AF352805

Sphingolipids

Sphingolipids: From Pathology to Therapeutic Perspectives—A Themed Honorary Issue to Prof. Lina Obeid

Editor

Gerhild van Echten-Deckert

MDPI • Basel • Beijing • Wuhan • Barcelona • Belgrade • Manchester • Tokyo • Cluj • Tianjin

Editor
Gerhild van Echten-Deckert
University of Bonn
Germany

Editorial Office
MDPI
St. Alban-Anlage 66
4052 Basel, Switzerland

This is a reprint of articles from the Special Issue published online in the open access journal *Cells* (ISSN 2073-4409) (available at: https://www.mdpi.com/journal/cells/special_issues/ Sphingolipids_Pathology).

For citation purposes, cite each article independently as indicated on the article page online and as indicated below:

LastName, A.A.; LastName, B.B.; LastName, C.C. Article Title. *Journal Name* **Year**, *Volume Number*, Page Range.

ISBN 978-3-03943-957-7 (Hbk)
ISBN 978-3-03943-958-4 (PDF)

Cover image courtesy of Sumaiya Yasmeen Afsar and Gerhild van Echten-Deckert.

Contents

About the Editor

Gerhild van Echten-Deckert conducted her PhD (Dr. rer. nat.) studies at the Diabetes Research Center at the University of Düsseldorf under the supervision of Dr. Jürgen Eckel. Her studies focused on the regulation of the cardiac insulin receptor. During these studies, Dr. van Echten-Deckert established a method for primary cultured adult cardiomyocytes. As a postdoc, she joined Professor Konrad Sandhoff at the University of Bonn where she intensively studied glycosphingolipid biosynthesis in the brain. She was granted the Habilitation Award of the German Society of Biochemistry and Molecular Biology for her contributions to the organization, regulation, and physiological impact of neuronal glycosphingolipid biosynthesis. After setting up her own research group, her main scientific interest was directed towards the role of sphingolipids in brain health and disease. Her work is highly interdisciplinary and she is involved in several collaborative projects. She initiated the generation of a mouse model in which sphingosine-1-phosphate (S1P)-lyase, the enzyme catalyzing the final step of sphingolipid catabolism, is specifically inactivated in neural cells. This model enabled her to show, for the first time, the impact of this enzyme for synaptic architecture, neuronal autophagy, and cognitive and motor skills. A deeper characterization of this model uncovered the cell type specific role of S1P in the brain, including activation of astroglia and microglia and thus its involvement in neuro-inflammation. Since 2013, she has been an Honorary Member of the Sphingolipid Club. She has participated in International Advisory Groups and chaired several sessions at international conferences including GRCs, FASEB conferences, ICC (International Ceramide Conference), and the International Conference on Bioactive Lipids.

Editorial

Special Issue on "Sphingolipids: From Pathology to Therapeutic Perspectives"

Gerhild van Echten-Deckert

Life & Medical Science (LIMES) Institute for Membrane Biology and Lipid, Biochemistry at the Kekulé-Institute, University of Bonn, 53121 Bonn, Germany; g.echten.deckert@uni-bonn.de

Received: 29 October 2020; Accepted: 30 October 2020; Published: 3 November 2020

It is an honor for us to dedicate this Special Issue to our dearest friend Lina Obeid, who was not only a pioneer in the field of sphingolipids, but also a remarkable personality. We lost her on November 19, 2019. There are no words to describe the deep pain we feel in losing Lina Obeid, an internationally recognized scientist, a wonderful mentor, ardent mother, and last but not least an empathic and cheerful friend. Among her numerous outstanding scientific achievements, I want to highlight her study published in 1993 in *Science*, where she showed for the first time that ceramide is a bioactive lipid that plays a role in apoptotic cell death [1]. This publication opened up a new era in sphingolipid research. Lina was also one of the first to recognize the key role of sphingosine kinase 1 (SK1) in cancer and to envision this enzyme as an anticancer target in p53-dependent tumors [2,3]. Last but not least, I want to mention the numerous excellent and helpful reviews she wrote, often together with her husband Yusuf Hannun, such as the one from 2018, which depicts and updates our knowledge on the critical roles of sphingolipid metabolism in physiology and disease [4]. Thank you, Lina.

Sphingolipids are ubiquitous components of cellular membranes [5]. Following their discovery in the brain and first description in 1884 by J.L.W. Thudichum [6], sphingolipids were largely overlooked for almost a century, perhaps due to their complexity and enigmatic nature [7,8]. It was with the discovery of sphingolipidoses, a series of inherited diseases caused by mutations of enzymes involved in sphingolipid degradation that sphingolipids returned to the limelight [9–11]. The essential breakthrough came decades later in the 1990s with the discovery that sphingolipids are not just structural elements of cellular membranes, but they are also intra- and extracellular signaling molecules [12]. It turned out that not only their complex carbohydrate head-groups, but especially their lipid backbones, including ceramide [1,13] and sphingosine-1-phosphate (S1P) [14] have selective physiological functions. As a result of this new concept, sphingolipids emerged as essential players in many pathologies including cancer [15,16], diabetes [17], neurodegenerative disorders [18], and autoimmune diseases [19]. The present issue reflects the evolvement of sphingolipids as bioactive signaling molecules that have unexpectedly eclectic functions in health, disease, and therapy.

In cells, sphingolipids are generally lower in abundance than glycerolipids or cholesterol, representing less than 20% of total lipid mass [20]. The mammalian brain contains the largest amounts of sphingolipids in the body. Thus, it is not surprising that a rather large number of the contributions to this Special Issue refer to the role of sphingolipids in brain health and disease.

The review by Lucaciu et al. highlights the role of S1P and its receptors in neurological disorders. This contribution provides a detailed picture of S1P metabolism and function in different cellular compartments. Most interestingly, the authors present an array of drugs targeting the S1P signaling pathway, which are being tested in clinical trials.

Related to this review, the study of Alam et al. shows that S1P accumulation in neural cells exerts pathological effects that are receptor-independent. This group of researchers have previously shown that S1P is not always neuroprotective [21,22]. Interestingly, the harmful effects of S1P were also shown in beta-cells of the pancreas [23]. Of note, neurons and pancreatic beta-cells share important developmental transcriptional programs [24].

Rose-Mary Boustany's research group studied the impact of sphingolipid metabolism in juvenile neuronal ceroid lipofuscinosis (now classified as CLN3 disease), which represents the most common form of neuronal ceroid lipofuscinoses (NCLs), a family of fatal, inherited pediatric neurodegenerative disorders [25]. The present contribution by Maloouf et al. proposes that flupirtine could be of use as a therapy for CLN3 disease and represents a first approach to define the mechanisms through which it may exert its actions.

Two other contributions focus on sphingolipid metabolism and its role in the central nervous system. Zoicas et al. elucidate that in major depression and comorbid anxiety, the activity of enzymes that catalyze both ceramide generation and degradation, including sphingomyelinases and ceramidases, respectively, are elevated in particular brain regions. In a related context, Cosima Rhein's research group has presented a novel mouse model of depression. In this model, overexpression of acid sphingomyelinase is restricted to the forebrain (see the contribution by Zoicas, Schumacher et al.).

However, the metabolism of ceramide is not only central to the brain, it has also been shown to play a major role in cellular and physiological processes including apoptosis, senescence, and inflammation [26]. The review by Duarte et al. focuses on ceramidases but also gives a detailed picture of ceramide metabolism and its involvement in numerous pathologies. Last, but not least, the authors discuss the usefulness of these enzymes as therapeutic tools.

As mentioned above, diabetes and the metabolic syndrome are other human pathologies in which the role of sphingolipids is continuously being unraveled. Intriguingly, inhibition of ceramide synthesis in models of metabolic diseases prevents insulin resistance and diabetes and hence complications related to this disease [27]. In addition, C16-ceramide accumulation has been directly correlated with insulin resistance [28]. Guitton et al. present a detailed overview of the role of S1P in obesity and type 2 diabetes, whereas the contribution of Gurgul-Convey discusses in depth what is known so far about the involvement of sphingolipids in beta-cell physiology and pathophysiology, and their contribution to type 1 diabetes development and progression.

Sphingolipids and cancer represent another important chapter with regard to the versatile functions of these molecules. In this volume, the review by Carrié et al. provides a very useful and important piece of information on the putative role of alterations of sphingolipid metabolism in tumorigenesis, progression and therapy of malignant melanoma.

The key role of SK1 in cancer was a core area of Lina Obeid's research [3]. Indeed, SK1 activity directly impacts the level of 3 bioactive sphingolipids; on the one hand it drives the formation of S1P while simultaneously clearing sphingosine, hence, impeding the generation of ceramide via the salvage pathway. It is therefore important to understand the mechanism of its regulation. The present Issue offers two contributions regarding this question. The contribution by Bonica et al. from Lina's group provides a comprehensive overview of the transcriptional regulation of SK1, indicating the role of different transcriptions factors as well as of microRNAs in different systems in healthy and diseased states. Post-transcriptional and post-translational control of SK1 by G-protein coupled receptors (GPCRs) is one of the research topics examined by Meyer zu Heringdorf's group. The latest findings of Blankenbach et al. on the Gq/11-mediated plasma membrane translocation and the activation of SK1 are part of this Special Issue.

The study by Gräler's group (Paul et al.) uncovers a new aspect of the role of S1P as an effective endothelial cell barrier stabilizing signaling molecule, which involves its secretion via the S1P-transporter Spinster homolog 2 (Spns2).

Studies on the role of S1P and its receptors in myocardial function are presented by Wafa et al., where it appears that S1P acts as a double-edged sword. As shown and discussed by the authors, depending on the physiological conditions, S1P may exert a protective or a deleterious effect on cardiac function.

Paola Signorelli's research group has extensive experience regarding the role of sphingolipids in cystic fibrosis. Here, Mingione et al. provide evidence on how the controlled modulation of sphingolipid metabolism can function as a useful therapeutic strategy to overcome this inherited disease.

The articles in this Special Issue are written by leading experts and cover many of the various roles played by sphingolipids in pathologies that are driven by perturbed sphingolipid metabolism. I am convinced that this fascinating lipid class will continue to be the subject of up-and-coming future discoveries, especially with regard to new therapeutic strategies.

Funding: Ongoing work in the van Echten-Deckert Lab is funded by the German Research Foundation (Deutsche Forschungsgemeinschaft, grant EC 118/10-1).

Acknowledgments: I want to particularly thank all the contributors of this Special Issue. In addition, I acknowledge the timely and constructive assistance of the expert reviewers.

Conflicts of Interest: The author declares no conflict of interest.

References

1. Obeid, L.M.; Linardic, C.M.; Karolak, L.A.; Hannun, Y.A. Programmed cell death induced by ceramide. *Science* **1993**, *259*, 1769–1771. [CrossRef] [PubMed]
2. Heffernan-Stroud, L.A.; Helke, K.L.; Jenkins, R.W.; De Costa, A.M.; Hannun, Y.A.; Obeid, L.M. Defining a role for sphingosine kinase 1 in P53-dependent tumors. *Oncogene* **2012**, *31*, 1166–1175. [CrossRef] [PubMed]
3. Heffernan-Stroud, L.A.; Obeid, L.M. Sphingosine kinase 1 in cancer. *Adv. Cancer Res.* **2013**, *117*, 201–235. [CrossRef] [PubMed]
4. Hannun, Y.A.; Obeid, L.M. Sphingolipids and their metabolism in physiology and disease. *Nat. Rev. Mol. Cell Biol.* **2018**, *19*, 175–191. [CrossRef]
5. Sonnino, S.; Prinetti, A.; Mauri, L.; Chigorno, V.; Tettamanti, G. Dynamic and structural properties of sphingolipids as driving forces for the formation of membrane domains. *Chem. Rev.* **2006**, *106*, 2111–2125. [CrossRef]
6. Thudichum, J.L.W. *Treatise on the Chemical Constitution of the Brain*; Baillière, Tindal and Cox: London, UK, 1884.
7. Merrill, A.H., Jr. Sphingolipid and glycosphingolipid metabolic pathways in the era of sphingolipidomics. *Chem. Rev.* **2011**, *111*, 6387–6422. [CrossRef] [PubMed]
8. van Echten, G.; Sandhoff, K. Ganglioside metabolism. Enzymology, topology, and regulation. *J. Biol. Chem.* **1993**, *268*, 5341–5344.
9. Futerman, A.H.; van Meer, G. The cell biology of lysosomal storage disorders. *Nat. Rev. Mol. Cell Biol.* **2004**, *5*, 554–565. [CrossRef]
10. Klein, A.D.; Futerman, A.H. Lysosomal storage disorders: Old diseases, present and future challenges. *Pediatr. Endocrinol. Rev.* **2013**, *11*, 59–63.
11. Kolter, T.; Sandhoff, K. Sphingolipid metabolism diseases. *Biochim. Biophys. Acta* **2006**, *1758*, 2057–2079. [CrossRef] [PubMed]
12. Hannun, Y.A.; Obeid, L.M. Principles of bioactive lipid signalling: Lessons from sphingolipids. *Nat. Rev. Mol. Cell Biol.* **2008**, *9*, 139–150. [CrossRef] [PubMed]
13. Hannun, Y.A. The sphingomyelin cycle and the second messenger function of ceramide. *J. Biol. Chem.* **1994**, *269*, 3125–3128.
14. Spiegel, S.; Milstien, S. Sphingosine-1-Phosphate: An enigmatic signalling lipid. *Nat. Rev. Mol. Cell Biol.* **2003**, *4*, 397–407. [CrossRef] [PubMed]
15. Ogretmen, B. Sphingolipid metabolism in cancer signalling and therapy. *Nat. Rev. Cancer* **2018**, *18*, 33–50. [CrossRef]
16. Segui, B.; Andrieu-Abadie, N.; Jaffrezou, J.P.; Benoist, H.; Levade, T. Sphingolipids as modulators of cancer cell death: Potential therapeutic targets. *Biochim. Biophys. Acta* **2006**, *1758*, 2104–2120. [CrossRef]
17. Summers, S.A. Ceramides in insulin resistance and lipotoxicity. *Prog. Lipid Res.* **2006**, *45*, 42–72. [CrossRef]
18. van Echten-Deckert, G.; Walter, J. Sphingolipids: Critical players in Alzheimer's disease. *Prog. Lipid Res.* **2012**, *51*, 378–393. [CrossRef]
19. Maceyka, M.; Spiegel, S. Sphingolipid metabolites in inflammatory disease. *Nature* **2014**, *510*, 58–67. [CrossRef]
20. Futerman, A.H. Sphingolipids. In *Biochemistry of Lipids and Membranes*, 6th ed.; Elsevier: Amsterdam, The Netherlands, 2016; Volume 5, pp. 297–326.

21. Karunakaran, I.; van Echten-Deckert, G. Sphingosine 1-phosphate—A double edged sword in the brain. *Biochim. Biophys. Acta* **2017**, *1859*, 1573–1582. [CrossRef]
22. van Echten-Deckert, G.; Hagen-Euteneuer, N.; Karaca, I.; Walter, J. Sphingosine-1-Phosphate: Boon and bane for the brain. *Cell. Physiol. Biochem.* **2014**, *34*, 148–157. [CrossRef]
23. Hahn, C.; Tyka, K.; Saba, J.D.; Lenzen, S.; Gurgul-Convey, E. Overexpression of sphingosine-1-phosphate lyase protects insulin-secreting cells against cytokine toxicity. *J. Biol. Chem.* **2017**, *292*, 20292–20304. [CrossRef]
24. Arntfield, M.E.; van der Kooy, D. βCell evolution: How the pancreas borrowed from the brain: The shared toolbox of genes expressed by neural and pancreatic endocrine cells may reflect their evolutionary relationship. *BioEssays* **2011**, *33*, 582–587. [CrossRef] [PubMed]
25. Wang, S. Juvenile neuronal ceroid lipofuscinoses. *Adv. Exp. Med. Biol.* **2012**, *724*, 138–142. [CrossRef]
26. Hannun, Y.A.; Obeid, L.M. The ceramide-centric universe of lipid-mediated cell regulation: Stress encounters of the lipid kind. *J. Biol. Chem.* **2002**, *277*, 25847–25850. [CrossRef]
27. Holland, W.L.; Brozinick, J.T.; Wang, L.P.; Hawkins, E.D.; Sargent, K.M.; Liu, Y.; Narra, K.; Hoehn, K.L.; Knotts, T.A.; Siesky, A.; et al. Inhibition of ceramide synthesis ameliorates glucocorticoid-, saturated-fat-, and obesity-induced insulin resistance. *Cell Metab.* **2007**, *5*, 167–179. [CrossRef]
28. Turpin, S.M.; Nicholls, H.T.; Willmes, D.M.; Mourier, A.; Brodesser, S.; Wunderlich, C.M.; Mauer, J.; Xu, E.; Hammerschmidt, P.; Bronneke, H.S.; et al. Obesity-Induced Cers6-dependent C16:0 ceramide production promotes weight gain and glucose intolerance. *Cell Metab.* **2014**, *20*, 678–686. [CrossRef]

Publisher's Note: MDPI stays neutral with regard to jurisdictional claims in published maps and institutional affiliations.

Review

The S1P–S1PR Axis in Neurological Disorders—Insights into Current and Future Therapeutic Perspectives

Alexandra Lucaciu [1,*], Robert Brunkhorst [2], Josef M. Pfeilschifter [3], Waltraud Pfeilschifter [1] and Julien Subburayalu [4,*]

[1] Department of Neurology, University Hospital Frankfurt, Goethe University Frankfurt, 60528 Frankfurt am Main, Germany
[2] Department of Neurology, RWTH Aachen University, 52074 Aachen, Germany
[3] Institute of General Pharmacology and Toxicology, Pharmazentrum Frankfurt, Goethe University Frankfurt, 60528 Frankfurt am Main, Germany
[4] Department of Medicine, Addenbrooke's Hospital, University of Cambridge, Cambridge CB2 0QQ, UK
* Correspondence: Alexandra.Lucaciu@kgu.de (A.L.); js2380@cam.ac.uk (J.S.)

Received: 14 May 2020; Accepted: 19 June 2020; Published: 22 June 2020

Abstract: Sphingosine 1-phosphate (S1P), derived from membrane sphingolipids, is a pleiotropic bioactive lipid mediator capable of evoking complex immune phenomena. Studies have highlighted its importance regarding intracellular signaling cascades as well as membrane-bound S1P receptor (S1PR) engagement in various clinical conditions. In neurological disorders, the S1P–S1PR axis is acknowledged in neurodegenerative, neuroinflammatory, and cerebrovascular disorders. Modulators of S1P signaling have enabled an immense insight into fundamental pathological pathways, which were pivotal in identifying and improving the treatment of human diseases. However, its intricate molecular signaling pathways initiated upon receptor ligation are still poorly elucidated. In this review, the authors highlight the current evidence for S1P signaling in neurodegenerative and neuroinflammatory disorders as well as stroke and present an array of drugs targeting the S1P signaling pathway, which are being tested in clinical trials. Further insights on how the S1P–S1PR axis orchestrates disease initiation, progression, and recovery may hold a remarkable potential regarding therapeutic options in these neurological disorders.

Keywords: sphingosine 1-phoshate; sphingosine 1-phosphate receptor; $S1P_{1-5}$; sphingosine 1-phosphate metabolism; sphingosine 1-phosphate antagonistst/inhibitors; sphingosine 1-phosphate signaling; stroke; multiple sclerosis; neurodegeneration; fingolimod

1. Introduction—S1P Metabolism and Signaling

Three decades ago, sphingosine 1-phosphate (S1P) was identified as an intracellular signaling agent in relation to calcium release from intracellular stores and metabolic adaptations [1]. The balance between sphingosine and S1P, both metabolites of its precursor ceramide, and their subsequent activation of effector kinases were shown to matter in imposing regulatory effects in the determination of whether a cell is destined for cell death or proliferation [2]. The sphingolipid metabolism is almost as complex as its protean intricacies to signaling pathways.

1.1. De Novo Sphingolipid Synthesis and Signaling at the Endoplasmic Reticulum

De novo sphingolipid biosynthesis is initiated in the smooth endoplasmic reticulum (sER) (Figure 1). Here, the α-aminocarbonic acid serine and the lipid palmitoyl-CoA (PalCoA) are enzymatically processed by the key enzyme serine palmitoyltransferase (SPT)—which is negatively regulated by ORM1-like protein 3 (ORMDL3) [3]—to 3-ketosphinganine [4–6]. Subsequent conversion of 3-ketosphinganine to S1P is promoted by enzymatic reactions including a reduction to sphinganine, a synthase reaction to dihydroceramide, and a desaturase reaction to ceramide followed by deacylation by ceramidase (CDase) and a phosphorylation by sphingosine kinase (SphK), which exists in two isoforms (SphK1 and SphK2) [6–8]. In general, the formation of the ceramide and the sphingoid bases represents the backbone of the sphingolipid metabolic pathway, as they can be utilized for the synthesis of complex glycosphingolipids. Glycosphingolipids are crucial components of cellular membranes [4,9], such as glucosylceramide or sphingomyelin manufactured by glucosylceramide synthase (GCS) or sphingomyelin synthase (SMS), respectively [10,11]. These various enzymatic reactions are not irreversible per se, since ceramide can be generated by sphingomyelin hydrolysis and/or recycling of complex sphingolipids [10,12,13]. Ultimately, S1P can irreversibly be degraded by S1P lyase into phosphoethanolamine (PE) and hexadecenal, both of which are being further processed [14,15]; PE is used for the synthesis of phosphatidylethanolamine and hexadecenal is used to replenish the PalCoA pool [15–18]. This cycle of de novo sphingolipid synthesis is tightly controlled by NOGO-B, a protein located within the membrane of the endoplasmic reticulum, which inhibits SPT [19,20]. Alternatively, S1P can be converted back to ceramides by dephosphorylation through sphingosine 1-phosphate phosphatase (SGPP) 1 or SGPP2 [21,22], both of which are members of the lipid phosphate phosphohydrolase (LPP) family [23]. This pathway can substantially contribute to the synthesis of complex sphingolipids within a cell subject to cell type and metabolic demand [24,25].

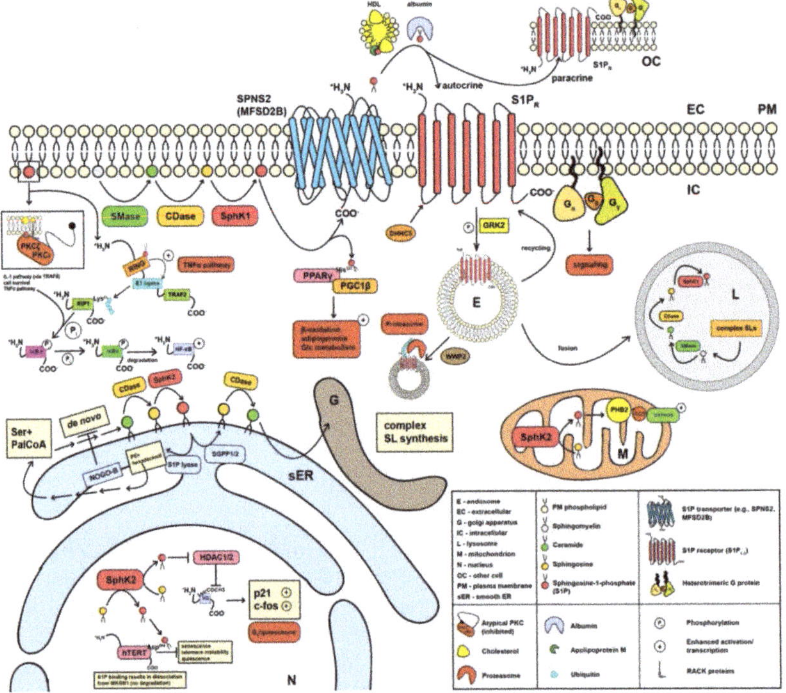

Figure 1. Sphingolipid biosynthesis, sphingosine 1-phosphate (S1P) release, and signaling. S1P is generated in different compartments within a cell. Nuclear S1P influences the balance between chromosome density by histones and telomere length impacting on metabolic adaptations and cell proliferation. De novo S1P synthesized at the smooth endoplasmic reticulum may be utilized for complex

sphingolipid synthesis, crucial components of cellular membranes. Mitochondrial S1P influences mitochondrial respiration by activating complex IV. S1P generated at the intracellular leaflet of the plasma membrane (PM) is either used for intracellular signaling converging on the TNFα or the PPARγ pathways or is exported to induce autocrine or paracrine stimulation transported by apolipoprotein (ApoM)-containing high-density lipoprotein (HDL) or albumin and signaling via membrane-bound S1P receptor (S1PR). Intracellularly, S1PR recruit different heterotrimeric G proteins to initiate different signaling pathways, which results in the down-regulation of S1PR via β-arrestin dependent recruitment of G protein receptor β-arrestin-regulated kinase 2 (GRK2), allowing dynamin and moesin-dependent endosome recruitment. Endosomal S1PR are either recruited to the PM or polyubiquitinylated by NEDD4-like E3 ubiquitin ligase WWP2 (WWP2) targeting S1PR for proteasomal degradation. Endosomal remnants are fused with the lysosome (the place where complex sphingolipids are degraded) to fully degrade proteinaceous or lipid cargo, ultimately replenishing the S1P pool. The figure is a modified version of Cartier and Hla [26] and Kunkel et al. [23].

1.2. Synthesis and Signaling of Sphingosine 1-Phosphate in Mitochondria and at the Plasma Membrane

Mitochondrial S1P (Figure 1), produced by SphK2, facilitates oxidative phosphorylation (OXPHOS). This effect was shown to be mediated by stabilizing the scaffolding protein prohibitin-2 (PHB2), which, in turn, eases the recruitment of cytochrome c oxidase (CCO) and thus the assembly of complex IV of the respiratory chain [23,27]. The most important site of S1P production is the plasma membrane (PM) itself (Figure 1), which is composed of a lipid bilayer, predominantly consisting of an extracellular (EC) and an intracellular (IC) phospholipid leaflet. Here, S1P is derived from sphingomyelin, an integral component of cellular membranes [4]. Sphingomyelin is metabolized to S1P via the enzymes sphingomyelinase (SMase), CDase, and sphingosine kinase 1 (SphK1) [28,29]. At this point, S1P can either be exported using the multi-pass membrane proteins spinster homolog 2 (SPNS2) [30–33] or major facilitator superfamily domain-containing protein 2B (MFSD2B) [30–32], respectively, or employed for further immediate intracellular signaling cascades. In terms of intracellular signaling, S1P can engage in tumor necrosis factor-α (TNF-α) receptor-associated factor 2 (TRAF2) [34] and TRAF6-dependent TNF-α signaling [35]. S1P can either recruit the atypical protein kinase C (PKC) subtypes ζ (PKCζ) or ι (PKCι) to sites of distinct membrane microdomains using receptor for activated C kinase (RACK) proteins [36–38] or associate with the TRAF2 complex [34]. This, in turn, allows the activation of the interleukin-1 (IL-1) pathway via TRAF6, cell survival, and convergence on the TNF-α pathway downstream of TRAF2 [28,34,39–41]. The association with the TRAF2 complex is said to occur by engaging with the N-terminal adjacent really interesting new group (RING) domain, which allows activation of the intramolecular E3 ligase domain of TRAF2 [28,34,41]. The activation of TRAF2's E3 ligase lures receptor-interacting serine/threonine-protein kinase 1 (RIP1) in close proximity, allowing polyubiquitination of its Lys^{63} residue. Polyubiquitination of Lys^{63} stimulates RIP1's kinase activity, which results in the phosphorylation of the inhibitor of nuclear factor κ-B kinase (IκB-k). As a consequence, the nuclear factor κ-light-chain-enhancer of activated B cells (NF-κB) signaling pathway is engaged, since IκB-k is now able to facilitate the down-regulation of the NF-κB inhibitor α (IκBα). This culminates in the uncoupling of IκBα from NF-κB revoking its inhibitory effect and allowing NF-κB's nuclear translocation [28,34,35]. Crosstalk between TRAF2 and TRAF6 was previously reported, as both can engage with atypical PKCs via protein p62 [39], are able to recruit TGF-β-activated kinase 1 and MAP3K7-binding protein 3 (TAB3) [40,42], or can be polyubiquitinylated given the presence of a RING domain in TRAF6 [43]. Moreover, a disruption of TRAF6 binding sites, for example, only mildly impacts NF-κB signaling in the presence of TRAF2 and TRAF3 [44]. Beyond the TNF-α signaling pathway, S1P signaling can vitalize adipogenesis, glucose (Glc) metabolism, and β-oxidation via peroxisome proliferator activated receptor γ (PPARγ) [15,45,46]. By binding to its His^{323} residue, S1P activates PPARγ, increasing the likelihood of an association with the PPARγ co-activator 1β (PGC1β), a necessary co-transcription factor for nuclear translocation [46,47].

1.3. Sphingosine 1-Phosphate Signaling in the Nucleus

SphK2, the predominant isoform in the sER and mitochondria [27,48], is also predominant in the nucleus (Figure 1) [49–53]. Nuclear S1P was shown to be critically involved in influencing the balance between cellular quiescence and proliferation. Hait et al., showed that S1P produced in the nucleus binds to the class I histone deacetylases (HDAC) 1 and 2, which results in their inhibition [53]. In general, HDAC1 and HDAC2's function lies in removing acetyl residues coupled to the α-aminocarbonic acid lysine close to the amino-terminal end of the histone protein H3 [54–56]. Therefore, the removal of these negatively charged residues culminates in a net positive charge of H3, increasing its tight association with the negatively charged deoxyribonucleic acid (DNA). Ultimately, the inhibition of HDAC1/2 enhances transcription of cyclin-dependent kinase inhibitor 1 (p21) and the proto-oncogene c-Fos (c-fos) [53]. Conversely, S1P may bind to human telomerase reverse transcriptase (hTERT) [57], which may allow makorin ring finger protein 1 (MKRN1) to dissociate due to competitive binding sites [57,58]. This signaling cascade results in telomere maintenance, cell proliferation, and tumor growth [57].

1.4. Sphingosine 1-Phosphate in Autocrine and Paracrine Signaling

Detailed experimental evidence is available for the mechanisms of "inside-out" autocrine and paracrine S1P signaling. After release into the extracellular compartment via SPNS2 [30–33] or MFSD2B [30–32], S1P is swiftly bound by its chaperones due to its hydrophobic character. These chaperones are apolipoprotein M (ApoM)-containing high-density lipoprotein (HDL)—to some extent also to very low-density lipoprotein (VLDL) and low-density lipoprotein (LDL) [59,60]—or albumin, respectively [61–63]. Subsequently, upon release and chaperoning by HDL or albumin, ligation of the five known heptameric G protein-coupled S1P receptors (S1PR) 1–5 (S1P$_{1–5}$) (Figure 1) [64–69] can result both in autocrine and paracrine signaling [70–72]. Signaling via S1PR is tightly regulated. Fine tuning of S1P-S1PR signaling may occur via post-translational modifications, e.g., through palmitoylation by the palmitoyltransferase DHHC5 (DHHC5) [73] and, at some point, termination of the signaling cascade may be achieved by β-arrestin-dependent recruitment of G protein-coupled receptor kinase 2 (GRK2), which phosphorylates S1PR, resulting in dynamin and moesin dependent establishment of the endosome [74–79]. At this point, re-routing, i.e., recycling of the receptor to the PM [80], polyubiquitination by the NEDD4-like E3 ubiquitin protein ligase WWP2 (WWP2) resulting in proteasomal degradation [81], or fusion with the lysosome in order for complete proteinaceous and lipid residue degradation can occur (Figure 1) [82–84]. The latter endolysosomal salvage pathway is of particular importance for cellular homeostasis and disassembly of complex sphingolipids to ceramides or allowing endolysosomal SphK1 to produce S1P, respectively [85–87]. The herein discussed mechanisms are briefly summarized in Figure 1.

1.5. External Action of Sphingosine 1-Phosphate through S1PR

With respect to S1PR ligation, the de facto signal triggered is dependent upon the S1PR subtype, the presence of co-regulatory agents, and the heterotrimeric G protein recruited. To date, five bona fide cognate receptors for S1P are known, namely S1P$_{1–5}$ (Figure 2) [88]. S1P$_1$, the most commonly expressed S1P receptor in the brain [89], appears to be most selective as it binds only to G$\alpha_{i/0}$ [66,90]. S1P$_2$, also binding to G$\alpha_{i/0}$, is capable of associating with Gα_q, G$_{12/13}$, and Gα_s [66,88,90], however, it couples most efficiently with G$_{12/13}$, subsequently activating the small GTPase Rho [91–93]. S1P$_3$ is said to couple with G$\alpha_{i/0}$, Gα_q, and G$_{12/13}$, although a higher affinity/likelihood for association with Gα_q was observed, ultimately resulting in intracellular Ca^{2+} enrichment and activation of PKC [92,94]. S1P$_4$ and S1P$_5$ can couple to Gα_s, Gα_q, and G$_{12/13}$ [88,95]. Regarding the intracellular signaling pathways triggered, the interested reader is referred to reviews entirely dedicated to detailing molecular signaling and transcriptional cascades triggered in appreciation of the heterotrimeric G protein recruited [96–99]. Regarding receptor activation or inhibition due to the presence of co-regulatory agents, S1P$_1$ can be activated by cluster of differentiation molecule (CD) 44 (CD44) (hyaluronic acid receptor) or activated Protein C (aPC) [100,101], whilst inhibition by CD69, S1P$_2$ (in dermal γδ T cells), or LPA$_1$ was reported previously [102–104]. Unlike NOGO-B, NOGO-A, a multi-pass PM and ER protein whose expression is confined to the central nervous system (CNS), was shown to activate the S1P binding domain Δ20 of S1P$_2$, thereby restricting neurite outgrowth via engagement with the G$_{13}$-RhoA signaling pathway [105]. Moreover, conjugated bile acids (CBAs) and FAM19A5 were other activators of S1P$_2$ [106,107]. Figure 2 gives a synopsis signaling cascades upon S1P$_{1-5}$ ligation.

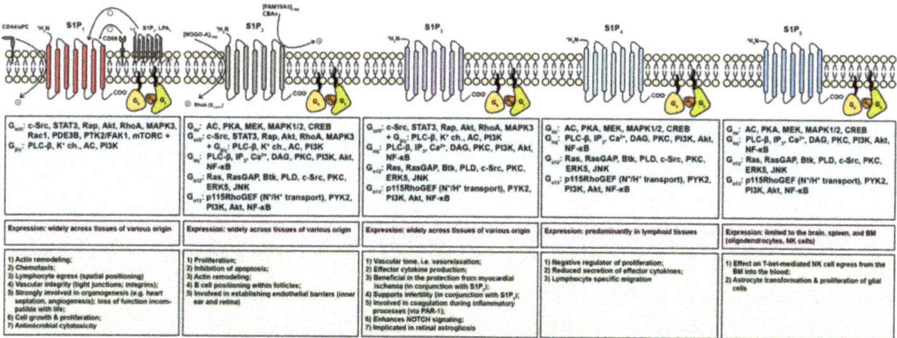

Figure 2. Sphingosine 1-phosphate receptors, canonical pathways, and functions triggered. The currently discovered S1P receptors 1–5 (S1P$_{1-5}$) are displayed. S1PR are naturally activated by S1P and to some extent by dihydro-S1P (sphinganine 1-phosphate) and phyto-S1P (4-hydroxysphinganine 1-phosphate), but also competitive/allosteric activation and inhibition by other molecules are described. Upon activation, S1PR can recruit various heterotrimeric G proteins, which, in turn, allow a finely tuned intracellular signaling cascade to be evoked by means of both Gα and Gβγ. Thus, differential S1PR expression in response to varying environmental lipid contexts, i.e., S1P, dihydro-S1P, and phyto-S1P, respectively, may result in a context- and cell type-dependent function triggered.

2. Implications of Sphingolipids in Neurological Disorders

2.1. The Sphingolipid Metabolism in Neurodegenerative Disorders

Neurodegenerative diseases are commonly characterized by intracellular or extracellular aggregation of misfolded proteins. These diseases most commonly comprise Alzheimer's disease (AD), characterized by the proteins amyloid-β and tau, Parkinson's disease (PD; α-synuclein), and amyotrophic lateral sclerosis (ALS), in which TAR DNA-binding protein 43 deposition is observed [108]. With respect to the accumulation of the ontogeny of the protein misfolded, different classifications of neurodegenerative conditions were established, denoted as tauopathies [109], synucleinopathies [110], or prion diseases [111] (Figure 3).

Regarding AD, increasing evidence supports the crosstalk between sphingolipids and aberrant protein aggregation [112,113]. Amyloid-β-peptide (Aβ) is cleaved from amyloid precursor protein (APP) by the β- and the γ-secretase enzymes, while α-secretase acts within the Aβ sequence [114]. APP cleavage and the release of Aβ from the PM subsequent to its production in lipid rafts are influenced by lipid composition [115,116]. Alterations in membrane lipid composition have a key role in the subsequent subcellular transport and trafficking of these proteins [114,116,117].

Perturbations in the neurovascular unit (NVU) result in a compromised barrier function and dysregulation and reduction in cerebral blood flow (CBF), which is implied to be involved in the pathogenesis of AD [118–127]. Vascular tightness via tight junctions is influenced by the sphingolipid metabolism. In that regard, acid SMase activity and ceramide production in endothelial cells were linked to vascular permeability [128]. Conversely, the acid SMase inhibition maintained enhanced tight junction regulation [128]. Similar mechanisms were observed to happen in astrocytes [129]. In brain tissue of AD patients', studies showed increased ceramide levels and decreased sphingomyelin and S1P levels [130–134]. A study on AD by Katsel et al., found a significant up-regulation of messenger ribonucleic acid (mRNA) of phospholipid phosphatase 3 (PLPP3) and S1P lyase 1 (SGPL1) at early stages after diagnosis, suggesting a lack of S1P as a spatiotemporal function may contribute to the degeneration of neurons [135]. Besides, Ceccom et al., reported in an immunohistochemical study of AD a reduction of SphK1 accompanied by enhanced S1P lyase expression in frontal and entorhinal human cortices to be accountable for the perturbed S1P metabolism observed, contributing to the deposition of amyloid and ultimately to neuronal damage [136]. More recently, Dominguez et al., described that the subcellular localization of S1P's production, e.g., by a disrupted equilibrium between cytosolic and nuclear SphK2, conferred pathogenic effects of S1P in AD [137].

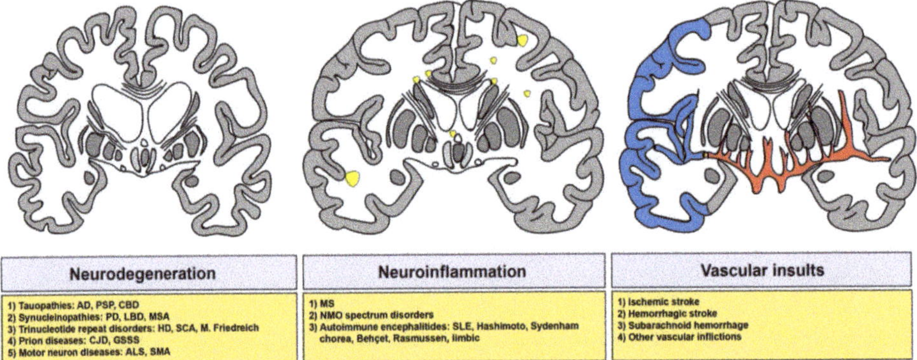

Figure 3. Clinical significance of distorted sphingosine 1-phosphate signaling in neurological disorders. Perturbed S1P signaling has been reported in various clinical conditions ranging from autoimmunity, infection, and cancer. S1P signaling was also shown to play a detrimental role in several neurological diseases, including neurodegenerative and neuroinflammatory conditions, but equally in cerebrovascular insults resulting in stroke or subarachnoid hemorrhage. The yellow patches refer to demyelinated/inflamed areas in the white matter. The area of the grey matter (grey) highlighted in blue denotes the infarcted region. This region is perfused by the middle cerebral artery (here blocked by an embolus), a branch of the cerebral vessel system (red).

With respect to PD, a chronic progressive disorder characterized by the degeneration of dopaminergic neurons in the pars compacta of the substantia nigra, emerging evidence has elucidated the role of mitochondrial and endolysosomal pathways and their interplay with ceramides in its pathogenesis. Xilouri et al., focused on α-synuclein degradation and suggested, in 2008, an impairment of neural autophagy-lysosomal pathways to be responsible for α-synuclein accumulation, unraveling a causal link between the pathogenic event and the initiation as well as the progression of the disease [138]. Later, further studies confirmed this hypothesis [139,140].

Glucocerebrosidase (GBA) mutations were found in subjects with parkinsonism [141,142], which were previously reported to predispose to the development of Lewy body disorders (LBD) [143]. Concerning LBD, Bras et al., investigated the neuronal ceramide metabolism and reported that several genes known to confer the risk of LBD development converge on the ceramide metabolism, although this remains to be confirmed neurohistologically [144]. In addition, implications in the pathogenesis of synucleinopathies were supported by descriptions of mutations in the GCase gene (*GBA1*) and altered sphingolipid pathways [145,146].

Mazzulli et al., could delineate that glucosylceramidase (GlcCer), the GCase substrates, enhanced the rate of α-synuclein oligomerization [147]. Recently, it was shown that the actual sphingolipid subspecies carry various potentials to cause formation of oligomeric α-synuclein, particularly in reflection of comparing in vitro with in vivo data [148]. Here, glucosylceramide, glucosylsphingosine, sphingosine, and S1P were shown to promote β-sheeted (oligomerized) structures of α-synuclein [148].

Somewhat aside of the molecular establishment of the aforementioned classifications, neurodegeneration is associated with the accumulation of most commonly CAG trinucleotide repeats, which encode for multiple glutamine residues to be translated, inevitably causing a toxic gain of function of the mutant protein [149]. The best-known condition for trinucleotide repeat disorders is Huntington's disease (HD) [150]. In HD, Di Pardo et al., showed an increase in SGPL1 in the striatum and the cortex and a decrease of SphK1 in the striatum of human post-mortem tissues, which were reflected by similar changes in mouse models of HD [151]. Moreover, the R6/2 mouse model revealed reduced levels of S1P [151,152] despite an up-regulation of SphK2. In contrast, no change was seen in either SphK2 in the YAC128 model or in humans [151]. Unfortunately, no data are currently available regarding S1P levels in YAC128 mice nor in human post-mortem tissues [153]. These findings warrant further investigations into the usability of druggable targets within the sphingolipid metabolism in HD.

Regarding ALS, a study by Henriquez et al., demonstrated a link between ALS severity and gene expressions or metabolite levels for sphingosine, ceramide (d18:1/26:0), SGPP2, SphK1, and UDP galactosyltransferase 8A (UGT8A) [154]. Shedding light on the therapeutic potential of the sphingolipid metabolism in ALS, Potenza et al., reported an improved neurological phenotype and an extended survival after fingolimod, a prodrug that becomes phosphorylated after application in vivo and acts as a receptor agonist against almost unanimously all S1PR—except for $S1P_2$ [155,156], administration in $mSOD1^{G93A}$ mice [157].

2.2. The Sphingolipid Metabolism in Neuroinflammatory Disorders

Multiple sclerosis (MS) represents an inflammatory disorder of the brain and the spinal cord featuring inflammation, demyelination, and neurodegeneration [158] (Figure 3). Over the last decades, murine experimental autoimmune encephalomyelitis (EAE) models have been used to decipher the mechanisms responsible for disease pathogenesis and progression and to identify druggable targets in order to develop therapeutics for multiple sclerosis [159–161]. It has been known for some time that the S1P metabolism can be exploited to slow disease progression in MS, e.g., by fingolimod, causing lymphocyte sequestration and ultimately preventing auto-reactive immune cell infiltration into the CNS [162]. The potency of exploiting the S1P S1PR axis by fingolimod in EAE was first shown in rats by Brinkmann et al. [155], implicating the feasibility to exploit S1PRs to influence lymphocyte egress. Subsequently, other studies have added to this observation [163–165]. Another report unveiled that prophylactic and therapeutic treatment with fingolimod resulted in suppression of EAE [166]. Choi et al., reported, in 2011, that a decline in disease severity of EAE by fingolimod involved astrocytic $S1P_1$ modulation as well, thus a loss of $S1P_1$ in astrocytes reduced disease severity, demyelination, axonal loss, and astrogliosis [167], arguing for additional CNS-specific effects of fingolimod in addition to lymphocyte redistribution. A recent study identified potential long-term effects caused by S1PR ligation. These long-term effects, according to Eken et al., confer an impact not only on lymphoid sequestration but similarly on non-lymphoid tissue regulatory T cell (T_{REG}) distribution, and, more importantly, on reducing the memory T_{REG} pool in favor of effector T_{REG} [168]. This could have implications for appropriate T cell zone access in lymph nodes via C-C chemokine receptor type 7 (CCR7) and subsequently their ability to control auto-reactive T cells in vivo [168,169]. These findings warrant further investigation into the precise mechanisms by which enzymes and lipids involved in the generation of S1P and their effects were linked to disease progression and treatment.

Cruz-Orengo et al., identified $S1P_2$ in the inbred SJL mouse strain as a sex- and strain-specific, disease-modifying molecule promoting the breakdown of adherens junctions, thus leading to blood-brain barrier (BBB) leakage, while antagonism of $S1P_2$ signaling led to an amelioration of disease severity in female mice [170]. BBB disruption could also be induced by ceramides, resulting in an increased migration of monocytes [129]. Moreover, Lopes et al., demonstrated that acid

SMase-derived ceramide regulates intracellular adhesion molecule 1 (ICAM-1) function during T cell transmigration across brain endothelial cells [171].

Concerning S1P receptor expression profiles in the disease model of EAE, mRNA for $S1P_1$ and $S1P_5$ in the spinal cord was down-regulated, and an up-regulation of $S1P_3$ and $S1P_4$ mRNAs occurred in the EAE model, which was reversable by fingolimod in accordance with structural restoration of the CNS parenchyma given a restriction to autoimmune T cell infiltration [166]. $S1P_3$ was shown to be involved in promoting systemic inflammation via activation of dendritic cells [172]. Concerning $S1P_3$ signaling in MS, Fischer et al., suggested that an increased expression of $S1P_3$ in EAE was likely due to astrocyte activation; however, its actual sequelae regarding detrimental effects (e.g., astrogliosis) and beneficial effects (e.g., remyelination) could not be established [173]. Apart from astrocytes, the same group reported enhanced SphK1 expression in macrophages of MS lesions [173]. A more definitive proof of $S1P_3$'s importance regarding an inflammatory cascade triggered in astrocytes was reported by Dusaban et al. [174]. The authors demonstrated $S1P_3$ to be up-regulated in astrocytes and to be able to engage with transforming protein RhoA (RhoA), and $S1P_3$ ligation was shown to promote IL-6, vascular endothelial growth factor A (VEGFa), and cyclooxygenase-2 (COX-2), which was accompanied by an increase of SphK1 and $S1P_3$ in vitro [174].

Neuromyelitis optica (NMO) spectrum disorders can also be classified as pertaining to the group of inflammatory brain disorders. Their hallmark feature was initially introduced by Devic and Gault [175,176] and characteristically consisted of a severe complement-mediated damage to the optic nerves and the spinal cord [177]. The discovery of highly specific serum autoantibody marker (NMO-IgG) eventually helped to differentiate this spectrum of disorders from MS and the prior interpretation as one entity [178,179]. Several reports have suggested that treatment with fingolimod in NMO may be contraindicated due to adverse events and worsening of disease severity [180–183]. However, exploitation of the sphingolipid metabolism to treat patients with NMO should not be excluded prematurely. Matsushita et al., demonstrated significantly higher levels of T_H1-related, i.e., C-C motif chemokine 4-like (CCL4) and C-X-C motif chemokine (CXC) 10 (CXCL10), and the T_H-17-related (and neutrophil-related) chemokine CXCL8 (IL-8) in NMO patients [184]. STAT3, which was recently shown to be linked to S1P signaling [185–187], is known to control the expression of chemokines and chemokine receptors in the recruitment of neutrophils [188] and T cells [189]. With respect to the mechanisms by which S1P signaling is tied to chemokine production and immune cell migration, studies revealed an interplay between S1PR and chemokine-driven migration of non-activated and naïve T cells [190]. In addition, binding of S1P produced by SphK1 to TRAF2 and cellular inhibitor of apoptosis 2 (cIAP2) in response to IL-1 signaling results in NF-κB activation [34]. This represents a relevant step in the recruitment of mononuclear cells to sites of sterile inflammation by means of interferon regulatory factor 1 (IRF1) expression and the resultant availability of the chemokines CXCL10 and CCL5 [191]. Therefore, mediation of the complex pathways of immune cell recruitment/trafficking, potentially resulting in favorable T_{REG} recruitment without disrupting T_H-17 or follicular T-helper cell (T_{FH}) sequestration, may hold the potential for future S1P metabolism-associated therapeutic perspectives in NMO [192].

Autoimmune conditions, such as systemic lupus erythematosus (SLE) or Hashimoto's disease, may present themselves with neurological pathology [193–195]. Therefore, due to their clinical heterogeneity, affecting potentially any organ of the body such as renal involvement may advise future neurological therapeutic perspectives. In that regard, studies have shown elevated S1P serum levels in patients with juvenile onset SLE [196] as well as in MRL-lpr/lpr mice [197]. In light of therapeutic targets for SLE, previous studies have concentrated on lupus nephritis. Here, fingolimod showed positive effects on survival, suppressing the continuation of autoimmunity [197]. In the context of murine lupus nephritis, in the NZB/W mouse model as well as in BXSB mice, fingolimod also proved to be beneficial [198–200]. However, inhibition of SphK2 in the MRL-lpr/lpr model could not convey protection from SLE [196]. These promising results prompted testing of cenerimod, a selective $S1P_1$ modulator (NCT02472795).

Other conditions denoted by aberrant inflammatory immune cell activation with a potential to cause immune encephalitis have been reported recently. For example, in autoimmune thyroiditis an enhanced expression of SphK1, S1P, and $S1P_1$ converging on STAT3 activation in $CD4^+$ T cells was demonstrated in mice by Han et al. [201]. Conversely, administration of fingolimod to these NOD.H-2^{h4} mice conferred the potential to reduce disease severity accompanied by a reduction of STAT3-related cell types, i.e., T_H1, T_H17, and T_{FH} cells [201].

2.3. The Sphingolipid Metabolism in Cerebrovascular Diseases

Lipid signaling plays pleiotropic roles in cerebral ischemia. In recent years, mounting evidence has emerged depicting the relationship between the sphingolipid metabolism and stroke (Figure 3). Studies have demonstrated that the driving force of neuroinflammation following cerebral ischemia are T cells. They migrate into the brain and amplify the initial detrimental damage [202–205]. Conversely, lymphocyte-deficient mice were shown to be protected from ischemic damage [206,207].

The S1P analogue and the $S1P_1$ functional antagonist fingolimod, originally derived from the fungal natural product ISP-1, was first synthesized in 1992 [208]. It impairs the egress of lymphocytes from primary and secondary lymphoid organs [155] and exerts immunomodulatory effects and non-immunological mechanisms [65,167,209–211]. Fingolimod was shown to provide protection from ischemic stroke and intracerebral hemorrhage [89,204,211–217], leading to the initiation of clinical studies demonstrating the efficacy of fingolimod for patients with acute ischemic stroke and improving clinical outcomes in patients with intracerebral hemorrhage [218]. In addition to reduced infarct volumes and improved neurological scores at 24 and 72 h after middle cerebral artery occlusion (MCAO; a commonly used animal model for ischemic stroke), fingolimod showed a deactivation of caspase-3, a reduction of terminal deoxynucleotidyl transferase-mediated uridine 5′-triphosphate-biotin nick end-labeling (TUNEL-) positive neurons, an activation of RAC-alpha serine/threonine-protein kinase (Akt) and extracellular-regulated kinase (ERK), and a Bcl-2 up-regulation, delineating an anti-apoptotic effect in neurons [211].

Moreover, studies have reported a role of S1PR in the preservation of endothelial barrier integrity [64,219], and phosphorylated fingolimod promotes the establishment of adherens junction in endothelial cells, i.e., an enhanced endothelial barrier function [65,209].

In contrast, Liesz et al., investigated the effect of fingolimod in permanent murine cerebral ischemia without achieving a significant reduction of infarct volumes and behavioral dysfunction despite effective lymphopenia [220]. In addition, Cai et al., unveiled no improvement in functional outcome and BBB integrity in large hemispheric infarctions and administration of fingolimod, either alone or in conjunction with recombinant-tissue plasminogen activator (rt-PA) [221]. Sanchez suggested that $S1P_1$ desensitization and/or degradation would potentially evoke detrimental effects on neurons and/or endothelial cells in the context of stroke. Therefore, the dosing and the timing of fingolimod administration seemed to be pivotal for its protective effects [222]. This is in accordance with a previous study by Brait et al., who showed that $S1P_1$ fosters protective effects regarding infarct volume after MCAO, however, only if the associated lymphopenia persists for at least 24 h [223].

SphK2 appears to wield an ambiguous nature in various disorders. SphK2 is the predominant S1P-synthesizing isoform in normal brain parenchyma [224] and particularly in cerebral microvascular endothelial cells [225]. SphK2 was recently shown to induce ischemic tolerance to stroke in C57BL/6 mice [226]. SphK2 is preferentially utilized to confer the neuroprotective effects of fingolimod, as it has a 30-fold higher affinity to the prodrug compared to SphK1 [225]. Mice lacking the SphK2 show larger ischemic lesions 24 h after 2 h of MCAO in comparison with wild-type animals [216], thus reinforcing the importance of extracellular signaling of S1PR. Moreover, studies have demonstrated that SphK2 predominates SphK1 in the phosphorylation of fingolimod in vitro [225] and in vivo [227]. In addition, hypoxia increases the expression and the activity levels of the SphK2 isoform in brain microvasculature, subsequently promoting ischemic tolerance [228]. In contrast to SphK2, $S1P_2$ is characterized as a key regulator of the pro-inflammatory phenotype of the endothelium [229] and promotes ischemia-induced vascular dysfunction [230]. Conversely, S1P generated by SphK1 potently facilitates the expression of IL-17A in activated microglia, thereby supporting neuronal apoptosis in cerebral ischemia [231]. This is in support of a study by Zheng et al., who found an enhanced expression of SphK1 in microglia 96 h after MCAO [232]. Subsequently, a cortical knockdown of SphK1 resulted in reduced infarct areas and less severe neurological deficits were observed [232].

Studies have highlighted the critical role of $S1P_2$ in ischemia-reperfusion injury, confirming that genetic deletion or inhibition of $S1P_2$ could block the development of hemorrhagic transformation and cerebral edema by inhibiting the matrix metalloproteinase-9 (MMP-9) activation in endothelial cells [230]. It was shown previously that the use of fingolimod conveyed a reduced risk of hemorrhagic transformation after thromboembolic occlusion [233]. In regard to these findings, the benefit of fingolimod in relation to hemorrhagic transformation was tested in randomized open-label multi-center trials [234,235]. Wan et al., focused on microRNA-149-5p and demonstrated its regulatory function on the permeability of the BBB after transient MCAO in rats by targeting $S1P_2$ of pericytes [236]. In their study, the expression of $S1P_2$ in pericytes increased at an early stage during ischemia/reperfusion, which was associated with an aggravation of BBB permeability in vivo and in vitro [236]. An engineered S1P chaperone, ApoM-Fc, maintained sustained S1P-S1PR signaling, resulting in a promoted function of the BBB after MCAO [237]. Another study puts emphasis on the importance of S1P in ameliorating the effects of stroke, as they reported reduced S1P lyase activity and a preferential synthesis of S1P and other sphingolipids in response to hypoxia [238].

In opposition to $S1P_2$, pathogenic mechanisms of $S1P_1$ and $S1P_3$ in cerebral ischemia rely on microglial activation [239,240]. Moreover, the same group elucidated the importance of $S1P_1$-regulation in promoting a pro-inflammatory M1 polarization of astrocytes, which was brought about by the intracellular signal transducers ERK1/2, p38, and JNK MAPK favoring brain damage after cerebral ischemia [241]. Furthermore, Zamanian et al., examined reactive astrogliosis in response to either MCAO or LPS and showed Pentraxin-related protein PTX3 (PTX3), tumor necrosis factor receptor superfamily member 12A (TNFRSF12A), and $S1P_3$ to be markers of reactive astrocytes after MCAO [242]. Liddelow et al., termed them "A1" and "A2" in analogy to the macrophage nomenclature [243]. Interestingly, $S1P_3$ was induced 46-fold after MCAO but only 6.4-fold by LPS. Under physiological circumstances, astrocytes were reported to express mainly $S1P_1$ and $S1P_3$, contrasting with very low levels for $S1P_2$ and $S1P_5$ [244–246]. However, in a recent study by Karunakaran et al., the authors demonstrated the importance of $S1P_2$ in microglial activation conferring impaired autophagy and propagating the inflammatory response in the BV2 microglial cell line [247]. Interestingly, similar effects were observed by the group after exogenous S1P administration or genetic knock out of SGPL1 [247].

S1P signaling is also functionally linked to influencing the pathophysiology during subarachnoid hemorrhage (SAH). In accordance with the detrimental effects caused by $S1P_2$ ligation in ischemia-reperfusion mentioned before, Yagi et al., demonstrated that S1P signaling increases vascular tone in the context of SAH, thus worsening neurological scores [248]. By employing a selective $S1P_2$ antagonistic treatment systemically using JTE013, they were able to set bounds to the extent of myogenic reactivity and to restore neurological scores to sham levels when administered instantly after SAH induction [248].

The various conditions that perturbed S1P signaling has been linked with are summed up in Figure 3.

3. Insights into Current and Future Therapeutic Perspectives

Almost a decade ago, the FDA approved the first drug aimed to interfere with the S1P-S1PR signaling cascade, fingolimod, in 2010 [162]. Due to fingolimod's preference for $S1P_1$ and the strong activation of this receptor subtype, $S1P_1$ eventually becomes down-regulated, resulting in a long-lasting functional antagonism that accounts for fingolimod-induced lymphocyte trapping in primary and secondary lymphoid organs. Consequently, this lymphocyte sequestration prevents auto-reactive T cells to migrate to the brain and therefore reduces the ferocious neurotoxic damage to myelin-associated proteins in patients with multiple sclerosis [158,249]. Nevertheless, due to its lack of specificity, a range of adverse effects (e.g., first-dose bradycardia [250], macular edema [251], elevated liver enzymes [252], and lymphopenia warranting vigilance regarding occurrence of infections [253–255]) is inevitable [256]. The endeavor to circumvent these unwanted drug effects led to the development of more tailored drugs aimed at selectively activating or inhibiting checkpoints within the sphingolipid metabolism on demand. To date, there is a huge number of clinical trials either completed (C), terminated (T), or momentarily being conducted in various clinical conditions examining the pharmacological exploitability of drugs designed to beneficially influence targets within the sphingolipid metabolism. A selection of these studies is presented below (Table 1). It is advisable to conceive that potentially all molecules/agents addressed in Figures 1 and 2 may represent future therapeutic targets.

Safingol, which targets PKC and non-selectively sphingosine kinases (SphK) [257], was identified in 1995 [258]. Safingol is now being tested in various cancer settings, since it is safe to co-administer with cisplatin and exerts tumoricidal effects [259,260]. Similarly, another drug targets SphK. The compound 3-(4-chlorophenyl)-adamantane-1-carboxylic acid (pyridine-4-ylmethyl)amide (ABC294-640) selectively inhibits SphK2. ABC294640 was identified in 2010 [261] and mechanistically competes with sphingosine for binding sites at SphK2. This, in turn, allows sphingosine and ceramides levels to increase (due to inability/slower rate of enzymatical conversion to S1P), facilitating apoptosis-inducing pathways [2,6,262]—a mechanism that was recently studied in various cancer therapies [263–266]. In that regard, conceivably, the exploitation of PKC and SphK subtypes in neurological conditions where their homeostasis of molecular activation, proliferation, and apoptosis is perturbed by crucially diminishing the S1P level within a cell and its immediate effects via PKC appears intuitive. This remains apprehensible, since S1P levels were previously shown to be enhanced to the disadvantage of its pro-differentiative and pro-apoptotic precursor ceramide [267,268].

Previously, a S1P-directed therapeutic agent was introduced [269]. The S1P-specific monoclonal antibody sonepcizumab (LT1009) is being tested in conditions where pathologies of the vasculature system occur [270–272].

Table 1. Trials of drugs interfering with the sphingosine 1-phosphate metabolism in clinical conditions. (C), completed; (T), terminated; (S), suspended (one study currently on halt due to COVID-19-related recruitment stop); SPHK1/2, sphingosine kinase 1/2; PKC, protein kinase C; S1P, sphingosine 1-phosphate; $S1P_n$, S1P receptor 'n'.

Target	Compounds (Mechanism of Action)	Indications	ClinicalTrials.gov Identifier	Phase
SPHK1/2, PKC inhibitor	Safingol	Solid tumor	NCT01553071	I
		Adult solid tumor (unspec.)	NCT00084812	I (C)
SPHK2	ABC294640 (SPHK2 inhibition)	Non-resectable, perihilar cholangiocarcinoma (extra- and intrahepatic)	NCT03377179	II
			NCT03414489	n/a
		Pancreatic cancer, adult solid tumor (unspec.)	NCT01488513	I (C)
		Multiple myeloma	NCT02757326	I, II (T)
S1P	n/a	Bacterial pneumonia	NCT04007328	II, III
		Food Allergy, anaphylaxis	NCT01776489	n/a
		Asthma	NCT04134351	n/a
		Pneumonia, chronic obstructive pulmonary disease, asthma	NCT03473119	n/a
	Sonepcizumab [LT1009] (S1P-specific mAb)	Solid tumors	NCT00661414	I (C)
		Neovascular age-related macular degeneration	NCT00767949	I
		Exudative age-related macular degeneration	NCT01414153	II (C)
		Pigment epithelial detachment	NCT01334255	I (T)
		Renal cell carcinoma	NCT01762033	II (T)
S1P lyase	LX3305 (S1P lyase inhibition)	Rheumatoid arthritis	NCT00847886, NCT01417052	I (C)
			NCT00903383	II (C)

Table 1. *Cont.*

Target	Compounds (Mechanism of Action)	Indications	ClinicalTrials.gov Identifier	Phase
S1P$_1$	n/a	Interstitial cystitis	NCT03003845	n/a
		Endometriosis	NCT02973854	n/a
		Vulvodynia	NCT02981433	n/a
	AKP11	Atopic dermatitis	ACTRN12617000763347	II
		Rheumatoid arthritis	ACTRN12617001223325	II
	BMS-986104	Rheumatoid arthritis (healthy volunteers)	NCT02211469	II (C)
	Cenerimod [ACT-334441] (S1P$_1$ agonist)	Systemic lupus erythematosus	NCT02472795	I, II (C)
		Healthy volunteers	NCT04052360	I (C)
			NCT04255277	I
	CS-077 (S1P$_1$ agonist)	Multiple sclerosis	NCT00616733	I (C)
	GSK2018682 (S1P$_1$ agonist)	Multiple sclerosis (healthy volunteers)	NCT01387217	I (C)
		Relapsing–remitting multiple sclerosis (healthy volunteers)	NCT01466322, NCT01431937	I (C)
	KRP203 (S1P$_1$ agonist)	Subacute cutaneous lupus erythematosus	NCT01294774	II (C)
		Hematological malignancies	NCT01830010	I (C)
		Ulcerative colitis	NCT01375179	II (T)
	Ponesimod [ACT-128800] (S1P$_1$ agonist)	Multiple sclerosis	NCT02425644	III (C)
			NCT03232073	III
			NCT02907177	III (T)
		Relapsing–remitting multiple sclerosis	NCT01093326	II
			NCT01006265	II (C)
		Plaque psoriasis	NCT00852670, NCT01208090	II (C)
		Chronic graft versus host disease	NCT02461134	II (T)
		Healthy volunteers	NCT02136888, NCT02068235, NCT03882255, NCT02223832	I (C)
		Pharmacokinetics	NCT02126956	I (C)
		Safety and tolerability	NCT02029482	I (C)

Table 1. *Cont.*

Target	Compounds (Mechanism of Action)	Indications	ClinicalTrials.gov Identifier	Phase
$S1P_1$, $S1P_3$, $S1P_4$, $S1P_5$	Fingolimod [FTY720] (S1PR modulator, $S1P_1$ functional antagonist)	Healthy volunteers	NCT00416845, NCT03757338	I (C)
		Multiple sclerosis	NCT00537082, NCT00670449, NCT00333138	II (C)
		Multiple sclerosis	NCT01892722	III
		Multiple sclerosis	NCT00662649, NCT00355134, NCT02939079, NCT00340834	III (C)
		Multiple sclerosis	NCT01647880	III (T)
		Multiple sclerosis	NCT01585298, NCT01333501	IV (C)
		Multiple sclerosis	NCT02232061, NCT01981161, NCT02769689	IV
		Multiple sclerosis	NCT02139696, NCT01592097, NCT01285479, NCT01281657, NCT02225977, NCT02408380, NCT03216915, NCT01811290, NCT02799199, NCT02776072, NCT01442194, NCT02021162, NCT02307877, NCT03243721	n/a
		Multiple sclerosis (autonomic nervous system dysfunction)	NCT02048072	IV (C)
		Multiple sclerosis (fatigue)	NCT01490840	IV (T)
		Multiple sclerosis (cognitive deficits)	NCT02141022	n/a
		Primary progressive multiple sclerosis	NCT00731692	III (T)

Table 1. *Cont.*

Target	Compounds (Mechanism of Action)	Indications	ClinicalTrials.gov Identifier	Phase
$S1P_1$, $S1P_3$, $S1P_4$, $S1P_5$	Fingolimod [FTY720] (S1PR modulator, $S1P_1$ functional antagonist)	Relapsing–remitting multiple sclerosis	NCT00289978, NCT01127750, NCT01201356, NCT01497262, NCT01199861	III (C)
			NCT01499667, NCT01633112	III (T)
			NCT01310166, NCT02325440	IV
			NCT02137707, NCT02307838, NCT01420055, NCT02720107, NCT03257358, NCT02373098, NCT01578330, NCT01623596, NCT01534182, NCT01317004, NCT01498887, NCT01216072, NCT01705236	IV (C)
			NCT01755871, NCT02342704, NCT03345940	IV (T)
			NCT01790269, NCT01704183, NCT02335892, NCT02277964	n/a
		Relapsing–remitting multiple sclerosis (cognition, brain volume loss)	NCT02575365	IV (T)
		Relapsing–remitting multiple sclerosis (depression)	NCT01436643	IV (T)
		Acute demyelinating optic neuritis	NCT01757691	II (T)
		Amyotrophic lateral sclerosis	NCT01786174	II (C)
		Intracerebral hemorrhage (hypertensive, intraparenchymal), cerebral edema	NCT04088630	I
		(Acute) stroke, (cerebro-) vascular accident, cerebral stroke	NCT02002390	II (C)
		Chemotherapy-induced peripheral neuropathy (numbness, pain, tingling)	NCT03943498	I

Table 1. *Cont.*

Target	Compounds (Mechanism of Action)	Indications	ClinicalTrials.gov Identifier	Phase
$S1P_1$, $S1P_3$, $S1P_4$, $S1P_5$	Fingolimod [FTY720] (S1PR modulator, $S1P_1$ functional antagonist)	Chronic inflammatory demyelinating polyradiculoneuropathy	NCT01625182	III (C)
		Breast carcinoma	NCT03941743	I
		Glioblastoma, anaplastic astrocytoma	NCT02490930	I (C)
		Coronavirus disease (COVID-19)	NCT04280588	II
		Asthma	NCT00785083	II (C)
		Rett syndrome	NCT02061137	II (C)
		Schizophrenia	NCT01779700	II (C)
		Renal insufficiency	NCT00731523	I (C)
		Kidney transplantation	NCT00239902, NCT00239798	II (C)
		Kidney transplantation	NCT00099736, NCT00239876, NCT00239811, NCT00099801, NCT00099749, NCT00239863, NCT00239785, NCT00098735	III (C)
$S1P_1$, $S1P_5$	Ceralifimod [ONO-4641] ($S1P_{1,5}$ agonist)	Multiple sclerosis	NCT01081782	II (C)
		Multiple sclerosis	NCT01226745	II (T)
	Ozanimod [RPC1063] ($S1P_{1,5}$ agonist)	Multiple sclerosis	NCT02797015	I (C)
		Multiple sclerosis	NCT02576717, NCT04140305	III
		Multiple sclerosis	NCT02294058	III (C)
		Relapsing–remitting multiple sclerosis	NCT01628393, NCT02047734	III (C)
		Ulcerative colitis	NCT01647516	II
		Ulcerative colitis	NCT02435992, NCT02531126, NCT03915769	III

Table 1. *Cont.*

Target	Compounds (Mechanism of Action)	Indications	ClinicalTrials.gov Identifier	Phase
$S1P_1$, $S1P_5$	Ozanimod [RPC1063] ($S1P_{1,5}$ agonist)		NCT02531113	II (C)
		Crohn's disease	NCT03467958, NCT03464097, NCT03440372, NCT03440385	III
		Healthy volunteers	NCT02994381, NCT03694119, NCT03644576, NCT03624959, NCT03665610	I (C)
			NCT04149678, NCT04211558	I
	Siponimod [BAF312] ($S1P_{1,5}$ modulator)	Healthy volunteers	NCT00422175	I (C)
		Multiple sclerosis	NCT03623243	III
		Relapsing-remitting multiple sclerosis	NCT01185821, NCT00879658	II (C)
		Secondary progressive multiple sclerosis	NCT01665144	III
			NCT02330965	n/a
		Polymyositis (, dermato-myositis)	NCT01801917, NCT01148810, NCT02029274	II (T)
		Hepatic impairment	NCT01565902	I (C)
		Hemorrhagic stroke, intracerebral hemorrhage (ICH)	NCT03338998	II (S)
		Renal impairment	NCT01904214	I (C)

Table 1. *Cont.*

Target	Compounds (Mechanism of Action)	Indications	ClinicalTrials.gov Identifier	Phase
$S1P_1, S1P_5, (S1P_4)$	Amiselimod [MT-1303] (S1PR modulator, $S1P_1$ functional antagonist)	Relapsing–remitting multiple sclerosis	NCT02193217, NCT02310048, NCT02293967	I (C)
		Crohn's disease	NCT01890655, NCT01742052	II (C)
			NCT02148185	I (C)
			NCT02389790, NCT02378688	II (C)
		Systemic lupus erythematosus	NCT02307643	I (C)
		Plaque psoriasis	NCT01987843	II (C)
		Inflammatory bowel disease	NCT01666327	I (C)
		Primary biliary cholangitis	NCT03155932	II (T)
		Inflammatory bowel disease(extra-int. skin manifestations)	NCT03139032	II (T)
	Etrasimod [APD334]	Ulcerative colitis	NCT02447302, NCT02536404	II (C)
			NCT03950232, NCT03945188, NCT03996369, NCT04176588	III
		Pyoderma gangrenosum	NCT03072953	II (T)
		Crohn's disease	NCT04173273	II
		Atopic dermatitis	NCT04162769	II

Conversely to the aforementioned mechanisms inevitably reducing the amount of S1P available, drugs either mimicking S1P effects at the receptor site or actually increasing S1P levels have been designed. The S1P lyase inhibitor LX3305 is currently being investigated in rheumatoid arthritis as an alternative to therapies with biologicals [273]. Conceptually, LX3305's tentative application in neurological conditions where S1P is reduced/disturbed appears undoubtedly apprehensible, e.g., in neurodegenerative disorders such as Alzheimer's disease [134], Parkinson's disease [274], Huntington's disease [151,153], or amyotrophic lateral sclerosis [154]. Several diseases have been implicated in aberrant S1PR-specific signaling pathways. In that regard, several drugs specifically designed to interact with $S1P_1$ are available to date. AKP11, for example, was compared against fingolimod in a rodent model of multiple sclerosis and was associated with a higher degree of endosomal receptor recycling upon withdrawal, lesser extent of proteasomal degradation, and milder and more easily reversible lymphopenia [275]. Despite similar therapeutic effects, an almost complete absence of adverse events was observed [275]. Similarly, but more recently, BMS-986104 was shown to act equivalently efficient to fingolimod in a T cell transfer colitis model, although not conveying as many cardiovascular and pulmonary adverse events in in vitro settings [276]. Moreover, cenerimod (ACT-334441) could also be confirmed as a potent and selective $S1P_1$ agonistic signaling properties, whilst broncho- and vasoconstrictive effects were not clinically relevant [277]. In humans, cenerimod showed an improvement in disease activity scores for systemic lupus without constraining an acceptable safety profile [278]. In contrast to these specific and well-tolerable agents, GSK2018682, another $S1P_1$ agonist, did bring about bradycardia and subsequent AV-block [279]. In contrast, BMS-986104 and cenerimod seem to have a favorable risk profile in comparison to fingolimod, which, of course, warrants further investigation in in vivo studies to determine its safety and efficacy in other conditions. Interestingly, in pancreatic islet transplantation, which at least in humans is a definitive treatment for type 1 diabetes mellitus denoted by high mortality and morbidity in the early phase after transplantation [280], KRP203 was shown to cause a marked increase in viable pancreatic islet transplants in C57BL/10 mice [281]. Nevertheless, KRP203 is not an entirely selective $S1P_1$ agonist, as it does bind to $S1P_3$ with 5-fold and to $S1P_2$ and $S1P_5$ 100-fold lesser selectivity [281]. Thus, concerns regarding adverse effects, particularly with potentially increased doses necessary and depending on pharmacogenetics, should govern careful investigations in humans. Lastly with respect to $S1P_1$ specific compounds, ponesimod (ACT-128800) was previously reported to display therapeutic efficacy in psoriasis whilst also distinguishing itself by swift reversibility upon discontinuation; however, some degree of cardiac effects was detected in clinical trials [282].

It may be advised, under some conditions, to exploit multiple S1PR pathways. $S1P_5$, for example, is implicated in having pro-fibrotic effects to act on proliferation and its involvement in early transforming-growth-factor-β (TGF-β)-signaling [283]. Therefore, fibrotic conditions or tissue scarring might be well-suited for treatment with an agent synthesized to evoke agonistic effects both against $S1P_1$ and $S1P_5$. Three compounds are currently being tested, which act in this pharmacological manner, namely: ceralifimod (ONO-4641), ozanimod (RPC1063), and siponimod (BAF312).

Ceralifimod (ONO-4641), 1-((6-(2-methoxy-4-propylbenzyl)oxy)-1-methyl-3,4-dihydronaphthalen -2-yl)methyl)azetidine-3-carboxylic acid 13n was recently synthesized in 2017 [284]. Beside ceralifimod, ozanimod (RPC1063), which acts similarly, was shown to cause beneficial effects in patients with multiple sclerosis and ulcerative colitis alike [285]. In addition, the authors found that ozanimod wielded strong influence on the innate immune cells. Here, plasmacytoid dendritic cells were lowered (potentially by means of sequestration), which, in turn, reduced interferon alpha (IFN-α) in lupus patients in addition to reducing the entirety of B cell and T cell subsets in the spleen [285].

Lastly, agents that not only depicted $S1P_1$ and $S1P_5$ agonistic features but that also were partially able to engage with $S1P_4$ were identified. This cohort of compounds currently comprises amiselimod (MT-1303) and etrasimod (APD334). The potential of being able to engage with $S1P_4$ may hold promising therapeutic benefits. In this regard, $S1P_4$, as discussed previously in this review, is predominantly expressed in lymphoid tissues [88,95], and ligation and its subsequent signaling are involved in marking time regarding proliferation [286,287], a reduction of effector cytokines secreted [286,287], and migration of lymphocytes [288,289]. It is worth noting that amiselimod displays a very safe risk profile [290–292], while it is too early to consistently assess this for etrasimod [293]. Their application and/or investigation regarding their future therapeutic exploitability in neurological conditions should find due consideration soon.

4. Conclusions

In this review, we described the complexities of the sphingosine 1-phosphate (S1P) signaling in neurological conditions in reflection of currently available S1P signaling targeted drugs. Starting with the de novo synthesis of S1P either at the smooth endoplasmic reticulum or other subcellular microdomains, we highlighted the currently established signaling pathways. However, S1P signaling also occurs after secretion and transportation by its chaperones HDL or albumin in the extracellular compartment, allowing either autocrine or paracrine signaling upon ligation to $S1P_{1-5}$. It was highlighted that each S1PR subtype is capable of coupling to a variety of heterotrimeric G proteins, subsequently allowing a tailored intracellular signaling cascade to be incited. However, under perturbed circumstances, the presence of co-activators, inhibitors of S1PR, or simply skewed S1PR patterns may predispose for disease onset and progression. Despite remarkable advances in understanding the contribution of sphingolipid signaling to neurological disorders, the field has yet a lot to learn. In this review, we highlighted the currently available literature regarding perturbations of the sphingolipid metabolism in the context of neurodegenerative, neuroinflammatory, and cerebrovascular diseases. A considerable number of clinical trials are being carried out testing S1P signaling targeted drugs in conditions linked to activation of the immune system. These trials may enhance our understanding of the importance of the S1P–S1PR axis and ultimately help to inform us about future therapeutic usability of these compounds in various neurological disorders. We reported on the importance of $S1P_1$ for vascular and other barrier functions. Activation of $S1P_1$ causes a significant improvement of vascular barrier properties and prevents microvascular leakage. Currently available drugs interacting with $S1P_1$ initially act as agonists but then may cause a profound and long-lasting desensitization and degradation of $S1P_1$. As outlined above, they finally act as functional antagonists with, in the long term, negative impact on vascular integrity. Currently, there is no pure $S1P_1$ receptor agonist available that does not desensitize the receptor. The compounds described thus far may indeed have a varying degree of agonistic and antagonistic properties. However, such a "true agonist" would be highly desirable and unique in order to protect from vascular leakage.

Author Contributions: A.L. and J.S. contributed equally to the design and the conception of this article and wrote the entire manuscript. R.B., J.M.P., and W.P. critically edited the manuscript and supervised the writing. All authors have read and agreed to the published version of the manuscript.

Funding: The authors would like to acknowledge the support by the Open Access Publication Fund of the Goethe University Frankfurt. R.B., J.M.P., and W.P. are funded by the German Research Organization (DFG; SFB1039). J.M.P. and W.P. are funded by the Fondation Leducq (SphingoNet). J.M.P. is funded by the Uniscientia Foundation Vaduz. J.S. is funded by a scholarship from the German Academic Scholarship Foundation (Studienstiftung des deutschen Volkes).

Conflicts of Interest: The authors declare no conflict of interest.

Abbreviations

AD	Alzheimer's disease
Akt	RAC-alpha serine/threonine-protein kinase
ALS	Amyotrophic lateral sclerosis
aPC	Activated protein C
ApoM	Apolipoprotein M

APP	Amyloid precursor protein
Aβ	Amyloid-β-peptide
BBB	Blood brain barrier
CBAs	Conjugated bile acids
CBF	Cerebral blood flow
CCL4	C-C motif chemokine 4-like
CCL5	C-C motif chemokine 5-like
CCO	Cytochrome c oxidase
CCR7	C-C chemokine receptor type 7
CD	Cluster of differentiation molecule
CDase	Ceramidase
c-fos	Proto-oncogene c-Fos
cIAP2	Cellular inhibitor of apoptosis 2
CNS	Central nervous system
COX-2	Cyclooxygenase-2
CXCL8	C-X-C motif chemokine 8, IL-8
CXCL10	C-X-C motif chemokine 10
DHHC5	Palmitoyltransferase DHHC5
DNA	Desoxyribonucleic acid
EAE	Experimental autoimmune encephalomyelitis
EC	Extracellular
ERK	Extracellular-regulated kinase
GBA	Glucocerebrosidase
GCS	Glucosylceramide synthase
Glc	Glucose
GlcCer	Glucosylceramidase
GRK2	G protein-coupled receptor kinase 2
$G\alpha_{i/o}$	Inhibitory Gα subunit
$G\alpha_q$	q subtype of Gα proteins
$G\alpha_s$	Stimulatory Gα subunit
$G_{12/13}$	12/13 subtype of Gα proteins
HD	Huntigton's disease
HDAC1/2	Histone deacetylases 1/2
HDL	High-density lipoprotein
hTERT	Human telomerase reverse transcriptase
IC	Intracellular
ICAM1	Intracellular adhesion molecule 1
IFN-α	Interferon α
IL-1	Interleukin-1
IL-6	Interleukin-6
IL-17A	Interleukin-17A
IRF1	Interferon regulatory factor 1
IκB	Inhibitor of nuclear factor κ-B kinase
IκBα	NF-κB inhibitor α
LBD	Lewy body disorders
LDL	Low-density lipoprotein
LPP	Lipid phosphate phosphohydrolase
MCAO	Middle cerebral artery occlusion
MFSD2B	Major facilitator superfamily domain-containing protein 2B
MKRN1	Makorin ring finger protein 1
MMP-9	Matrix metalloproteinase-9
mRNA	Messenger ribonucleic acid
MS	Multiple slerosis
NF-κB	Nuclear factor κ-light-chain-enhancer of activated B cells
NMO	Neuromyelitis optica

NVU	Neurovascular unit
ORMDL3	ORM1-like protein 3
OXPHOS	Oxidative phosphorylation
PalCoA	Palmitoyl-CoA
PD	Parkinson's disease
PE	Phosphoethanolamine
PGC1β	PPARγ co-activator 1β
PHB2	Prohibitin-2
PKC	Protein kinase C
PKCζ	Atypical PKC subtype ζ
PKCι	Atypical PKC subtype ι
PLPP3	Phospholipid phosphatase 3
PM	Plasma membrane
PPARγ	Peroxisome proliferator activated receptor γ
PTX3	Pentraxin-related protein 3
p21	Cyclin-dependent kinase inhibitor 1
RACK	Receptor for activated C kinase proteins
RhoA	Transforming protein RhoA
RING	Really interesting new group
RIP1	Receptor-interacting serine/threonine-protein kinase 1
rt-PA	Recombinant-tissue plasminogen activator
SAH	Subarachnoid hemorrhage
sER	Smooth endoplasmic reticulum
SGPL1	S1P lyase 1
SGPP1/2	Sphingosine 1-phosphate phosphohydrolase 1/2
SLE	Systemic lupus erythematosus
SMase	Sphingomyelinase
SMS	Sphingomyelin synthase
SphK1	Sphingosine kinase 1
SphK2	Sphingosine kinase 2
SPNS2	Spinster homolog 2
SPT	Serine palmitoyltransferase
STAT3	Signal transducer and activator of transcription 3
S1P	Sphingosine 1-phosphate
S1PR	Sphingosine 1-phosphate receptor
S1P$_{1-5}$	S1PR subtype 1–5
TAB3	TGF-β-activated kinase 1 and MAP3K7-binding protein3
T$_{FH}$	Follicular T$_H$ cell
TGF-β	Transforming-growth-factor-β
T$_H$-17	T$_H$-17 type cell
TNFRSF12A	Tumor necrosis factor receptor superfamily member 12A
TNF-α	Tumor necrosis factor-α
TRAF2	TNF-α receptor-associated factor 2
TRAF6	TNF-α receptor-associated factor 6
T$_{REG}$	Regulatory T cell
TUNEL	Terminal deoxynucleotidyl transferase-mediated, uridine 5′ triphosphate-biotin nick end-labeling
UGT8A	UDP galactosyltransferase 8A
VEGFa	Vascular endothelial growth factor A
VLDL	Very low-density lipoprotein
WWP2	NEDD4-like E3 ubiquitin ligase WWP2

References

1. Zhang, H.; Desai, N.N.; Olivera, A.; Seki, T.; Brooker, G.; Spiegel, S. Sphingosine-1-phosphate, a novel lipid, involved in cellular proliferation. *J. Cell Biol.* **1991**, *114*, 155–167. [CrossRef]
2. Cuvillier, O.; Pirianov, G.; Kleuser, B.; Vanek, P.G.; Coso, O.A.; Gutkind, S.; Spiegel, S. Suppression of cermide-mediated programmed cell death by sphingosine-1-phosphate. *Nature* **1996**, *381*, 800–803. [CrossRef]
3. Breslow, D.K.; Collins, S.R.; Bodenmiller, B.; Aebersold, R.; Simons, K.; Shevchenko, A.; Ejsing, C.S.; Weissman, J.S. Orm family proteins mediate sphingolipid homeostasis. *Nature* **2010**, *463*, 1048–1053. [CrossRef] [PubMed]
4. Breslow, D.K.; Weissman, J.S. Membranes in Balance: Mechanisms of Sphingolipid Homeostasis. *Mol. Cell.* **2010**, *40*, 267–279. [CrossRef] [PubMed]
5. Hannun, Y.A.; Obeid, L.M. The ceramide-centric universe of lipid-mediated cell regulation: Stress encounters of the lipid kind. *J. Biol. Chem.* **2002**, *277*, 25847–25850. [CrossRef]
6. Linn, S.C.; Kim, H.S.; Keane, E.M.; Andras, L.M.; Wang, E.; Merrill, A.H., Jr. Regulation of de novo sphingolipid biosynthesis and the toxic consequences of its disruption. *Biochem. Soc. Trans.* **2001**, *29*, 831–835. [CrossRef]
7. Xu, R.; Jin, J.; Hu, W.; Sun, W.; Bielawski, J.; Szulc, Z.; Taha, T.; Obeid, L.M.; Mao, C. Golgi alkaline ceramidase regulates cell proliferation and survival by controlling levels of sphingosine and S1P. *FASEB J.* **2006**, *20*, 1813–1825. [CrossRef] [PubMed]
8. Galadari, S.; Wu, B.X.; Mao, C.; Roddy, P.; El Bawab, S.; Hannun, Y.A. Identification of a novel amidase motif in neutral ceramidase. *Biochem. J.* **2006**, *393*, 687–695. [CrossRef] [PubMed]
9. Prinetti, A.; Loberto, N.; Chigorno, V.; Sonnino, S. Glycosphingolipid behaviour in complex membranes. *Biochim. Biophys. Acta* **2009**, *1788*, 184–193. [CrossRef]
10. Maceyka, M.; Spiegel, S. Sphingolipid metabolites in inflammatory disease. *Nature* **2014**, *510*, 58–67. [CrossRef]
11. Tafesse, F.G.; Ternes, P.; Holthuis, J.C.M. The multigenic sphingomyelin synthase family. *J. Biol. Chem.* **2006**, *281*, 29421–29425. [CrossRef] [PubMed]
12. Hakomori, S. Traveling for the glycosphingolipid path. *Glycoconj. J.* **2000**, *17*, 627–647. [CrossRef]
13. Ichikawa, S.; Hirabayashi, Y. Glucosylceramide synthase and glycosphingolipid synthesis. *Trends Cell Biol.* **1998**, *8*, 198–202. [CrossRef]
14. Zhou, J.; Saba, J.D. Identification of the first mammalian sphingosine phosphate lyase gene and its functional expression in yeast. *Biochem. Biophys. Res. Commun.* **1998**, *242*, 502–507. [CrossRef]
15. Bektas, M.; Allende, M.L.; Lee, B.G.; Chen, W.; Amar, M.J.; Remaley, A.T.; Saba, J.D.; Proia, R.L. Sphingosine 1-phosphate lyase deficiency disrupts lipid homeostasis in liver. *J. Biol. Chem.* **2010**, *285*, 10880–10889. [CrossRef] [PubMed]
16. Bandhuvula, P.; Saba, J.D. Sphingosine-1-phosphate lyase in immunity and cancer: Silencing the siren. *Trends Mol. Med.* **2007**, *13*, 210–217. [CrossRef]
17. Nakahara, K.; Ohkuni, A.; Kitamura, T.; Abe, K.; Naganuma, T.; Ohno, J.; Zoeller, R.A.; Kihara, A. The Sjögren-Larsson syndrome gene encodes a hexadecenal dehydrogenase of the sphingosine 1-phosphate degradation pathway. *Mol. Cell.* **2012**, *46*, 461–471. [CrossRef]
18. Dobrosotskaya, I.Y.; Seegmiller, A.C.; Brown, M.S.; Goldstein, J.L.; Rawson, R.B. Regulation of SREBP processing and membrane lipid production by phospholipids in Drosophila. *Science* **2002**, *29*, 879–883. [CrossRef] [PubMed]
19. Cantalupo, A.; Zhang, Y.; Kothiya, M.; Galvani, S.; Obinata, H.; Bucci, M.; Giordano, F.J.; Jiang, X.-C.; Hla, T.; Di Lorenzo, A. Nogo-B regulates endothelial sphingolipid homeostasis to control vascular function and blood pressure. *Nat. Med.* **2015**, *21*, 1028–1037. [CrossRef]
20. Zhang, Y.; Huang, Y.; Cantalupo, A.; Azevedo, P.S.; Siragusa, M.; Bielawski, J.; Giordano, F.J.; Di Lorenzo, A. Endothelial Nogo-B regulates sphingolipid biosynthesis to promote pathological cardiac hypertrophy during chronic pressure overload. *JCI Insight* **2016**, *1*, e85484. [CrossRef]
21. Le Stunff, H.; Giussani, P.; Maceyka, M.; Lépine, S.; Milstien, S.; Spiegel, S. Recycling of sphingosine is regulated by the concerted actions of sphingosine-1-phosphate phosphohydrolase 1 and sphingosine kinase 2. *J. Biol. Chem.* **2007**, *282*, 34372–34380. [CrossRef] [PubMed]

22. Ogawa, C.; Kihara, A.; Gokoh, M.; Igarashi, Y. Identification and characterization of a novel human sphingosine-1-phosphate phosphohydrolase, hSPP2. *J. Biol. Chem.* **2003**, *278*, 1268–1272. [CrossRef] [PubMed]

23. Kunkel, G.T.; Maceyka, M.; Milstien, S.; Spiegel, S. Targeting the sphingosine-1-phosphate axis in cancer, inflammation and beyond. *Nat. Rev. Drug. Discov.* **2013**, *12*, 688–702. [CrossRef] [PubMed]

24. Gillard, B.K.; Clement, R.G.; Marcus, D.M. Variations among cell lines in the synthesis of sphingolipids in de novo and recycling pathways. *Glycobiology* **1998**, *8*, 885–890. [CrossRef] [PubMed]

25. Tettamanti, G.; Bassi, R.; Viani, P.; Riboni, L. Salvage pathways in glycosphingolipid metabolism. *Biochimie* **2003**, *85*, 423–437. [CrossRef]

26. Cartier, A.; Hla, T. Sphingosine 1-phosphate: Lipid signaling in pathology and therapy. *Science* **2019**, *366*, eaar5551. [CrossRef]

27. Strub, G.M.; Paillard, M.; Liang, J.; Gomez, L.; Allegood, J.C.; Hait, N.C.; Maceyka, M.; Price, M.M.; Chen, Q.; Simpson, D.C; et al. Sphingosine-1-phosphate produced by sphingosine kinase 2 in mitochondria interacts with prohibitin 2 to regulate complex IV assembly and respiration. *FASEB J.* **2011**, *25*, 600–612. [CrossRef]

28. Maceyka, M.; Harikumar, K.B.; Milstien, S.; Spiegel, S. Sphingosine-1-phosphate signaling and its role in disease. *Trends Cell Biol.* **2012**, *22*, 50–60. [CrossRef]

29. Huwiler, A.; Kolter, T.; Pfeilschifter, J.; Sandhoff, K. Physiology and pathophysiology of sphingolipid metabolism and signaling. *Biochim. Biophys. ACTA* **2000**, *1485*, 63–99. [CrossRef]

30. Pappu, R.; Schwab, S.R.; Cornelissen, I.; Pereira, J.P.; Regard, J.B.; Xu, Y.; Camerer, E.; Zheng, Y.-W.; Huang, Y.; Cyster, J.G.; et al. Promotion of lymphocyte egress into blood and lymph by distinct sources of sphingosine-1-phosphate. *Science* **2007**, *316*, 295–298. [CrossRef]

31. Venkataraman, K.; Lee, Y.-M.; Michaud, J.; Thangada, S.; Ai, Y.; Bonkovsky, H.L.; Parikh, N.S.; Habrukowich, C.; Hla, T. Vascular endothelium as a contributor of plasma sphingosine 1-phosphate. *Circ. Res.* **2008**, *102*, 669–676. [CrossRef] [PubMed]

32. Hisano, Y.; Kobayashi, N.; Yamaguchi, A.; Nishi, T. Mouse SPNS2 functions as a Sphingosine-1-phosphate transporter in vascular endothelial cells. *PLoS One* **2012**, *7*, e38941. [CrossRef]

33. Fukuhara, S.; Simmons, S.; Kawamura, S.; Inoue, A.; Orba, Y.; Tokudome, T.; Sunden, Y.; Arai, Y.; Moriwaki, K.; Ishida, J.; et al. The sphingosine-1-phosphate transporter Spns2 expressed on endothelial cells regulates lymphocyte trafficking in mice. *J. Clin. Investig.* **2012**, *122*, 1416–1426. [CrossRef] [PubMed]

34. Alvarez, S.E.; Harikumar, K.B.; Hait, N.C.; Allegood, J.; Strub, G.M.; Kim, E.Y.; Maceyka, M.; Jiang, H.; Luo, C.; Kordula, T.; et al. Sphingosine-1-phosphate is a missing cofactor for the E3 ubiquitin ligase TRAF2. *Nature* **2010**, *465*, 1084–1088. [CrossRef]

35. Spiegel, S.; Milstien, S. The outs and the ins of sphingosine-1-phosphate in immunity. *Nat. Rev. Immunol.* **2011**, *11*, 403–415. [CrossRef]

36. Kajimoto, T.; Caliman, A.D.; Tobias, I.S.; Okada, T.; Pilo, C.A.; Van, A.-A. N.; McCammon, J.A.; Nakamura, S.-I.; Newton, A.C. Activation of atypical protein kinase C by sphingosine 1-phosphate revealed by an aPKC-specific activity reporter. *Sci. Signal.* **2019**, *12*, eaat6662. [CrossRef] [PubMed]

37. Adams, D.R.; Ron, D.; Kiely, P.A. RACK1, a multifaceted scaffolding protein: Structure and function. *Cell Commun. Signal.* **2011**, *9*, 22. [CrossRef]

38. Steinberg, S.F. Structural basis of protein kinase C isoform function. *Physiol. Rev.* **2008**, *88*, 1341–1378. [CrossRef]

39. Moscat, J.; Diaz-Meco, M.T. The atypical protein kinase Cs: Functional specificity mediated by specific protein adapters. *EMBO Rep.* **2000**, *1*, 399–403. [CrossRef]

40. Li, S.; Wang, L.; Dorf, M.E. PKC phosphorylation of TRAF2 mediates IKKα/β recruitment and K63-linked polyubiquitination. *Mol. Cell.* **2009**, *33*, 30–42. [CrossRef]

41. Etemadi, N.; Chopin, M.; Anderton, H.; Tanzer, M.C.; Rickard, J.A.; Abeysekera, W.; Hall, C.; Spall, S.; Wang, B.; Xiong, Y.; et al. TRAF2 regulates TNF and NF-κB signalling to suppress apoptosis and skin inflammation independently of sphingosine kinase. *Elife* **2015**, *4*, e10592. [CrossRef] [PubMed]

42. Kanayama, A.; Seth, R.B.; Sun, L.; Ea, C.-K.; Hong, M.; Shaito, A.; Chiu, Y.-H.; Deng, L.; Chen, Z.J. TAB2 and TAB3 activate the NF-κB pathway through binding to polyubiquitin chains. *Mol. Cell.* **2004**, *15*, 535–548. [CrossRef] [PubMed]

43. Deng, L.; Wang, C.; Spencer, E.; Yang, L.; Braun, A.; You, J.; Slaughter, C.; Pickart, C.; Chen, Z.J. Activation of the Iκb kinase complex by TRAF6 requires a dimeric ubiquitin-conjugating enzyme complex and a unique polyubiquitin chain. *Cell* **2000**, *103*, 351–361. [CrossRef]

44. Manning, E.; Pullen, S.S.; Souza, D.J.; Kehry, M.; Noelle, R.J. Cellular responses to murine CD40 in a mouse B cell line may be TRAF dependent or independent. *Eur. J. Immunol.* **2002**, *32*, 39–49. [CrossRef]

45. Tontonoz, P.; Spiegelman, B.M. Fat and beyond: The diverse biology of PPARγ. *Annu. Rev. Biochem.* **2008**, *77*, 289–312. [CrossRef]

46. Parham, K.A.; Zebol, J.R.; Tooley, K.L.; Sun, W.Y.; Moldenhauer, L.M.; Cockshell, M.P.; Gliddon, B.L.; Moretti, P.A.; Tigyi, G.; Pitson, S.M.; et al. Sphingosine 1-phosphate is a ligand for peroxisome proliferator-activated receptor-γ that regulates neoangiogenesis. *FASEB J.* **2015**, *29*, 3638–3653. [CrossRef]

47. Ogretmen, B. Sphingolipid metabolism in cancer signalling and therapy. *Nat. Rev. Cancer* **2018**, *18*, 33–50. [CrossRef]

48. Maceyka, M.; Sankala, H.; Hait, N.C.; Le Stunff, H.; Liu, H.; Toman, R.; Collier, C.; Zhang, M.; Satin, L.S.; Merrill, A.H., Jr.; et al. SphK1 and SphK2, sphingosine kinase isoenzymes with opposing functions in sphingolipid metabolism. *J. Biol. Chem.* **2005**, *280*, 37118–37129. [CrossRef]

49. Igarashi, N.; Okada, T.; Hayashi, S.; Fujita, T.; Jahangeer, S.; Nakamura, S.-I. Sphingosine kinase 2 is a nuclear protein and inhibits DNA synthesis. *J. Biol. Chem.* **2003**, *278*, 46832–46839. [CrossRef]

50. Sankala, H.M.; Hait, N.C.; Paugh, S.W.; Shida, D.; Lépine, S.; Elmore, L.W.; Dent, P.; Milstien, S.; Spiegel, S. Involvement of sphingosine kinase 2 in p53-independent induction of p21 by the chemotherapeutic drug doxorubicin. *Cancer Res.* **2007**, *67*, 10466–10474. [CrossRef]

51. Ding, G.; Sonoda, H.; Yu, H.; Kajimoto, T.; Goparaju, S.K.; Jahangeer, S.; Okada, T.; Nakamura, S.-I. Protein kinase D-mediated phosphorylation and nuclear export of sphingosine kinase 2. *J. Biol. Chem.* **2007**, *282*, 27493–27502. [CrossRef] [PubMed]

52. Spiegel, S.; Maczis, M.A.; Maceyka, M.; Milstien, S. New insights into functions of the sphingosine-1-phosphate transporter SPNS2. *J. Lipid Res.* **2019**, *60*, 484–489. [CrossRef] [PubMed]

53. Hait, N.C.; Allegood, J.; Maceyka, M.; Strub, G.M.; Harikumar, K.B.; Singh, S.K.; Luo, C.; Marmorstein, R.; Kordula, T.; Milstien, S.; et al. Regulation of histone acetylation in the nucleus by sphingosine-1-phosphate. *Science* **2009**, *325*, 1254–1257. [CrossRef]

54. Jamaladdin, S.; Kelly, R.D.W.; O'Regan, L.; Dovey, O.M.; Hodson, G.E.; Millard, C.J.; Portolano, N.; Fry, A.M.; Schwabe, J.W.R.; Cowley, S.M. Histone deacetylase (HDAC) 1 and 2 are essential for accurate cell division and the pluripotency of embryonic stem cells. *Proc. Natl. Acad. Sci. USA* **2014**, *111*, 9840–9845. [CrossRef] [PubMed]

55. Choudhary, C.; Kumar, C.; Gnad, F.; Nielsen, M.L.; Rehman, M.; Walther, T.C.; Olsen, J.V.; Mann, M. Lysine acetylation targets protein complexes and co-regulates major cellular functions. *Science* **2009**, *325*, 834–840. [CrossRef]

56. Kelly, R.D.W.; Chandru, A.; Watson, P.J.; Song, Y.; Blades, M.; Robertson, N.S.; Jamieson, A.G.; Schwabe, J.W.R.; Cowley, S.M. Histone deacetylase (HDAC) 1 and 2 complexes regulate both histone acetylation and crotonylation in vivo. *Sci. Rep.* **2018**, *8*, 14690. [CrossRef] [PubMed]

57. Panneer Selvam, S.; De Palma, R.M.; Oaks, J.J.; Oleinik, N.; Peterson, Y.K.; Stahelin, R.V.; Skoralakes, E.; Ponnusamy, S.; Garrett-Mayer, E.; Smith, C.D.; et al. Binding of the sphingolipid S1P to hTERT stabilizes telomerase at the nuclear periphery by allosterically mimicking protein phosphorylation. *Sci. Signal.* **2015**, *8*, ra58. [CrossRef]

58. Kim, J.H.; Park, S.-M.; Kang, M.R.; Oh, S.-Y.; Lee, T.H.; Muller, M.T.; Chung, I.K. Ubiquitin ligase MKRN1 modulates telomere length homeostasis through a proteolysis of hTERT. *Genes Dev.* **2005**, *19*, 776–781. [CrossRef]

59. Murata, N.; Sato, K.; Kon, J.; Tomura, H.; Yanagita, M.; Kuwabara, A.; Ui, M.; Okajima, F. Interaction of sphingosine 1-phosphate with plasma components, including lipoproteins, regulates the lipid receptor-mediated actions. *Biochem. J.* **2000**, *352*, 809–815. [CrossRef]

60. Argraves, K.M.; Argraves, W.S. HDL serves as a S1P signaling platform mediating a multitude of cardiovascular effects. *J. Lipid Res.* **2007**, *48*, 2325–2333. [CrossRef]

61. Christoffersen, C.; Obinata, H.; Kumaraswamy, S.B.; Galvani, S.; Ahnström, J.; Sevvana, M.; Egerer-Sieber, C.; Muller, Y.A.; Hla, T.; Nielsen, L.B.; et al. Endothelium-protective sphingosine-1-phosphate provided by HDL-associated apolipoprotein M. *Proc. Natl. Acad. Sci. USA* **2011**, *108*, 9613–9618. [CrossRef] [PubMed]

62. Kurano, M.; Tsukamoto, K.; Ohkawa, R.; Hara, M.; Iino, J.; Kageyama, Y.; Ikeda, H.; Yatomi, Y. Liver involvement in sphingosine 1-phosphate dynamism revealed by adenoviral hepatic overexpression of apolipoprotein M. *Atherosclerosis.* **2013**, *229*, 102–109. [CrossRef] [PubMed]

63. Liu, M.; Seo, J.; Allegood, J.; Bi, X.; Zhu, X.; Boudyguina, E.; Gebre, A.K.; Avni, D.; Shah, D.; Sorci-Thomas, M.G.; et al. Hepatic apolipoprotein M (ApoM) overexpression stimulates formation of larger ApoM/sphingosine 1-phosphate-enriched plasma high density lipoprotein. *J. Biol. Chem.* **2014**, *289*, 2801–2814. [CrossRef]

64. Lee, M.J.; Thangada, S.; Claffey, K.P.; Ancellin, N.; Liu, C.H.; Kluk, M.; Volpi, M.; Sha'afi, R.I.; Hla, T. Vascular endothelial cell adherens junction assembly and morphogenesis induced by sphingosine-1-phosphate. *Cell* **1999**, *99*, 301–312. [CrossRef]

65. Sanchez, T.; Estrada-Hernandez, T.; Paik, J.-H.; Wu, M.-T.; Venkataraman, K.; Brinkmann, V.; Claffey, K.; Hla, T. Phosphorylation and action of the immunomodulator FTY720 inhibits vascular endothelial cell growth factor-induced vascular permeability. *J. Biol. Chem.* **2003**, *278*, 47281–47290. [CrossRef] [PubMed]

66. Cyster, J.G.; Schwab, S.R. Sphingosine-1-phosphate and lymphocyte egress from lymphoid organs. *Annu. Rev. Immunol.* **2012**, *30*, 69–94. [CrossRef]

67. Blaho, V.A.; Hla, T. An update on the biology of sphingosine 1-phosphate receptors. *J. Lipid Res.* **2014**, *55*, 1596–1608. [CrossRef]

68. Regard, J.B.; Sato, I.T.; Coughlin, S.R. Anatomical profiling of G protein-coupled receptor expression. *Cell* **2008**, *135*, 561–571. [CrossRef]

69. Garcia, J.G.; Liu, F.; Verin, A.D.; Birukova, A.; Dechert, M.A.; Gerthoffer, W.T.; Bamberg, J.R.; English, D. Sphingosine 1-phosphate promotes endothelial cell barrier integrity by Edg-dependent cytoskeletal rearrangement. *J. Clin. Investig.* **2001**, *108*, 689–701. [CrossRef]

70. Rosen, H.; Goetzl, E.J. Sphingosine 1-phosphate and its receptors: An autocrine and paracrine network. *Nat. Rev. Immunol.* **2005**, *5*, 560–570. [CrossRef]

71. Alvarez, S.E.; Milstien, S.; Spiegel, S. Autocrine and paracrine roles of sphingosine-1-phosphate. *Trends Endocrinol. Metab.* **2007**, *18*, 300–307. [CrossRef]

72. Rosen, H.; Stevens, R.C.; Hanson, M.; Roberts, E.; Oldstone, M.B.A. Sphingosine-1-phosphate and its receptors: Structure, signaling, and influence. *Annu. Rev. Biochem.* **2013**, *82*, 637–662. [CrossRef]

73. Badawy, S.M.M.; Okada, T.; Kajimoto, T.; Ijuin, T.; Nakamura, S.I. DHHC5-mediated palmitoylation of S1P receptor subtype 1 determines G-protein coupling. *Sci. Rep.* **2017**, *7*, 16552. [CrossRef] [PubMed]

74. Kono, M.; Tucker, A.E.; Tran, J.; Bergner, J.B.; Turner, E.M.; Proia, R.L. Sphingosine-1-phosphate receptor 1 reporter mice reveal receptor activation sites in vivo. *J. Clin. Investig.* **2014**, *124*, 2076–2086. [CrossRef]

75. Kono, M.; Conlon, E.G.; Lux, S.Y.; Yanagida, K.; Hla, T.; Proia, R.L. Bioluminescence imaging of G protein-coupled receptor activation in living mice. *Nat. Commun.* **2017**, *8*, 1163. [CrossRef] [PubMed]

76. Arnon, T.I.; Xu, Y.; Lo, C.; Pham, T.; An, J.; Coughlin, S.; Dorn, G.W.; Cyster, J.G. GRK2-dependent S1PR1 desensitization is required for lymphocytes to overcome their attraction to blood. *Science* **2011**, *333*, 1898–1903. [CrossRef] [PubMed]

77. Thangada, S.; Khanna, K.M.; Blaho, V.A.; Oo, M.L.; Im, D.-S.; Guo, C.; Lefrancois, L.; Hla, T. Cell-surface residence of sphingosine 1-phosphate receptor 1 on lymphocytes determines lymphocyte egress kinetics. *J. Exp. Med.* **2010**, *207*, 1475–1483. [CrossRef]

78. Willinger, T.; Ferguson, S.M.; Pereira, J.P.; De Camilli, P.; Flavell, R.A. Dynamin 2-dependent endocytosis is required for sustained S1PR1 signaling. *J. Exp. Med.* **2014**, *211*, 685–700. [CrossRef]

79. Nomachi, A.; Yoshinaga, M.; Liu, J.; Kanchanawong, P.; Tohyama, K.; Thumkeo, D.; Watanabe, T.; Narumiya, S.; Hirata, T. Moesin controls clathrin-mediated S1PR1 internalization in T cells. *PLoS ONe* **2013**, *8*, e82590. [CrossRef]

80. Liu, C.H.; Thangada, S.; Lee, M.J.; Van Brooklyn, J.R.; Spiegel, S.; Hla, T. Ligand-induced trafficking of the sphingosine-1-phosphate receptor EDG- 1. *Mol. Biol. Cell.* **1999**, *10*, 1179–1190. [CrossRef]

81. Oo, M.L.; Chang, S.H.; Thangada, S.; Wu, M.-T.; Rezaul, K.; Blaho, V.; Hwang, S.-I.; Han, D.K.; Hla, T. Engagement of S1P1-degradative mechanisms leads to vascular leak in mice. *J. Clin. Investig.* **2011**, *121*, 2290–2300. [CrossRef] [PubMed]

82. Schulze, H.; Kolter, T.; Sandhoff, K. Principles of lysosomal membrane degradation: Cellular topology and biochemistry of lysosomal lipid degradation. *Biochim. Biophys. Acta* **2009**, *1793*, 674–683. [CrossRef] [PubMed]

83. Luzio, J.P.; Gray, S.R.; Bright, N.A. Endosome-lysosome fusion. *Biochem. Soc. Trans.* **2010**, *38*, 1413–1416. [CrossRef] [PubMed]

84. Lawrence, R.E.; Zoncu, R. The lysosome as a cellular centre for signalling, metabolism and quality control. *Nat. Cell Biol.* **2019**, *21*, 133–142. [CrossRef]

85. Kitatani, K.; Idkowiak-Baldys, J.; Hannun, Y.A. The sphingolipid salvage pathway in ceramide metabolism and signaling. *Cell Signal.* **2008**, *20*, 1010–1018. [CrossRef]

86. Proia, R.L.; Hla, T. Emerging biology of sphingosine-1-phosphate: Its role in pathogenesis and therapy. *J. Clin. Investig.* **2015**, *125*, 1379–1387. [CrossRef] [PubMed]

87. Young, M.M.; Takahashi, Y.; Fox, T.E.; Yun, J.K.; Kester, M.; Wang, H.-G. Sphingosine kinase 1 cooperates with autophagy to maintain endocytic membrane trafficking. *Cell Rep.* **2016**, *17*, 1532–1545. [CrossRef] [PubMed]

88. Chun, J.; Hla, T.; Lynch, K.R.; Spiegel, S.; Moolenaar, W.H. International Union of Basic and Clinical Pharmacology. LXXVIII. Lysophospholipid receptor nomenclature. *Pharmacol. Rev.* **2010**, *62*, 579–587. [CrossRef] [PubMed]

89. Moon, E.; Han, J.E.; Jeon, S.; Ryu, J.H.; Choi, J.W.; Chun, J. Exogenous S1P exposure potentiates ischemic stroke damage that is reduced possibly by inhibiting S1P receptor signaling. *Mediators Inflamm.* **2015**, *2015*, 492659. [CrossRef]

90. Siehler, S.; Manning, D.R. Pathways of transduction engaged by sphingosine 1-phosphate through G protein-coupled receptors. *Biochim. Biophys. Acta* **2002**, *1582*, 94–99. [CrossRef]

91. Gonda, K.; Okamoto, H.; Takuwa, N.; Yatomi, Y.; Okazaki, H.; Sakurai, T.; Kimura, S.; Sillard, R.; Harii, K.; Takuwa, Y. The novel sphingosine 1-phosphate receptor AGR16 is coupled via pertussis toxin-sensitive and -insensitive G-proteins to multiple signalling pathways. *Biochem. J.* **1999**, *337*, 67–75. [CrossRef] [PubMed]

92. Windh, R.T.; Lee, M.J.; Hla, T.; An, S.; Barr, A.J.; Manning, D.R. Differential coupling of the sphingosine 1-phosphate receptors Edg-1, Edg-3, and H218/Edg-5 to the G(i), G(q), and G12 families of heterotrimeric G proteins. *J. Biol. Chem.* **1999**, *274*, 27351–27358. [CrossRef] [PubMed]

93. Okamoto, H.; Takuwa, N.; Yokomizo, T.; Sugimoto, N.; Sakurada, S.; Shigematsu, H.; Takuwa, Y. Inhibitory regulation of Rac activation, membrane ruffling, and cell migration by the G protein-coupled sphingosine-1-phosphate receptor EDG5 but not EDG1 or EDG3. *Mol. Cell Biol.* **2000**, *20*, 9247–9261. [CrossRef] [PubMed]

94. Ancellin, N.; Hla, T. Differential pharmacological properties and signal transduction of the sphingosine 1-phosphate receptors EDG-1, EDG-3, and EDG-5. *J. Biol. Chem.* **1999**, *274*, 18997–19002. [CrossRef]

95. Graeler, M.; Goetzl, E.J. Activation-regulated expression and chemotactic function of sphingosine 1-phosphate receptors in mouse splenic T cells. *FASEB J.* **2002**, *16*, 1874–1878. [CrossRef]

96. Neves, S.R.; Ram, P.T.; Iyengar, R. G protein pathways. *Science* **2002**, *296*, 1636–1639. [CrossRef]

97. Ghosh, P.; Rangamani, P.; Kufareva, I. The GAPs, GEFs, GDIs and … now, GEMs: New kids on the heterotrimeric G protein signaling block. *Cell Cycle.* **2017**, *16*, 607–612. [CrossRef]

98. Oldham, W.M.; Hamm, H.E. Heterotrimeric G protein activation by G-protein-coupled receptors. *Nat. Rev. Mol. Cell Biol.* **2008**, *9*, 60–71. [CrossRef]

99. Smith, J.S.; Lefkowitz, R.J.; Rajagopal, S. Biased signalling: From simple switches to allosteric microprocessors. *Nat. Rev. Drug Discov.* **2018**, *17*, 243–260. [CrossRef]

100. Feistritzer, C.; Riewald, M. Endothelial barrier protection by activated protein C through PAR1-dependent sphingosine 1-phosphate receptor-1 crossactivation. *Blood* **2005**, *105*, 3178–3184. [CrossRef]

101. Singleton, P.A.; Dudek, S.M.; Ma, S.-F.; Garcia, J.G.N. Transactivation of sphingosine 1-phosphate receptors is essential for vascular barrier regulation: Novel role for hyaluronan and CD44 receptor family. *J. Biol. Chem.* **2006**, *281*, 34381–34393. [CrossRef] [PubMed]

102. Shiow, L.R.; Rosen, D.B.; Brdičková, N.; Xu, Y.; An, J.; Lanier, L.L.; Cyster, J.G.; Matloubian, M. CD69 acts downstream of interferon-α/β to inhibit S1P1 and lymphocyte egress from lymphoid organs. *Nature* **2006**, *440*, 540–544. [CrossRef] [PubMed]

103. Hisano, Y.; Kono, M.; Cartier, A.; Engelbrecht, E.; Kano, K.; Kawakami, K.; Xiong, Y.; Piao, W.; Galvani, S.; Yanagida, K.; et al. Lysolipid receptor cross-talk regulates lymphatic endothelial junctions in lymph nodes. *J. Exp. Med.* **2019**, *216*, 1582–1598. [CrossRef] [PubMed]

104. Laidlaw, B.J.; Gray, E.E.; Zhang, Y.; Ramírez-Valle, F.; Cyster, J.G. Sphingosine-1-phosphate receptor 2 restrains egress of γδ T cells from the skin. *J. Exp. Med.* **2019**, *216*, 1487–1496. [CrossRef] [PubMed]

105. Kempf, A.; Tews, B.; Arzt, M.E.; Weinmann, O.; Obermair, F.J.; Pernet, V.; Zagrebelsky, M.; Delekate, A.; Iobbi, C.; Zemmar, A.; et al. The sphingolipid receptor S1PR2 is a receptor for Nogo-A repressing synaptic plasticity. *PLoS Biol.* **2014**, *12*, e1001763. [CrossRef]

106. Studer, E.; Zhou, X.; Zhao, R.; Wang, Y.; Takabe, K.; Nagahashi, M.; Pandak, W.M.; Dent, P.; Spiegel, S.; Shi, R.; et al. Conjugated bile acids activate the sphingosine-1-phosphate receptor 2 in primary rodent hepatocytes. *Hepatology* **2012**, *55*, 267–276. [CrossRef]

107. Wang, Y.; Chen, D.; Zhang, Y.; Wang, P.; Zheng, C.; Zhang, S.; Yu, B.; Zhang, L.; Zhao, G.; Ma, B.; et al. Novel adipokine, FAM19A5, inhibits neointima formation after injury through sphingosine-1-phosphate receptor 2. *Circulation* **2018**, *138*, 48–63. [CrossRef]

108. Peng, C.; Trojanowski, J.Q.; Lee, V.M.-Y. Protein transmission in neurodegenerative disease. *Nat. Rev. Neurol.* **2020**, *16*, 199–212. [CrossRef]

109. Goedert, M.; Masuda-Suzukake, M.; Falcon, B. Like prions: The propagation of aggregated tau and α-synuclein in neurodegeneration. *Brain* **2017**, *140*, 266–278. [CrossRef]

110. Angot, E.; Steiner, J.A.; Hansen, C.; Li, J.-Y.; Brundin, P. Are synucleinopathies prion-like disorders? *Lancet Neurol.* **2010**, *9*, 1128–1138. [CrossRef]

111. Aguzzi, A.; Nuvolone, M.; Zhu, C. The immunobiology of prion diseases. *Nat. Rev. Immunol.* **2013**, *13*, 888–902. [CrossRef] [PubMed]

112. Shubhra Chakrabarti, S.; Bir, A.; Poddar, J.; Sinha, M.; Ganguly, A.; Chakrabarti, S. Ceramide and sphingosine-1-phosphate in cell death pathways: Relevance to the pathogenesis of Alzheimer's disease. *Curr. Alzheimer Res.* **2016**, *13*, 1232–1248. [CrossRef] [PubMed]

113. Di Paolo, G.; Kim, T.-W. Linking lipids to Alzheimer's disease: Cholesterol and beyond. *Nat. Rev. Neurosci.* **2011**, *12*, 284–296. [CrossRef] [PubMed]

114. Van Echten-Deckert, G.; Walter, J. Sphingolipids: Critical players in Alzheimer's disease. *Prog. Lipid Res.* **2012**, *51*, 378–393. [CrossRef]

115. Lemkul, J.A.; Bevan, D.R. Lipid composition influences the release of Alzheimer's amyloid β-peptide from membranes. *Protein Sci.* **2011**, *20*, 1530–1545. [CrossRef] [PubMed]

116. Grassi, S.; Giussani, P.; Mauri, L.; Prioni, S.; Sonnino, S.; Prinetti, A. Lipid rafts and neurodegeneration: Structural and functional roles in physiologic aging and neurodegenerative diseases. *J. Lipid Res.* **2020**, *61*, 636–654. [CrossRef] [PubMed]

117. Crivelli, S.M.; Giovagnoni, C.; Visseren, L.; Scheithauer, A.-L.; de Wit, N.; den Hoedt, S.; Losen, M.; Mulder, M.T.; Walter, J.; de Vries, H.E. et al.; et al. Sphingolipids in Alzheimer's disease, how can we target them? *Adv. Drug Deliv. Rev* **2020**, *S0169-409X(20)30002-8*. [CrossRef] [PubMed]

118. Kisler, K.; Nelson, A.R.; Montagne, A.; Zlokovic, B.V. Cerebral blood flow regulation and neurovascular dysfunction in Alzheimer disease. *Nat. Rev. Neurosci.* **2017**, *18*, 419–434. [CrossRef] [PubMed]

119. Iadecola, C. The pathobiology of vascular dementia. *Neuron* **2013**, *80*, 844–866. [CrossRef]

120. Iadecola, C. Neurovascular regulation in the normal brain and in Alzheimer's disease. *Nat. Rev. Neurosci.* **2004**, *5*, 347–360. [CrossRef]

121. Zlokovic, B.V. Neurovascular pathways to neurodegeneration in Alzheimer's disease and other disorders. *Nat. Rev. Neurosci.* **2011**, *12*, 723–738. [CrossRef] [PubMed]

122. Toledo, J.B.; Cairns, N.J.; Da, X.; Chen, K.; Carter, D.; Fleisher, A.; Householder, E.; Ayutyanont, N.; Roontiva, A.; Bauer, R.J.; et al. Clinical and multimodal biomarker correlates of ADNI neuropathological findings. *Acta Neuropathol. Commun.* **2013**, *1*, 65. [CrossRef] [PubMed]

123. Montagne, A.; Barnes, S.R.; Sweeney, M.D.; Halliday, M.R.; Sagare, A.P.; Zhao, Z.; Toga, A.W.; Jacobs, R.E.; Liu, C.Y.; Amezcua, L.; et al. Blood-brain barrier breakdown in the aging human hippocampus. *Neuron* **2015**, *85*, 296–302. [CrossRef] [PubMed]

124. Sweeney, M.D.; Sagare, A.P.; Zlokovic, B.V. Cerebrospinal fluid biomarkers of neurovascular dysfunction in mild dementia and Alzheimer's disease. *J. Cereb. Blood Flow Metab.* **2015**, *35*, 1055–1068. [CrossRef]

125. Arvanitakis, Z.; Capuano, A.W.; Leurgans, S.E.; Bennett, D.A.; Schneider. JA. Relation of cerebral vessel disease to Alzheimer's disease dementia and cognitive function in elderly people: A cross-sectional study. *Lancet Neurol.* **2016**, *15*, 934–943. [CrossRef]

126. Iturria-Medina, Y.; Sotero, R.C.; Toussaint, P.J.; Mateos-Pérez, J.M.; Evans, A.C. Early role of vascular dysregulation on late-onset Alzheimer's disease based on multifactorial data-driven analysis. *Nat. Commun.* **2016**, *7*, 11934. [CrossRef]

127. Nelson, A.R.; Sweeney, M.D.; Sagare, A.P.; Zlokovic, B.V. Neurovascular dysfunction and neurodegeneration in dementia and Alzheimer's disease. *Biochim. Biophys. Acta* **2016**, *1862*, 887–900. [CrossRef]
128. Becker, K.A.; Fahsel, B.; Kemper, H.; Mayeres, J.; Li, C.; Wilker, B.; Keitsch, S.; Soddemann, M.; Sehl, C.; Kohnen, M.; et al. Staphylococcus aureus alpha-toxin disrupts endothelial-cell tight junctions via acid sphingomyelinase and ceramide. *Infect. Immun.* **2017**, *86*, e00606–17. [CrossRef]
129. van Doorn, R.; Nijland, P.G.; Dekker, N.; Witte, M.E.; Lopes-Pinheiro, M.A.; van het Hof, B.; Kooij, G.; Reijerkerk, A.; Dijkstra, C.; van van der Valk, P.; et al. Fingolimod attenuates ceramide-induced blood-brain barrier dysfunction in multiple sclerosis by targeting reactive astrocytes. *Acta Neuropathol.* **2012**, *124*, 397–410. [CrossRef]
130. Cutler, R.G.; Kelly, J.; Storie, K.; Pedersen, W.A.; Tammara, A.; Hatanpaa, K.; Troncoso, J.C.; Mattson, M.P. Involvement of oxidative stress-induced abnormalities in ceramide and cholesterol metabolism in brain aging and Alzheimer's disease. *Proc. Natl. Acad. Sci. USA* **2004**, *101*, 2070–2075. [CrossRef]
131. Han, X.; Holtzman, D.M.; McKeel, D.W., Jr.; Kelley, J.; Morris, J.C. Substantial sulfatide deficiency and ceramide elevation in very early Alzheimer's disease: Potential role in disease pathogenesis. *J. Neurochem.* **2002**, *82*, 809–818. [CrossRef]
132. He, X.; Huang, Y.; Li, B.; Gong, C.-X.; Schuchman, E.H. Deregulation of sphingolipid metabolism in Alzheimer's disease. *Neurobiol. Aging* **2010**, *31*, 398–408. [CrossRef]
133. Filippov, V.; Song, M.A.; Zhang, K.; Vinters, H.V.; Tung, S.; Kirsch, W.M.; Yang, J.; Duerksen-Hughes, P.J. Increased ceramide in brains with alzheimer's and other neurodegenerative diseases. *J. Alzheimer's Dis.* **2012**, *29*, 537–547. [CrossRef] [PubMed]
134. Couttas, T.A.; Kain, N.; Daniels, B.; Lim, X.Y.; Shepherd, C.; Kril, J.; Pickford, R.; Li, H.; Garner, B.; Don, A.S. Loss of the neuroprotective factor sphingosine 1-phosphate early in Alzheimer's disease pathogenesis. *Acta Neuropathol. Commun.* **2014**, *2*, 9. [CrossRef]
135. Katsel, P.; Li, C.; Haroutunian, V. Gene expression alterations in the sphingolipid metabolism pathways during progression of dementia and Alzheimer's disease: A shift toward ceramide accumulation at the earliest recognizable stages of Alzheimer's disease? *Neurochem. Res.* **2007**, *32*, 845–856. [CrossRef] [PubMed]
136. Ceccom, J.; Loukh, N.; Lauwers-Cances, V.; Touriol, C.; Nicaise, Y.; Gentil, C.; Uro-Coste, E.; Pitson, S.; Maurage, C.A.; Duyckaerts, C.; et al. Reduced sphingosine kinase-1 and enhanced sphingosine 1-phosphate lyase expression demonstrate deregulated sphingosine 1-phosphate signaling in Alzheimer's disease. *Acta Neuropathol. Commun.* **2014**, *2*, 12. [CrossRef]
137. Dominguez, G.; Maddelein, M.-L.; Pucelle, M.; Nicaise, Y.; Maurage, C.-A.; Duyckaerts, C.; Cuvillier, O.; Delisle, M.-B. Neuronal sphingosine kinase 2 subcellular localization is altered in Alzheimer's disease brain. *Acta Neuropathol. Commun.* **2018**, *6*, 25. [CrossRef]
138. Xilouri, M.; Vogiatzi, T.; Vekrellis, K.; Stefanis, L. α-synuclein degradation by autophagic pathways: A potential key to Parkinson's disease pathogenesis. *Autophagy* **2008**, *4*, 917–919. [CrossRef] [PubMed]
139. Dehay, B.; Martinez-Vicente, M.; Caldwell, G.A.; Caldwell, K.A.; Yue, Z.; Cookson, M.R.; Klein, C.; Vila, M.; Bezard, E. Lysosomal impairment in Parkinson's disease. *Mov. Disord.* **2013**, *28*, 725–732. [CrossRef]
140. Manzoni, C.; Lewis, P.A. Dysfunction of the autophagy/lysosomal degradation pathway is a shared feature of the genetic synucleinopathies. *FASEB J.* **2013**, *27*, 3424–3429. [CrossRef]
141. Lwin, A.; Orvisky, E.; Goker-Alpan, O.; LaMarca, M.E.; Sidransky, E. Glucocerebrosidase mutations in subjects with parkinsonism. *Mol. Genet. Metab.* **2004**, *81*, 70–73. [CrossRef] [PubMed]
142. Neumann, J.; Bras, J.; Deas, E.; O'Sullivan, S.S.; Parkkinen, L.; Lachmann, R.H.; Li, A.; Holton, J.; Guerreiro, R.; Paudel, R.; et al. Glucocerebrosidase mutations in clinical and pathologically proven Parkinson's disease. *Brain* **2009**, *132*, 1783–1794. [CrossRef] [PubMed]
143. Mata, I.F.; Samii, A.; Schneer, S.H.; Roberts, J.W.; Griffith, A.; Leis, B.C.; Schellenberg, G.D.; Sidransky, E.; Bird, T.D.; Leverenz, J.B.; et al. Glucocerebrosidase gene mutations: A risk factor for Lewy body disorders. *Arch. Neurol.* **2008**, *65*, 379–382. [CrossRef] [PubMed]
144. Bras, J.; Singleton, A.; Cookson, M.R.; Hardy, J. Emerging pathways in genetic Parkinson's disease: Potential role of ceramide metabolism in Lewy body disease. *FEBS J.* **2008**, *275*, 5767–5773. [CrossRef] [PubMed]
145. Sidransky, E. Gaucher disease and parkinsonism. *Mol. Genet. Metab.* **2005**, *84*, 302–304. [CrossRef]
146. Sidransky, E.; Lopez, G. The link between the GBA gene and parkinsonism. *Lancet Neurol.* **2012**, *11*, 986–998. [CrossRef]

147. Mazzulli, J.R.; Xu, Y.-H.; Sun, Y.; Knight, A.L.; McLean, P.J.; Caldwell, G.A.; Sidransky, E.; Grabowski, G.A.; Krainc, D. Gaucher disease glucocerebrosidase and α-synuclein form a bidirectional pathogenic loop in synucleinopathies. *Cell* **2011**, *146*, 37–52. [CrossRef]

148. Taguchi, Y.V.; Liu, J.; Ruan, J.; Pacheco, J.; Zhang, X.; Abbasi, J.; Keutzer, J.; Mistry, P.K.; Chandra, S.S. Glucosylsphingosine promotes α-synuclein pathology in mutant GBA-associated parkinson's disease. *J. Neurosci.* **2017**, *37*, 9617–9631. [CrossRef]

149. Everett, C.M.; Wood, N.W. Trinucleotide repeats and neurodegenerative disease. *Brain* **2004**, *127*, 2385–2405. [CrossRef]

150. Jimenez-Sanchez, M.; Licitra, F.; Underwood, B.R.; Rubinsztein, D.C. Huntington's disease: Mechanisms of pathogenesis and therapeutic strategies. *Cold Spring Harb. Perspect. Med.* **2017**, *7*, a024240. [CrossRef]

151. Di Pardo, A.; Amico, E.; Basit, A.; Armirotti, A.; Joshi, P.; Neely, M.D.; Vuono, R.; Castaldo, S.; Digilio, A.F.; Scalabrì, F.; et al. Defective sphingosine-1-phosphate metabolism is a druggable target in Huntington's disease. *Sci. Rep.* **2017**, *7*, 5280. [CrossRef] [PubMed]

152. Pirhaji, L.; Milani, P.; Dalin, S.; Wassie, B.T.; Dunn, D.E.; Fenster, R.J.; Avila-Pacheco, J.; Greengard, P.; Clish, C.B.; Heiman, M.; et al. Identifying therapeutic targets by combining transcriptional data with ordinal clinical measurements. *Nat. Commun.* **2017**, *8*, 623. [CrossRef] [PubMed]

153. Di Pardo, A.; Maglione, V. The S1P axis: New exciting route for treating Huntington's disease. *Trends Pharmacol. Sci.* **2018**, *39*, 468–480. [CrossRef] [PubMed]

154. Henriques, A.; Croixmarie, V.; Bouscary, A.; Mosbach, A.; Keime, C.; Boursier-Neyret, C.; Walter, B.; Spedding, M.; Loeffler, J.-P. Sphingolipid metabolism is dysregulated at transcriptomic and metabolic levels in the spinal cord of an animal model of amyotrophic lateral sclerosis. *Front. Mol. Neurosci.* **2018**, *10*, 433. [CrossRef] [PubMed]

155. Brinkmann, V.; Davis, M.D.; Heise, C.E.; Albert, R.; Cottens, S.; Hof, R.; Bruns, C.; Prieschl, E.; Baumruker, T.; Hiestand, P.; et al. The immune modulator FTY720 targets sphingosine 1-phosphate receptors. *J. Biol. Chem.* **2002**, *277*, 21453–21457. [CrossRef] [PubMed]

156. Mandala, S.; Hajdu, R.; Bergstrom, J.; Quackenbush, E.; Xie, J.; Milligan, J.; Thornton, R.; Shei, G.-J.; Card, D.; Keohane, C.; et al. Alteration of lymphocyte trafficking by sphingosine-1-phosphate receptor agonists. *Science* **2002**, *296*, 346–349. [CrossRef]

157. Potenza, R.L.; De Simone, R.; Armida, M.; Mazziotti, V.; Pèzzola, A.; Popoli, P.; Minghetti, L. Fingolimod: A disease-modifier drug in a mouse model of amyotrophic lateral sclerosis. *Neurotherapeutics* **2016**, *13*, 918–927. [CrossRef]

158. Compston, A.; Coles, A. Multiple sclerosis. *Lancet* **2008**, *372*, 1502–1517. [CrossRef]

159. Constantinescu, C.S.; Farooqi, N.; O'Brien, K.; Gran, B. Experimental autoimmune encephalomyelitis (EAE) as a model for multiple sclerosis (MS). *Br. J. Pharmacol.* **2011**, *164*, 1079–1106. [CrossRef]

160. Grant, J.L.; Ghosn, E.E.B.; Axtell, R.C.; Herges, K.; Kuipers, H.F.; Woodling, N.S.; Andreasson, K.; Herzenberg, L.A.; Herzenberg, L.A.; Steinman, L. Reversal of paralysis and reduced inflammation from peripheral administration of β-amyloid in TH1 and TH17 versions of experimental autoimmune encephalomyelitis. *Sci. Transl. Med.* **2012**, *4*, 145ra105. [CrossRef]

161. Gijbels, K.; Engelborghs, S.; De Deyn, P.P. Experimental autoimmune encephalomyelitis: An animal model for multiple sclerosis. *Neurosci. Res. Commun.* **2000**, *26*, 193–206. [CrossRef]

162. Brinkmann, V.; Billich, A.; Baumruker, T.; Heining, P.; Schmouder, R.; Francis, G.; Aradhye, S.; Burtin, P. Fingolimod (FTY720): Discovery and development of an oral drug to treat multiple sclerosis. *Nat. Rev. Drug Discov.* **2010**, *9*, 883–897. [CrossRef] [PubMed]

163. Fujino, M.; Funeshima, N.; Kitazawa, Y.; Kimura, H.; Amemiya, H.; Suzuki, S.; Li, X.-K. Amelioration of experimental autoimmune encephalomyelitis in Lewis rats by FTY720 treatment. *J. Pharmacol. Exp. Ther.* **2003**, *305*, 70–77. [CrossRef] [PubMed]

164. Webb, M.; Tham, C.-S.; Lin, F.-F.; Lariosa-Willingham, K.; Yu, N.; Hale, J.; Mandala, S.; Chun, J.; Rao, T.S. Sphingosine 1-phosphate receptor agonists attenuate relapsing-remitting experimental autoimmune encephalitis in SJL mice. *J. Neuroimmunol.* **2004**, *153*, 108–121. [CrossRef] [PubMed]

165. Kataoka, H.; Sugahara, K.; Shimano, K.; Teshima, K.; Koyama, M.; Fukunari, A.; Chiba, K. FTY720, sphingosine 1-phosphate receptor modulator, ameliorates experimental autoimmune encephalomyelitis by inhibition of T cell infiltration. *Cell Mol. Immunol.* **2005**, *2*, 439–448.

166. Foster, C.A.; Mechtcheriakova, D.; Storch, M.K.; Balatoni, B.; Howard, L.M.; Bornancin, F.; Wlachos, A.; Sobanov, J.; Kinnunen, A.; Baumruker, T. FTY720 rescue therapy in the dark agouti rat model of experimental autoimmune encephalomyelitis: Expression of central nervous system genes and reversal of blood-brain-barrier damage. *Brain Pathol.* **2009**, *19*, 254–266. [CrossRef]

167. Choi, J.W.; Gardell, S.E.; Herr, D.R.; Rivera, R.; Lee, C.-W.; Noguchi, K.; Teo, S.T.; Yung, Y.C.; Lu, M.; Kennedy, G.; et al. FTY720 (fingolimod) efficacy in an animal model of multiple sclerosis requires astrocyte sphingosine 1-phosphate receptor 1 (S1P1) modulation. *Proc. Natl. Acad. Sci. USA* **2011**, *108*, 751–756. [CrossRef]

168. Eken, A.; Duhen, R.; Singh, A.K.; Fry, M.; Buckner, J.H.; Kita, M.; Bettelli, E.; Oukka, M. S1P1 deletion differentially affects TH17 and regulatory T cells. *Sci. Rep.* **2017**, *7*, 12905. [CrossRef]

169. Smigiel, K.S.; Richards, E.; Srivastava, S.; Thomas, K.R.; Dudda, J.C.; Klonowski, K.D.; Campbell, D.J. CCR7 provides localized access to IL-2 and defines homeostatically distinct regulatory T cell subsets. *J. Exp. Med.* **2014**, *211*, 121–136. [CrossRef]

170. Cruz-Orengo, L.; Daniels, B.P.; Dorsey, D.; Basak, S.A.; Grajales-Reyes, J.G.; McCandless, E.E.; Piccio, L.; Schmidt, R.E.; Cross, A.H.; Crosby, S.D.; et al. Enhanced sphingosine-1-phosphate receptor 2 expression underlies female CNS autoimmunity susceptibility. *J. Clin. Investig.* **2014**, *124*, 2571–2584. [CrossRef]

171. Lopes Pinheiro, M.A.; Kroon, J.; Hoogenboezem, M.; Geerts, D.; van Het Hof, B.; van der Pol, S.M.A.; van Buul, J.D.; de Vries, H.E. Acid sphingomyelinase–derived ceramide regulates ICAM-1 function during T cell transmigration across brain endothelial cells. *J. Immunol.* **2016**, *196*, 72–79. [CrossRef] [PubMed]

172. Niessen, F.; Schaffner, F.; Furlan-Freguia, C.; Pawlinski, R.; Bhattacharjee, G.; Chun, J.; Derian, C.K.; Andrade-Gordon, P.; Rosen, H.; Ruf, W. Dendritic cell PAR1-S1P3 signalling couples coagulation and inflammation. *Nature* **2008**, *452*, 654–658. [CrossRef] [PubMed]

173. Fischer, I.; Alliod, C.; Martinier, N.; Newcombe, J.; Brana, C.; Pouly, S. Sphingosine kinase 1 and sphingosine 1-phosphate receptor 3 are functionally upregulated on astrocytes under pro-inflammatory conditions. *PLoS ONE* **2011**, *6*, e23905. [CrossRef] [PubMed]

174. Dusaban, S.S.; Chun, J.; Rosen, H.; Purcell, N.H.; Brown, J.H. Sphingosine 1-phosphate receptor 3 and RhoA signaling mediate inflammatory gene expression in astrocytes. *J. Neuroinflammation* **2017**, *14*, 111. [CrossRef]

175. Devic, E. *Congrès français de médecine (Premiere Session; Lyon, 1894; procès-verbaux, mémoires et discussions; publiés par M. le Dr, L. Bard)*; Asselin et Houzeau: Lyon, France, 1895.

176. Gault, F. De la neuromyélite optique aiguë. Ph.D. Thesis, Faculté de Médecine Lyon Est, Lyon, France, 1894.

177. Wingerchuk, D.M.; Lennon, V.A.; Lucchinetti, C.F.; Pittock, S.J.; Weinshenker, B.G. The spectrum of neuromyelitis optica. *Lancet Neurol.* **2007**, *6*, 805–815. [CrossRef]

178. Weinshenker, B.G.; Wingerchuk, D.M.; Pittock, S.J.; Lucchinetti, C.F.; Lennon, V.A. NMO-IgG: A specific biomarker for neuromyelitis optica. *Dis. Markers* **2006**, *22*, 197–206. [CrossRef]

179. Jarius, S.; Franciotta, D.; Bergamaschi, R.; Wright, H.; Littleton, E.; Palace, J.; Hohlfeld, R.; Vincent, A. NMO-IgG in the diagnosis of neuromyelitis optica. *Neurology* **2007**, *68*, 1076–1077. [CrossRef]

180. Min, J.H.; Kim, B.J.; Lee, K.H. Development of extensive brain lesions following fingolimod (FTY720) treatment in a patient with neuromyelitis optica spectrum disorder. *Mult. Scler.* **2012**, *18*, 113–115. [CrossRef]

181. Yoshii, F.; Moriya, Y.; Ohnuki, T.; Ryo, M.; Takahashi, W. Fingolimod-induced leukoencephalopathy in a patient with neuromyelitis optica spectrum disorder. *Mult. Scler. Relat Disord.* **2016**, *7*, 53–57. [CrossRef]

182. Izaki, S.; Narukawa, S.; Kubota, A.; Mitsui, T.; Fukaura, H.; Nomura, K. A case of neuromyelitis optica spectrum disorder developing a fulminant course with multiple white-matter lesions following fingolimod treatment. *Clin. Neurol.* **2013**, *53*, 513–517. [CrossRef]

183. Tanaka, M.; Oono, M.; Motoyama, R.; Tanaka, K. Longitudinally extensive spinal cord lesion after initiation, and multiple extensive brain lesions after cessation of fingolimod treatment in a patient with recurrent myelitis and anti-aquaporin 4 antibodies. *Clin. Exp. Neuroimmunol.* **2013**, *4*, 239–240. [CrossRef]

184. Matsushita, T.; Tateishi, T.; Isobe, N.; Yonekawa, T.; Yamasaki, R.; Matsuse, D.; Murai, H.; Kira, J.-I. Characteristic cerebrospinal fluid cytokine/chemokine profiles in neuromyelitis optica, relapsing remitting or primary progressive multiple sclerosis. *PLoS One* **2013**, *8*, e61835. [CrossRef] [PubMed]

185. Lee, H.; Deng, J.; Kujawski, M.; Yang, C.; Liu, Y.; Herrmann, A.; Kortylewski, M.; Horne, D.; Somlo, G.; Forman, S.; et al. STAT3-induced S1PR1 expression is crucial for persistent STAT3 activation in tumors. *Nat. Med.* **2010**, *16*, 1421–1428. [CrossRef] [PubMed]

186. Liang, J.; Nagahashi, M.; Kim, E.Y.; Harikumar, K.B.; Yamada, A.; Huang, W.-C.; Hait, N.C.; Allegood, J.C.; Price, M.M.; Avni, D.; et al. Sphingosine-1-phosphate links persistent STAT3 activation, chronic intestinal inflammation, and development of colitis-associated cancer. *Cancer Cell* **2013**, *23*, 107–120. [CrossRef] [PubMed]

187. Nguyen, A.V.; Wu, Y.Y.; Liu, Q.; Wang, D.; Nguyen, S.; Loh, R.; Pang, J.; Friedman, K.; Orlofsky, A.; Augenlicht, L.; et al. STAT3 in epithelial cells regulates inflammation and tumor progression to malignant state in colon. *Neoplasia* **2013**, *15*, 998–1008. [CrossRef]

188. Nguyen-Jackson, H.; Panopoulos, A.D.; Zhang, H.; Li, H.S.; Watowich, S.S. STAT3 controls the neutrophil migratory response to CXCR2 ligands by direct activation of G-CSF-induced CXCR2 expression and via modulation of CXCR2 signal transduction. *Blood* **2010**, *115*, 3354–3363. [CrossRef]

189. McLoughlin, R.M.; Jenkins, B.J.; Grail, D.; Williams, A.S.; Fielding, C.A.; Parker, C.R.; Ernst, M.; Topley, N.; Jones, S.A. IL-6 trans-signaling via STAT3 directs T cell infiltration in acute inflammation. *Proc. Natl. Acad. Sci. USA* **2005**, *102*, 9589–9594. [CrossRef]

190. Yopp, A.C.; Ochando, J.C.; Mao, M.; Ledgerwood, L.; Ding, Y.; Bromberg, J.S. Sphingosine 1-phosphate receptors regulate chemokine-driven transendothelial migration of lymph node but not splenic T cells. *J. Immunol.* **2005**, *175*, 2913–2924. [CrossRef]

191. Harikumar, K.B.; Yester, J.W.; Surace, M.J.; Oyeniran, C.; Price, M.M.; Huang, W.-C.; Hait, N.C.; Allegood, J.C.; Yamada, A.; Kong, X.; et al. K63-linked polyubiquitination of transcription factor IRF1 is essential for IL-1-induced production of chemokines CXCL10 and CCL5. *Nat. Immunol.* **2014**, *15*, 231–238. [CrossRef]

192. Uzawa, A.; Mori, M.; Arai, K.; Sato, Y.; Hayakawa, S.; Masuda, S.; Taniguchi, J.; Kuwabara, S. Cytokine and chemokine profiles in neuromyelitis optica: Significance of interleukin-6. *Mult. Scler.* **2010**, *16*, 1443–1452. [CrossRef]

193. Muscal, E.; Brey, R.L. Neurologic manifestations of systemic lupus erythematosus in children and adults. *Neurol. Clin.* **2010**, *28*, 61–73. [CrossRef] [PubMed]

194. Kirshner, H.S. Hashimoto's encephalopathy: A brief review. *Curr. Neurol. Neurosci. Rep.* **2014**, *14*, 476. [CrossRef]

195. Sanna, G.; Piga, M.; Terryberry, J.W.; Peltz, M.T.; Giagheddu, S.; Satta, L.; Ahmed, A.; Cauli, A.; Montaldo, C.; Passiu, G.; et al. Central nervous system involvement in systemic lupus erythematosus: Cerebral imaging and serological profile in patients with and without overt neuropsychiatric manifestations. *Lupus* **2000**, *9*, 573–583. [CrossRef] [PubMed]

196. Snider, A.J. Sphingosine kinase and sphingosine-1-phosphate: Regulators in autoimmune and inflammatory disease. *Int. J. Clin. Rheumtol.* **2013**, *8*, 453–463. [CrossRef] [PubMed]

197. Okazaki, H.; Hirata, D.; Kamimura, T.; Sato, H.; Iwamoto, M.; Yoshio, T.; Masuyama, J.; Fujimura, A.; Kobayashi, E.; Kano, S.; et al. Effects of FTY720 in MRL-lpr/lpr mice: Therapeutic potential in systemic lupus erythematosus. *J. Rheumatol.* **2002**, *29*, 707–716. [PubMed]

198. Alperovich, G.; Rama, I.; Lloberas, N.; Franquesa, M.; Poveda, R.; Gomà, M.; Herrero-Fresneda, I.; Cruzado, J.M.; Bolaños, N.; Carrera, M.; et al. New immunosuppresor strategies in the treatment of murine lupus nephritis. *Lupus* **2007**, *16*, 18–24. [CrossRef]

199. Ando, S.; Amano, H.; Amano, E.; Minowa, K.; Watanabe, T.; Nakano, S.; Nakiri, Y.; Morimoto, S.; Tokano, Y.; Lin, Q.; et al. FTY720 exerts a survival advantage through the prevention of end-stage glomerular inflammation in lupus-prone BXSB mice. *Biochem. Biophys. Res. Commun.* **2010**, *394*, 804–810. [CrossRef]

200. Wenderfer, S.E.; Stepkowski, S.M.; Braun, M.C. Increased survival and reduced renal injury in MRL/lpr mice treated with a novel sphingosine-1-phosphate receptor agonist. *Kidney Int.* **2008**, *74*, 1319–1326. [CrossRef]

201. Han, C.; He, X.; Xia, X.; Guo, J.; Liu, A.; Liu, X.; Wang, X.; Li, C.; Peng, S.; Zhao, W.; et al. Sphk1/S1P/S1PR1 Signaling is involved in the development of autoimmune thyroiditis in patients and NOD.H-2^{h4} mice. *Thyroid* **2019**, *29*, 700–713. [CrossRef]

202. Yilmaz, G.; Arumugam, T.V.; Stokes, K.Y.; Granger, D.N. Role of T lymphocytes and interferon-γ in ischemic stroke. *Circulation* **2006**, *113*, 2105–2112. [CrossRef]

203. Lo, E.H. T time in the brain. *Nat. Med.* **2009**, *15*, 844–846. [CrossRef] [PubMed]

204. Shichita, T.; Sugiyama, Y.; Ooboshi, H.; Sugimori, H.; Nakagawa, R.; Takada, I.; Iwaki, T.; Okada, Y.; Iida, M.; Cua, D.J.; et al. Pivotal role of cerebral interleukin-17-producing γδT cells in the delayed phase of ischemic brain injury. *Nat. Med.* **2009**, *15*, 946–950. [CrossRef] [PubMed]

205. Iadecola, C.; Anrather, J. The immunology of stroke: From mechanisms to translation. *Nat. Med.* **2011**, *17*, 796–808. [CrossRef] [PubMed]
206. Hurn, P.D.; Subramanian, S.; Parker, S.M.; Afentoulis, M.E.; Kaler, L.J.; Vandenbark, A.A.; Offner, H. T- and B-cell-deficient mice with experimental stroke have reduced lesion size and inflammation. *J. Cereb. Blood Flow Metab.* **2007**, *27*, 1798–1805. [CrossRef] [PubMed]
207. Kleinschnitz, C.; Schwab, N.; Kraft, P.; Hagedor, I.; Dreykluft, A.; Schwarz, T.; Austinat, M.; Nieswandt, B.; Wiendl, H.; Stoll, G. Early detrimental T-cell effects in experimental cerebral ischemia are neither related to adaptive immunity nor thrombus formation. *Blood* **2010**, *115*, 3835–3842. [CrossRef]
208. Adachi, K.; Kohara, T.; Nakao, N.; Arita, M.; Chiba, K.; Mishina, T.; Sazaki, S.; Fujita, T. Design, synthesis, and structure-activity relationships of 2-substituted-2-amino-1,3-propanediols: Discovery of a novel immunosuppressant, FTY720. *Bioorganic Med. Chem. Lett.* **1995**, *5*, 853–856. [CrossRef]
209. Brinkmann, V.; Cyster, J.G.; Hla, T. FTY720: Sphingosine 1-phosphate receptor-1 in the control of lymphocyte egress and endothelial barrier function. *Am. J. Transplant.* **2004**, *4*, 1019–1025. [CrossRef]
210. Rolland, W.B., II.; Manaenko, A.; Lekic, T.; Hasegawa, Y.; Ostrowski, R.; Tang, J.; Zhang, J.H. FTY720 is neuroprotective and improves functional outcomes after intracerebral hemorrhage in mice. *Acta Neurochir. Suppl.* **2011**, *111*, 213–217. [CrossRef]
211. Hasegawa, Y.; Suzuki, H.; Sozen, T.; Rolland, W.; Zhang, J.H. Activation of sphingosine 1-phosphate receptor-1 by FTY720 is neuroprotective after ischemic stroke in rats. *Stroke* **2010**, *41*, 368–374. [CrossRef] [PubMed]
212. Czech, B.; Pfeilschifter, W.; Mazaheri-Omrani, N.; Strobel, M.A.; Kahles, T.; Neumann-Haefelin, T.; Rami, A.; Huwiler, A.; Pfeilschifter, J. The immunomodulatory sphingosine 1-phosphate analog FTY720 reduces lesion size and improves neurological outcome in a mouse model of cerebral ischemia. *Biochem. Biophys. Res. Commun.* **2009**, *389*, 251–256. [CrossRef]
213. Kraft, P.; Göb, E.; Schuhmann, M.K.; Göbel, K.; Deppermann, C.; Thielmann, I.; Herrmann, A.M.; Lorenz, K.; Brede, M.; Stoll, G.; et al. FTY720 ameliorates acute ischemic stroke in mice by reducing thrombo-inflammation but not by direct neuroprotection. *Stroke* **2013**, *44*, 3202–3210. [CrossRef] [PubMed]
214. Nazari, M.; Keshavarz, S.; Rafati, A.; Namavar, M.R.; Haghani, M. Fingolimod (FTY720) improves hippocampal synaptic plasticity and memory deficit in rats following focal cerebral ischemia. *Brain Res. Bull.* **2016**, *124*, 95–102. [CrossRef] [PubMed]
215. Wei, Y.; Yemisci, M.; Kim, H.-H.; Yung, L.M.; Shin, H.K.; Hwang, S.-K.; Guo, S.; Qin, T.; Alsharif, N.; Brinkmann, V.; et al. Fingolimod provides long-term protection in rodent models of cerebral ischemia. *Ann Neurol.* **2011**, *69*, 119–129. [CrossRef] [PubMed]
216. Pfeilschifter, W.; Czech-Zechmeister, B.; Sujak, M.; Mirceska, A.; Koch, A.; Rami, A.; Steinmetz, H.; Foerch, C.; Huwiler, A.; Pfeilschifter, J. Activation of sphingosine kinase 2 is an endogenous protective mechanism in cerebral ischemia. *Biochem. Biophys. Res. Commun.* **2011**, *413*, 212–217. [CrossRef]
217. Rolland, W.B.; Lekic, T.; Krafft, P.R.; Hasegawa, Y.; Altay, O.; Hartman, R.; Ostrowski, R.; Manaenko, A.; Tang, J.; Zhang, J.H. Fingolimod reduces cerebral lymphocyte infiltration in experimental models of rodent intracerebral hemorrhage. *Exp. Neurol.* **2013**, *241*, 45–55. [CrossRef] [PubMed]
218. Fu, Y.; Zhang, N.; Ren, L.; Yan, Y.; Sun, N.; Li, Y.-J.; Han, W.; Xue, R.; Liu, Q.; Hao, J.; et al. Impact of an immune modulator fingolimod on acute ischemic stroke. *Proc. Natl. Acad. Sci. USA* **2014**, *111*, 18315–18320. [CrossRef]
219. Schaphorst, K.L.; Chiang, E.; Jacobs, K.N.; Zaiman, A.; Natarajan, V.; Wigley, F.; Garcia, J.G.N. Role of sphingosine-1 phosphate in the enhancement of endothelial barrier integrity by platelet-released products. *Am. J. Physiol. Lung Cell Mol. Physiol.* **2003**, *285*, L258–L267. [CrossRef]
220. Liesz, A.; Zhou, W.; Mracskó, É.; Karcher, S.; Bauer, H.; Schwarting, S.; Sun, L.; Bruder, D.; Stegemann, S.; Cerwenka, A.; et al. Inhibition of lymphocyte trafficking shields the brain against deleterious neuroinflammation after stroke. *Brain* **2011**, *134*, 704–720. [CrossRef]
221. Cai, A.; Schlunk, F.; Bohmann, F.; Kahefiolasl, S.; Brunkhorst, R.; Foerch, C.; Pfeilschifter, W. Coadministration of FTY720 and rt-PA in an experimental model of large hemispheric stroke-No influence on functional outcome and blood-brain barrier disruption. *Exp. Transl. Stroke Med.* **2013**, *5*, 11. [CrossRef]
222. Sanchez, T. Sphingosine-1-phosphate signaling in endothelial disorders. *Curr. Atheroscler. Rep.* **2016**, *18*, 31. [CrossRef]

223. Brait, V.H.; Tarrasón, G.; Gavaldà, A.; Godessart, N.; Planas, A.M. Selective sphingosine 1-phosphate receptor 1 agonist is protective against ischemia/reperfusion in mice. *Stroke* **2016**, *47*, 3053–3056. [CrossRef] [PubMed]

224. Blondeau, N.; Lai, Y.; Tyndall, S.; Popolo, M.; Topalkara, K.; Pru, J.K.; Zhang, L.; Kim, H.; Liao, J.K.; Ding, K.; et al. Distribution of sphingosine kinase activity and mRNA in rodent brain. *J. Neurochem.* **2007**, *103*, 509–517. [CrossRef] [PubMed]

225. Billich, A.; Bornancin, F.; Dévay, P.; Mechtcheriakova, D.; Urtz, N.; Baumruker, T. Phosphorylation of the immunomodulatory drug FTY720 by sphingosine kinases. *J. Biol. Chem.* **2003**, *278*, 47408–47415. [CrossRef]

226. Wacker, B.K.; Perfater, J.L.; Gidday, J.M. Hypoxic preconditioning induces stroke tolerance in mice via a cascading HIF, sphingosine kinase, and CCL2 signaling pathway. *J. Neurochem.* **2012**, *123*, 954–962. [CrossRef] [PubMed]

227. Zemann, B.; Kinzel, B.; Müller, M.; Reuschel, R.; Mechtcheriakova, D.; Urtz, N.; Bornancin, F.; Baumruker, T.; Billich, A. Sphingosine kinase type 2 is essential for lymphopenia induced by the immunomodulatory drug FTY720. *Blood* **2006**, *107*, 1454–1458. [CrossRef]

228. Wacker, B.K.; Park, T.S.; Gidday, J.M. Hypoxic preconditioning-induced cerebral ischemic tolerance: Role of microvascular sphingosine kinase 2. *Stroke* **2009**, *40*, 3342–3348. [CrossRef]

229. Zhang, W.; An, J.; Jawadi, H.; Siow, D.L.; Lee, J.-F.; Zhao, J.; Gartung, A.; Maddipati, K.R.; Honn, K.V.; Wattenberg, B.W.; et al. Sphingosine-1-phosphate receptor-2 mediated NFκB activation contributes to tumor necrosis factor-α induced VCAM-1 and ICAM-1 expression in endothelial cells. *Prostaglandins Other Lipid Mediat.* **2013**, *106*, 62–71. [CrossRef]

230. Kim, G.S.; Yang, L.; Zhang, G.; Zhao, H.; Selim, M.; McCullough, L.D.; Kluk, M.J.; Sanchez, T. Critical role of sphingosine-1-phosphate receptor-2 in the disruption of cerebrovascular integrity in experimental stroke. *Nat. Commun.* **2015**, *6*, 7893. [CrossRef] [PubMed]

231. Lv, M.; Zhang, D.; Dai, D.; Zhang, W.; Zhang, L. Sphingosine kinase 1/sphingosine-1-phosphate regulates the expression of interleukin-17A in activated microglia in cerebral ischemia/reperfusion. *Inflamm. Res.* **2016**, *65*, 551–562. [CrossRef]

232. Zheng, S.; Wei, S.; Wang, X.; Xu, Y.; Xiao, Y.; Liu, H.; Jia, J.; Cheng, J. Sphingosine kinase 1 mediates neuroinflammation following cerebral ischemia. *Exp. Neurol.* **2015**, *272*, 160–169. [CrossRef] [PubMed]

233. Campos, F.; Qin, T.; Castillo, J.; Seo, J.H.; Arai, K.; Lo, E.H.; Waeber, C. Fingolimod reduces hemorrhagic transformation associated with delayed tissue plasminogen activator treatment in a mouse thromboembolic model. *Stroke* **2013**, *44*, 505–511. [CrossRef] [PubMed]

234. Zhu, Z.; Fu, Y.; Tian, D.; Sun, N.; Han, W.; Chang, G.; Dong, Y.; Xu, X.; Liu, Q.; Huang, D.; et al. Combination of the immune modulator fingolimod with alteplase in acute ischemic stroke: A pilot trial. *Circulation* **2015**, *132*, 1104–1112. [CrossRef] [PubMed]

235. Zhang, S.; Zhou, Y.; Zhang, R.; Zhang, M.; Campbell, B.; Lin, L.; Shi, F.-D.; Lou, M. Rationale and design of combination of an immune modulator fingolimod with alteplase bridging with mechanical thrombectomy in acute ischemic stroke (FAMTAIS) trial. *Int. J. Stroke* **2017**, *12*, 906–909. [CrossRef] [PubMed]

236. Wan, Y.; Jin, H.-J.; Zhu, Y.-Y.; Fang, Z.; Mao, L.; He, Q.; Xia, J.-P.; Li, M.; Li, Y.; Chen, X.; et al. MicroRNA-149-5p regulates blood-brain barrier permeability after transient middle cerebral artery occlusion in rats by targeting S1PR2 of pericytes. *FASEB J.* **2018**, *32*, 3133–3148. [CrossRef] [PubMed]

237. Swendeman, S.L.; Xiong, Y.; Cantalupo, A.; Yuan, A.; Burg, N.; Hisano, Y.; Cartier, A.; Liu, C.H.; Engelbrecht, E.; Blaho, V.; et al. An engineered S1P chaperone attenuates hypertension and ischemic injury. *Sci. Signal.* **2017**, *10*, eaal2722. [CrossRef] [PubMed]

238. Testai, F.D.; Kilkus, J.P.; Berdyshev, E.; Gorshkova, I.; Natarajan, V.; Dawson, G. Multiple sphingolipid abnormalities following cerebral microendothelial hypoxia. *J. Neurochem.* **2014**, *131*, 530–540. [CrossRef] [PubMed]

239. Gaire, B.P.; Lee, C.H.; Sapkota, A.; Lee, S.Y.; Chun, J.; Cho, H.J.; Nam, T.-G.; Choi, J.W. Identification of sphingosine 1-phosphate receptor subtype 1 (S1P1) as a pathogenic factor in transient focal cerebral ischemia. *Mol. Neurobiol.* **2018**, *55*, 2320–2332. [CrossRef]

240. Gaire, B.P.; Song, M.-R.; Choi, J.W. Sphingosine 1-phosphate receptor subtype 3 (S1P3) contributes to brain injury after transient focal cerebral ischemia via modulating microglial activation and their M1 polarization. *J. Neuroinflammation* **2018**, *15*, 284. [CrossRef]

241. Gaire, B.P.; Bae, Y.J.; Choi, J.W. S1P1 regulates M1/M2 polarization toward brain injury after transient focal cerebral ischemia. *Biomol. Ther.* **2019**, *27*, 522–529. [CrossRef]

242. Zamanian, J.L.; Xu, L.; Foo, L.C.; Nouri, N.; Zhou, L.; Giffard, R.G.; Barres, B.A. Genomic analysis of reactive astrogliosis. *J. Neurosci.* **2012**, *32*, 6391–6410. [CrossRef]

243. Liddelow, S.A.; Guttenplan, K.A.; Clarke, L.E.; Bennett, F.C.; Bohlen, C.J.; Schirmer, L.; Bennett, M.L.; Münch, A.E.; Chung, W.-S.; Peterson, T.C.; et al. Neurotoxic reactive astrocytes are induced by activated microglia. *Nature* **2017**, *541*, 481–487. [CrossRef] [PubMed]

244. Anelli, V.; Bassi, R.; Tettamanti, G.; Viani, P.; Riboni, L. Extracellular release of newly synthesized sphingosine-1-phosphate by cerebellar granule cells and astrocytes. *J. Neurochem.* **2005**, *92*, 1204–1215. [CrossRef] [PubMed]

245. Mullershausen, F.; Craveiro, L.M.; Shin, Y.; Cortes-Cros, M.; Bassilana, F.; Osinde, M.; Wishart, W.L.; Guerini, D.; Thallmair, M.; Schwab, M.E.; et al. Phosphorylated FTY720 promotes astrocyte migration through sphingosine-1-phosphate receptors. *J. Neurochem.* **2007**, *102*, 1151–1161. [CrossRef] [PubMed]

246. Herr, D.R.; Chun, J. Effects of LPA and S1P on the nervous system and implications for their involvement in disease. *Curr. Drug Targets* **2007**, *8*, 155–167. [CrossRef] [PubMed]

247. Karunakaran, I.; Alam, S.; Jayagopi, S.; Frohberger, S.J.; Hansen, J.N.; Kuehlwein, J.; Hölbling, B.V.; Schumak, B.; Hübner, M.P.; Gräler, M.H.; et al. Neural sphingosine 1-phosphate accumulation activates microglia and links impaired autophagy and inflammation. *Glia* **2019**, *67*, 1859–1872. [CrossRef]

248. Yagi, K.; Lidington, D.; Wan, H.; Fares, J.C.; Meissner, A.; Sumiyoshi, M.; Ai, J.; Foltz, W.D.; Nedospasov, S.A.; Offermanns, S.; et al. Therapeutically targeting tumor necrosis factor-α/sphingosine-1-phosphate signaling corrects myogenic reactivity in subarachnoid hemorrhage. *Stroke* **2015**, *46*, 2260–2270. [CrossRef]

249. Olsson, T.; Zhi, W.W.; Hojeberg, B.; Kostulas, V.; Jiang, Y.P.; Anderson, G.; Ekre, H.P.; Link, H. Autoreactive T lymphocytes in multiple sclerosis determined by antigen-induced secretion of interferon-γ. *J. Clin. Investig.* **1990**, *86*, 981–985. [CrossRef]

250. Comi, G. Position and practical use of fingolimod in Europe. *Clin. Exp. Neuroimmunol.* **2014**, *5*, 19–33. [CrossRef]

251. Camm, J.; Hla, T.; Bakshi, R.; Brinkmann, V. Cardiac and vascular effects of fingolimod: Mechanistic basis and clinical implications. *Am. Heart J.* **2014**, *168*, 632–644. [CrossRef]

252. Cohen, J.A.; Barkhof, F.; Comi, G.; Hartung, H.-P.; Khatri, B.O.; Montalba, Y.; Pelletier, J.; Capra, R.; Gallo, P.; Izquierdo, G.; et al. Oral fingolimod or intramuscular interferon for relapsing multiple sclerosis. *N. Engl. J. Med.* **2010**, *362*, 402–415. [CrossRef]

253. Arvin, A.M.; Wolinsky, J.S.; Kappos, L.; Morris, M.I.; Reder, A.T.; Tornatore, C.; Gershon, A.; Gershon, M.; Levin, M.J.; Bezuidenhoudt, M.; et al. Varicella-zoster virus infections in patients treated with fingolimod: Risk assessment and consensus recommendations for management. *JAMA Neurol.* **2015**, *72*, 31–39. [CrossRef] [PubMed]

254. Kappos, L.; Radue, E.-W.; O'Connor, P.; Polman, C.; Hohlfeld, R.; Calabresi, P.; Selmaj, K.; Agoropoulou, C.; Leyk, M.; Zhang-Auberson, L.; et al. A placebo-controlled trial of oral fingolimod in relapsing multiple sclerosis. *N. Engl. J. Med.* **2010**, *362*, 387–401. [CrossRef] [PubMed]

255. Foerch, C.; Friedauer, L.; Bauer, B.; Wolf, T.; Adam, E.H. Severe COVID-19 infection in a patient with multiple sclerosis treated with fingolimod. *Mult. Scler. Relat. Disord.* **2020**, *42*, 102180. [CrossRef]

256. Brinkmann, V. Sphingosine 1-phosphate receptors in health and disease: Mechanistic insights from gene deletion studies and reverse pharmacology. *Pharmacol. Ther.* **2007**, *115*, 84–105. [CrossRef]

257. Carvajal, R.D.; Merrill, A.H., Jr.; Dials, H.; Barbi, A.; Schwartz, G.K. A phase I clinical study of safingol followed by cisplatin: Promising activity in refractory adrenocortical cancer with novel pharmacology. *J. Clin. Oncol.* **2006**, *24*, 13044–13044. [CrossRef]

258. Schwartz, G.K.; Haimovitz-Friedman, A.; Dhupar, S.K.; Ehleiter, D.; Maslak, P.; Lai, L.; Loganzo, F., Jr.; Kelsen, D.P.; Fuks, Z.; Albino, A.P. Potentiation of apoptosis by treatment with the protein kinase C-specific inhibitor safingol in mitomycin C- treated gastric cancer cells. *J. Natl. Cancer Inst.* **1995**, *87*, 1394–1399. [CrossRef]

259. Dickson, M.A.; Carvajal, R.D.; Merrill, A.H., Jr.; Gonen, M.; Cane, L.M.; Schwartz, G.K. A phase I clinical trial of safingol in combination with cisplatin in advanced solid tumors. *Clin. Cancer Res.* **2011**, *17*, 2484–2492. [CrossRef]

260. Ling, L.-U.; Tan, K.-B.; Lin, H.; Chiu, G.N.C. The role of reactive oxygen species and autophagy in safingol-induced cell death. *Cell Death Dis.* **2011**, *2*, e129. [CrossRef]

261. French, K.J.; Zhuang, Y.; Maines, L.W.; Gao, P.; Wang, W.; Beljanski, V.; Upson, J.J.; Green, C.L.; Keller, S.N.; Smith, C.D. Pharmacology and antitumor activity of ABC294640, a selective inhibitor of sphingosine kinase-2. *J. Pharmacol. Exp. Ther.* **2010**, *333*, 129–139. [CrossRef]

262. Chaurasia, B.; Summers, S.A. Ceramides - Lipotoxic inducers of metabolic disorders. *Trends Endocrinol. Metab.* **2015**, *26*, 538–550. [CrossRef]

263. Britten, C.D.; Garrett-Mayer, E.; Chin, S.H.; Shirai, K.; Ogretmen, B.; Bentz, T.A.; Brisendine, A.; Anderton, K.; Cusack, S.L.; Maines, L.W.; et al. A phase I study of ABC294640, a first-in-class sphingosine kinase-2 inhibitor, in patients with advanced solid tumors. *Clin. Cancer Res.* **2017**, *23*, 4642–4650. [CrossRef]

264. Xun, C.; Chen, M.-B.; Qi, L.; Tie-Ning, Z.; Peng, X.; Ning, L.; Zhi-Xiao, C.; Li-Wei, W. Targeting sphingosine kinase 2 (SphK2) by ABC294640 inhibits colorectal cancer cell growth in vitro and in vivo. *J. Exp. Clin. Cancer Res.* **2015**, *34*, 94. [CrossRef] [PubMed]

265. Xu, L.; Jin, L.; Yang, B.; Wang, L.; Xia, Z.; Zhang, Q.; Xu, J. The sphingosine kinase 2 inhibitor ABC294640 inhibits cervical carcinoma cell growth. *Oncotarget* **2018**, *9*, 2384–2394. [CrossRef] [PubMed]

266. Dai, L.; Smith, C.D.; Foroozesh, M.; Miele, L.; Qin, Z. The sphingosine kinase 2 inhibitor ABC294640 displays anti-non-small cell lung cancer activities in vitro and in vivo. *Int. J. Cancer* **2018**, *142*, 2153–2162. [CrossRef] [PubMed]

267. Abuhusain, H.J.; Matin, A.; Qiao, Q.; Shen, H.; Kain, N.; Day, B.W.; Stringer, B.W.; Daniels, B.; Laaksonen, M.A.; Teo, C.; et al. A metabolic shift favoring sphingosine 1-phosphate at the expense of ceramide controls glioblastoma angiogenesis. *J. Biol. Chem.* **2013**, *288*, 37355–37364. [CrossRef]

268. Mahajan-Thakur, S.; Bien-Möller, S.; Marx, S.; Schroeder, H.; Rauch, B.H. Sphingosine 1-phosphate (S1P) signaling in glioblastoma multiforme—A systematic review. *Int. J. Mol. Sci.* **2017**, *18*, 2448. [CrossRef]

269. Visentin, B.; Vekich, J.A.; Sibbald, B.J.; Cavalli, A.L.; Moreno, K.M.; Matteo, R.G.; Garland, W.A.; Lu, Y.; Hall, H.S.; et al. Validation of an anti-sphingosine-1-phosphate antibody as a potential therapeutic in reducing growth, invasion, and angiogenesis in multiple tumor lineages. *Cancer Cell* **2006**, *9*, 225–238. [CrossRef]

270. Caballero, S.; Swaney, J.; Moreno, K.; Afzal, A.; Kielczewski, J.; Stoller, G.; Cavalli, A.; Garland, W.; Hansen, G.; Sabbadini, R.; et al. Anti-sphingosine-1-phosphate monoclonal antibodies inhibit angiogenesis and sub-retinal fibrosis in a murine model of laser-induced choroidal neovascularization. *Exp. Eye Res.* **2009**, *88*, 367–377. [CrossRef]

271. Sabbadini, R.A. Sphingosine-1-phosphate antibodies as potential agents in the treatment of cancer and age-related macular degeneration. *Br. J. Pharmacol.* **2011**, *162*, 1225–1238. [CrossRef]

272. Pal, S.K.; Drabkin, H.A.; Reeves, J.A.; Hainsworth, J.D.; Hazel, S.E.; Paggiarino, D.A.; Wojciak, J.; Woodnutt, G.; Bhatt, R.S. A phase 2 study of the sphingosine-1-phosphate antibody sonepcizumab in patients with metastatic renal cell carcinoma. *Cancer* **2017**, *123*, 576–582. [CrossRef]

273. Fleischmann, R. Novel small-molecular therapeutics for rheumatoid arthritis. *Curr. Opin. Rheumatol.* **2012**, *24*, 335–341. [CrossRef] [PubMed]

274. Pyszko, J.A.; Strosznajder, J.B. The key role of sphingosine kinases in the molecular mechanism of neuronal cell survival and death in an experimental model of Parkinson's disease. *Folia Neuropathol.* **2014**, *52*, 260–269. [CrossRef]

275. Samuvel, D.J.; Saxena, N.; Dhindsa, J.S.; Singh, A.K.; Gill, G.S.; Grobelny, D.W.; Singh, I. AKP-11 - A novel S1P1 agonist with favorable safety profile attenuates experimental autoimmune encephalomyelitis in rat model of multiple sclerosis. *PLoS ONE* **2015**, *10*, e0141781. [CrossRef] [PubMed]

276. Dhar, T.G.M.; Xiao, H.-Y.; Xie, J.; Lehman-McKeeman, L.D.; Wu, D.-R.; Dabros, M.; Yang, X.; Taylor, T.L.; Zhou, X.D.; Heimrich, E.M.; et al. Identification and preclinical pharmacology of BMS-986104: A differentiated S1P1 receptor modulator in clinical trials. *ACS Med. Chem. Lett.* **2016**, *7*, 283–288. [CrossRef] [PubMed]

277. Piali, L.; Birker-Robaczewska, M.; Lescop, C.; Froidevaux, S.; Schmitz, N.; Morrison, K.; Kohl, C.; Rey, M.; Studer, R.; Vezzali, E.; et al. Cenerimod, a novel selective S1P1 receptor modulator with unique signaling properties. *Pharmacol. Res. Perspect.* **2017**, *5*, e00370. [CrossRef] [PubMed]

278. Hermann, V.; Batalov, A.; Smakotina, S.; Juif, P.E.; Cornelisse, P. First use of cenerimod, a selective S1P 1 receptor modulator, for the treatment of SLE: A double-blind, randomised, placebo-controlled, proof-of-concept study. *Lupus Sci. Med.* **2019**, *6*, e000354. [CrossRef]

279. Xu, J.; Gray, F.; Henderson, A.; Hicks, K.; Yang, J.; Thompson, P.; Oliver, J. Safety, pharmacokinetics, pharmacodynamics, and bioavailability of GSK2018682, a sphingosine-1-phosphate receptor modulator, in healthy volunteers. *Clin. Pharmacol. Drug Dev.* **2014**, *3*, 170–178. [CrossRef]

280. Gruessner, R.W.; Sutherland, D.E.; Troppmann, C.; Benedetti, E.; Hakim, N.; Dunn, D.L.; Gruessner, A.C. The surgical risk of pancreas transplantation in the cyclosporine era: An overview. *J. Am. Coll. Surg.* **1997**, *185*, 128–144. [CrossRef]
281. Khattar, M.; Deng, R.; Kahan, B.D.; Schroder, P.M.; Phan, T.; Rutzky, L.P.; Stepkowski, S.M. Novel sphingosine-1-phosphate receptor modulator KRP203 combined with locally delivered regulatory T cells induces permanent acceptance of pancreatic islet allografts. *Transplantation* **2013**, *95*, 919–927. [CrossRef]
282. D'Ambrosio, D.; Freedman, M.S.; Prinz, J. Ponesimod, a selective S1P1 receptor modulator: A potential treatment for multiple sclerosis and other immune-mediated diseases. *Ther. Adv. Chronic Dis.* **2016**, *7*, 18–33. [CrossRef]
283. Schmidt, K.G.; Herrero San Juan, M.; Trautmann, S.; Berninger, L.; Schwiebs, A.; Ottenlinger, F.M.; Thomas, D.; Zaucke, F.; Pfeilschifter, J.M.; Radeke, H.H. Sphingosine-1-phosphate receptor 5 modulates early-stage processes during fibrogenesis in a mouse model of systemic sclerosis: A pilot study. *Front. Immunol.* **2017**, *8*, 1242. [CrossRef] [PubMed]
284. Kurata, H.; Kusumi, K.; Otsuki, K.; Suzuki, R.; Kurono, M.; Komiya, T.; Hagiya, H.; Mizuno, H.; Shioya, H.; Ono, T.; et al. Discovery of a 1-Methyl-3,4-dihydronaphthalene-based sphingosine-1-phosphate (S1P) receptor agonist ceralifimod (ONO-4641). A S1P1 and S1P5 selective agonist for the treatment of autoimmune diseases. *J. Med. Chem.* **2017**, *60*, 9508–9530. [CrossRef] [PubMed]
285. Meadows, K.R.T.; Steinberg, M.W.; Clemons, B.; Stokes, M.E.; Opiteck, G.J.; Peach, R.; Scott, F.L. Ozanimod (RPC1063), a selective S1PR1 and S1PR5 modulator, reduces chronic inflammation and alleviates kidney pathology in murine systemic lupus erythematosus. *PLoS ONE* **2018**, *13*, e0193236. [CrossRef]
286. Wang, W.; Graeler, M.H.; Goetzl, E.J. Type 4 sphingosine 1-phosphate G protein-coupled receptor (S1P4) transduces S1P effects on T cell proliferation and cytokine secretion without signaling migration. *FASEB J.* **2005**, *19*, 1731–1733. [CrossRef]
287. Wang, W.; Huang, M.-C.; Goetzl, E.J. Type 1 sphingosine 1-phosphate G protein-coupled receptor (S1P1) mediation of enhanced IL-4 generation by CD4 T cells from S1P1 transgenic mice. *J. Immunol.* **2007**, *178*, 4885–4890. [CrossRef]
288. Gräler, M.H.; Grosse, R.; Kusch, A.; Kremmer, E.; Gudermann, T.; Lipp, M. The sphingosine 1-phosphate receptor S1P4 regulates cell shape and motility via coupling to Gi and G12/13. *J. Cell Biochem.* **2003**, *89*, 507–519. [CrossRef]
289. Schulze, T.; Golfier, S.; Tabeling, C.; Räbel, K.; Gräler, M.H.; Witzenrath, M.; Lipp, M. Sphingosine-1-phospate receptor 4 (S1P4) deficiency profoundly affects dendritic cell function and TH 17-cell differentiation in a murine model. *FASEB J.* **2011**, *25*, 4024–4036. [CrossRef]
290. Sugahara, K.; Maeda, Y.; Shimano, K.; Mogami, A.; Kataoka, H.; Ogawa, K.; Hikida, K.; Kumagai, H.; Asayama, M.; Yamamoto, T.; et al. Amiselimod, a novel sphingosine 1-phosphate receptor-1 modulator, has potent therapeutic efficacy for autoimmune diseases, with low bradycardia risk. *Br. J. Pharmacol.* **2017**, *174*, 15–27. [CrossRef]
291. Kappos, L.; Arnold, D.L.; Bar-Or, A.; Camm, A.J.; Derfuss, T.; Sprenger, T.; Davies, M.; Piotrowska, A.; Ni, P.; Harada, T. Two-year results from a phase 2 extension study of oral amiselimod in relapsing multiple sclerosis. *Mult. Scler.* **2018**, *24*, 1605–1616. [CrossRef]
292. Shimano, K.; Maeda, Y.; Kataoka, H.; Murase, M.; Mochizuki, S.; Utsumi, H.; Oshita, K.; Sugahara, K. Amiselimod (MT-1303), a novel sphingosine 1-phosphate receptor-1 functional antagonist, inhibits progress of chronic colitis induced by transfer of $CD4^+CD45RBhigh$ T cells. *PLoS ONE* **2019**, *14*, e0226154. [CrossRef]
293. Sandborn, W.J.; Peyrin-Biroulet, L.; Zhang, J.; Chiorean, M.; Vermeire, S.; Lee, S.D.; Kühlbacher, T.; Yacyshyn, B.; Cabell, C.H.; Naik, S.U.; et al. Efficacy and safety of etrasimod in a phase 2 randomized trial of patients with ulcerative colitis. *Gastroenterology* **2020**, *158*, 550–561. [CrossRef] [PubMed]

Article

Neurodegeneration Caused by S1P-Lyase Deficiency Involves Calcium-Dependent Tau Pathology and Abnormal Histone Acetylation

Shah Alam [1], Antonia Piazzesi [1], Mariam Abd El Fatah [1], Maren Raucamp [2] and Gerhild van Echten-Deckert [1],*

[1] LIMES Institute for Membrane Biology and Lipid Biochemistry, Kekulé-Institute, University of Bonn, 53121 Bonn, Germany; shah.bonn@outlook.com (S.A.); piazzesa@hushmail.com (A.P.); mariam@uni-bonn.de (M.A.E.F.)

[2] Institute of Physiology, University Bonn, 53111 Bonn, Germany; Maren.Raucamp@ukb.uni-bonn.de

* Correspondence: g.echten.deckert@uni-bonn.de; Tel.: +49-228-73-2703; Fax: +49-228-73-4845

Received: 29 June 2020; Accepted: 23 September 2020; Published: 28 September 2020

Abstract: We have shown that sphingosine 1-phosphate (S1P) generated by sphingosine kinase 2 (SK2) is toxic in neurons lacking S1P-lyase (SGPL1), the enzyme that catalyzes its irreversible cleavage. Interestingly, patients harboring mutations in the gene encoding this enzyme (*SGPL1*) often present with neurological pathologies. Studies in a mouse model with a developmental neural-specific ablation of SGPL1 (SGPL1$^{fl/fl/Nes}$) confirmed the importance of S1P metabolism for the presynaptic architecture and neuronal autophagy, known to be essential for brain health. We now investigated in SGPL1-deficient murine brains two other factors involved in neurodegenerative processes, namely tau phosphorylation and histone acetylation. In hippocampal and cortical slices SGPL1 deficiency and hence S1P accumulation are accompanied by hyperphosphorylation of tau and an elevated acetylation of histone3 (H3) and histone4 (H4). Calcium chelation with BAPTA-AM rescued both tau hyperphosphorylation and histone acetylation, designating calcium as an essential mediator of these (patho)physiological functions of S1P in the brain. Studies in primary cultured neurons and astrocytes derived from SGPL1$^{fl/fl/Nes}$ mice revealed hyperphosphorylated tau only in SGPL1-deficient neurons and increased histone acetylation only in SGPL1-deficient astrocytes. Both could be reversed to control values with BAPTA-AM, indicating the close interdependence of S1P metabolism, calcium homeostasis, and brain health.

Keywords: Sphingosine 1-phosphate (S1P); S1P-lyase (SGPL1); tau; calcium; histone acetylation; hippocampus; cortex; astrocytes; neurons

---

1. Introduction

Sphingosine 1-phosphate (S1P), an evolutionarily conserved catabolic intermediate of sphingolipid metabolism, regulates diverse biological processes in the brain including neural development, differentiation, and survival [1,2]. S1P exerts its functions either as a ligand of five specific G-protein coupled receptors (S1PR1-5) or alternatively as an intracellular second messenger [3,4]. Notably, one of the first described intracellular, receptor-independent effects of S1P is its involvement in calcium homeostasis [5,6].

S1P-lyase (SGPL1) irreversibly cleaves S1P in the final step of sphingolipid catabolism, generating ethanolamine phosphate and a long-chain aldehyde [7]. Of interest, in 2017 different research groups reported patients and their relatives harboring autosomal recessive mutations in *SGPL1* and exhibiting a variety of pathologies including congenital steroid-resistant nephrotic syndrome, primary adrenal insufficiency, and last but not least central and peripheral neurological defects [8].

The essential role of S1P in brain development became clear years ago, when elimination of S1P production was shown to severely disturb neurogenesis including neural tube closure, and angiogenesis leading to embryonic death [9]. Yet, reports on the involvement of S1P in the pathology of neurodegenerative diseases including Alzheimer's disease (AD) are rather conflicting [10]. On the one hand, it is assumed that loss of the neuroprotective factor S1P occurs early in AD pathogenesis [11]. Indeed reduced expression of sphingosine kinase 1, one of the two sphingosine kinases known to generate S1P, and a simultaneous augmented expression of SGPL1 were detected in AD brains [12]. On the other hand, S1P was shown to stimulate neuronal beta-site amyloid precursor protein (APP) cleaving enzyme (BACE1) that catalyzes the rate-limiting step of the formation of amyloid beta peptide (Aβ), the major component of senile plaques in AD [13]. In addition, increased S1P levels were shown to induce death of terminally differentiated post-mitotic neurons [14,15]. Moreover, S1P was found to be increased in cerebrospinal fluid during early stages of AD [16].

In an attempt to clarify the function of S1P in the brain, we generated a mouse model in which SGPL1 was inactivated specifically in neural cells (SGPL1$^{fl/fl/Nes}$). As expected, SGPL1 ablation leads to S1P accumulation in the brain which was found to affect presynaptic architecture and function [17]. In addition, we demonstrated that SGPL1 deficiency blocks neuronal autophagy at its early stages because of reduced phosphatidylethanolamine (PE) production [18]. Consequently, an accumulation of aggregate-prone proteins such as APP and α-synuclein (SNCA) was detected. All these molecular changes in neurons were accompanied by deficits in motor coordination as well as in spatial and associative learning and memory [17,18].

We have also shown that S1P promotes excessive phosphorylation of tau in neurons generated from SGPL1 systemic knockout mice [14]. Note that tau is the major neuronal microtubule assembly activator protein and there is no doubt regarding its essential involvement in the etiopathogenesis of AD and a family of related neurodegenerative disorders known as tauopathies [19]. Tau neurotoxicity has been linked to heterochromatin relaxation and hence to aberrant gene expression in tauopathies [20]. Studies in primary cultured neurons revealed that nuclear tau directly regulates pericentromericheterochtomatin integrity that appears disrupted in AD neurons [21]. Recently, an epigenome-wide study in which acetylation of lysine 9 of histone3 (H3K9ac) was used as a marker for transcriptionally active open chromatin, also led to the conclusion that in aging and AD brains tau pathology drives chromatin rearrangement [22]. Furthermore, in post-mortem AD brains, increased levels of acetylated H3 and H4 were detected and correlated with the load of hyperphosphorylated tau [23].

Remarkably, in a tumorigenic cell line, S1P generated by sphingosine kinase 2 (SK2) was reported to specifically enhance acetylation of H3 and H4 at K9 and K5, respectively, by directly inhibiting histone deacetylases 1 and 2 (HDACs 1, 2) [24]. In the present study, we demonstrate that accumulation of S1P as a result of SGPL1 deficiency increases tau phosphorylation and histone acetylation also in brain slices. Furthermore, we found that both effects can be rescued in the presence of the calcium chelator BAPTA-AM, indicating that this process is calcium-dependent. Notably, these effects were cell-type specific, with increases in tau phosphorylation and histone acetylation found in neurons and astrocytes, respectively. Taken together, our results further elucidate the extensive and complex interrelation of S1P metabolism and brain health.

2. Materials and Methods

2.1. List of Abbreviations

AD: Alzheimer's disease; APP: amyloid precursor protein; ac: acetylated; H3: histone3; H4: histone4; H2B, histone2B; K5, lysine residue 5; K9, lysine residue 9; K12, lysine residue 12; HDAC: histone deacetylase; S1P: sphingosine 1-phosphate; SGPL1: S1P-lyase; SK: sphingosine kinase; S1PR: S1P receptor; S396/404, serine residue 396 and serine residue 404; S262/356, serine residue 262 and serine residue 356.

2.2. Antibodies and Chemicals

Monoclonal antibody against phosphorylated tau PHF1 (S396/404), 12E8(S262/356) and against total tau (K9JA) was a kind gift from Prof. Dr. Eckhard Mandelkow and Prof. Dr. Eva-Maria Mandelkow (DZNE, University of Bonn, Germany). Acetyl-Histone H3 antibody sampler kit comprising acetylation of K9, K14, K18, K27, and K56, anti-H4K5ac, and anti-H2BK12ac antibody were from Cell Signaling Technology Danvers, MA, USA (9927, 8647 and 5410). Anti-HDAC1, -HDAC2, -HDAC3, and -HDAC6 antibodies were from Cell Signaling Technology (antibody Sampler Kit #9928). Anti-SGPL1 antibody was from abcam (Cambridge, UK; ab56183) and Anti-glial fibrillary acidic protein (GFAP) antibody from Cell Signaling Technology (12389).Secondary antibodies were HRP linked anti-rabbit and anti-mouse IgG (Cell Signaling Technology, 7074 and 7076). 5,5′-Dimethyl-BAPTA-AM was from Sigma- Aldrich, Munich, Germany (16609).

2.3. Animals

The SGPL1$^{flox/flox}$ lines were generated as recently described [17]. SGPL1$^{flox/flox}$ mice, harboring "floxed" exons 10–12 on both Sgpl1 alleles were crossbred with mice expressing the Nes (nestin) - Cre transgene. Thus SGPL1$^{fl/fl/Nes}$mice (nKO) in which "floxed" exons are excised by Cre recombinase were obtained. For all the experiments, the floxed mice (SGPL1$^{fl/fl}$) served as controls. Brain tissue was taken from mice housed in standard conditions at the University of Bonn.

2.4. Ethical Statement

All animal experiments were conducted in accordance with the guidelines of the Animal Care Committee of the University of Bonn. The experimental protocols were approved by Landesamt für Natur, Umwelt und Verbraucherschutz (LANUV) Nordrhein-Westfalen (NRW) (LANUV NRW, Az. 81–02.05.40.19.013).

2.5. Cell Culture

Primary neuronal culture: Granular cells were cultured from the cerebella of 6-days old mice as described previously [25]. Briefly, neurons were isolated by mild trypsinization (0.05%, w/v; Sigma-Aldrich, Munich, Germany P6567) and dissociated by passing them repeatedly through a constricted Pasteur pipette in a DNase solution (0.1%, w/v; Roche, Basel, Switzerland 04716728001). The cells were then suspended in Dulbecco's modified Eagle's medium (Thermo Fisher Scientific, Waltham, MA, USA 10566032) containing 10% heat-inactivated horse serum (Thermo Fisher Scientific, 16050130) supplemented with 100 units/mL penicillin and 100 mg/mL streptomycin (Gibco™ Thermo Scientific 15140122) and plated onto precoated poly-L-lysine (Sigma-Aldrich, P6282) 6-well plates, 35 mm in diameter (Sarstedt, Nümbrecht, Germany (83.3920.300). Twenty-four h after plating, 1% cytosine ß-D-arabinofuranoside hydrochloride (Sigma-Aldrich, C6645) was added to the medium to arrest the division of non-neuronal cells. After 10 days in culture, cells were used for experiments as indicated.

Primary astrocyte culture: Mixed cortical cell isolation for astrocyte culture was performed using P1 to P4 mouse pups as described previously [26]. Briefly, cerebral cortices were dissected in Ca^{2+}- and Mg^{2+}-free HBSS (Gibco™, Thermo Scientific, 14185652) and incubated in 0.125% trypsin for 10 min at 37 °C. The resulting cell suspension was diluted in complete media Dulbecco's modified Eagle's medium supplemented 10% fetal bovine serum (PAN biotech, Aidenbach, Germany, P40-47100) and 1% penicillin/ streptomycin. The cell suspension was plated on poly-L-Lysine (P-1399) coated T75 cell culture flasks and kept at 37 °C in a humidified 5% CO_2 incubator. Medium was renewed every 2 days. After about 21 days, flasks were shaken horizontally, and the medium containing detached microglia and oligodendrocyte precursor cells (OPC) was removed. Later, astrocytes were collected and seeded onto 6-well cell culture dishes (35 mm diameter) and used for experiments after 24 h, as indicated.

2.6. BAPTA- AM Treatment

Hippocampal and cortical slices of 200 μm thickness were prepared in ice-cold high sucrose solution (220 mM sucrose, 26 mM $NaHCO_3$, 10 mM glucose, 6 mM $MgSO_4.7H_2O$, 3 mM KCL solid, 1.25 mM NaH_2PO_4. H_2O, 0.43 mM $CaCl_2$) gassed with carbogen. Then, both hippocampal and cortical slices were incubated in artificial cerebrospinal fluid (119 mMNaCl, 26.2 mM $NaHCO_3$, 2.5 mMKCl, 1 mM NaH_2PO_4, 1.3 mM $MgCl_2$, 10 mM glucose) with and without 150 μM BAPTA-AM for 2 h and kept at -80 °C until use.

2.7. Western Immunoblotting

Tissue and cell samples were homogenized in RIPA buffer (20 mMTris-HCl, pH 7.5, 150 mMNaCl, 1 mM EDTA, 1 mM EGTA, 1% NP-40 (Thermo Fisher Scientific, Waltham, MA, USA, FNN0021), 1% Nadcap (Sigma-Aldrich, Munich, Germany, D6750), 2.5 mM $Na_4P_2O_7$, 1 mM b-glycerophosphate, 1 mM Na_3VO_4, 1 mg/mL leupeptin (Thermo Fisher Scientific, 78435). Samples were kept on ice for 1 h followed by centrifugation at 14,000× rpm at 4 °C for 45 min. The protein concentration of the supernatants was determined using the BCA assay (Sigma–Aldrich). Samples were stored at -20 °C until use. Laemmli Sample Buffer (Bio-rad Laboratories, Munich, Germany, 1610747) was added to lysates and samples were heated for 10 min at 95 °C before loading on SDS-PAGE gel. Proteins were separated by SDS-PAGE in running buffer (25 mM Tris, pH 8.3, 192 mM glycine, 0.1% SDS) at 50 V for 15 min, then 1 h at 150 V. Transfer onto nitrocellulose membranes (Porablot NCL; Macherey-Nagel, Thermo Fisher Scientific, 741290) was performed at 4 °C and 400 mA for 2 h in blotting buffer (50 mMTris, pH 9.2, 40 mM glycine, 20% methanol). Membranes were blocked with 5% milk powder (Bio-Rad Laboratories, 1706404) in TBS-Tween 20 (20 mM Tris, pH 7.5, 150 mM NaCl, 0.1% Tween 20, Sigma-Aldrich, P9416) for 1 h, washed 3 times (10 min each) and incubated at 4 °C overnight with the primary antibody. Then membranes were washed again and incubated for 1 h at room temperature with an HRP-conjugated secondary antibody. Western BLoT Chemiluminescence HRP Substrate (TAKARA Bio, Saint-Germain-en-Laye, France, T7101B) was used for detection with the VersaDoc 5000 imaging system (Bio-Rad, Hercules, CA, USA). β-actin was used as loading control. Quantification was performed using ImageJ and Prism GraphPad program.

2.8. RNA Isolation and Real-Time PCR

Up to 1 μg of total RNA (isolated with EXTRAzol from Blirt, 7Bioscience, Hartheim/Rhein, Germany, EM30-200) was used for reverse transcription with the ProtoScript® II First Strand cDNA Synthesis kit (New England Biolabs, Frankfurt/Main, Germany, E6560L). The resulting total cDNA was then applied to real-time PCR (CFX96-real time PCR, Bio-Rad Laboratories, Munich, Germany) using β-actin and 18S RNA as housekeeping genes. The primers for real-time PCR were designed using the online tool from NCBI BLAST primer and obtained from Invitrogen, Carlsbad, CA, USA. They are listed as follows: name: forward primer (for), reverse primer (rev): β-actin, 5′-CTTTGCAGCTCCTTCGTTGC (for) and 5′-CCTTCTGACCCATTCCCACC (rev); 18S RNA, 5′-CCCCTCGATGCTCTTAGCTG (for) and 5′-CTTTCGCTCTGGTCCGTCTT (rev); HDAC1, 5′-AGCTGGGCTTTCCAAGTTACC (for) and 5′-TGGTCCACACCCTTCTCGTA (rev); HDAC2, 5′-CGGCCAAGCCTGACTTAGAT (for) and 5′-TTTTCAGCTGTCCTCGGTGG (rev); HDAC3, 5′-TGCCCCAGATTTCACACTCC (for) and 5′-TGGTCCAGATACTGGCGTGA (rev); HDAC6, 5′- GGCGCAGATTAGAGAGCCTT (for) and 5′-GAAGGGGTGACTGGGGATTG (rev); SGPL1, 5′-TTTCCTCATGGTGTGATGGA (for) and 5′-C CCCAGACAAGCATCCAC3(rev).The reactions were performed at 95 °C for 30 s, 95 °C for 10 s, and 60 °C for 1 min. Relative normalized mRNA expression was obtained from real-time qPCR. Statistical significance of the relative normalized mRNA expression was determined by *t*-test in Prism GraphPad program.

2.9. Immunohistochemistry

Brains were removed and quick-frozen in liquid nitrogen. Cryo-sectioning was used to produce 10 μm sagittal sections, which were placed on Superfrost Plus positively charged microscope slides. Brain sections were fixed for 5 min in ice-cold 4% (*v/v*) paraformaldehyde in phosphate-buffered saline (PBS). Sections were then permeabilized with 0.1% (*v/v*) Triton X-100 in PBS for 30 min at room temperature (RT). Tissue sections were blocked in 20% (*v/v*) normal goat serum in PBS for 30 min and incubated overnight at 4 °C with primary antibody (PHF1 and 12E8). The primary antibodies were diluted 1:200 in PBS containing 0.5% lambda-carrageenan (Sigma Aldrich, Munich, Germany, 22049) and 0.02% sodium azide and applied overnight to the sections at 4 °C. Following a washing step, brain sections were incubated with Cy3-conjugated anti-rabbit antibody diluted 1:300 in PBS with the same additions as above for 1 h at RT. Finally, antibody-labeled brain sections were embedded in Fluoromount G medium with DAPI (Electron Microscopy Sciences, Hatfield, PA, USA for microscopic analysis (Zeiss Axioskop 2 epi-fluorescence microscope equipped with a digital Zeiss AxioCamHRc camera, Carl Zeiss Jena, Jena, Germany).

2.10. Immunocytochemistry

Cover slips with astrocytes were rinsed 3 times with PBS at room temperature (RT) and then fixed in methanol (−20 °C, 5 min). Between each incubation step cells were always rinsed 3 times with PBS. Then cells were blocked in 20% (*v/v*) normal goat serum in PBS for 30 min. and incubated overnight with anti-GFAP antibody diluted 1:200 with PBS at 4 °C and thenwith anti-rabbit Alexa Fluor 488 (1:300)-conjugated secondary antibodies (Invitrogen, Carlsbad, CA, USA) for 50 min at RT. Finally, cells were embedded in Fluoromount G medium with DAPI for microscopic analyses (Zeiss Axioskop 2 epi-fluorescence microscope equipped with a digital Zeiss AxioCamHRc camera, Carl Zeiss Jena, Jena, Germany).

2.11. Statistical Analysis

The GraphPad Prism 5 software was used for statistical analysis.All values are expressed as means ± SEMobtained from at least 3 independent experiments. The significanceof differences between the experimental groups and controls was assessed by either Student's *t*-test with false discovery rate (FDR) correction or One-Way ANOVA, as appropriate. $p/q < 0.05$ was considered statistically significant (* $p/q < 0.05$; ** $p/q < 0.01$; *** $p/q < 0.001$; compared with the respective control group).

3. Results

3.1. Elevated Phosphorylation of Tau in SGPL1-Deficient Brains Is Cell Type Specific

We have previously shown that tau phosphorylation is elevated in primary cultured neurons derived from brains of systemic SGPL1-knockout (KO) mice [14]. Here, we generated a neural-specific Sgpl1 knockout (SGPL1$^{fl/fl/Nes}$; nKO) mouse and performed our analysis in brain slices, primary neuronal and astrocyte cultures (Supplementary Figure S1). We have found that tau phosphorylation at disease-relevant sites is also significantly increased in hippocampal and cortical slices from SGPL1$^{fl/fl/Nes}$mice consistent with our previously reported findings in systemic KO mice [14] (Figure 1A,B). Furthermore, this increase in tau phosphorylation at disease-relevant sites was not accompanied by changes in total tau levels (Figure 1B). A more refined analysis in primary cultured neurons and astrocytes from SGPL1$^{fl/fl/Nes}$mice further revealed that this effect is primarily attributed to neurons, as tau phosphorylation remained unaffected in astrocytes (Figure 1C). Accordingly, phosphorylated tau was increased by about 40% in SGPL1-deficient neurons whereas no changes were detectable in astrocytes lacking SGPL1 when compared with the respective control cells (Figure 1C). This indicates that the increase in tau phosphorylation in hippocampus and cortex is due to hyperphosphorylation of tau in neurons.

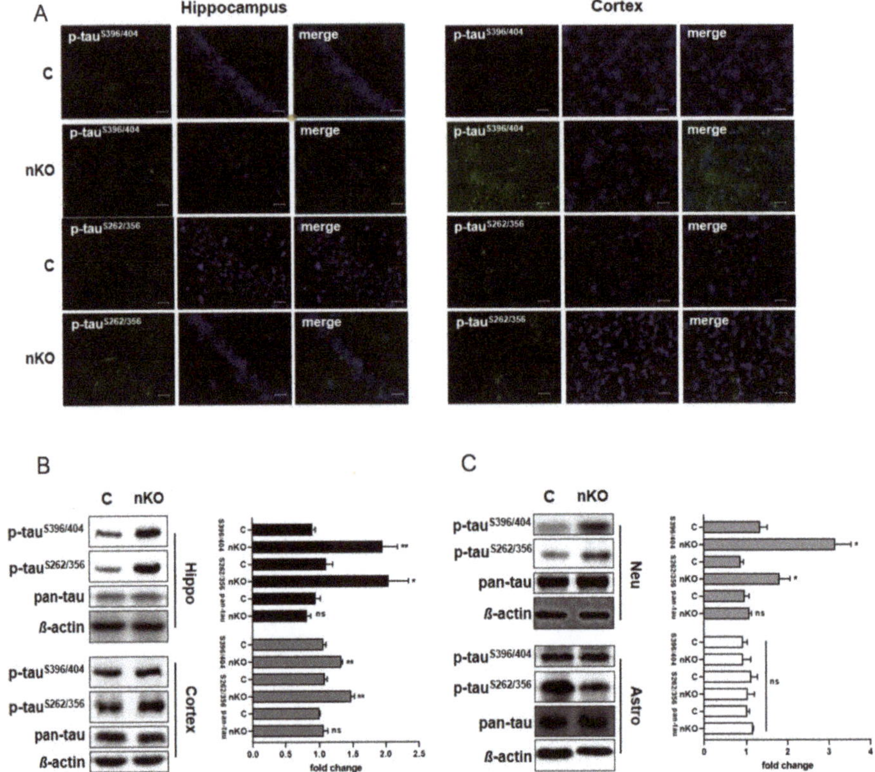

Figure 1. SGPL1 deficiency results in tau hyperphosphorylation in the brain. (**A**) Representative images of hippocampal and cortical slices stained for phospho-tau$^{S396/404}$, phospho-tau$^{S262/356}$, and DAPI from control (**C**) and SGPL1$^{fl/fl/Nes}$ (nKO) mice. Scale bar: 200 μm. (**B,C**) Protein quantification of phospho-tau$^{S396/404}$, phospho-tau$^{S262/356}$, and total tau (pan-tau) in the hippocampus (Hippo, black), cortex (dark grey), neurons (Neu, light grey) and astrocytes (Astro, white) in control (**C**) and SGPL1$^{fl/fl/Nes}$ (nKO) mice. Bars mean +/− S.E.M, Student's *t*-test with false discovery rate (FDR) correction, n = 3–5, * *q* < 0.05, ** *q* < 0.01, ns = not significant.

3.2. Histone Acetylation Levels Vary in Different Cell Types Derived from Brains Lacking SGPL1

Based on reports that tau stimulates chromatin relaxation [20] and that S1P accumulation elevates histone acetylation in tumorigenic cells [24], we analyzed histone acetylation in hippocampal and cortical slices from SGPL1$^{fl/fl/Nes}$ mice. We found that, upon S1P accumulation in the brain, acetylation of H3 is also significantly increased in both hippocampal and cortical slices by about 36% and 32%, respectively (Figure 2A). An investigation of specific acetylation sites revealed that K9 of H3 was significantly increased by about 45% in the hippocampus, and to a lesser extent (about 35%) in the cortex, as compared to the respective controls (Figure 2A). Interestingly, acetylation of all other lysine residues examined, including K14 and K18 was not affected by SGPL1 deficiency (Supplementary Figure S2). Next, we investigated acetylation of H3 in primary cultured cells derived from SGPL1$^{fl/fl/Nes}$ mice. We found that acetylation of H3 was significantly increased in astrocytes generated from SGPL1$^{fl/fl/Nes}$ mice (Figure 2B), whereas no changes of H3 acetylation were observed in primary cultured neurons of these mice (Figure 2C). Notably, the acetylation of H3 (H3ac) largely resembled that of lysine 9 of H3 (H3K9ac), whereas the total H3 level was only slightly increased (Figure 2B). These results strongly suggest that the increases in histone acetylation observed in the brain upon SGPL1 deficiency is due to epigenetic changes in astrocytes, rather than neurons. Interestingly, we also found that acetylation of H4 at K5 was significantly increased in astrocytes of SGPL1$^{fl/fl/Nes}$ mice, similar to what was shown previously in breast cancer cells (Figure 2B) [24]. Furthermore, we observed an increase in acetylation of H2B at K12, also consistent with a report of S1P accumulation in breast cancer cells (Figure 2B). Given the differences in histone acetylation observed, and given that S1P can act as an inhibitor of histone deacetylases [24], we next investigated whether the expression of histone deacetylases was affected in SGPL1$^{fl/fl/Nes}$ mice. However, we found that both mRNA and protein levels of HDAC1, 2, 3 and 6 remained unaffected in SGPL1$^{fl/fl/Nes}$ mice, compared to controls (Figure 2D–G). These results show that S1P accumulation in the brain has a cell type-specific effect on protein posttranslational modifications, without affecting the expression levels of the deacetylases responsible.

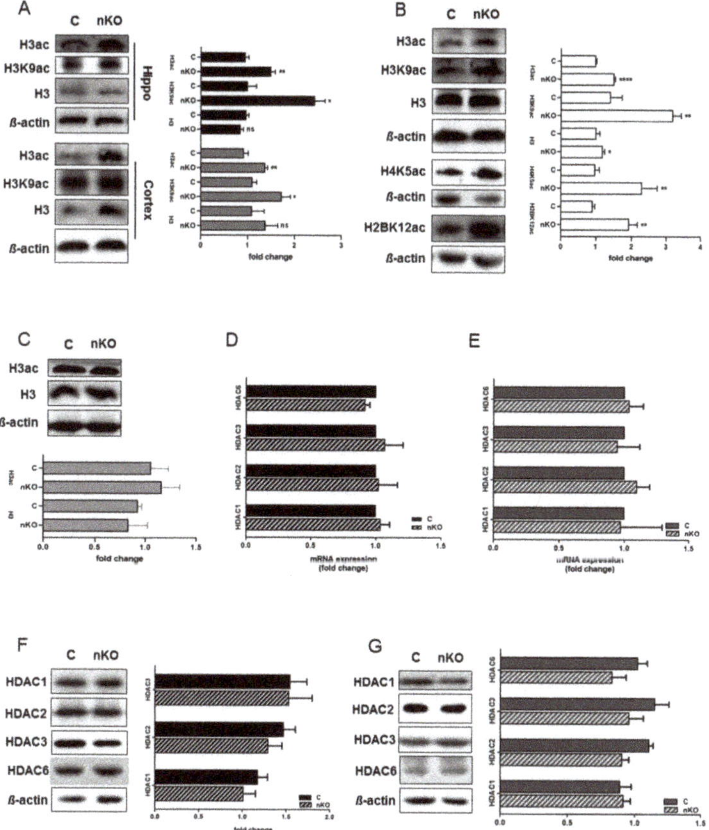

Figure 2. SGPL1 deficiency affects histone acetylation in the brain without affecting histone deacetylases (HDAC) expression. (**A**) Protein quantification of H3 pan-acetylation (H3ac), H3K9 acetylation (H3K9ac), and total H3 in the hippocampus (Hippo, black) and cortex (dark grey) from control (C) and SGPL1$^{fl/fl/Nes}$ (nKO) mice. (**B**) Protein quantification of H3 pan-acetylation (H3ac), H3K9 acetylation (H3K9ac), total H3, H4K5 acetylation (H4K5ac), and H2BK12 acetylation (H2BK12ac) in astrocytes from control (C) and SGPL1$^{fl/fl/Nes}$ (nKO) mice. (**C**) Protein quantification of H3 pan-acetylation (H3ac) and total H3 in primary neuronal culture from control (C) and SGPL1$^{fl/fl/Nes}$ (nKO) mice. (**D–E**) qRT-PCR of HDAC1, 2, 3, and 6 in the hippocampus (**D**) and cortex (**E**) of control (C) and SGPL1$^{fl/fl/Nes}$ (nKO) mice. (**F–G**) Protein quantification of HDAC1, 2, 3, and 6 in the hippocampus (**F**) and cortex (**G**) of control (C) and SGPL1$^{fl/fl/Nes}$ (nKO) mice. For all: Bars mean +/− S.E.M, Student's *t*-test with false discovery rate (FDR) correction, n = 3–7, * $q < 0.05$, ** $q < 0.01$, *** $q < 0.001$, ns = not significant.

Together, these results indicate that different cell types can be responsible for interrelated effects detected when studying certain brain regions.

3.3. Calcium Chelation Reverses Both Tau Phosphorylation and Histone Acetylation in the Brain of SGPL1$^{fl/fl/Nes}$ mice

We have previously suggested that increased calcium concentrations might account for the neurotoxic effect of S1P in SGPL1-deficient neurons [14]. Calcium measurements in hippocampal slices of SGPL1$^{fl/fl/Nes}$ mice revealed a persistent elevation of basal calcium concentration in pyramidal neurons of the CA1 region amounting about 223 nM, a value that exceeds control concentrations by about 2.5-fold [27]. To find out whether elevated basal calcium concentration is linked to tau phosphorylation, we subjected hippocampal as well as cortical slices to BAPTA-AM treatment. Notably, calcium chelation by BAPTA reversed tau phosphorylation in SGPL1-deficient hippocampal and cortical slices to control values, specifically at the pathological phosphoepitope at serine residue S396/404 (Figure 3A), while phosphorylation of serine residues S262/356 was not affected by BAPTA (Supplementary Figure S3). Furthermore, we found that histone acetylation in the same samples also returned to control levels following BAPTA-AM treatment (Figure 3B).

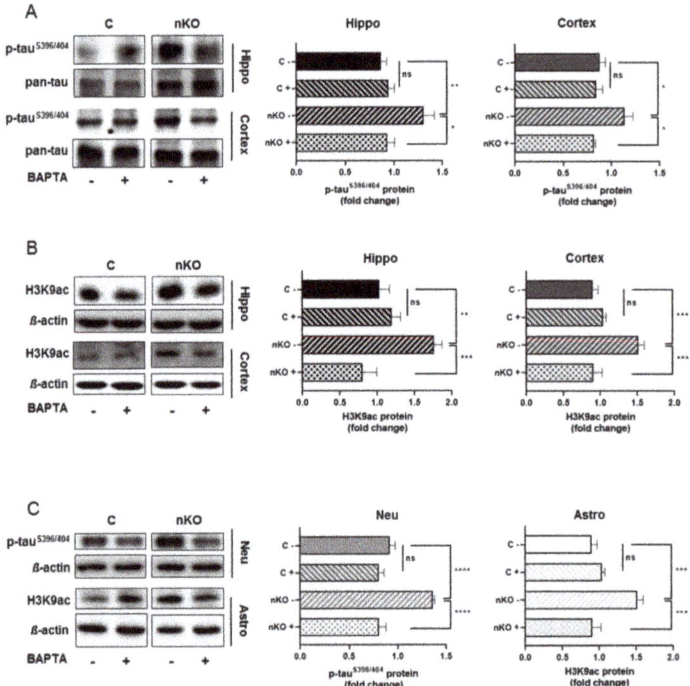

Figure 3. BAPTA-AM treatment reversed tau$^{S396/404}$ hyperphosphorylation and H3K9 acetylation in the brain of SGPL1$^{fl/fl/Nes}$ mice. (**A**) Protein quantification of phospho-tau$^{S396/404}$ and total tau (pan-tau) in the hippocampus (Hippo, black) and cortex (dark grey) of control (**C**) and SGPL1$^{fl/fl/Nes}$ (nKO) mice with (+) and without (-) BAPTA-AM treatment. (**B**) Protein quantification of H3K9 acetylation (H3K9ac) and total H3 in the hippocampus (Hippo, black) and cortex (dark grey) of control (**C**) and SGPL1$^{fl/fl/Nes}$ (nKO) mice with (+) and without (-) BAPTA-AM treatment. (**C**) Protein quantification of phospho-tau$^{S396/404}$ and H3K9 acetylation (H3K9ac) in neurons (Neu, light grey) and astrocytes (Astro, white) in control (**C**) and SGPL1$^{fl/fl/Nes}$ (nKO) mice with (+) and without (-) BAPTA-AM treatment. For all: Bars mean +/− S.E.M, One-Way ANOVA with Tukey's post-hoc correction, n = 3–7, * $p < 0.05$, ** $p < 0.01$, *** $p < 0.001$, **** $p < 0.0001$, ns = not significant.

3.4. Neuronal Tau Pathology and Augmented Histone Acetylation in Astrocytes of SGPL1$^{fl/fl/Nes}$ Mice Are Calcium Dependent

As shown above, tau hyperphosphorylation appeared to be neuron specific, whereas the unusual increase of histone acetylation was restricted to astrocytes. The fact that, in hippocampal and cortical slices, BAPTA-AM reversed both parameters to control values prompted us to investigate the effect of calcium chelation in primary cultured neurons and astrocytes, respectively. Treatment of primary cultured neurons from SGPL1$^{fl/fl/Nes}$ mice with BAPTA-AM recapitulated the effect on tau phosphorylation described above in hippocampal and in cortical slices. Thus, the expression of phosphorylated tau at serineS396/404 returned to control values following calcium chelation (Figure 3C). Likewise, the treatment of primary cultured astrocytes derived from SGPL1$^{fl/fl/Nes}$ mice reversed the acetylation of H3 at K9 (H3K9ac) to control amounts (Figure 3C).

4. Discussion

SGPL1 deficiency in brain has been shown to affect neuronal health and to cause neuroinflammation accompanied by impairment of cognitive and motor skills in mice [17,18,28]. The central aim of the present study was to further unravel the molecular bases of the neurological phenotype of SGPL1$^{fl/fl/Nes}$ mice. We therefore investigated tau expression and phosphorylation in SGPL1-deficient animals. While the amount of total tau remained unaffected in the brains of SGPL1$^{fl/fl/Nes}$ mice, phosphorylation of tau at disease-relevant sites was significantly increased in brain slices and neuronal cultures derived from these mice. Based on reports from the early nineties that altered calcium homeostasis may be a key event leading to altered tau disposition and neuronal degeneration [29], we assumed that increase of phosphorylated tau in SGPL1-deficient brains could be linked to the increase of neuronal calcium in SGPL1$^{fl/fl/Nes}$ mice. Our results confirmed that at least one of the investigated disease-relevant phosphorylation sites can be reversed by BAPTA-AM treatment, indicating that the effect of SPGL1 deficiency on tau phosphorylation is also calcium-dependent. Although several changes of tau expression, mutation and posttranslational modifications were described in diverse pathologies of the central nervous system, hyperphosphorylation appears to be of particular importance for its pathologic function [19,30]. Our results show that SPGL1 deficiency in the brain can lead to a neuronal-specific, calcium-dependent hyperphosphorylation of tau at a site relevant to tauopathies.

We have previously shown that SPGL1 deficiency leads to an accumulation of aggregate-prone proteins in the brain, along with deficits in motor coordination, learning and memory [17,18]. Furthermore, we have shown that SGPL1 deficiency on the one hand triggers the ubiquitin-proteasome system (UPS) in brains of SGPL1$^{fl/fl/Nes}$ mice [17], while on the other hand, it blocks neuronal autophagy flux at an early stage [18]. The implication of both systems in neuronal health and death is well established [31–34]. Notably, both systems are also responsible for tau clearance [35]. While numerous reports suggested that tau is a proteasomal substrate, other studies found that the autophagy/lysosomal pathway is the primary degradation machinery for tau [35]. Our results regarding unchanged amounts of total tau could be explained by these opposing effects of SGPL1 deficiency on the two degradation systems in neurons [17,18]. Further studies into the molecular mechanisms by which SGPL1 deficiency and/or S1P accumulation in the brain affect tau hyperphosphorylation, clearance and pathology could therefore be of great interest to the field of tauopathies.

Although tau is predominantly produced by neurons in the brain, tau pathology is not restricted to neurons [30,36]. However, astrocytes derived from SGPL1-deficient brains did not show any changes in tau expression or phosphorylation. One explanation might be the fact that many studies regarding tau pathology were performed in transgenic animals harboring the wild-type or mutant human tau transgene [30,37]. On the other hand, apart from hyperphosphorylation, microglial activation and thus neuroinflammation appear to be essential for astrocytic tau pathology [30]. We have recently shown that in SGPL1$^{fl/fl/Nes}$ mice, microglial activation and hence release of pro-inflammatory cytokines including interleukin (IL) 6 (IL-6) and tumor necrosis factor alpha (TNF-α), is caused by S1P released from SGPL1-deficient astrocytes [28]. Whether and how these factors are involved in the tau hyperphosphorylation of SGPL1-deficient astrocytes has to be clarified in future studies.

It is well-established that mobilization of intracellular calcium stores is a universal signaling mechanism that cells employ for responding to a wide range of external stimuli [38]. Notably in SGPL1-deficient neurons, this stimulus is S1P [5]. We also found it rather interesting that both S1P [24] and tau, downstream of its pathologic accumulation [22] have been shown to affect histone acetylation, though in different cellular contexts. Specifically, S1P produced by overexpression of SK2 in the nucleus of breast cancer cells was shown to inhibit HDAC 1 and 2, thus increasing acetylation of H3 at K9, of H4 at K5, and rather weakly that of H2B at K12 [24]. Moreover, a compromised acetylation homeostasis has been suggested to be intimately coupled with neurodegeneration [39]. We therefore decided to investigate whether SGPL1 deficiency in the brain could exert an analogous effect. Intriguingly, we also found calcium-dependent increases in H3K9 and H2BK12 acetylation in brain slices and primary cultured astrocytes, but not in neurons derived from SGPL1$^{fl/fl/Nes}$ mice, without affecting the overall expression levels of HDAC1, 2, 3, or 6. These results indicate that SGPL1 deficiency also plays a role in histone posttranslational modifications in astrocytes, though further studies are needed to elucidate the role of this epigenetic disruption on overall brain health.

In SGPL1-deficient mouse embryonic fibroblasts (MEFs) an increase of acetylated H3K9 was reported, but no changes in the level of H4 and H2B acetylation could be detected [40]. Despite this similarity, the underlying molecular mechanisms appear to diverge in the three cell types. In the SK2-overexpressing cancer cell lines, nuclear S1P was shown to directly interact and thus inhibit HDACs1 and 2 [24] while in SGPL1-deficient MEFs a reduced expression of HDACs1, 2, and 3 was reported [40]. Intriguingly, the decreased expression of HDACs was correlated with an elevated basal calcium level in SGPL1-deficient MEFs [40]. By contrast, in hippocampal and cortical slices derived from SGPL1-deficient murine brains, no changes in the expression of HDACs could be observed. However, calcium chelation by BAPTA-AM restored histone acetylation, suggesting that calcium mediates the effect of S1P on histone acetylation independent of HDAC expression. Given these differences between models, we cannot make assumptions as to the specific molecular mechanism responsible for altered histone acetylation in our system.

Interestingly, epigenetic dysregulation currently attracts much attention as a pivotal player in aging and age-related neurodegenerative disorders, such as AD, Parkinson's disease and Huntington's disease, where it may mediate interactions between genetic and environmental risk factors, or directly interact with disease-specific pathological factors [23,41]. Furthermore, a recent epigenome-wide association study using H3K9ac as a marker for transcriptionally active open chromatin revealed that tau pathology is associated with broad changes in the brain's epigenome [22]. This study was conducted in aged human dorsolateral prefrontal cortices. Also, in post-mortem human brains from AD patients, tau pathology was correlated with augmented H3 and H4 acetylation [23]. Given our findings that SPGL1 deficiency results in both altered tau homeostasis and histone acetylation, further studies into how SPGL1 deficiency and/or S1P accumulation fit into these complex systems are needed. While a closer look in primary cultured neurons and astrocytes uncovered a cell type specificity to the effect of SGPL1 deficiency, this does not exclude their association in the nervous tissue. At present, the association of tau pathology and histone acetylation appears rather conflicting and far from clear. It has been reported that specific inhibition of HDAC3 and consequently an increased acetylation of H3 and H4 was shown to reduce tau phosphorylation at disease- associated sites, including serine 396, and was proposed as a novel neuroprotective mechanism [42]. Similarly, the specific inhibition of HDAC6 caused a significant reduction of tau phosphorylation [43]. HDAC6 inhibition also increased acetylation of Hsp90 which caused ubiquitination of phosphorylated tau thus alleviating abnormal tau accumulation [43]. Despite many open questions, we feel that our results show that SGPL1 deficiency impacts brain health and might help explain the potential molecular mechanisms underlying the diverse neuropathological phenotypes in humans harboring mutations in the *SGPL1* gene [8].

5. Conclusions

Our results indicate that SGPL1 depletion augments tau phosphorylation in neurons and simultaneously enhances histone acetylation in astrocytes suggesting a negative impact on brain health. On the other hand, immunohistochemical analysis in frontal and entorhinal cortices from 56 human AD brains revealed an augmented SGPL1 expression correlating with amyloid deposits [12]. The same study also reported a decreased expression of sphingosine kinase 1 as well as of S1PR1 suggesting a global deregulation of S1P signaling in human AD brains [12]. Our results rely primarily on SGPL1 deficiency. Hence, although sites relevant to tauopathies including AD are hyperphosphorylated, we feel that, due to the multitude of phosphorylation sites in tau and the complexity of this phenomenon [44], more studies are needed to finally understand the function of SGPL1 in tauopathies. At present, our findings are first and foremost interesting for a better understanding of the phenotype of humans with insufficient SGPL1 activity [8].

Supplementary Materials: The following are available online at http://www.mdpi.com/2073-4409/9/10/2189/s1, Figure S1: Neural-specific depletion of SGPL1; Figure S2: SGPL1 deficiency has no effect on acetylaztion of Histone3 at lysine 14 and 18; Figure S3: BAPTA-AM treatment has no effect on tau$^{S262/356}$ hyperphosphorylation in the brain of SGPL1$^{fl/fl/Nes}$ mice.

Author Contributions: Conceptualization: S.A. and G.v.E.-D.; methodology: S.A., M.A.E.F., and M.R.; Formal analysis and investigation: S.A. and A.P.; supervision and writing, G.v.E.-D. and A.P. All authors have read and agreed to the published version of the manuscript.

Funding: This work was supported by a grant EC 118/10-1 to G.v.E.D. from the German Research Foundation (Deutsche Forschungsgemeinschaft, DFG).

Acknowledgments: We thank Dieter Swandulla and the LIMES Institute for financial support and Margit Zweyer for excellent technical assistance.

Conflicts of Interest: The authors declare no conflict of interest.

References

1. Grassi, S.; Mauri, L.; Prioni, S.; Cabitta, L.; Sonnino, S.; Prinetti, A.; Giussani, P. Sphingosine 1-Phosphate Receptors and Metabolic Enzymes as Druggable Targets for Brain Diseases. *Front. Pharmacol.* **2019**, *10*. [CrossRef] [PubMed]

2. Karunakaran, I.; van Echten-Deckert, G. Sphingosine 1-Phosphate—A Double Edged Sword in the Brain. *Biochim. Biophys. Acta* **2017**, *1859*, 1573–1582. [CrossRef] [PubMed]

3. Spiegel, S.; Milstien, S. The Outs and the Ins of Sphingosine-1-Phosphate in Immunity. *Nat. Rev. Immunol.* **2011**, *11*, 403–415. [CrossRef] [PubMed]

4. Spiegel, S.; Milstien, S. Sphingosine-1-Phosphate: An Enigmatic Signalling Lipid. *Nat. Rev. Mol. Cell. Biol.* **2003**, *4*, 397–407. [CrossRef]

5. Ghosh, T.K.; Bian, J.; Gill, D.L. Sphingosine 1-Phosphate Generated in the Endoplasmic Reticulum Membrane Activates Release of Stored Calcium. *J. Biol. Chem.* **1994**, *269*, 22628–22635.

6. Meyer zu Heringdorf, D.; Liliom, K.; Schaefer, M.; Danneberg, K.; Jaggar, J.H.; Tigyi, G.; Jakobs, K.H. Photolysis of Intracellular Caged Sphingosine-1-Phosphate Causes Ca2+ Mobilization Independently of G-Protein-Coupled Receptors. *FEBS Lett.* **2003**, *554*, 443–449. [CrossRef]

7. Saba, J.D. Fifty Years of Lyase and a Moment of Truth: Sphingosine Phosphate Lyase from Discovery to Disease. *J. Lipid Res.* **2019**, *60*, 456–463. [CrossRef]

8. Choi, Y.J.; Saba, J.D. Sphingosine Phosphate Lyase Insufficiency Syndrome (Splis): A Novel Inborn Error of Sphingolipid Metabolism. *Adv. Biol. Regul.* **2019**, *71*, 128–140. [CrossRef]

9. Mizugishi, K.; Yamashita, T.; Olivera, A.; Miller, G.F.; Spiegel, S.; Proia, R.L. Essential Role for Sphingosine Kinases in Neural and Vascular Development. *Mol. Cell Biol.* **2005**, *25*, 11113–11121. [CrossRef]

10. Van Echten-Deckert, G.; Hagen-Euteneuer, N.; Karaca, I.; Walter, J. Sphingosine-1-Phosphate: Boon and Bane for the Brain. *Cell Physiol. Biochem.* **2014**, *34*, 148–157. [CrossRef]

11. Couttas, T.A.; Kain, N.; Daniels, B.; Lim, X.Y.; Shepherd, C.; Kril, J.; Pickford, R.; Li, H.; Garner, B.; Don, A.S. Loss of the Neuroprotective Factor Sphingosine 1-Phosphate Early in Alzheimer's Disease Pathogenesis. *Acta Neuropathol. Commun.* **2014**, *2*, 9. [CrossRef] [PubMed]

12. Ceccom, J.; Loukh, N.; Lauwers-Cances, V.; Touriol, C.; Nicaise, Y.; Gentil, C.; Uro-Coste, E.; Pitson, S.; Maurage, C.A.; Duyckaerts, C.; et al. Reduced Sphingosine Kinase-1 and Enhanced Sphingosine 1-Phosphate Lyase Expression Demonstrate Deregulated Sphingosine 1-Phosphate Signaling in Alzheimer's Disease. *Acta Neuropathol. Commun.* **2014**, *2*, 12. [CrossRef] [PubMed]

13. Takasugi, N.; Sasaki, T.; Suzuki, K.; Osawa, S.; Isshiki, H.; Hori, Y.; Shimada, N.; Higo, T.; Yokoshima, S.; Fukuyama, T.; et al. Bace1 Activity Is Modulated by Cell-Associated Sphingosine-1-Phosphate. *J. Neurosci.* **2011**, *31*, 6850–6857. [CrossRef] [PubMed]

14. Hagen, N.; Hans, M.; Hartmann, D.; Swandulla, D.; van Echten-Deckert, G. Sphingosine-1-Phosphate Links Glycosphingolipid Metabolism to Neurodegeneration Via a Calpain-Mediated Mechanism. *Cell Death Differ.* **2011**, *18*, 1356–1365. [CrossRef] [PubMed]

15. Hagen, N.; Van Veldhoven, P.P.; Proia, R.L.; Park, H.; Merrill, A.H., Jr.; van Echten-Deckert, G. Subcellular Origin of Sphingosine 1-Phosphate Is Essential for Its Toxic Effect in Lyase-Deficient Neurons. *J. Biol. Chem.* **2009**, *284*, 11346–11353. [CrossRef]

16. Ibanez, C.; Simo, C.; Barupal, D.K.; Fiehn, O.; Kivipelto, M.; Cedazo-Minguez, A.; Cifuentes, A. A New Metabolomic Workflow for Early Detection of Alzheimer's Disease. *J. Chromatogr. A* **2013**, *1302*, 65–71. [CrossRef]

17. Mitroi, D.N.; Deutschmann, A.U.; Raucamp, M.; Karunakaran, I.; Glebov, K.; Hans, M.; Walter, J.; Saba, J.; Graler, M.; Ehninger, D.; et al. Sphingosine 1-Phosphate Lyase Ablation Disrupts Presynaptic Architecture and Function Via an Ubiquitin- Proteasome Mediated Mechanism. *Sci. Rep.* **2016**, *6*, 37064. [CrossRef]

18. Mitroi, D.N.; Karunakaran, I.; Graler, M.; Saba, J.D.; Ehninger, D.; Ledesma, M.D.; van Echten-Deckert, G. Sgpl1 (Sphingosine Phosphate Lyase 1) Modulates Neuronal Autophagy Via Phosphatidylethanolamine Production. *Autophagy* **2017**, *13*, 885–899. [CrossRef]

19. Iqbal, K.; Liu, F.; Gong, C.X. Tau and Neurodegenerative Disease: The Story So Far. *Nat. Rev. Neurol.* **2016**, *12*, 15–27. [CrossRef]

20. Frost, B.; Hemberg, M.; Lewis, J.; Feany, M.B. Tau Promotes Neurodegeneration through Global Chromatin Relaxation. *Nat. Neurosci.* **2014**, *17*, 357–366. [CrossRef]

21. Mansuroglu, Z.; Benhelli-Mokrani, H.; Marcato, V.; Sultan, A.; Violet, M.; Chauderlier, A.; Delattre, L.; Loyens, A.; Talahari, S.; Begard, S.; et al. Loss of Tau Protein Affects the Structure, Transcription and Repair of Neuronal Pericentromeric Heterochromatin. *Sci. Rep.* **2016**, *6*, 33047. [CrossRef] [PubMed]

22. Klein, H.U.; McCabe, C.; Gjoneska, E.; Sullivan, S.E.; Kaskow, B.J.; Tang, A.; Smith, R.V.; Xu, J.; Pfenning, A.R.; Bernstein, B.E.; et al. Epigenome-Wide Study Uncovers Large-Scale Changes in Histone Acetylation Driven by Tau Pathology in Aging and Alzheimer's Human Brains. *Nat. Neurosci.* **2019**, *22*, 37–46. [CrossRef] [PubMed]

23. Narayan, P.J.; Lill, C.; Faull, R.; Curtis, M.A.; Dragunow, M. Increased Acetyl and Total Histone Levels in Post-Mortem Alzheimer's Disease Brain. *Neurobiol. Dis.* **2015**, *74*, 281–294. [CrossRef] [PubMed]

24. Hait, N.C.; Allegood, J.; Maceyka, M.; Strub, G.M.; Harikumar, K.B.; Singh, S.K.; Luo, C.; Marmorstein, R.; Kordula, T.; Milstien, S.; et al. Regulation of Histone Acetylation in the Nucleus by Sphingosine-1-Phosphate. *Science* **2009**, *325*, 1254–1257. [CrossRef]

25. Van Echten-Deckert, G.; Zschoche, A.; Bar, T.; Schmidt, R.R.; Raths, A.; Heinemann, T.; Sandhoff, K. Cis-4-Methylsphingosine Decreases Sphingolipid Biosynthesis by Specifically Interfering with Serine Palmitoyltransferase Activity in Primary Cultured Neurons. *J. Biol. Chem.* **1997**, *272*, 15825–15833. [CrossRef]

26. Schildge, S.; Bohrer, C.; Beck, K.; Schachtrup, C. Isolation and Culture of Mouse Cortical Astrocytes. *J. Vis. Exp.* **2013**, *71*, e50079. [CrossRef]

27. Raucamp, M. Impact of Sphingosine 1-Phosphate Lyase on the Synaptic Transmission in Mouse Hippocampus. Ph.D Thesis, University Bonn, Bonn, Germany, 2019.

28. Karunakaran, I.; Alam, S.; Jayagopi, S.; Frohberger, S.J.; Hansen, J.N.; Kuehlwein, J.; Holbling, B.V.; Schumak, B.; Hubner, M.P.; Graler, M.H.; et al. Neural Sphingosine 1-Phosphate Accumulation Activates Microglia and Links Impaired Autophagy and Inflammation. *Glia* **2019**, *67*, 1859–1872. [CrossRef]

29. Mattson, M.P.; Engle, M.G.; Rychlik, B. Effects of Elevated Intracellular Calcium Levels on the Cytoskeleton and Tau in Cultured Human Cortical Neurons. *Mol. Chem. Neuropathol.* **1991**, *15*, 117–142. [CrossRef]

30. Leyns, C.E.G.; Holtzman, D.M. Glial Contributions to Neurodegeneration in Tauopathies. *Mol. Neurodegener.* **2017**, *12*, 50. [CrossRef]

31. Dantuma, N.P.; Bott, L.C. The Ubiquitin-Proteasome System in Neurodegenerative Diseases: Precipitating Factor, yet Part of the Solution. *Front. Mol. Neurosci.* **2014**, *7*, 70. [CrossRef]

32. Menzies, F.M.; Fleming, A.; Rubinsztein, D.C. Compromised Autophagy and Neurodegenerative Diseases. *Nat. Rev. Neurosci.* **2015**, *16*, 345–357. [CrossRef] [PubMed]

33. Nixon, R.A. The Role of Autophagy in Neurodegenerative Disease. *Nat. Med.* **2013**, *19*, 983–997. [CrossRef] [PubMed]

34. Zheng, Q.; Huang, T.; Zhang, L.; Zhou, Y.; Luo, H.; Xu, H.; Wang, X. Dysregulation of Ubiquitin-Proteasome System in Neurodegenerative Diseases. *Front. Aging Neurosci.* **2016**, *8*, 303. [CrossRef] [PubMed]

35. Lee, M.J.; Lee, J.H.; Rubinsztein, D.C. Tau Degradation: The Ubiquitin-Proteasome System Versus the Autophagy-Lysosome System. *Prog. Neurobiol.* **2013**, *105*, 49–59. [CrossRef]

36. Forman, M.S.; Lal, D.; Zhang, B.; Dabir, D.V.; Swanson, E.; Lee, V.M.; Trojanowski, J.Q. Transgenic Mouse Model of Tau Pathology in Astrocytes Leading to Nervous System Degeneration. *J. Neurosci.* **2005**, *25*, 3539–3550. [CrossRef]

37. Ikeda, M.; Shoji, M.; Kawarai, T.; Kawarabayashi, T.; Matsubara, E.; Murakami, T.; Sasaki, A.; Tomidokoro, Y.; Ikarashi, Y.; Kuribara, H.; et al. Accumulation of Filamentous Tau in the Cerebral Cortex of Human Tau R406w Transgenic Mice. *Am. J. Pathol.* **2005**, *166*, 521–531. [CrossRef]

38. Lee, H.C.; Zhao, Y.J. Resolving the Topological Enigma in Ca(2+) Signaling by Cyclic Adp-Ribose and Naadp. *J. Biol. Chem.* **2019**, *294*, 19831–19843. [CrossRef]

39. Saha, R.N.; Pahan, K. Hats and Hdacs in Neurodegeneration: A Tale of Disconcerted Acetylation Homeostasis. *Cell Death Differ.* **2006**, *13*, 539–550. [CrossRef]

40. Ihlefeld, K.; Claas, R.F.; Koch, A.; Pfeilschifter, J.M.; Meyer Zu Heringdorf, D. Evidence for a Link between Histone Deacetylation and Ca(2)+ Homoeostasis in Sphingosine-1-Phosphate Lyase-Deficient Fibroblasts. *Biochem. J.* **2012**, *447*, 457–464. [CrossRef]

41. Lardenoije, R.; Iatrou, A.; Kenis, G.; Kompotis, K.; Steinbusch, H.W.; Mastroeni, D.; Coleman, P.; Lemere, C.A.; Hof, P.R.; van den Hove, D.L.; et al. The Epigenetics of Aging and Neurodegeneration. *Prog. Neurobiol.* **2015**, *131*, 21–64. [CrossRef]

42. Janczura, K.J.; Volmar, C.H.; Sartor, G.C.; Rao, S.J.; Ricciardi, N.R.; Lambert, G.; Brothers, S.P.; Wahlestedt, C. Inhibition of Hdac3 Reverses Alzheimer's Disease-Related Pathologies in Vitro and in the 3xtg-Ad Mouse Model. *Proc. Natl. Acad. Sci. USA* **2018**, *115*, E11148–E11157. [CrossRef] [PubMed]

43. Fan, S.J.; Huang, F.I.; Liou, J.P.; Yang, C.R. The Novel Histone De Acetylase 6 Inhibitor, Mpt0g211, Ameliorates Tau Phosphorylation and Cognitive Deficits in an Alzheimer's Disease Model. *Cell Death Dis.* **2018**, *9*, 655. [CrossRef] [PubMed]

44. Kimura, T.; Sharma, G.; Ishiguro, K.; Hisanaga, S.I. Phospho-Tau Bar Code: Analysis of Phosphoisotypes of Tau and Its Application to Tauopathy. *Front. Neurosci.* **2018**, *12*, 44. [CrossRef] [PubMed]

Article

Exogenous Flupirtine as Potential Treatment for CLN3 Disease

Katia Maalouf [1,†], Joelle Makoukji [1,†], Sara Saab [1], Nadine J. Makhoul [1], Angelica V. Carmona [2], Nihar Kinarivala [3], Noël Ghanem [4], Paul C. Trippier [2,5,6] and Rose-Mary Boustany [1,7,8,*]

[1] Department of Biochemistry and Molecular Genetics, American University of Beirut Medical Center, Beirut 1107 2020, Lebanon; kg21@aub.edu.lb (K.M.); jm53@aub.edu.lb (J.M.); sas85@mail.aub.edu (S.S.); nm36@aub.edu.lb (N.J.M.)

[2] Department of Pharmaceutical Sciences, College of Pharmacy, University of Nebraska Medical Center, Omaha, NE 68198-6120, USA; angelica.carmona@unmc.edu (A.V.C.); paul.trippier@unmc.edu (P.C.T.)

[3] Department of Pharmaceutical Sciences, Jerry H. Hodge School of Pharmacy, Texas Tech University Health Sciences Center, Amarillo, TX 79106, USA; Nihar.kinarivala@gmail.com

[4] Department of Biology, American University of Beirut, Beirut 1107 2020, Lebanon; ng13@aub.edu.lb

[5] Fred & Pamela Buffett Cancer Center, University of Nebraska Medical Center, Omaha, NE 68198-6120, USA

[6] UNMC Center for Drug Discovery, University of Nebraska Medical Center, Omaha, NE 68198-6120, USA

[7] Neurogenetics Program, AUBMC Special Kids Clinic, Division of Pediatric Neurology, Department of Pediatrics and Adolescent Medicine, American University of Beirut Medical Center, Beirut 1107 2020, Lebanon

[8] Neurogenetics Program and Pediatric Neurology, Departments of Pediatrics, Adolescent Medicine and Biochemistry, American University of Beirut, P.O. Box 11-0236 Riad El Solh 1107 2020, Beirut 1107 2020, Lebanon

* Correspondence: rb50@aub.edu.lb; Tel.: +961-350-000 (ext. 5640/1/2)

† K.M. and J.M. contributed equally to this work.

Received: 16 June 2020; Accepted: 6 July 2020; Published: 11 August 2020

Abstract: CLN3 disease is a fatal neurodegenerative disorder affecting children. Hallmarks include brain atrophy, accelerated neuronal apoptosis, and ceramide elevation. Treatment regimens are supportive, highlighting the importance of novel, disease-modifying drugs. Flupirtine and its new allyl carbamate derivative (compound 6) confer neuroprotective effects in CLN3-deficient cells. This study lays the groundwork for investigating beneficial effects in $Cln3^{\Delta ex7/8}$ mice. WT/$Cln3^{\Delta ex7/8}$ mice received flupirtine/compound 6/vehicle for 14 weeks. Short-term effect of flupirtine or compound 6 was tested using a battery of behavioral testing. For flupirtine, gene expression profiles, astrogliosis, and neuronal cell counts were determined. Flupirtine improved neurobehavioral parameters in open field, pole climbing, and Morris water maze tests in $Cln3^{\Delta ex7/8}$ mice. Several anti-apoptotic markers and ceramide synthesis/degradation enzymes expression was dysregulated in $Cln3^{\Delta ex7/8}$ mice. Flupirtine reduced astrogliosis in hippocampus and motor cortex of male and female $Cln3^{\Delta ex7/8}$ mice. Flupirtine increased neuronal cell counts in male mice. The newly synthesized compound 6 showed promising results in open field and pole climbing. In conclusion, flupirtine improved behavioral, neuropathological and biochemical parameters in $Cln3^{\Delta ex7/8}$ mice, paving the way for potential therapies for CLN3 disease.

Keywords: sphingolipids; neurodegeneration; ceramide; CLN3 disease; $Cln3^{\Delta ex7/8}$ mice; flupirtine; allyl carbamate derivative; apoptosis

---

1. Introduction

The neuronal ceroid lipofuscinoses (NCLs) constitute a family of fatal pediatric neurodegenerative disorders that primarily affect the central nervous system (CNS) [1]. NCLs are atypical lysosomal storage disorders that manifest accumulation of lipopigments in the lysosomes of neurons and other cell types [2]. CLN3 disease arises due to mutations in the CLN3 gene and is the most common variant of the NCL group [3]. This neurological disease manifests at four to eight years of age with progressive visual deterioration, seizures, blindness, motor and cognitive decline, mental retardation, epilepsy and early death during the second or third decade of life [4]. Massive cortical neuronal cell loss due to neuronal apoptosis within the cortex [5], and neuronal loss in hippocampus and microglial activation in this region are documented [6]. Eighty five per cent of patients with CLN3 disease harbor a 1.02 kb deletion eliminating exons 7/8 and creating a truncated CLN3 protein [7,8].

CLN3 protein influences major cellular functions, including apoptosis and cell growth 9. Apoptosis is the mechanism of neuronal and photoreceptor cell loss in human brain from patients with CLN3 disease 3 [9]. Ceramide, a pro-apoptotic lipid second messenger, mediates anti-proliferative events of apoptosis, growth inhibition, cell differentiation, and senescence [10].

Ceramide levels are increased in CLN3-deficient cells and in brain of CLN3 patients [11]. Studies confirm that CLN3 protein expression modulates brain ceramide levels. Levels of lipids ceramide, SM, GalCer, GluCer, and globoside are elevated in human CLN3-deficient fibroblasts. Ceramide levels normalized following restoration of CLN3 function, but not following caspase inhibition by zVAD, a pan-inhibitor of caspases [8,11,12]. Overexpressing CLN3 protein results in a drop in ceramide levels [13]. Increased ceramide levels and neuronal cell loss are evident in brain sections from post-mortem CLN3 disease patients and in brains and sera of $Cln3^{\Delta ex7/8}$ mice [14]. Treatment regimens for CLN3 disease are largely supportive, not curative, and do not target the underlying causes of the disease.

Flupirtine is a centrally acting non-opioid drug previously widely used in clinics as an analgesic [15,16]. Flupirtine maleate is the salt of this drug, henceforth, referred to as just flupirtine. It is neuroprotective, has muscle relaxant and anticonvulsant properties [17] and suppresses neuronal hyper-excitability [18]. Flupirtine protects photoreceptor and neuronal cells from apoptosis induced by various insults [13]. There is evidence suggesting that flupirtine reduces brain injury, induces remodeling of brain tissue, and diminishes cognitive impairment in in vivo animal models of ischemic stroke [15]. Flupirtine protects lymphoblasts, differentiated human post-mitotic hNT neurons and PC12 neuronal precursor cells from apoptosis induced by etoposide [13,19]. A newly synthesized allyl carbamate derivative of flupirtine (compound 6) has shown potential neuroprotective effects in vitro, as one of nine flupirtine aromatic carbamate derivative [19–21]. Compound 6 imparted a 150% increase in Bcl-2/Bax ratio in vitro which is protective [19]. Flupirtine and its allyl carbamate derivative (compound 6) increased cell viability in human CLN3 patient-derived lymphoblasts and in neuronal precursor PC12 cells transfected with siRNA directed against CLN3, exhibiting significant anti-apoptotic and neuroprotective effects [19].

This study tests, in vivo, oral supplementation of flupirtine for a period of 14 weeks in $Cln3^{\Delta ex7/8}$ knock-in mice. Outcomes of efficacy include improved behavioral measures, altered gene expression profiles, decreased glial immunoreactivity, and increased neuronal cell numbers in specific brain regions. Supplementation of compound 6 assessed effectiveness in several parameters, as a first step in also developing it as potential treatment for CLN3 disease.

2. Materials and Methods

2.1. Animals

This mouse work was conducted in accordance with an approved American University of Beirut (AUB) Institutional Animal Care and Use Committee (IACUC) protocol (IACUC approval #18-08-496). Animal testing was carried out at the AUB Animal Care Facility where animals were housed. C57BL/6J (JAX stock number: 000664) and homozygous $Cln3^{\Delta ex7/8}$ (JAX stock number: 029471) mice were obtained from the Jackson laboratory, kept in a 12-h light/dark cycle (lights onset at 6:00 am) and supplied with access to food and water ad libitum. Room temperature was maintained between 18–26 °C, and relative humidity between 30–70%. Mice were housed in groups of 3–4/cage. All efforts to minimize number of animals and animal suffering were applied. Mice were monitored for weekly weights, basic behavior and general health throughout the study, and were bred in-house.

2.2. Flupirtine and Compound 6 Treatment

Flupirtine and compound 6 were dissolved in vehicle (0.5% di-methyl sulphoxide (DMSO) in 10% phosphate-buffered saline (PBS)), at a dose of 30 mg/kg body weight for a period of 14 weeks starting at 4 weeks of age. Vehicle treatment consisted of 0.5% DMSO in 10% PBS. Mice were treated 'per os' by drinking water with a consistent supply in a volume of ≈ 8 mL/day/mouse. Both drugs were synthesized by Dr. Paul Trippier at the department of pharmaceutical sciences in the School of Pharmacy at Texas Tech University Health Sciences Center (Figure 1). Mice were divided into five groups, consisting of 16 mice each (eight males and females) and consisted of C57BL/6J vehicle-treated WT mice, C57BL/6J compound 6-treated WT mice, vehicle-treated $Cln3^{\Delta ex7/8}$ mice, flupirtine-treated $Cln3^{\Delta ex7/8}$ mice and compound 6-treated $Cln3^{\Delta ex7/8}$ mice. Genotype was confirmed by PCR of DNA mouse blood.

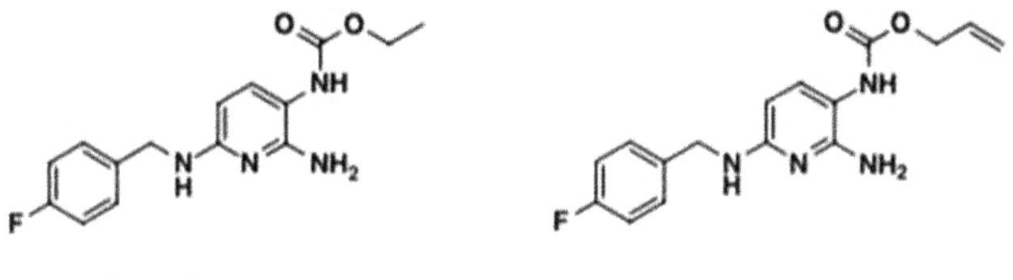

Flupirtine

Compound 6

Figure 1. Chemical structure of flupirtine and its allyl carbamate derivative, compound 6.

2.3. Behavioral Studies

Mice were held in their cages in the behavioral room for testing, with lights dimmed for 60 min prior to onset of tests, for habituation. All behavioral assays were performed during the light cycle. Each test was performed at n, the same time of the day and within the same hour, when possible, to minimize variability between cohorts. Comparison between groups was carried out on males and females separately.

Open field: A mouse was placed on the periphery of a transparent Plexiglas cubic box (dimensions: 50 cm width, 50 cm length and 30 cm in height, with the center 16.4 cm^2) so that locomotion would be apparent to the operator and exploratory behavior videotaped. Specific parameters were recorded by a top-mounted video-recorder using EthoVision software (Noldus Information Technology, Wageningen, The Netherlands) for each animal including total distance traveled, average speed, mobility duration, rearing and walling frequencies. Each mouse was allowed to move in the arena freely for 5 min. The box was wiped with 50% ethanol between each mouse to avoid olfactory cuing.

Pole climbing: Mice were habituated to the task in five trials/day for 1 day. On the next test day, five trial measures/mouse were performed by placing the mouse head upward on top of a rough-surfaced pole (1 cm in diameter and 60 cm in height) wrapped with tape to prevent slipping. The time for the mouse to completely turn its head down (t_{turn}), time to reach the middle of the pole ($t_{1/2}$), time it takes to descend and settle on the floor (t_{total}), and time the mouse spent freezing during descent (t_{stop}) were recorded. Each mouse had a maximum time of two minutes to climb down to avoid exhaustion.

Morris water maze (MWZ): The spatial learning abilities of mice were assessed in a MWM task. The apparatus consisted of a circular pool 100 cm diameter, filled up to 50 cm with water made opaque by addition of a small amount of non-toxic white paint and maintained at 21–22 °C. The pool is virtually subdivided into four quadrants using the software. A circular escape platform was placed in a fixed south-west (SW) location hidden 0.5 cm below the surface of the water, and five fixed-position geometric visual cues were kept in the brightly lit room throughout the period of testing. A digital camera was positioned above the center of the tank and linked to a tracking system (ANY-Maze behavioral tracking software, version 6.3, USA) in order to record escape latencies, path distance (m), percentage of thigmotaxis path and swim speed for each trial, together with time spent swimming in each of the four quadrants of the maze.

Mice (eight males and females from each treatment group) were given four consecutive days of acquisition training sessions that consisted of four trials per day. Throughout the course of this acquisition period the position of the hidden platform remained fixed (SW) and the entry point was varied from trial to trial, but the sequence remained fixed for all mice within that day. We used four different entry points (north, south, east, and west). The sequence of starting points was modified from one day to the other. Mice were given 60 s to find the platform, and if the mouse failed to locate the platform within this period, it was guided onto it. All mice were allowed to rest on the platform for a 30-s interval after each trial. At the end of the training block, mice were put in a drying cage and allowed to dry prior to being returned to their experimental cages. The last day of the test, the 'probe trial' was performed with the platform removed from the maze and the rodent released from the North entry point of the pool to find the previous location of the hidden platform.

2.4. Corticosterone Immunoassay Kit

Blood was collected from the inferior vena cava, left to clot, centrifuged and spun for 15 min at 10,000 rpm at 4 °C. Serum corticosterone levels were determined using DetectX® Corticosterone Enzyme Immunoassay Kit (Cat. No K014-H5, Arbor Assays, Ann Arbor, MI, USA), according to manufacturer's instructions. Five µL of standards or mouse serum samples were assayed in duplicate and run altogether on one single plate simultaneously. The absorbance was measured using TriStar2S microplate reader (Berthold Technologies, Bad Wildbad, Germany) at 450 nm.

2.5. RNA Extraction from Brain Tissue

Mice were deeply anesthetized with a mixture of xylazine and ketamine (10 mg/kg and 100 mg/kg, respectively) and brains were rapidly dissected and "snap" frozen in liquid nitrogen to preserve RNA integrity, and stored at −80 °C. A total of 30 mg ground fresh brain tissue (from four males and females in each treatment group) were homogenized using a motorized rotor-stator homogenizer and RNA extracted using RNeasyPlus Mini Kits (Cat No. 74134, Qiagen, Germantown, MD, USA) according to manufacturer's instructions. For assessing RNA quality, A260/A280 and A260/A230 ratios for RNA are analyzed with the ExperionTM Automated Electrophoresis System (BioRad, Hercules, CA, USA). RNA concentrations are determined by absorption at 260 nm wavelength with a ND-1000 spectrometer (Nanodrop Technologies LLC, Wilmington, DE, USA).

2.6. Quantitative Real-Time PCR (qRT-PCR)

Total RNA extracted from fresh brain tissues was reverse transcribed using RevertAid Reverse Transcriptase (Thermo Fisher Scientific, Waltham, MA, USA) with 2 µg of input RNA and random primers (Thermo Fisher Scientific, USA). qRT-PCR reactions were performed in 384-well plates using specific primers (Tm = 60 °C) (TIB MOLBIOL, Berlin, Germany) (Table 1) and iTaq SYBR Green Supermix (BioRad, Hercules, CA, USA) as a fluorescent detection dye, in CFX384TM Real-Time PCR (BioRad, USA), in a final volume of 10 µL. To characterize generated amplicons and to control contamination by unspecific by-products, melt curve analysis was applied. Results were normalized to β-actin or Gapdh mRNA level. All reactions were performed in duplicate, and results were calculated using the ΔΔCt method.

Table 1. Mouse cDNA primer sequences.

Gene Name	Primer Sequence (F = forward, R = reverse)
Bcl-xL F	AGGTTCCTAAGCTTCGCAATTC
Bcl-xL R	TGTTTAGCGATTCTCTTCCAGG
BCL-2 F	TGTGTGTGGAGAGCGTCAAC
BCL-2 R	TGAGCAGAGTCTTCAGAGAC
Sptlc3 F	TGATTTCTCTCCGGTGATCC
Sptlc3 R	GGAAATCCAACAACCACCAC
Samd8 F	ATCACATTGCTCACGCTGAC
Samd8 R	GCAATTTTCGGACTGAGAGC
β-actin F	ACACTGTGCCCATCTACGAG
β-actin R	ATTTCCCTCTCAGCTGTGGT
Gapdh F	TGTTCCTACCCCCAATGTGT
Gapdh R	AGTTGCTGTTGAAGTCGCAG

2.7. Immunohistochemistry

For morphological and immunohistochemical sectioning, 4 mice/treatment group were deeply anesthetized with a mixture of xylazine and ketamine (10 mg/kg and 100 mg/kg, respectively) and fixed by cardiac puncture with 30 mL of 4% paraformaldehyde (PFA) in PBS. Brains were carefully isolated and fixed with 4% PFA solution (pH of 7.4) for 2 h at 4 °C, then cryoprotected in a solution of 20% sucrose overnight. The following day, brains were processed and frozen using embedding medium, Optimal Cutting Temperature (OCT) compound, for later tangential sectioning on glass slides. Brains were cut in coronal sections (20 μm) using a cryostat and sections stored at −20 °C for further analysis. Brain coronal cryosections were treated with PBS for 5 min twice, then incubated with PBST (0.1% Triton X-100 in PBS) for 10 min twice. Sections were incubated with blocking solution (PBST 0.1%-FBS 10%) for 1 h at room temperature (RT). They were then incubated with each of the primary antibodies: anti-GFAP (1:500, Abcam, Cambridge, UK, catalogue #ab7260) and anti-NeuN antibody (1/300, Abcam, Cambridge, UK, catalogue #ab104225) in antibody solution (PBST 0.1%-FBS 1%) overnight at 4 °C. After washing in PBST, slides were treated with Sudan black for 40 min, then washed with PBS three times. Brain cryosections were then incubated with biotinylated secondary antibody diluted in antibody solution at RT for 1 h. Samples were counterstained with 1:10,000 Hoechst (Sigma, St. Louis, MO, USA) and then mounted in Fluoromount (Sigma, St. Louis, MO, USA).

Signal quantification was assessed using Leica microscope software imaging. For microscopic imaging, three sections/mouse were selected. Primary motor cortex was viewed at 40× magnification with three to four photos/section for motor cortex layers (I–VI). Hippocampus images for GFAP were obtained at 40× magnification. Intensity quantification was depicted by the ratio of integrated density over total area of image using Image J software, version 1.52a. NeuN positive cells were quantified manually using Image J software. Number of NeuN positive cells divided by total area of image.

2.8. Statistical Analysis

Basic statistical analysis was conducted using GraphPad Prism 6 statistical package (GraphPad Software version 6.04, San Diego, CA, USA). Data was expressed as mean ± standard error of the mean (SEM). For two group comparisons, Student's *t*-test was used with quantitative continuous variables. Comparisons between different groups were statistically tested with either one-way analysis of variants (ANOVA) or two-factorial ANOVA followed by Tukey's post-hoc test for multiple group comparisons. All tests are two sided and a p-value < 0.05 is considered as statistically significant.

3. Results

3.1. Impact of Flupirtine on Motor Behavior of Homozygous Cln3$^{\Delta ex7/8}$ Mice

The open field behavioral test assesses novel environment exploration, general locomotor activity, and provides an initial screen for anxiety-related behavior in rodents. Open field behavioral testing showed that vehicle-treated Cln3$^{\Delta ex7/8}$ mice exhibit increased spontaneous locomotor activity compared to vehicle-treated WT controls in both genders at 15 weeks of age. Vehicle-treated Cln3$^{\Delta ex7/8}$ mice were significantly more mobile than their vehicle-treated WT littermates (Figure 2A,B). These results indicate that WT mice display increased anxiety-like behavior when put in the novel test environment compared with Cln3$^{\Delta ex7/8}$ mice. Normally, rodents display distinct aversion to large, brightly lit, open and unknown environments. They have been phylogenetically conditioned to see these types of environments as dangerous [22]. To confirm whether Cln3$^{\Delta ex7/8}$ mice have less anxiety than WT mice, we analyzed serum levels of the predominant murine glucocorticoid, corticosterone, in these animals. Male Cln3$^{\Delta ex7/8}$ mice experienced significantly lower corticosterone levels than those measured in WT animals (Figure 2E). Similarly, female Cln3$^{\Delta ex7/8}$ mice experienced lower corticosterone levels than WT animals, that was very close to significance ($p = 0.09$) (Figure 2F). These results confirm that WT mice have an enhanced physiologic response to stress, characterized by increased hypothalamic-pituitary-adrenal axis activity. Flupirtine treatment slowed down significantly the locomotor hyperactivity of male and female Cln3$^{\Delta ex7/8}$ mice (Figure 2A,B).

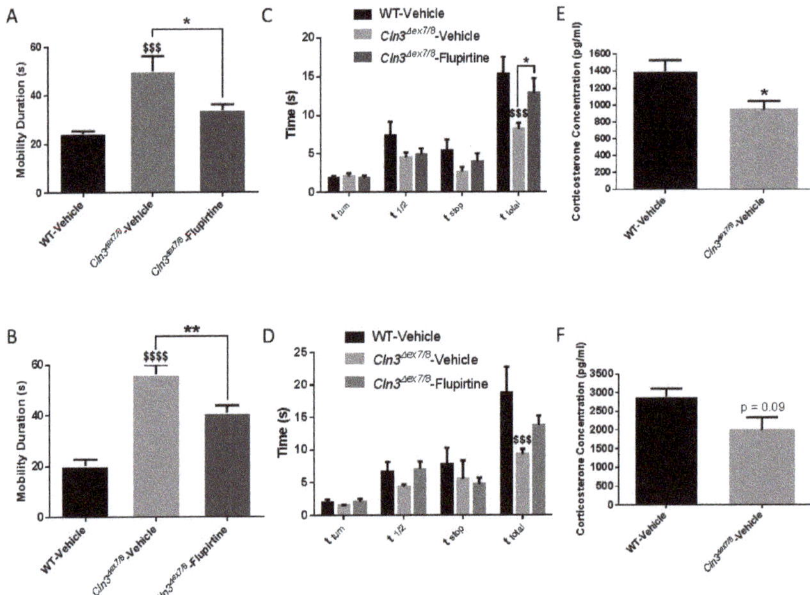

Figure 2. Impact of flupirtine on motor behavior of male and female Cln3$^{\Delta ex7/8}$ mice. Open field behavioral parameter (mobility duration) in vehicle-treated WT, vehicle-treated Cln3$^{\Delta ex7/8}$, and flupirtine-treated Cln3$^{\Delta ex7/8}$ (**A**) male and (**B**) female mice ($n = 8$ per group). Pole climbing test in vehicle-treated WT, vehicle-treated Cln3$^{\Delta ex7/8}$, and flupirtine-treated Cln3$^{\Delta ex7/8}$ (**C**) males and (**D**) female mice ($n = 8$ per group). Time to turn (t_{turn}), time needed to reach center of pole ($t_{1/2}$), amount of time stopped (t_{stop}), and total time to descend (t_{total}) were recorded. All data are expressed as mean $\pm$ SEM. $^\$$ p: compared to vehicle-treated WT mice; and * p: compared to vehicle-treated Cln3$^{\Delta ex7/8}$ mice. * $p < 0.05$, ** $p < 0.01$, $^{\$\$\$}$ $p < 0.001$, and $^{\$\$\$\$}$ $p < 0.0001$. Corticosterone levels in vehicle-treated WT and Cln3$^{\Delta ex7/8}$ (**E**) male and (**F**) female mice ($n = 8$ per group). Data are expressed as mean $\pm$ SEM.

The pole climbing test measures motor coordination, vertical orientation capability, and balance of mice. Vehicle-treated $Cln3^{\Delta ex7/8}$ male and female mice descended the pole faster than vehicle-treated WT mice (Figure 2C,D), in line with the hyperactivity of $Cln3^{\Delta ex7/8}$ mice observed in the open field experiment. In males, flupirtine supplementation had a significant impact on increasing and delaying the descent compared to vehicle-treated $Cln3^{\Delta ex7/8}$ mice (Figure 2C). In females, flupirtine supplementation also trended to increase and delay the descent compared to vehicle-treated $Cln3^{\Delta ex7/8}$ mice, however it did not reach significance (Figure 2D).

3.2. Impact of Flupirtine on Learning and Memory of Homozygous $Cln3^{\Delta ex7/8}$ Mice

The Morris water maze (MWM) is a test used to assess cognitive function, more specifically, spatial learning and memory. The 'probe trial', in which the platform was removed, was used to assess spatial memory for the previously learned platform location. The time spent in the target quadrant compared to the average time spent in the non-target quadrants is an indicator of the animal's recall of platform location. WT male and female mice spent significantly more time in the target quadrant as compared to the average time spent in the non-target quadrants, suggestive of recall of platform location (Figure 3A–C). Vehicle-treated $Cln3^{\Delta ex7/8}$ male and female mice spent significantly less time than their WT male littermates in the target quadrant (Figure 3A–C). Flupirtine treatment improved memory retention in $Cln3^{\Delta ex7/8}$ male and female mice, with a significant increase in time spent in the target quadrant compared to vehicle-treated $Cln3^{\Delta ex7/8}$ mice, suggestive of recall of platform location (Figure 3A–C).

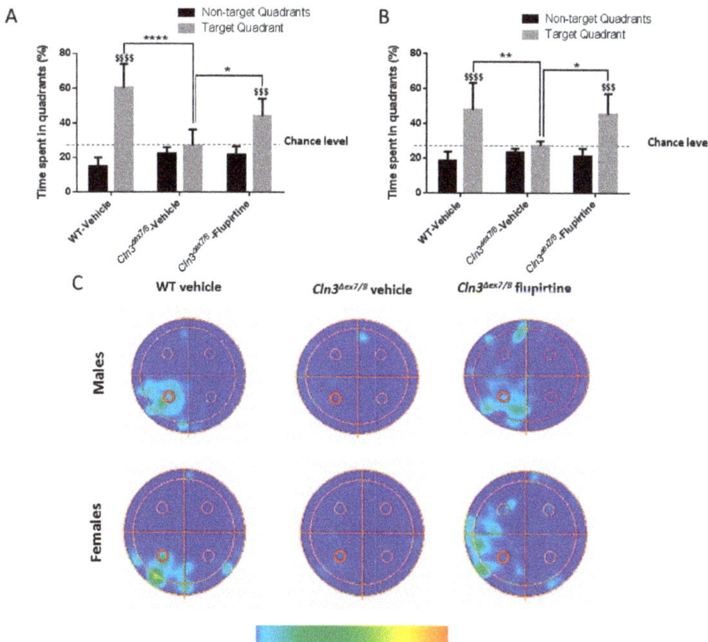

Figure 3. Impact of flupirtine on learning and memory of male and female $Cln3^{\Delta ex7/8}$ mice. Percentage of time spent in target vs. non-target quadrant of (**A**) male and (**B**) female mice in the different groups (vehicle-treated WT, vehicle-treated $Cln3^{\Delta ex7/8}$, and flupirtine-treated $Cln3^{\Delta ex7/8}$) during the probe test (n = 8 per group). All data are expressed as mean ± SEM. $^{\$}$ p: Target quadrant compared to non-target quadrant; and * p: compared to vehicle-treated WT mice. * $p < 0.05$, ** $p < 0.01$, **** $p < 0.0001$, $^{\$\$\$}$ $p < 0.001$, and $^{\$\$\$\$}$ $p < 0.0001$. (**C**) Representative heat maps of swimming paths of one mouse from different groups during the probe test in male and female mice. An empty, bolded red circle indicates location of the target platform (o).

3.3. Impact of Flupirtine on Anti-Apoptotic Gene Expression in Male and Female Cln3$^{\Delta ex7/8}$ Mice

The expression of several anti-apoptotic (*Bcl-2, Bcl-xL, Akt, Xiap*) and pro-apoptotic genes (*Fadd, Cytochrome C, Caspase 3, Caspase 6, Caspase 9, Apaf-1, Bad, Bax*) were measured in male and female mice. The anti-apoptotic gene B-cell lymphoma extra-large (*Bcl-xl*) gene expression level was downregulated in vehicle-treated Cln3$^{\Delta ex7/8}$ mice compared to vehicle-treated WT male mice (Figure 4A). Flupirtine treatment had a significant effect only on *Bcl-xl* expression levels, by increasing mRNA expression level in Cln3$^{\Delta ex7/8}$ male mice (Figure 4A). In female mice, B-cell lymphoma 2 (*Bcl-2*) gene expression was slightly higher in vehicle-treated Cln3$^{\Delta ex7/8}$ mice compared to vehicle-treated WT female mice (Figure 4B). Flupirtine, however, significantly increased only the mRNA expression level of *Bcl-2* in Cln3$^{\Delta ex7/8}$ compared to vehicle-treated Cln3$^{\Delta ex7/8}$ mice (Figure 4B). The expression level of other anti-apotic and pro-apotic genes remained unchanged (data not shown).

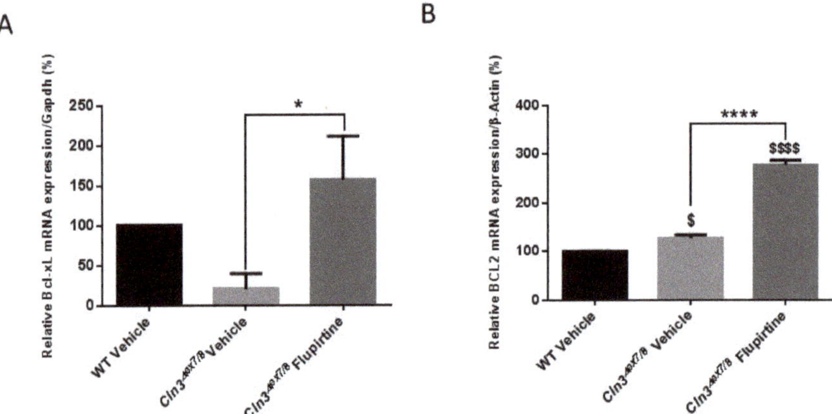

Figure 4. Mouse brain gene expression of anti-apoptotic markers. (**A**) *Bcl-xl* gene expression levels in the brain of vehicle-treated WT, vehicle-treated Cln3$^{\Delta ex7/8}$, and flupirtine-treated Cln3$^{\Delta ex7/8}$ male mice ($n = 4$ per group). Data are expressed as mean ± SEM, * p: compared to Cln3$^{\Delta ex7/8}$ vehicle-treated mice. * $p < 0.05$; (**B**) Bcl-2 gene expression levels in the brain of vehicle-treated WT, vehicle-treated Cln3$^{\Delta ex7/8}$, and flupirtine-treated Cln3$^{\Delta ex7/8}$ female mice ($n = 4$ per group). Data are expressed as mean ± SEM. $^{\$}$ p: compared to vehicle-treated WT mice, and, * p: compared to vehicle-treated Cln3$^{\Delta ex7/8}$ mice. **** $p < 0.0001$, $^{\$}$ $p < 0.05$, and $^{\$\$\$\$}$ $p < 0.0001$.

3.4. Impact of Flupirtine on Gene Expression of Enzymes of Ceramide Metabolism in Cln3$^{\Delta ex7/8}$ Mice

Several ceramide synthesis enzymes (*Sptlc2, Sptlc3, Kdsr, CerS1-6, Degs1, Degs2, Smpd2, Smpd3, Gba, GalC*) and ceramide degradation enzymes (*Asah1, Asah2, Samd8, Ugcg, Ugt8*) were investigated in male and female mice. Serine palmitoyltransferase 3 (*Sptlc3*) levels were significantly upregulated in vehicle-treated Cln3$^{\Delta ex7/8}$ mice compared to vehicle-treated WT male mice (Figure 5A). Flupirtine treatment had a significant impact only on the expression level of key ceramide synthesis enzyme, Sptlc3, in the de novo pathway, by reducing mRNA expression levels compared to Cln3$^{\Delta ex7/8}$ vehicle-treated male mice (Figure 5A). In female mice, sterile alpha motif domain containing 8 (*Samd8*) gene expression level did not differ between vehicle-treated WT and vehicle-treated Cln3$^{\Delta ex7/8}$ female mice. *Samd8* is an endoplasmic reticulum (ER) transferase that converts phosphatidylethanolamine (PE) and ceramide to ceramide phosphoethanolamine (CPE). Flupirtine significantly increased expression levels of only *Samd8* compared to vehicle-treated WT and to Cln3$^{\Delta ex7/8}$ female mice (Figure 5B). The expression level of other ceramide synthesis/degradation enzymes remained unchanged (data not shown).

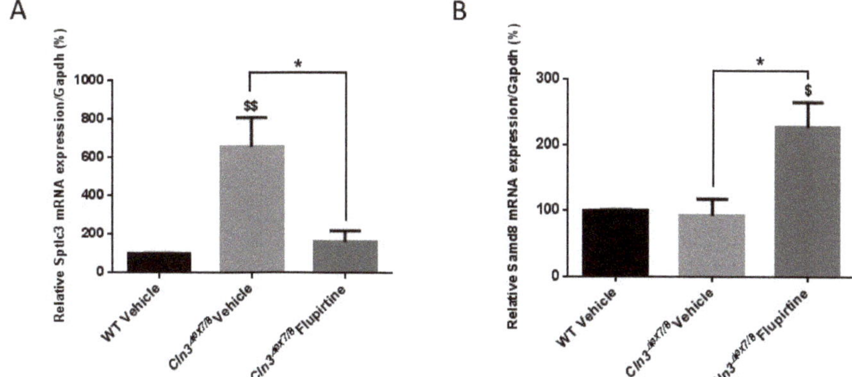

Figure 5. Mouse brain gene expression of ceramide synthesis/degradation enzymes. (**A**) Sptlc3 gene expression levels in the brain of vehicle-treated WT, vehicle-treated $Cln3^{\Delta ex7/8}$, and flupirtine-treated $Cln3^{\Delta ex7/8}$ male mice ($n = 4$ per group). Data are expressed as mean ± SEM. $^{\$}$ p: compared to WT vehicle-treated mice, and, * p: compared to vehicle-treated $Cln3^{\Delta ex7/8}$ mice. * $p < 0.05$, and $^{\$\$}$ $p < 0.01$; (**B**) Samd8 gene expression levels in the brain of vehicle-treated WT, vehicle-treated $Cln3^{\Delta ex7/8}$, and flupirtine-treated $Cln3^{\Delta ex7/8}$ female mice ($n = 4$ per group). Data are expressed as mean ± SEM. $^{\$}$ p: compared to vehicle-treated WT mice, and, * p: compared to vehicle-treated $Cln3^{\Delta ex7/8}$ mice. * $p < 0.05$ and $^{\$}$ $p < 0.05$.

3.5. Effect of Flupirtine Supplementation on Astrocytosis in $Cln3^{\Delta ex7/8}$ Mouse Brains

Fluorescence microscopy demonstrated significant enhanced GFAP immunostaining (green) in vehicle-treated $Cln3^{\Delta ex7/8}$ male mice relative to WT littermates in CA1/2 and CA3 hippocampus, dentate gyrus (DG), and motor cortex (MC) (Figure 6A,B). Treatment with flupirtine decreased glial activation in all brain regions studied in $Cln3^{\Delta ex7/8}$ male mice (Figure 6A,B). Results were statistically significant only in CA1/2 hippocampus, and motor cortex (MC). Hoechst staining (blue) indicates diminished number of hippocampal neurons in vehicle-treated $Cln3^{\Delta ex7/8}$ versus WT, and also an increase in neurons in flupirtine-treated male mice versus vehicle-treated $Cln3^{\Delta ex7/8}$ male mice (Figure 6A).

In females, GFAP immunostaining was significantly enhanced in vehicle-treated $Cln3^{\Delta ex7/8}$ mice relative to their WT female littermates in CA1/2 and CA3 hippocampus, as well as in DG (Figure 7A,B). Treatment with flupirtine significantly decreased astrogliosis in CA1/2 and CA3 hippocampus, and DG in $Cln3^{\Delta ex7/8}$ female mice (Figure 7A,B). Also, note an increase in the number of blue, Hoechst-stained neurons in hippocampus and dentate gyrus in both wild type and flupirtine-treated $Cln3^{\Delta ex7/8}$ female mice compared to vehicle-treated $Cln3^{\Delta ex7/8}$ female mice (Figure 7A). In motor cortex (MC) of females, no significant difference in GFAP levels was observed (data not shown).

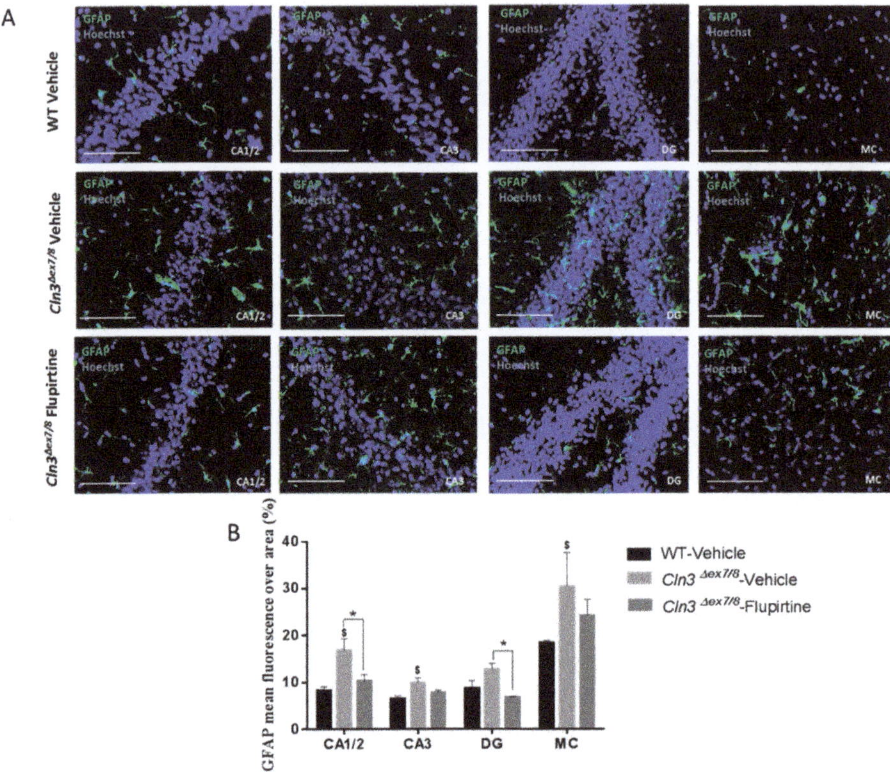

Figure 6. Impact of flupirtine supplementation on astrocytosis in hippocampus and motor cortex of male $Cln3^{\Delta ex7/8}$ mice. (**A**) Representative images of hippocampus regions CA1, CA2, CA3 and dentate gyrus (DG), as well as primary motor cortex from vehicle-treated WT, vehicle-treated $Cln3^{\Delta ex7/8}$, and flupirtine-treated $Cln3^{\Delta ex7/8}$ male mice stained with GFAP in green and counterstained with Hoechst in blue ($n = 4$; Scale bars = 20 μm).; (**B**) Glial fibrillary acidic protein (GFAP) mean fluorescence over area in vehicle-treated WT, vehicle-treated $Cln3^{\Delta ex7/8}$, and flupirtine-treated $Cln3^{\Delta ex7/8}$ male mice in hippocampus regions CA1, CA2, CA3 and dentate gyrus (DG), as well as in primary motor cortex (MC). Data are expressed as mean ± SEM. $^{\$}$ $p < 0.05$ compared to vehicle-treated WT, * $p < 0.05$ compared to vehicle-treated $Cln3^{\Delta ex7/8}$.

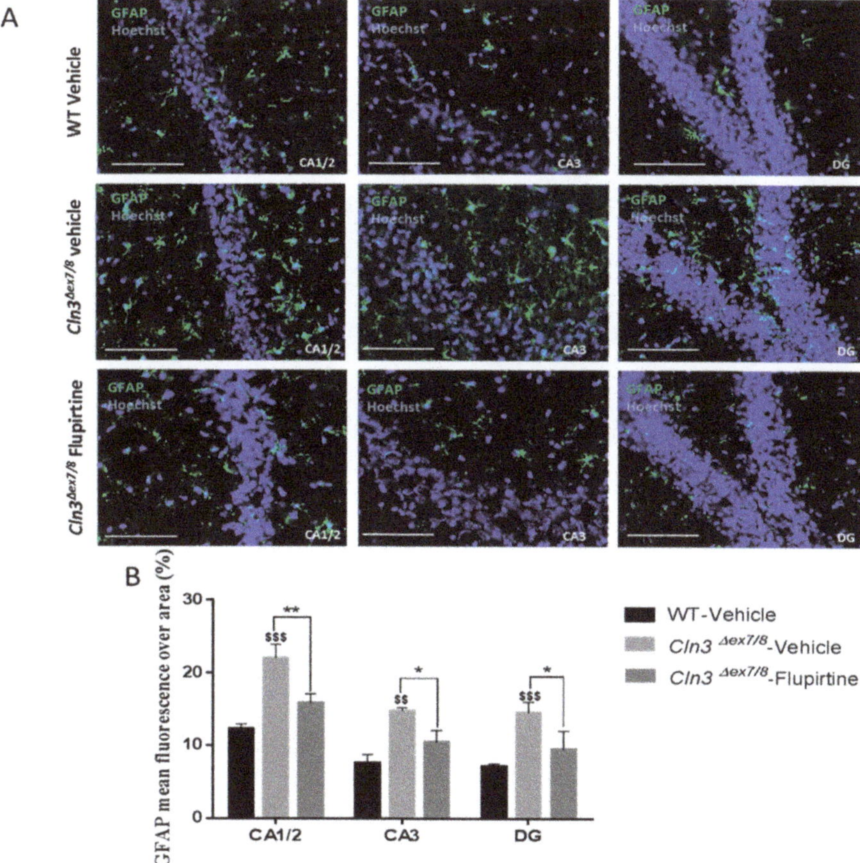

Figure 7. Impact of flupirtine supplementation on astrocytosis in hippocampus and motor cortex of female $Cln3^{\Delta ex7/8}$ mice. (**A**) Representative images of hippocampus regions CA1, CA2, CA3 and dentate gyrus (DG), from vehicle-treated WT, vehicle-treated $Cln3^{\Delta ex7/8}$, and flupirtine-treated $Cln3^{\Delta ex7/8}$ female mice stained with GFAP in green and counterstained with Hoechst in blue ($n = 4$; scale bars = 20 μm).; (**B**) glial fibrillary acidic protein (GFAP) mean fluorescence over area in vehicle-treated WT, vehicle-treated $Cln3^{\Delta ex7/8}$, and flupirtine-treated $Cln3^{\Delta ex7/8}$ male mice in hippocampus regions CA1, CA2, CA3 and dentate gyrus (DG), as well as in primary motor cortex (MC). Data are expressed as mean ± SEM. $^{\$}$ *p*: compared to vehicle-treated WT, * *p*: compared to vehicle-treated $Cln3^{\Delta ex7/8}$. * $p < 0.05$, ** $p < 0.01$, $^{\$\$}$ $p < 0.01$, and $^{\$\$\$}$ $p < 0.001$.

3.6. Impact of Flupirtine Supplementation on Neuronal Cell Counts in $Cln3^{\Delta ex7/8}$ Brains

The number of NeuN-positive cells decreased significantly in the MC of vehicle-treated $Cln3^{\Delta ex7/8}$ male mice versus vehicle-treated WT mice (Figure 8A,B). Although not significant, NeuN-stained cells in MC of flupirtine-treated $Cln3^{\Delta ex7/8}$ male mice compared to vehicle-treated $Cln3^{\Delta ex7/8}$ mice showed a slight increase, close to the level in WT mice (Figure 8A,B). No significant difference in the neuronal cell counts of female MC was seen (data not shown).

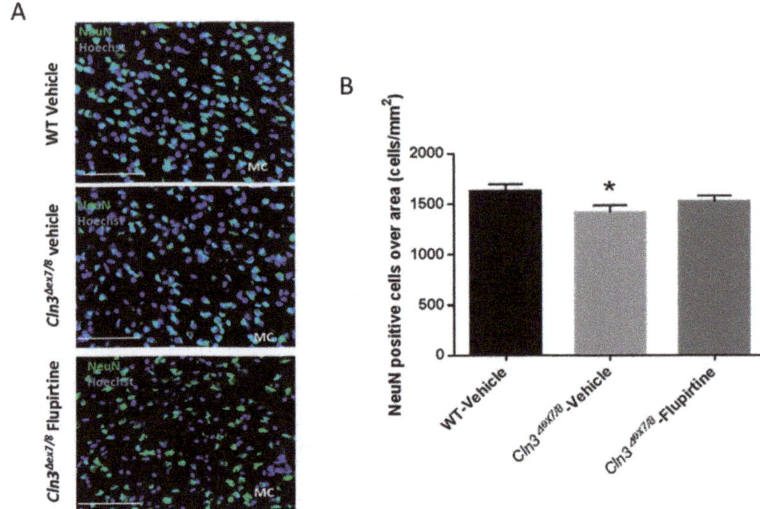

Figure 8. Effect of flupirtine treatment on motor cortex neuron numbers in $Cln3^{\Delta ex7/8}$ mice. (**A**) Representative images of primary motor cortex from vehicle-treated WT, vehicle-treated $Cln3^{\Delta ex7/8}$ and flupirtine-treated $Cln3^{\Delta ex7/8}$ male mice stained with NeuN (mature neuronal marker) in green and counterstained with Hoechst in blue ($n = 4$; Scale bars = 50 μm).; (**B**) NeuN-positive cells normalized to area in the primary motor cortex (MC) of vehicle-treated WT, vehicle-treated $Cln3^{\Delta ex7/8}$, and flupirtine-treated $Cln3^{\Delta ex7/8}$ male mice. Data are expressed as mean ± SEM. * $p < 0.05$ compared to vehicle-treated WT.

3.7. Impact of Compound 6 Supplementation on Behavioral Parameters

In the open field test, compound 6 significantly slowed down locomotor hyperactivity of male and female $Cln3^{\Delta ex7/8}$ mice (Figure 9A,B). Similarly, a similar effect was seen in the pole-climbing test, where compound 6 significantly increased time needed by male mice to descend the pole compared to vehicle treated mice (Figure 9C).

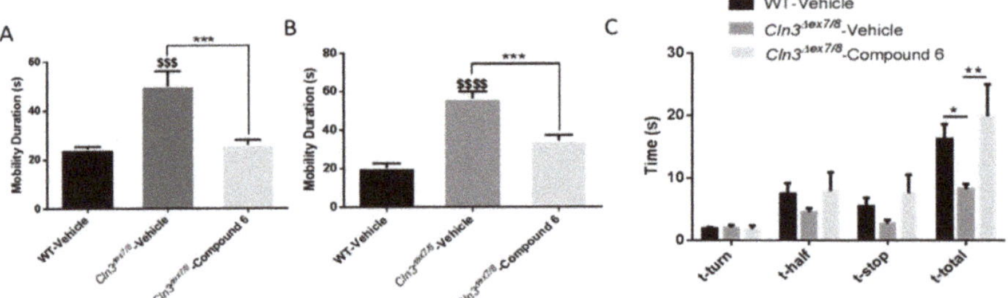

Figure 9. Impact of compound 6 treatment in $Cln3^{\Delta ex7/8}$ mice. Open field behavioral parameter (mobility duration) in vehicle-treated WT, vehicle-treated $Cln3^{\Delta ex7/8}$, and compound 6-treated $Cln3^{\Delta ex7/8}$ (**A**) male and (**B**) female mice ($n = 8$ per group). Data are expressed as mean ± SEM. $ p: compared to WT vehicle-treated mice, and, * p: compared to vehicle-treated $Cln3^{\Delta ex7/8}$ mice. *** $p < 0.001$, $$$ $p < 0.001$, and $$$$ $p < 0.0001$. (**C**) Pole climbing test in vehicle-treated WT, vehicle-treated $Cln3^{\Delta ex7/8}$, and flupirtine-treated $Cln3^{\Delta ex7/8}$ male mice ($n = 8$ per group). Time to turn (t_{turn}), time needed to reach center of pole ($t_{1/2}$), amount of time stopped (t_{stop}), and total time to descend (t_{total}) were recorded. Data are expressed as mean ± SEM. * $p < 0.05$, ** $p < 0.01$.

4. Discussion

This study addresses novel small molecule treatment strategies for CLN3 disease in a $Cln3^{\Delta ex7/8}$ knock-in mouse model. Hopefully, this will translate into future knowledge to improve the lives of CLN3 patients.

Flupirtine is well known for its significant powerful anti-oxidative, anti-apoptotic, and neuroprotective effects in vitro and in vivo. The effectiveness of the chosen daily dose of flupirtine (30 mg/kg per os) has been demonstrated in testing for anti-nociceptive, anticonvulsant, and anti-apoptotic activity in rodents [23–27]. This study is the first to evaluate the use of flupirtine as potential treatment for CLN3 disease in an animal model, i.e., in vivo. Previous studies proved that flupirtine reaches brain regions, including hippocampus and cortex, and that it rapidly crosses the blood brain barrier and enters other tissues. The liver is the primary organ responsible for its metabolism [15].

Behavioral tests assessed different aspects of $Cln3^{\Delta ex7/8}$ mouse motor strength, coordination, balance, as well as learning and spatial memory functions before and after flupirtine treatment. Male and female vehicle-treated $Cln3^{\Delta ex7/8}$ mice exhibit significantly increased mobility with respect to WT controls in the open field test. The hyperactive phenotype prominent in vehicle-treated $Cln3^{\Delta ex7/8}$ mice is described for the first time here in mice at of 18 weeks of age. This was not previously documented in $Cln3^{\Delta ex7/8}$ mice at 40 weeks of age [28]. This suggests that, at a young age, $Cln3^{\Delta ex7/8}$ mice express a distinct behavioral phenotype prior to onset of more severe CLN3 disease symptoms, including motor decline. $Cln3^{\Delta ex7/8}$ mice also showed increased random and chaotic exploratory activity compared to WT controls in a novel environment. This indicates inattentiveness and diminished executive function. Treatment with flupirtine significantly attenuated this abnormal mobility in male and female $Cln3^{\Delta ex7/8}$ mice. This phenotype may be consistent with the notion that diminished executive function of vehicle-treated $Cln3^{\Delta ex7/8}$ mice is driving their inattentive hyperactivity. The vertical pole test evaluates spatial and motor orientation and balance of mice [29]. Reduced latency of vehicle-treated $Cln3^{\Delta ex7/8}$ mice to descend the pole is explained by their quick and less controlled behavior compared to WT animals, consistent with open field behavioral assessments of excessive mobility. The aforementioned behavioral test reveals prominent hyperactivity and attentional deficits in vehicle-treated $Cln3^{\Delta ex7/8}$ mice, while treatment with flupirtine lessened this impulsive phenotype. These results are in line with other mouse models of neurodevelopmental psychiatric disorder of attention-deficit hyperactivity disorder (ADHD). Mice recapitulating this disease are characterized by impulsivity, inattentiveness, and hyperactivity [30]. Stress has a direct profound effect on rodent behavior and physical activity [31]. Chronic stress leads to increased corticosterone levels [31], as observed in WT male and female mice with respect to $Cln3^{\Delta ex7/8}$ mice. Several studies elucidate that animals who do not experience stress show higher exploration, locomotion, and physical activity in an open field test, as a reaction to an unknown environment [32,33]. Therefore, the markedly decreased corticosterone levels in serum of $Cln3^{\Delta ex7/8}$ mice with respect to WT animals explains their hyperactivity resulting from reduced stress levels in male and female $Cln3^{\Delta ex7/8}$ mice. Learning and cognitive ability of mice was tested at 16 weeks of age using the Morris water maze (MWM) test. Analysis of the swim paths of mice during the probe trial showed that flupirtine-treated male and female $Cln3^{\Delta ex7/8}$ mice adopted strategies and maintained spatial preference in the target quadrant, contrary to vehicle-treated $Cln3^{\Delta ex7/8}$ mice that show coverage of the whole maze. Flupirtine significantly enhanced spatial learning, navigation and memory retention in $Cln3^{\Delta ex7/8}$ male and female mice.

Apoptosis is a naturally-occurring mechanism of cell death and helps maintain tissue homeostasis [34]. Neuronal cell loss is evident in brain sections from post-mortem CLN3 disease patients [3]. Numerous apoptotic cells are present within cortical brain sections from CLN3 disease patients [3]. CLN3 patient-derived lymphoblasts have decreased growth rate compared to normal lymphoblasts, validating that apoptosis is one of the mechanisms implicated in CLN3 disease pathogenesis [35]. Other studies demonstrate damage and apoptosis of neuronal and glial cells in hippocampus and cortex of CLN3 patients in addition to marked loss of cortical neurons due to apoptotic cell death [6]. CLN3 defects also perturb calcium signaling, leading to a profound defects in neuronal survival [36].

Most neuronal death in CLN2 and CLN3 brains takes place via apoptosis, and the surviving neurons upregulate Bcl-2 [5]. Treatment with flupirtine significantly upregulated expression of anti-apoptotic BCL-2 in CLN3-deficient cells in vitro [20]. This is the case *in vivo* in this current study as flupirtine-treated $Cln3^{\Delta ex7/8}$ female mice show a remarkable increase in *Bcl-2* expression. In males, another anti-apoptotic protein, *Bcl-xl*, was upregulated following treatment with flupirtine. Different proteins were impacted in male versus female mice, yet the end result was upregulation of anti-apoptotic pathways and hence, reduction of cell death in brains of $Cln3^{\Delta ex7/8}$ mice given flupirtine. This variation among sexes is not a new observation in this disease. We documented this in another study using exogenous galactosylceramide as potential treatment for CLN3 disease [28].

CLN3 is directly implicated in apoptotic cell death signaling cascades by activating caspase-dependent and caspase-independent pathways [8]. Ceramide is a major sphingolipid second messenger implicated in several cell processes and impacts divergent pathways [37]. Sphingolipids are major bioactive lipids involved in homeostasis, growth, proliferation and cell death [38]. Ceramide mediates anti-proliferative events, such as apoptosis, growth inhibition, cell differentiation, and senescence [10]. This biomolecule possesses complex biophysical properties and acts as a central hub. Regulation of its levels affects catabolism and break-down of various sphingolipid species [10]. Ceramide is generated through several complex interrelated pathways either via the de novo pathway, sphingomyelin, or cerebroside catabolism. Ceramide is synthesized de novo in the endoplasmic reticulum or through breakdown of sphingomyelin in Golgi, plasma membrane, or mitochondrial membrane [39]. Defects in ceramide signaling pathways often result in augmenting programmed cell death in multiple cell types, including neurons [40]. Previous studies show that ceramide levels are increased in CLN3-deficient cells and brain of CLN3 patients [11]. Published reports demonstrate that 17 week-old $Cln3^{\Delta ex7/8}$ mice express higher levels of ceramide in brain compared to age-matched WT mice [14]. Sptlc3 catalyzes the initial steps in formation of ceramide via the de novo pathway by condensing serine and palmitoyl Co-A to generate 3-ketoshphinganine (3-KDS) [10]. We documented decreased *Sptlc3* levels in flupirtine-treated $Cln3^{\Delta ex7/8}$ mice versus vehicle-treated $Cln3^{\Delta ex7/8}$ male mice. Downregulation of *Sptlc3* leads to diminution of ceramide generation via the de novo ceramide pathway. As for the other enzymes of the de novo pathway, the expression of *Sptlc2* and *Degs1* also decreased with flupirtine treatment, but did not reach significance (data not shown). This supports our conclusion that flupirtine treatment in male mice impacts the de novo synthesis pathway. In females, however, a different pathway in ceramide signaling is at play. *Samd8* is an ER transferase that converts phosphatidylethanolamine (PE) and ceramide to ceramide phosphoethanolamine (CPE). *Samd8* levels are increased in flupirtine-treated $Cln3^{\Delta ex7/8}$ female mice. *Samd8* operates as a ceramide sensor to control ceramide homeostasis in the ER rather than a converter of ceramides. This implicates that the ceramide salvage pathway is modulated in flupirtine-treated $Cln3^{\Delta ex7/8}$ female mice with the end result of diminished ceramide levels. Empirical evidence from our study confirms reduced synthesis (decreased *Sptlc3* expression) in flupirtine-treated male mice, but increased degradation of ceramide (increased *Samd8* expression) in female flupirtine-treated mice. Physiologic gender differences affects drug activity and characteristics, including pharmacokinetics [41]. Differences in body size result in larger distribution volumes, faster total clearance, and less tissue absorption of some medications in men compared to women [42]. This may explain higher brain absorption in female compared to male mice. Moreover, sex hormones, in females, have a direct effect on drug absorption, distribution, metabolism, elimination and adverse effects [43].This study implies that sex-specific drug dosing regimens may be warranted for treatment of neurological diseases that affect the blood brain barrier, including CLN3 disease in mouse and man.

Activation of the glial cell population contributes to imbalance in CNS function and impacts cognitive function [44]. Hyperactive astrocytes are observed in neurodegenerative disorders and following brain injury documented by GFAP as biomarker. In males, microscopic inspection of vehicle-treated $Cln3^{\Delta ex7/8}$ mouse brain documents widespread intracellular GFAP staining in hippocampal regions (CA1, CA2, CA3, and the dentate gyrus) and in the MC relative to age-matched, vehicle-treated WT animals. Mice treated with flupirtine had significantly lower levels of GFAP staining in these regions. This data provides evidence that flupirtine attenuates astrogliosis at the level of the hippocampus and MC in $Cln3^{\Delta ex7/8}$ male mice. In females, flupirtine was able to attenuate GFAP immunostaining in CA1/2, CA3 and DG regions of the hippocampus. The motor cortex did not show any difference in vehicle-treated $Cln3^{\Delta ex7/8}$ female mice compared to WT. This may explain better performance in $Cln3^{\Delta ex7/8}$ female mice on the rotarod compared to males as it tests motor skills (data not shown).

The neuronal nuclear protein (NeuN) is a marker not detected in glial cells or other cells in the brain [28]. NeuN assesses neuronal health and loss of this protein is indicative of damage. Neuronal cell loss in CLN3 patients is at the root of CLN3 disease pathogenesis [45]. NeuN-positive cell were significantly ablated in vehicle-treated $Cln3^{\Delta ex7/8}$ mouse. Although it did not reach significance, flupirtine resulted in an increase in the neuronal population in motor cortex (MC) of male mice. This tsuggests that flupirtine conferred neuroprotection and reduced cell death in brains of $Cln3^{\Delta ex7/8}$ mice. In females, there was no difference in neuronal counts in motor cortex of vehicle-treated $Cln3^{\Delta ex7/8}$ female mice compared to WT. In a previous study, affected female mice 44 weeks of age did show a diminution in NeuN positive cells [28], suggesting that unlike males, neuronal loss in females starts at a later age.

The novel, flupirtine-like allyl carbamate derivative, compound 6, developed to possess physicochemical properties desirable for CNS therapeutics had an improved Multiparameter optimization (MPO) score. The latter predicts blood–brain barrier (BBB) penetration, an essential parameter for neuroprotective compounds, and was ≥ 4 more effective [19]. The newly synthesized allyl carbamate derivative of flupirtine, compound 6, showed potential for improved neuroprotection, after screening in vitro nine flupirtine derivatives [20]. Here, we report early promising behavioral in vivo results for compound 6 for treatment of $Cln3^{\Delta ex7/8}$ mice. Although pharmacokinetics and toxicological safety remain to be established for compound 6, the promising behavioral data obtained in this study are worth reporting. Treatment with compound 6 significantly attenuated the high mobility documented by open field and pole climbing in $Cln3^{\Delta ex7/8}$ mice, suggesting more work is necessary to determine optimal dosing for this compound.

5. Conclusions

In conclusion, these findings suggest that flupirtine, and compound 6, improve neurobehavioral measures. Flupirtine impacted ceramide biosynthesis and apoptotic signaling pathways. Flupirtine affected a broad-spectrum of actionable targets, providing insights into the pathobiology of CLN3 disease in humans, particularly uncovering impact on gender-specific signaling pathways. Flupirtine shows promise in males and females, and the allyl carbamate derivative, compound 6, needs further preclinical analyses and development. These findings extend our knowledge of the role of drugs in the treatment of a fatal pediatric neurodegenerative disease implying more work lies ahead for development into clinically applicable therapies for CLN3 patients.

6. Patents

R.-M.B. has an Application for Method of Treating Batten Disease. Inventor: Rose-Mary Boustany. Duke (File No. 5405-240 PR). US Patent and Trademark No. 10/148,859 (U.S. National Phase); Use Patent issued 11/23/2004 US Patent # 6 821 995, expired 11/23/2014.

N.K., P.C.T. and R.-M.B. are inventors on a patent application detailing the aromatic carbamates described herein: 'Functionalized Pyridine Carbamates with enhanced Neuroprotective Activity' PCT Int. Appl. (2019), WO2019014547 A8.

Author Contributions: K.M., J.M., S.S., and N.J.M. performed experiments, analyzed data, and interpreted the results. A.V.C., N.K., and P.C.T. synthesized the drugs. N.G. helped in analyzing data and interpreting results. K.M., J.M., and S.S. prepared the figures, and drafted the main manuscript. R.-M.B. conceived the study, obtained funding for the study, designed experiments, reviewed data and analyses, and revised and edited the manuscript. All authors have read and agreed to the published version of the manuscript.

Funding: We would like to thank OpenMinds for financial support (R.-M.B.) for this work (#620229), AUB—Collaborative Research Grant (N.G and R.-M.B.) (#24473), and University Research Board (URB) at AUB (N.G.) and the Lebanese National Council for Scientific Research (N.G.).

Acknowledgments: A very particular acknowledgement to my late and dear friend Lina Marie Obeid, who, together with her husband, Yusuf Hannun, introduced me to the importance of sphingolipid biology.

Conflicts of Interest: The authors declare no conflict of interest.

References

1. Jalanko, A.; Braulke, T. Neuronal ceroid lipofuscinoses. *Biochim. Biophys. Acta (BBA) Mol. Cell Res.* **2009**, *1793*, 697–709. [CrossRef] [PubMed]

2. Persaud-Sawin, D.-A.; Mousallem, T.; Wang, C.; Zucker, A.; Kominami, E.; Boustany, R.-M.N. Neuronal Ceroid Lipofuscinosis: A Common Pathway? *Pediat. Res.* **2007**, *61*, 146–152. [CrossRef] [PubMed]

3. Lane, S.C.; Jolly, R.D.; Schmechel, D.E.; Alroy, J.; Boustany, R.-M. Apoptosis as the Mechanism of Neurodegeneration in Batten's Disease. *J. Neurochem.* **1996**, *67*, 677–683. [CrossRef] [PubMed]

4. Munroe, P.B.; Mitchison, H.M.; O'Rawe, A.M.; Anderson, J.W.; Boustany, R.M.; Lerner, T.J.; Taschner, P.E.; de Vos, N.; Breuning, M.H.; Gardiner, R.M.; et al. Spectrum of mutations in the Batten disease gene, CLN3. *Am. J. Hum. Genet.* **1997**, *61*, 310–316. [CrossRef] [PubMed]

5. Puranam, K.; Qian, W.H.; Nikbakht, K.; Venable, M.; Obeid, L.; Hannun, Y.; Boustany, R.M. Upregulation of Bcl-2 and elevation of ceramide in Batten disease. *Neuropediatrics* **1997**, *28*, 37–41. [CrossRef] [PubMed]

6. Pontikis, C.C.; Cotman, S.L.; MacDonald, M.E.; Cooper, J.D. Thalamocortical neuron loss and localized astrocytosis in the Cln3Δex7/8 knock-in mouse model of Batten disease. *Neurobiol. Dis.* **2005**, *20*, 823–836. [CrossRef] [PubMed]

7. Lerner, T.J.; Boustany, R.-M.N.; Anderson, J.W.; D'Arigo, K.L.; Schlumpf, K.; Buckler, A.J.; Gusella, J.F.; Haines, J.L. Isolation of a novel gene underlying batten disease, *CLN3*. *Cell* **1995**, *82*, 949–957. [CrossRef]

8. Persaud-Sawin, D.A.; Boustany, R.M.N. Cell death pathways in juvenile Batten disease. *Apoptosis* **2005**, *10*, 973–985. [CrossRef]

9. Guo, W.-X.; Mao, C.; Obeid, L.M.; Boustany, R.-M. A Disrupted Homologue of the Human CLN3 or Juvenile Neuronal Ceroid Lipofuscinosis Gene in Saccharomyces cerevisiae: A Model to Study Batten Disease. *Cell. Mol. Neurobiol.* **1999**, *19*, 671–680. [CrossRef]

10. Ogretmen, B.; Hannun, Y.A. Biologically active sphingolipids in cancer pathogenesis and treatment. *Nat. Rev. Cancer* **2004**, *4*, 604–616. [CrossRef] [PubMed]

11. Puranam, K.L.; Guo, W.-X.; Qian, W.-H.; Nikbakht, K.; Boustany, R.-M. CLN3 Defines a Novel Antiapoptotic Pathway Operative in Neurodegeneration and Mediated by Ceramide. *Mol. Genet. Metab.* **1999**, *66*, 294–308. [CrossRef] [PubMed]

12. Rusyn, E.; Mousallem, T.; Persaud-Sawin, D.-A.; Miller, S.; Boustany, R.-M.N. CLN3p Impacts Galactosylceramide Transport, Raft Morphology, and Lipid Content. *Pediat. Res.* **2008**, *63*, 625–631. [CrossRef] [PubMed]

13. Dhar, S.; Bitting, R.L.; Rylova, S.N.; Jansen, P.J.; Lockhart, E.; Koeberl, D.D.; Amalfitano, A.; Boustany, R.-M.N. Flupirtine blocks apoptosis in batten patient lymphoblasts and in human postmitotic CLN3- and CLN2-deficient neurons. *Ann. Neurol.* **2002**, *51*, 448–466. [CrossRef] [PubMed]

14. El-Sitt, S.; Soueid, J.; Al Ali, J.; Makoukji, J.; Makhoul, N.J.; Harati, H.; Boustany, R.-M. Developmental Comparison of Ceramide in Wild-Type and Cln3 (Δex7/8) Mouse Brains and Sera. *Front. Neurol.* **2019**, *10*, 128. [CrossRef] [PubMed]

15. Patil, A.M.; Matter, A.B.; Raol, H.Y.; Bourne, W.A.D.; Kelley, A.R.; Kompella, B.U. Brain Distribution and Metabolism of Flupirtine, a Nonopioid Analgesic Drug with Antiseizure Effects, in Neonatal Rats. *Pharmaceutics* **2018**, *10*, 281. [CrossRef] [PubMed]

16. Perovic, S.; Pergande, G.; Ushijima, H.; Kelve, M.; Forrest, J.; Müller, W.E.G. Flupirtine Partially Prevents Neuronal Injury Induced by Prion Protein Fragment and Lead Acetate. *Neurodegeneration* **1995**, *4*, 369–374. [CrossRef]

17. Müller, W.E.G.; Laplanche, J.-L.; Ushijima, H.; Schröder, H.C. Novel approaches in diagnosis and therapy of Creutzfeldt–Jakob disease. *Mech. Ageing Dev.* **2000**, *116*, 193–218. [CrossRef]

18. Devulder, J. Flupirtine in Pain Management. *CNS Drugs* **2010**, *24*, 867–881. [CrossRef] [PubMed]

19. Kinarivala, N.; Patel, R.; Boustany, R.-M.; Al-Ahmad, A.; Trippier, P.C. Discovery of Aromatic Carbamates that Confer Neuroprotective Activity by Enhancing Autophagy and Inducing the Anti-Apoptotic Protein B-Cell Lymphoma 2 (Bcl-2). *J. Med. Chem.* **2017**, *60*, 9739–9756. [CrossRef]

20. Makoukji, J.; Saadeh, F.; Mansour, K.A.; El-Sitt, S.; Al Ali, J.; Kinarivala, N.; Trippier, P.C.; Boustany, R.-M. Flupirtine derivatives as potential treatment for the neuronal ceroid lipofuscinoses. *Ann. Clin. Transl. Neurol.* **2018**, *5*, 1089–1103. [CrossRef] [PubMed]

21. Kinarivala, N.; Trippier, P.C. Progress in the Development of Small Molecule Therapeutics for the Treatment of Neuronal Ceroid Lipofuscinoses (NCLs). *J. Med. Chem.* **2016**, *59*, 4415–4427. [CrossRef] [PubMed]

22. Seibenhener, M.L.; Wooten, M.C. Use of the Open Field Maze to measure locomotor and anxiety-like behavior in mice. *J. Vis. Exp.* **2015**, *96*, e52434. [CrossRef] [PubMed]

23. Huang, P.; Li, C.; Fu, T.; Zhao, D.; Yi, Z.; Lu, Q.; Guo, L.; Xu, X. Flupirtine attenuates chronic restraint stress-induced cognitive deficits and hippocampal apoptosis in male mice. *Behav. Brain Res.* **2015**, *288*, 1–10. [CrossRef]

24. Jaeger, H.M.; Pehlke, J.R.; Kaltwasser, B.; Kilic, E.; Bähr, M.; Hermann, D.M.; Doeppner, T.R. The indirect NMDAR inhibitor flupirtine induces sustained post-ischemic recovery, neuroprotection and angioneurogenesis. *Oncotarget* **2015**, *6*, 14033. [CrossRef]

25. Kumar, M.; Gokul, C.G.; Somashekar, H.S.; Adake, P.; Acharya, A.; Santhosh, R. Anticonvulsant activity of flupirtine in Albino mice. *Pharmacologyonline* **2011**, *3*, 860–867.

26. Morecroft, I.; Murray, A.; Nilsen, M.; Gurney, A.M.; MacLean, M.R. Treatment with the Kv7 potassium channel activator flupirtine is beneficial in two independent mouse models of pulmonary hypertension. *Br. J. Pharm.* **2009**, *157*, 1241–1249. [CrossRef]

27. Nickel, B. The antinociceptive activity of flupirtine: A structurally new analgesic. *Postgrad. Med. J.* **1987**, *63* Suppl 3, 19–28.

28. El-Sitt, S.; Soueid, J.; Maalouf, K.; Makhoul, N.; Al Ali, J.; Makoukji, J.; Asser, B.; Daou, D.; Harati, H.; Boustany, R.-M. Exogenous Galactosylceramide as Potential Treatment for CLN3 Disease. *Ann. Neurol.* **2019**, *86*, 729–742. [CrossRef] [PubMed]

29. Justice, J.N.; Carter, C.S.; Beck, H.J.; Gioscia-Ryan, R.A.; McQueen, M.; Enoka, R.M.; Seals, D.R. Battery of behavioral tests in mice that models age-associated changes in human motor function. *Age* **2014**, *36*, 583–592. [CrossRef]

30. Bouchatta, O.; Manouze, H.; Bouali-benazzouz, R.; Kerekes, N.; Ba-M'hamed, S.; Fossat, P.; Landry, M.; Bennis, M. Neonatal 6-OHDA lesion model in mouse induces Attention-Deficit/Hyperactivity Disorder (ADHD)-like behaviour. *Sci. Rep.* **2018**, *8*, 15349. [CrossRef] [PubMed]

31. DeVallance, E.; Riggs, D.; Jackson, B.; Parkulo, T.; Zaslau, S.; Chantler, P.D.; Olfert, I.M.; Bryner, R.W. Effect of chronic stress on running wheel activity in mice. *PLoS ONE* **2017**, *12*, e0184829. [CrossRef] [PubMed]

32. Bondar, N.P.; Lepeshko, A.A.; Reshetnikov, V.V. Effects of Early-Life Stress on Social and Anxiety-Like Behaviors in Adult Mice: Sex-Specific Effects. *Behav. Neurol.* **2018**, *2018*, 1538931. [CrossRef] [PubMed]

33. Borkar, C.D.; Dorofeikova, M.; Le, Q.E.; Vutukuri, R.; Vo, C.; Hereford, D.; Resendez, A.; Basavanhalli, S.; Sifnugel, N.; Fadok, J.P. Sex differences in behavioral responses during a conditioned flight paradigm. *Behav. Brain Res.* **2020**, *389*, 112623. [CrossRef]

34. Gelbard, H.A.; Boustany, R.M.; Schor, N.F. Apoptosis in development and disease of the nervous system: II. Apoptosis in childhood neurologic disease. *Pediat. Neurol.* **1997**, *16*, 93–97. [CrossRef]

35. Persaud-Sawin, D.A.; VanDongen, A.; Boustany, R.M. Motifs within the CLN3 protein: Modulation of cell growth rates and apoptosis. *Hum. Mol. Genet.* **2002**, *11*, 2129–2142. [CrossRef] [PubMed]

36. Bosch, M.E.; Kielian, T. Astrocytes in juvenile neuronal ceroid lipofuscinosis (CLN3) display metabolic and calcium signaling abnormalities. *J. Neurochem.* **2019**, *148*, 612–624. [CrossRef] [PubMed]

37. Jayadev, S.; Liu, B.; Bielawska, A.E.; Lee, J.Y.; Nazaire, F.; Pushkareva, M.Y.; Obeid, L.M.; Hannun, Y.A. Role or Ceramide in Cell Cycle Arrest. *J. Biol. Chem.* **1995**, *270*, 2047–2052. [CrossRef]

38. Bilal, F.; Montfort, A.; Gilhodes, J.; Garcia, V.; Riond, J.; Carpentier, S.; Filleron, T.; Colacios, C.; Levade, T.; Daher, A.; et al. Sphingomyelin Synthase 1 (SMS1) Downregulation Is Associated With Sphingolipid Reprogramming and a Worse Prognosis in Melanoma. *Front. Pharmacol.* **2019**, *10*. [CrossRef] [PubMed]

39. Persaud-Sawin, D.A.; McNamara, J.O., II; Rylova, S.; Vandongen, A.; Boustany, R.M. A galactosylceramide binding domain is involved in trafficking of CLN3 from Golgi to rafts via recycling endosomes. *Pediat. Res.* **2004**, *56*, 449–463. [CrossRef] [PubMed]

40. Di Pardo, A.; Maglione, V. Sphingolipid Metabolism: A New Therapeutic Opportunity for Brain Degenerative Disorders. *Front. Neurosci.* **2018**, *12*, 249. [CrossRef] [PubMed]

41. Whitley, H.; Lindsey, W. Sex-based differences in drug activity. *Am Fam Physician* **2009**, *80*, 1254–1258. [PubMed]

42. Schwartz, J.B. The influence of sex on pharmacokinetics. *Clin. Pharm.* **2003**, *42*, 107–121. [CrossRef] [PubMed]

43. Spoletini, I.; Vitale, C.; Malorni, W.; Rosano, G.M. Sex differences in drug effects: Interaction with sex hormones in adult life. *Handb. Exp. Pharm.* **2012**, *214*, 91–105. [CrossRef]

44. Becerra-Calixto, A.; Cardona-Gómez, G.P. The Role of Astrocytes in Neuroprotection after Brain Stroke: Potential in Cell Therapy. *Front. Mol. Neurosci.* **2017**, *10*, 88. [CrossRef] [PubMed]

45. Braak, H.; Goebel, H.H. Loss of pigment-laden stellate cells: A severe alteration of the isocortex in juvenile neuronal ceroid-lipofuscinosis. *Acta Neuropathol.* **1978**, *42*, 53–57. [CrossRef] [PubMed]

Article

Anxiety and Depression Are Related to Higher Activity of Sphingolipid Metabolizing Enzymes in the Rat Brain

Iulia Zoicas [1,*,†], Christiane Mühle [1,†], Anna K. Schmidtner [2,3], Erich Gulbins [4], Inga D. Neumann [2] and Johannes Kornhuber [1]

[1] Department of Psychiatry and Psychotherapy, Friedrich-Alexander University Erlangen-Nürnberg (FAU), 91054 Erlangen, Germany; Christiane.Muehle@uk-erlangen.de (C.M.); Johannes.Kornhuber@uk-erlangen.de (J.K.)

[2] Department of Behavioural and Molecular Neurobiology, University of Regensburg, 93040 Regensburg, Germany; Anna-Kri.Schmidtner@mail.huji.ac.il (A.K.S.); inga.neumann@ur.de (I.D.N.)

[3] Edmond and Lily Safra Center for Brain Sciences, Hebrew University of Jerusalem, Jerusalem 9190401, Israel

[4] Department of Molecular Biology, University of Duisburg-Essen, 45147 Essen, Germany; erich.gulbins@uni-due.de

* Correspondence: Iulia.Zoicas@uk-erlangen.de; Tel.: +49-9131-85-46005; Fax: +49-9131-85-36381

† These authors contributed equally to this work.

Received: 28 April 2020; Accepted: 15 May 2020; Published: 17 May 2020

Abstract: Changes in sphingolipid metabolism have been suggested to contribute to the pathophysiology of major depression. In this study, we investigated the activity of acid and neutral sphingomyelinases (ASM, NSM) and ceramidases (AC, NC), respectively, in twelve brain regions of female rats selectively bred for high (HAB) versus low (LAB) anxiety-like behavior. Concomitant with their highly anxious and depressive-like phenotype, HAB rats showed increased activity of ASM and NSM as well as of AC and NC in multiple brain regions associated with anxiety- and depressive-like behavior, including the lateral septum, hypothalamus, ventral hippocampus, ventral and dorsal mesencephalon. Strong correlations between anxiety-like behavior and ASM activity were found in female HAB rats in the amygdala, ventral hippocampus and dorsal mesencephalon, whereas NSM activity correlated with anxiety levels in the dorsal mesencephalon. These results provide novel information about the sphingolipid metabolism, especially about the sphingomyelinases and ceramidases, in major depression and comorbid anxiety.

Keywords: anxiety; depression; sphingolipids; sphingomyelinase; ceramidase

---

1. Introduction

Major depressive disorder (MDD) is a severe and chronic mood disorder, with a lifetime prevalence of approximately 11% in men and 18% in women [1]. MDD is highly comorbid with other psychiatric disorders, including generalized anxiety, social anxiety and alcohol use disorders, and is associated with increased suicidality [2]. The major symptoms of MDD are a depressed mood and loss of interest, anhedonia, feelings of worthlessness, weight loss and insomnia. Despite its severity and high prevalence, the pathogenesis of MDD is yet unclear. One of the mechanisms involved in the pathogenesis of MDD seems to relate to an altered metabolism of sphingolipids such as sphingomyelin and ceramide [3]. The enzyme sphingomyelinase catalyzes the hydrolysis of sphingomyelin to ceramide and phosphorylcholine [4]. Depending on the pH optimum of the enzyme, several isoforms are known, including acid sphingomyelinase (ASM), neutral sphingomyelinase (NSM) and alkaline sphingomyelinase [5]. As expected from their pH optimum, these enzymes are differentially localized throughout the body. As such, ASM is ubiquitously distributed in all tissues [6] with a lysosomal form but is also constitutively secreted [7]. NSM is predominantly localized in the central nervous system [8,9], and alkaline sphingomyelinase is active in the digestive system but not in the central nervous system [10]. Similar to sphingomyelinases, ceramidases differ in their pH optimum for the breakdown of ceramide to sphingosine and fatty acid. The acid ceramidase (AC) and neutral ceramidase (NC) show much higher activities in the brain compared to peripheral tissues [9]. An alkaline isoform was detected in the human cerebellum [11].

Several clinical studies reported an altered sphingomyelin and ceramide metabolism in MDD. As such, ASM activity was increased in peripheral blood mononuclear cells of patients experiencing a major depressive episode [12]. Moreover, secretory ASM activity was related to depression severity and predicted the improvement of depressive symptoms during therapy [13]. Such alterations in ASM activity might involve changes in alternative splicing of the gene coding for ASM, which differed between MDD patients and healthy controls [14,15]. Similarly, several plasma ceramide species were increased in patients experiencing a major depressive episode during the past two years compared with healthy controls and patients experiencing a major depressive episode for more than two years [16]. Higher plasma ceramide Cer16:0, Cer18:0, Cer20:0, Cer22:0, Cer24:0 and Cer24:1 levels were also observed in patients with MDD and bipolar disorder [17], and higher plasma levels of ceramide Cer16:0 and Cer18:0 were associated with higher severity of depression symptoms in patients with coronary artery disease [18]. In contrast, plasma sphingomyelin SM26:1 [19], SM39:1 and SM39:2 [20] levels were decreased in MDD patients, and the SM23:1/SM16:0 ratio was negatively correlated with the severity of depression symptoms in a Dutch family-based lipidomics study [21].

Similarly increased ASM activity resulting in decreased sphingomyelin and increased ceramide concentrations was described in rodent models of MDD. For example, exposure to chronic unpredictable stress, which was shown to induce a depressive-like and anxious phenotype [22,23], increased the levels of several ceramide species in the hippocampus and frontal cortex but not in the amygdala and cerebellum in mice [24]. In contrast, sphingomyelin concentration was reduced by chronic unpredictable stress [24], suggesting an altered sphingolipid metabolism. Chronic administration of corticosterone, another model known to induce a depressive-like and anxious phenotype [25–27], also increased ceramide concentrations in the dorsal and ventral hippocampus [28]. The direct involvement of ceramide in the pathogenesis of depression was demonstrated in naïve mice, which developed a depressive-like but not an anxious phenotype after infusion of Cer16:0 ceramide into the dorsal hippocampus, whereas infusion of Cer8:0 or Cer20:0 ceramide exhibited no effect [29,30]. Interestingly, infusion of Cer16:0 ceramide into the basolateral amygdala induced an anxious but not a depressive-like phenotype in mice [30], demonstrating the complex species- and brain-region-specific modulation of emotional behavior by ceramide. Significant associations between ASM activity and depressive-like behavior were also described in transgenic mice overexpressing ASM throughout the body (ASMtg). These ASMtg mice showed an increased serum and hippocampal ASM activity and an increased hippocampal ceramide concentration that was associated with a depressive-like and anxious phenotype [29,31,32]. Interestingly, female but not male ASMtg mice also showed a social anxious phenotype [31], suggesting an association between increased ASM activity and deficits in social behavior in females. Sex-specific and brain-regional effects of ASM activity on emotional behavior were also demonstrated in conditional transgenic mice in which the overexpression of ASM was restricted to the forebrain (ASMtg[fb]). In these ASMtg[fb] mice, males showed higher ASM activity in the frontal cortex, hippocampus, lateral septum and amygdala that resulted in a depressive-like phenotype, whereas females showed a higher ASM activity in the hypothalamus and a social anxious phenotype but not a depressive-like phenotype [33]. Despite this improved understanding of the effects of ASM and ceramide on MDD, little is known about the involvement of other sphingomyelinases and ceramidases in the pathophysiology of MDD or anxiety [34].

In this study, we characterized the brain activity of ASM and NSM as well as of AC and NC in an animal model of innate hyper-anxiety and MDD, namely, in Wistar rats selectively bred for extremely high (HAB) versus low (LAB) anxiety-like behavior, based on their performance on the elevated plus maze (EPM). This model is highly relevant for studying the sphingolipid metabolizing enzymes, as HAB rats show an anxious and depressive-like phenotype similar to ASMtg and ASMtg[fb] mice but not a social anxious phenotype [35,36]. On the other hand, LAB rats show a low level of anxiety and depressive-like behavior when compared with HAB rats and unselected Wistar rats [37]. Given that MDD is more prevalent in women [38] and the social anxious phenotype was observed only in female ASMtg [31] and ASMtg[fb] mice [33], experiments were performed in female HAB and LAB rats.

2. Materials and Methods

2.1. Animals

Female HAB and LAB rats were bred at the University of Regensburg. Rats were kept in colonies of 4 rats per cage under standard laboratory conditions (12:12 light-dark cycle, lights on at 06:00, 22 °C, 60% humidity, food and water ad libitum). Experiments were performed during the light phase in accordance to the Guide for the Care and Use of Laboratory Animals of the Government of Unterfranken (project identification code: 55.2-2532-2-384, approved on 06.04.2017) and the Guidelines of the NIH.

2.2. Experimental Design

At 9 weeks of age, the anxiety-like behavior of the rats was tested in the EPM. At 11 weeks of age, the naturally occurring social preference of the rats, as an indicator of social anxiety-like behavior, was tested in the social preference test (SPT). One week later, the depressive-like behavior of the rats was tested in the novelty-suppressed feeding paradigm (NSF). Twenty-four hours later, the rats were rapidly decapitated under CO_2 anesthesia and their brains were removed, snap frozen and stored at −80 °C until further analysis. Twelve brain regions (i.e., the frontal cortex, dorsal and ventral striatum, lateral septum, amygdala, dorsal and ventral hippocampus, thalamus, hypothalamus, dorsal and ventral mesencephalon and cerebellum) were dissected out of coronal brain slices based on previous studies [30,39]. The activities of ASM and NSM as well as of AC and NC were analyzed from one hemisphere, counterbalanced between rats.

Experiments were performed in two batches of rats. In the first batch (n = 14 HAB rats and n = 15 LAB rats), all behavioral tests were performed as described above. In the second batch (n = 8 HAB rats and n = 8 LAB rats), the anxiety-like behavior was tested on the EPM at 9 weeks of age, and brains were collected at a comparable time point to the first batch. The activities of ASM, NSM, AC and NC were analyzed from both batches combined.

2.3. Elevated Plus-Maze Test (EPM)

The anxiety-like behavior of the rats was tested in the EPM as previously described [36]. The EPM consisted of two closed arms (50 × 10 × 40 cm; 10 lx) and two open arms (50 × 10 cm; 40 lx) connected through a central neutral zone (10 × 10 cm) elevated 70 cm from the floor. Rats were placed into the neutral zone facing one closed arm, and the 5-min test was recorded. A decreased percentage of time spent on the open arms indicated increased anxiety-like behavior.

2.4. Social Preference Test (SPT)

The naturally occurring social preference of the rats, as an indicator of social anxiety-like behavior, and the preference for social novelty of the rats, as an indicator of social recognition, were tested in the SPT as previously described [31,40]. Rats were placed in a novel arena (80 × 40 × 40 cm) and allowed to habituate for 30 s. Two identical wire-mesh cages (20 × 9 × 9 cm) were simultaneously placed at opposite side-walls of the arena for 5 min. One cage remained empty, and one cage contained an age-matched unfamiliar female rat (same rat). The initial position of the same rat varied between experimental rats to prevent possible place preference. After 5 min, the empty cage was exchanged by an identical cage containing a novel female rat for an additional 5 min. Experiments were recorded, and the time spent investigating (sniffing) the empty cage, the same and the novel rat was analyzed using JWatcher (Version 1.0, Macquarie University and UCLA). A higher investigation time directed toward the same rat versus the empty cage during the first 5 min indicated social preference and thus a lack of social anxiety. A higher investigation time directed toward the novel versus the same rat during the second 5 min indicated social recognition and preference for social novelty.

2.5. Novelty-Suppressed Feeding Paradigm (NSF)

The depressive-like behavior of rats was tested in the NSF as previously described [30,31]. Rats were food-deprived for 24 h prior to testing with unlimited water access. Rats were placed in a novel arena ($80 \times 80 \times 40$ cm) with the head facing one of the corners. Immediately afterwards, a single food pellet (ssniff Spezialdiäten GmbH, Soest, Germany) was placed in the center of the arena. The latency to feed, defined as biting the food pellet for longer than 3 s, was manually analyzed. An increased feeding latency indicated depressive-like behavior. The test lasted maximally 20 min. HAB rats that did not feed within these 20 min (56% of rats) were not removed from the study but were allocated 1200 s as their feeding latency value.

2.6. Measurement of Sphingomyelinase and Ceramidase Activities

The activity of sphingolipid metabolizing enzymes was determined using the fluorescent substrate BODIPY-FL-C12-SM (N-(4,4-difluoro-5,7-dimethyl-4-bora-3a,4a-diaza-s-indacene-3-dodecanoyl) sphingosylphosphocholine, D-7711, Thermo Fisher Scientific, Waltham, MA, USA) for sphingomyelinases and NBD-C12-ceramide (Cayman, obtained from Biomol, Hamburg, Germany) for ceramidases, with four replicates for each sample based on a previously established method [41]. Tissues were homogenized in lysis reagent (250 mM sucrose, 1 mM EDTA, 0.2% Triton X-100, 1× Roche protease inhibitor cocktail; approximately 200 μL/10 mg tissue) using a Tissue Lyser LT bead mill (Qiagen) with steal beads followed by freezing at −80 °C to enhance lysis. Supernatants obtained after thawing, ultrasound treatment (water bath for 60 s) and centrifugation at $16,000 \times g$ at 4 °C for 10 min were diluted 1:4 in lysis reagent and used for activity assays and for protein determination (Bradford/Coomassie kit, Thermo Fisher Scientific, Waltham, MA, USA). A standard enzyme reaction in a 96-well polystyrene plate contained 58 pmol sphingomyelin or 50 pmol ceramide as a substrate in a total volume of 50 μL of reaction buffer of the following composition: 200 mM sodium acetate buffer (pH 5.0), 500 mM NaCl, 0.2% IGEPAL® CA-630 (NP 40) detergent for ASM, 200 mM HEPES buffer (pH 7.0), 200 mM MgCl$_2$, 0.05% IGEPAL® CA-630 (NP 40) for NSM; 200 mM sodium acetate buffer (pH 4.5), 100 mM NaCl, 0.03% IGEPAL® CA-630 (NP 40) for AC and 200 mM HEPES (pH 7.0), 100 mM NaCl, 0.03% IGEPAL® CA-630 (NP 40) for NC. The reaction was initiated by the addition of 2 μL of tissue lysate corresponding to 0.5–1 μg protein. After incubation at 37 °C for 1–18 h, depending on enzymatic activity, reactions were stopped by freezing at −20 °C and stored until further processing. To separate product and uncleaved substrate, 1.5 μL of each reaction were directly spotted on silica gel 60 thin layer chromatography plates (ALUGRAM SIL G, 818232, Macherey–Nagel, Düren, Germany) and separated using ethyl acetate with 1% (v/v) acetic acid as a solvent. Signals were detected on a Typhoon Trio scanner (488 nm excitation, 520 nm emission, 325–385 V, 100 μm resolution, GE Healthcare Life Sciences, Buckinghamshire, UK) and quantified with the ImageQuant software (GE Healthcare Life Sciences, Buckinghamshire, UK). Enzymatic activities were calculated as the hydrolysis rate of sphingomyelin or ceramide (pmol), respectively, per time (h) and per protein (μg).

2.7. Statistical Analysis

For statistical analysis, SPSS (Version 21, SPSS Inc., Chicago, IL, USA) was used. Data were analyzed using the Student t-test and two-way ANOVA for repeated measures, followed by a Bonferroni post-hoc analysis whenever appropriate. Spearman correlations were calculated to evaluate associations between behavior and enzyme activities within groups. Statistical significance was set at

$p < 0.05$. For each parameter, outliers deviating more than two standard deviations from the mean were excluded from analysis. Graphs were prepared using GraphPad Prism 7.00 (GraphPad Software Inc., San Diego, CA, USA).

3. Results

3.1. Behavioral Phenotype of HAB and LAB Females

The expected highly anxious and depressive-like phenotype of HAB females compared with LAB females was visible in both the EPM (T(40) = −40.02; $p < 0.001$; Figure 1a) and NSF (T(26) = 5.63; $p < 0.001$; Figure 1b), respectively. In the SPT, all rats displayed normal social preference and a lack of social anxiety (stimulus effect F(1,52) = 246.64; $p < 0.001$; group × stimulus effect F(1,52) = 1.90; $p = 0.174$; Figure 1c). Furthermore, all rats displayed normal social recognition and preference for social novelty (stimulus effect F(1,52) = 44.38; $p < 0.001$; group × stimulus effect F(1,52) = 0.43; $p = 0.514$; Figure 1d).

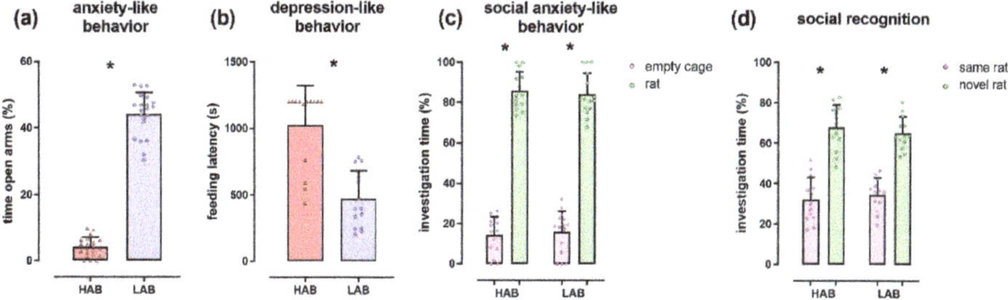

Figure 1. The behavioral phenotype of female HAB and LAB rats. (**a**) Percentage of time spent on the open arms of the elevated plus-maze, as an indicator of anxiety-like behavior; (**b**) Feeding latency shown in the novelty-suppressed feeding paradigm, as an indicator of depressive-like behavior; (**c,d**) Percentage of investigation time of the empty cage, the same and the novel female rat shown during the first (**c**) and second (**d**) 5 min of the social preference test, as indicators of social anxiety-like behavior and social recognition, respectively. Female HAB rats showed an anxious and depressive-like phenotype compared with LAB rats but unaltered social anxiety and social recognition. Data represent individual data points with means ± SEM. * $p < 0.05$.

3.2. Enzyme Activities in Selected Brain Regions of HAB and LAB Females

HAB females showed an increased ASM activity in the lateral septum (+40%, T(40) = 2.52; $p = 0.016$), hypothalamus (+15%, T(42) = 3.40; $p = 0.001$), ventral hippocampus (+10%, T(41) = 2.31; $p = 0.026$) and ventral mesencephalon (+11%, T(41) = 2.42; $p = 0.020$) compared with LAB females, whereas no differences in ASM activity were found in the frontal cortex, amygdala, dorsal hippocampus, dorsal and ventral striatum, dorsal mesencephalon, thalamus and cerebellum (Figure 2a). HAB females showed an increased NSM activity in the ventral mesencephalon (+10%, T(39) = 2.08; $p = 0.044$) but no significant differences in the other brain regions (Figure 2b). HAB females also showed an increased AC activity in the dorsal (+24%, T(43) = 3.69; $p < 0.001$) and ventral (+20%, T(42) = 2.98; $p = 0.005$) striatum, hypothalamus (+25%, T(41) = 3.50; $p = 0.001$), thalamus (+27%, T(43) = 2.18; $p = 0.035$) and ventral mesencephalon (+21%, T(41) = 2.10; $p = 0.042$) (Figure 2c) and an increased NC activity in the hypothalamus (+42%, T(41) = 2.84; $p = 0.007$) and dorsal mesencephalon (+13%, T(41) = 2.37; $p = 0.022$) compared with LAB females (Figure 2d). The only enzymatic activity that was decreased in HAB females compared with LAB females was the NC activity in the amygdala (−62%, T(42) = −2.47; $p = 0.019$) (Figure 2d).

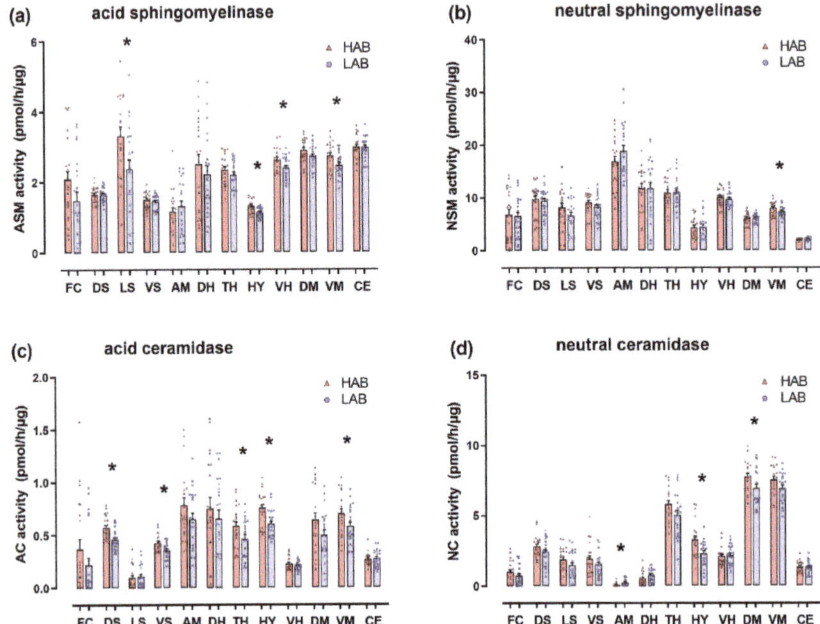

Figure 2. The activity of sphingolipid metabolizing enzymes in brains of female HAB and LAB rats. The activity of acid and neutral sphingomyelinases (ASM in (**a**), NSM in (**b**)) as well as of acid and neutral ceramidases (AC in (**c**), NC in (**d**)) was analyzed in the frontal cortex (FC), dorsal striatum (DS), lateral septum (LS), ventral striatum (VS), amygdala (AM), dorsal hippocampus (DH), thalamus (TH), hypothalamus (HY), ventral hippocampus (VH), dorsal mesencephalon (DM), ventral mesencephalon (VM) and cerebellum (CE). HAB rats showed a significant increase in the activity of these enzymes compared with LAB rats. Data represent individual data points with means ± SEM. * $p < 0.05$.

3.3. Correlations between Behavior and Enzyme Activities in HAB and LAB Females

In addition to the observed group differences, enzyme activities were also related to depression- and anxiety-like behavior within the groups. The anxiety-like behavior expressed as percentage time spent on the open arms of the EPM negatively correlated with ASM activity in the amygdala ($r = -0.48$; $p = 0.031$), ventral hippocampus ($r = -0.57$; $p = 0.008$) and dorsal mesencephalon ($r = -0.60$; $p = 0.006$) and with NSM activity in the dorsal mesencephalon ($r = -0.64$; $p = 0.002$) in HAB females but with none of the parameters in LAB females.

Depressive-like behavior expressed as latency to feed after a 24-h food deprivation period in the NSF negatively correlated with AC activity in the frontal cortex in LAB females ($r = -0.62$; $p = 0.015$) but with none of the parameters in HAB females. This lower number of correlations between enzyme activities and depressive- versus anxiety-like behavior might be due to the lower number of rats in which depressive-like behavior was assessed ($n = 29$ versus $n = 45$).

4. Discussion

Our study describes for the first time the activity of sphingomyelinases and ceramidases in the brain of rats with high and low levels of innate anxiety- and depressive-like behavior. Concomitant with their highly anxious and depressive-like phenotype, HAB rats showed increased activity of ASM and NSM as well as of AC and NC in multiple brain regions associated with anxiety- and depressive-like behavior, including the lateral septum, hypothalamus, ventral hippocampus and ventral and dorsal mesencephalon.

These findings extend previous reports of increased brain ASM activity in transgenic models of MDD, namely, in ASMtg [29] and ASMtg[fb] mice [33]. As such, increased ASM activity was previously described in the dorsal and ventral hippocampus in ASMtg mice overexpressing ASM in the whole body [29] (no other brain regions were tested in this study), whereas in ASMtg[fb] mice overexpressing ASM only in the forebrain [33], ASM activity was increased in multiple brain regions. As such, both male and female ASMtg[fb] mice showed increased ASM activity in the dorsal striatum, dorsal and ventral hippocampus and amygdala compared with wild-type controls. Furthermore, male ASMtg[fb] mice showed higher ASM activity compared with female ASMtg[fb] mice in all these regions except in the dorsal striatum where the ASM activity was similarly high in both male and female ASMtg[fb] mice. In the hypothalamus, the activity of ASM was increased in female but not in male ASMtg[fb] mice, whereas in the frontal cortex, lateral septum and ventral mesencephalon, the ASM activity was increased in male but not in female ASMtg[fb] mice. Despite this increased ASM activity, female ASMtg[fb] mice did not show an anxious and depressive-like phenotype [33]. In contrast, the behavior of the male ASMtg[fb] mice and female HAB rats clearly changed, displaying a significant anxious and depressive-like phenotype in the EPM and NSF, respectively. By directly comparing these two animal models of MDD with similar behavioral deficits, an important role of the lateral septum and ventral mesencephalon in anxiety- and depressive-like behavior might be suggested, as female HAB rats and male ASMtg[fb] mice but not female ASMtg[fb] mice showed an increased ASM activity within these brain areas. Given that the lateral septum plays a critical role in regulating processes related to mood and motivation [42,43] and is neuroanatomically connected with several brain regions known to regulate emotional behavior (e.g., the hippocampus, amygdala, hypothalamus and the mesolimbic system [44]), it is feasible that ASM within the lateral septum and ventral mesencephalon regulates anxiety- and depressive-like behavior. The ASM within the amygdala, ventral hippocampus and dorsal mesencephalon might also regulate anxiety- and depressive-like behavior given that a negative correlation between anxiety levels and ASM activity was found in these brain regions in female HAB rats, i.e., rats that spent less time on the open arms of the EPM and were thereby more anxious also showed higher levels of ASM activity in the amygdala, ventral hippocampus and dorsal mesencephalon. However, the increased ASM activity within the amygdala and ventral hippocampus in female ASMtg[fb] mice did not result in an anxious and depressive-like phenotype [33], suggesting a differential contribution of the amygdalar and ventral hippocampal ASM to anxiety- and depressive-like behavior in these two animal models of MDD. An indirect involvement of amygdalar ASM in anxiety-like behavior has been already suggested in a previous study that showed that infusion of Cer16:0 ceramide into the basolateral amygdala induced an anxious phenotype in mice [30]. As ceramide is generated by the activity of ASM, a higher ASM activity might be expected to increase ceramide levels. It is also possible that a very high level of ASM activity within the amygdala and ventral hippocampus, as seen in male but not in female ASMtg[fb] mice [33], might be necessary to elicit changes in anxiety- and depressive-like behavior in this transgenic mouse model.

We also aimed to identify the brain region/s contributing to ASM effects on social anxiety by comparing the alterations in ASM activity between female ASMtg[fb] mice with social anxiety and female HAB rats without social anxiety. However, for now we cannot suggest which brain region mediates these effects. Based on the study performed in ASMtg[fb] mice, the hypothalamus was a promising brain region given that the ASM activity was increased in female but not in male ASMtg[fb] mice and that only female ASMtg[fb] mice showed a social anxious phenotype [33]. Although the hypothalamus is highly relevant for several types of social behavior, including social anxiety [45], it is unlikely that the increased ASM activity within the hypothalamus contributes to social anxiety in general, as female HAB rats were not socially anxious despite the increased ASM activity in the hypothalamus. In female HAB rats, the increased ASM activity in the hypothalamus seems to relate more to their depressive-like behavior, given that a tendency towards a positive correlation between depressive-like behavior and ASM activity was found ($r = 0.54$; $p = 0.056$). As such, rats that showed higher feeding latencies in the NSF and thereby a more severe depressive-like behavior tended to show higher levels of ASM activity in the hypothalamus. A more detailed analysis of ASM activity within distinct hypothalamic nuclei might reveal whether specific nuclei mediate the effects of ASM on social anxiety. Alternatively, a high level of ASM activity in other brain regions such as the thalamus and/or ventral striatum might be necessary to induce social anxiety, given that several studies suggested an important role of the thalamus [46,47] and the ventral striatum [48,49] in the pathophysiology of social anxiety and that ASM activity was unaltered in these brain regions in female HAB rats.

Although previous studies investigated the activity of ASM in depressive patients [12] and in animal models of MDD [29,33], little is known about the contribution of other enzymes of the sphingolipid metabolism such as NSM, AC and NC in MDD. We showed that the activity of NSM was also increased in the ventral mesencephalon in female HAB rats, similar to ASM, suggesting that ASM and NSM might regulate anxiety- and depressive-like behavior together within this brain region. Similar to ASM, NSM within the dorsal mesencephalon might also regulate anxiety-like behavior in female HAB rats given that a negative correlation between anxiety levels and NSM activity was observed, i.e., rats that spent less time on the open arms of the EPM and were thereby more anxious also showed higher levels of NSM activity in the dorsal mesencephalon.

As ceramide is generated by the activity of ASM and NSM, an increase in ASM and NSM activity levels is expected to increase ceramide levels and therefore also increase the activity of the enzymes metabolizing ceramide, i.e., AC and NC. In agreement with this hypothesis, we found an increased AC activity in the dorsal and ventral striatum, thalamus, hypothalamus and ventral mesencephalon, whereas the activity of NC was increased in the hypothalamus and dorsal mesencephalon. However, whether this relatively small but significantly increased activity of ASM and NSM is physiologically relevant and leads to increased ceramide levels remains to be verified. Mass spectrometry of brain-specific tissue samples could reveal such alterations in sphingolipid and ceramide species of different chain length. Moreover, new label-free imaging mass spectrometry techniques could visualize the distribution and provide quantitative data on various subtypes of lipids in these tissue sections [32]. In addition, future studies could include mass spectrometry as well as enzyme activity assays of cerebrospinal fluid samples with detectable levels of both ASM [50] and NSM [51]. Further enzymes involved in regulating ceramide levels and known to be affected in neuropsychiatric disorders (e.g., alkaline ceramidase and sphingomyelin synthase [52]) might also play a role and could be altered in specific brain regions, counterbalancing or increasing the effects of the analyzed sphingomyelinases and ceramidases.

Taken together, our results add a novel piece of information to the complex regulation of sphingolipid metabolism, especially of the sphingomyelinases and ceramidases, in MDD and demonstrate that all investigated enzymes (i.e., ASM, NSM, AC and NC) show predominantly an increased activity in multiple brain regions associated with anxiety- and depressive-like behavior.

Author Contributions: Conceptualization, I.Z., I.D.N. and J.K.; Data curation, I.Z. and C.M.; Formal analysis, I.Z. and C.M.; Funding acquisition, E.G., I.D.N. and J.K.; Investigation, C.M. and A.K.S.; Methodology, I.Z. and C.M.; Project administration, I.Z. and C.M.; Resources, I.D.N. and J.K.; Supervision, I.Z.; Validation, I.Z. and C.M.; Visualization, C.M.; Writing—original draft, I.Z.; Writing—review and editing, I.Z., C.M., E.G., I.D.N. and J.K. All authors have read and agreed to the published version of the manuscript.

Funding: This research was founded by the German Federal Ministry of Education and Research (BMBF), Research Grants No. 01EE1401A to I.N. and 01EE1401C to J.K and by the Deutsche Forschungsgemeinschaft (DFG) grants KO 947/13-3 to J.K. and GU 335/29-3 to E.G. The work was also supported by intramural grants from the Universitätsklinikum of the Friedrich-Alexander University Erlangen-Nürnberg (FAU). C.M. is an associated fellow of the research training group 2162 "Neurodevelopment and Vulnerability of the Central Nervous System" funded by the DFG—270949263/GRK2162. In addition, we acknowledge support by the DFG and the FAU within the funding program Open Access Publishing.

Acknowledgments: We thank Sabine E. Huber, Sina Kirsten and Andrea Leicht (Friedrich-Alexander University Erlangen-Nürnberg) and Andrea Havasi (University of Regensburg) for excellent technical support.

Conflicts of Interest: The authors declare no conflict of interest. The funders had no role in the design of the study; in the collection, analyses, or interpretation of data; in the writing of the manuscript or in the decision to publish the results.

References

1. World Health Organization. *Depression and Other Common Mental Disorders: Global Health Estimates*; World Health Organization: Geneva, Switzerland, 2017.
2. Lenz, B.; Rother, M.; Bouna-Pyrrou, P.; Mühle, C.; Tektas, O.Y.; Kornhuber, J. The androgen model of suicide completion. *Prog. Neurobiol.* **2019**, *172*, 84–103. [CrossRef] [PubMed]
3. Müller, C.P.; Reichel, M.; Mühle, C.; Rhein, C.; Gulbins, E.; Kornhuber, J. Brain membrane lipids in major depression and anxiety disorders. *Biochim. Biophys. Acta* **2015**, *1851*, 1052–1065. [CrossRef] [PubMed]
4. Schneider, P.B.; Kennedy, E.P. Sphingomyelinase in normal human spleens and in spleens from subjects with Niemann-Pick disease. *J. Lipid Res.* **1967**, *8*, 202–209. [PubMed]
5. Goni, F.M.; Alonso, A. Sphingomyelinases: Enzymology and membrane activity. *FEBS Lett.* **2002**, *531*, 38–46. [CrossRef]
6. Weinreb, N.J.; Brady, R.O.; Tappel, A.L. The lysosomal localization of sphingolipid hydrolases. *Biochim. Biophys. Acta* **1968**, *159*, 141–146. [CrossRef]
7. Kornhuber, J.; Rhein, C.; Müller, C.P.; Mühle, C. Secretory sphingomyelinase in health and disease. *Biol. Chem.* **2015**, *396*, 707–736. [CrossRef]
8. Rao, B.G.; Spence, M.W. Sphingomyelinase activity at pH 7.4 in human brain and a comparison to activity at pH 5.0. *J. Lipid Res.* **1976**, *17*, 506–515.
9. Mühle, C.; Reichel, M.; Gulbins, E.; Kornhuber, J. Sphingolipids in psychiatric disorders and pain syndromes. *Handb. Exp. Pharm.* **2013**, 431–456. [CrossRef]
10. Duan, R.D.; Hertervig, E.; Nyberg, L.; Hauge, T.; Sternby, B.; Lillienau, J.; Farooqi, A.; Nilsson, A. Distribution of alkaline sphingomyelinase activity in human beings and animals. Tissue and species differences. *Dig. Dis. Sci.* **1996**, *41*, 1801–1806. [CrossRef]
11. Sugita, M.; Willians, M.; Dulaney, J.T.; Moser, H.W. Ceramidase and ceramide synthesis in human kidney and cerebellum. Description of a new alkaline ceramidase. *Biochim. Biophys. Acta* **1975**, *398*, 125–131. [CrossRef]
12. Kornhuber, J.; Medlin, A.; Bleich, S.; Jendrossek, V.; Henkel, A.W.; Wiltfang, J.; Gulbins, E. High activity of acid sphingomyelinase in major depression. *J. Neural Transm. (Vienna)* **2005**, *112*, 1583–1590. [CrossRef] [PubMed]
13. Mühle, C.; Wagner, C.J.; Färber, K.; Richter-Schmidinger, T.; Gulbins, E.; Lenz, B.; Kornhuber, J. Secretory acid sphingomyelinase in the serum of medicated patients predicts the prospective course of depression. *J. Clin. Med.* **2019**, *8*, 846. [CrossRef] [PubMed]
14. Rhein, C.; Reichel, M.; Kramer, M.; Rotter, A.; Lenz, B.; Mühle, C.; Gulbins, E.; Kornhuber, J. Alternative splicing of *SMPD1* coding for acid sphingomyelinase in major depression. *J. Affect. Disord.* **2017**, *209*, 10–15. [CrossRef] [PubMed]
15. Rhein, C.; Tripal, P.; Seebahn, A.; Konrad, A.; Kramer, M.; Nagel, C.; Kemper, J.; Bode, J.; Mühle, C.; Gulbins, E.; et al. Functional implications of novel human acid sphingomyelinase splice variants. *PLoS ONE* **2012**, *7*, e35467. [CrossRef] [PubMed]
16. Gracia-Garcia, P.; Rao, V.; Haughey, N.J.; Bandaru, V.V.; Smith, G.; Rosenberg, P.B.; Lobo, A.; Lyketsos, C.G.; Mielke, M.M. Elevated plasma ceramides in depression. *J. Neuropsychiatry Clin. Neurosci.* **2011**, *23*, 215–218. [CrossRef] [PubMed]
17. Brunkhorst-Kanaan, N.; Klatt-Schreiner, K.; Hackel, J.; Schroter, K.; Trautmann, S.; Hahnefeld, L.; Wicker, S.; Reif, A.; Thomas, D.; Geisslinger, G.; et al. Targeted lipidomics reveal derangement of ceramides in major depression and bipolar disorder. *Metabolism* **2019**, *95*, 65–76. [CrossRef] [PubMed]
18. Dinoff, A.; Saleem, M.; Herrmann, N.; Mielke, M.M.; Oh, P.I.; Venkata, S.L.V.; Haughey, N.J.; Lanctot, K.L. Plasma sphingolipids and depressive symptoms in coronary artery disease. *Brain Behav.* **2017**, *7*, e00836. [CrossRef]
19. Moaddel, R.; Shardell, M.; Khadeer, M.; Lovett, J.; Kadriu, B.; Ravichandran, S.; Morris, P.J.; Yuan, P.; Thomas, C.J.; Gould, T.D.; et al. Plasma metabolomic profiling of a ketamine and placebo crossover trial of major depressive disorder and healthy control subjects. *Psychopharmacol (Berlin)* **2018**, *235*, 3017–3030. [CrossRef]
20. Liu, X.; Li, J.; Zheng, P.; Zhao, X.; Zhou, C.; Hu, C.; Hou, X.; Wang, H.; Xie, P.; Xu, G. Plasma lipidomics reveals potential lipid markers of major depressive disorder. *Anal. Bioanal. Chem.* **2016**, *408*, 6497–6507. [CrossRef]

21. Demirkan, A.; Isaacs, A.; Ugocsai, P.; Liebisch, G.; Struchalin, M.; Rudan, I.; Wilson, J.F.; Pramstaller, P.P.; Gyllensten, U.; Campbell, H.; et al. Plasma phosphatidylcholine and sphingomyelin concentrations are associated with depression and anxiety symptoms in a Dutch family-based lipidomics study. *J. Psychiatr. Res.* **2013**, *47*, 357–362. [CrossRef]

22. Willner, P. The chronic mild stress (CMS) model of depression: History, evaluation and usage. *Neurobiol. Stress* **2017**, *6*, 78–93. [CrossRef]

23. Zhou, X.D.; Shi, D.D.; Zhang, Z.J. Antidepressant and anxiolytic effects of the proprietary Chinese medicine Shexiang Baoxin pill in mice with chronic unpredictable mild stress. *J. Food Drug Anal.* **2019**, *27*, 221–230. [CrossRef] [PubMed]

24. Oliveira, T.G.; Chan, R.B.; Bravo, F.V.; Miranda, A.; Silva, R.R.; Zhou, B.; Marques, F.; Pinto, V.; Cerqueira, J.J.; Di Paolo, G.; et al. The impact of chronic stress on the rat brain lipidome. *Mol. Psychiatry* **2016**, *21*, 80–88. [CrossRef] [PubMed]

25. Gregus, A.; Wintink, A.J.; Davis, A.C.; Kalynchuk, L.E. Effect of repeated corticosterone injections and restraint stress on anxiety and depression-like behavior in male rats. *Behav. Brain Res.* **2005**, *156*, 105–114. [CrossRef] [PubMed]

26. Murray, F.; Smith, D.W.; Hutson, P.H. Chronic low dose corticosterone exposure decreased hippocampal cell proliferation, volume and induced anxiety and depression like behaviours in mice. *Eur. J. Pharm.* **2008**, *583*, 115–127. [CrossRef] [PubMed]

27. David, D.J.; Samuels, B.A.; Rainer, Q.; Wang, J.W.; Marsteller, D.; Mendez, I.; Drew, M.; Craig, D.A.; Guiard, B.P.; Guilloux, J.P.; et al. Neurogenesis-dependent and -independent effects of fluoxetine in an animal model of anxiety/depression. *Neuron* **2009**, *62*, 479–493. [CrossRef]

28. Miranda, A.M.; Bravo, F.V.; Chan, R.B.; Sousa, N.; Di Paolo, G.; Oliveira, T.G. Differential lipid composition and regulation along the hippocampal longitudinal axis. *Transl. Psychiatry* **2019**, *9*, 144. [CrossRef] [PubMed]

29. Gulbins, E.; Palmada, M.; Reichel, M.; Luth, A.; Bohmer, C.; Amato, D.; Muller, C.P.; Tischbirek, C.H.; Groemer, T.W.; Tabatabai, G.; et al. Acid sphingomyelinase-ceramide system mediates effects of antidepressant drugs. *Nat. Med.* **2013**, *19*, 934–938. [CrossRef]

30. Zoicas, I.; Huber, S.E.; Kalinichenko, L.S.; Gulbins, E.; Müller, C.P.; Kornhuber, J. Ceramides affect alcohol consumption and depressive-like and anxiety-like behavior in a brain region- and ceramide species-specific way in male mice. *Addict. Biol.* **2019**. [CrossRef]

31. Zoicas, I.; Reichel, M.; Gulbins, E.; Kornhuber, J. Role of acid sphingomyelinase in the regulation of social behavior and memory. *PLoS ONE* **2016**, *11*, e0162498. [CrossRef]

32. Müller, C.P.; Kalinichenko, L.S.; Tiesel, J.; Witt, M.; Stockl, T.; Sprenger, E.; Fuchser, J.; Beckmann, J.; Praetner, M.; Huber, S.E.; et al. Paradoxical antidepressant effects of alcohol are related to acid sphingomyelinase and its control of sphingolipid homeostasis. *Acta Neuropathol.* **2017**, *133*, 463–483. [CrossRef] [PubMed]

33. Zoicas, I.; Schumacher, F.; Kleuser, B.; Reichel, M.; Gulbins, E.; Fejtova, A.; Kornhuber, J.; Rhein, C. The forebrain—Specific overexpression of acid sphingomyelinase induces depressive-like symptoms in mice. *Cells* **2020**, accepted.

34. Kalinichenko, L.S.; Mühle, C.; Eulenburg, V.; Praetner, M.; Reichel, M.; Gulbins, E.; Kornhuber, J.; Müller, C.P. Enhanced alcohol preference and anxiolytic alcohol effects in Niemann-Pick Disease model in mice. *Front. Neurol.* **2019**, *10*, 731. [CrossRef] [PubMed]

35. Liebsch, G.; Montkowski, A.; Holsboer, F.; Landgraf, R. Behavioural profiles of two Wistar rat lines selectively bred for high or low anxiety-related behaviour. *Behav. Brain Res.* **1998**, *94*, 301–310. [CrossRef]

36. Schmidtner, A.K.; Slattery, D.A.; Glasner, J.; Hiergeist, A.; Gryksa, K.; Malik, V.A.; Hellmann-Regen, J.; Heuser, I.; Baghai, T.C.; Gessner, A.; et al. Minocycline alters behavior, microglia and the gut microbiome in a trait-anxiety-dependent manner. *Transl. Psychiatry* **2019**, *9*, 223. [CrossRef]

37. Wegener, G.; Mathe, A.A.; Neumann, I.D. Selectively bred rodents as models of depression and anxiety. *Curr. Top. Behav. Neurosci.* **2012**, *12*, 139–187. [CrossRef]

38. Brivio, E.; Lopez, J.P.; Chen, A. Sex differences: Transcriptional signatures of stress exposure in male and female brains. *Genes Brain Behav.* **2020**, *19*, e12643. [CrossRef]

39. Huber, S.E.; Zoicas, I.; Reichel, M.; Mühle, C.; Buttner, C.; Ekici, A.B.; Eulenburg, V.; Lenz, B.; Kornhuber, J.; Müller, C.P. Prenatal androgen receptor activation determines adult alcohol and water drinking in a sex-specific way. *Addict. Biol.* **2018**, *23*, 904–920. [CrossRef]

40. Lukas, M.; Toth, I.; Reber, S.O.; Slattery, D.A.; Veenema, A.H.; Neumann, I.D. The neuropeptide oxytocin facilitates pro-social behavior and prevents social avoidance in rats and mice. *Neuropsychopharmacology* **2011**, *36*, 2159–2168. [CrossRef]

41. Mühle, C.; Kornhuber, J. Assay to measure sphingomyelinase and ceramidase activities efficiently and safely. *J. Chromatogr. A* **2017**, *1481*, 137–144. [CrossRef]

42. Sheehan, T.P.; Neve, R.L.; Duman, R.S.; Russell, D.S. Antidepressant effect of the calcium-activated tyrosine kinase Pyk2 in the lateral septum. *Biol. Psychiatry* **2003**, *54*, 540–551. [CrossRef]

43. Muigg, P.; Hoelzl, U.; Palfrader, K.; Neumann, I.; Wigger, A.; Landgraf, R.; Singewald, N. Altered brain activation pattern associated with drug-induced attenuation of enhanced depression-like behavior in rats bred for high anxiety. *Biol. Psychiatry* **2007**, *61*, 782–796. [CrossRef] [PubMed]

44. Sheehan, T.P.; Chambers, R.A.; Russell, D.S. Regulation of affect by the lateral septum: Implications for neuropsychiatry. *Brain Res. Brain Res. Rev.* **2004**, *46*, 71–117. [CrossRef] [PubMed]

45. Choleris, E.; Devidze, N.; Kavaliers, M.; Pfaff, D.W. Steroidal/neuropeptide interactions in hypothalamus and amygdala related to social anxiety. *Prog. Brain Res.* **2008**, *170*, 291–303. [CrossRef] [PubMed]

46. Vertes, R.P.; Linley, S.B.; Hoover, W.B. Limbic circuitry of the midline thalamus. *Neurosci. Biobehav. Rev.* **2015**, *54*, 89–107. [CrossRef]

47. Wang, X.; Cheng, B.; Luo, Q.; Qiu, L.; Wang, S. Gray matter structural alterations in social anxiety disorder: A voxel-based meta-analysis. *Front. Psychiatry* **2018**, *9*, 449. [CrossRef]

48. Becker, M.P.I.; Simon, D.; Miltner, W.H.R.; Straube, T. Altered activation of the ventral striatum under performance-related observation in social anxiety disorder. *Psychol. Med.* **2017**, *47*, 2502–2512. [CrossRef]

49. Gonzalez, M.Z.; Puglia, M.H.; Morris, J.P.; Connelly, J.J. Oxytocin receptor genotype and low economic privilege reverses ventral striatum-social anxiety association. *Soc. Neurosci.* **2019**, *14*, 67–79. [CrossRef]

50. Mühle, C.; Huttner, H.B.; Walter, S.; Reichel, M.; Canneva, F.; Lewczuk, P.; Gulbins, E.; Kornhuber, J. Characterization of acid sphingomyelinase activity in human cerebrospinal fluid. *PLoS ONE* **2013**, *8*, e62912. [CrossRef]

51. Sarrafpour, S.; Ormseth, C.; Chiang, A.; Arakaki, X.; Harrington, M.; Fonteh, A. Lipid metabolism in late-onset Alzheimer's disease differs from patients presenting with other dementia phenotypes. *Int. J. Environ. Res. Public Health* **2019**, *16*, 1995. [CrossRef]

52. Mühle, C.; Bilbao Canalejas, R.D.; Kornhuber, J. Sphingomyelin synthases in neuropsychiatric health and disease. *Neurosignals* **2019**, *27*, 54–76. [CrossRef] [PubMed]

Article

The Forebrain-Specific Overexpression of Acid Sphingomyelinase Induces Depressive-Like Symptoms in Mice

Iulia Zoicas [1], Fabian Schumacher [2,3], Burkhard Kleuser [2], Martin Reichel [1,†], Erich Gulbins [3], Anna Fejtova [1], Johannes Kornhuber [1,‡] and Cosima Rhein [1,4,*,‡]

[1] Department of Psychiatry and Psychotherapy, Friedrich-Alexander Universität Erlangen-Nürnberg, 91054 Erlangen, Germany; iulia.zoicas@uk-erlangen.de (I.Z.); martin.reichel@charite.de (M.R.); anna.fejtova@uk-erlangen.de (A.F.); johannes.kornhuber@uk-erlangen.de (J.K.)

[2] Department of Toxicology, University of Potsdam, 14558 Nuthetal, Germany; fabian.schumacher@uni-potsdam.de (F.S.); kleuser@uni-potsdam.de (B.K.)

[3] Department of Molecular Biology, University of Duisburg-Essen, 45147 Essen, Germany; erich.gulbins@uni-due.de

[4] Department of Psychosomatic Medicine and Psychotherapy, Friedrich-Alexander Universität Erlangen-Nürnberg, 91054 Erlangen, Germany

* Correspondence: Cosima.Rhein@uk-erlangen.de; Tel.: +49-9131-85-44542

† Current Address: Department of Nephrology and Medical Intensive Care, Charité Universitätsmedizin Berlin, 10115 Berlin, Germany.

‡ These authors contributed equally to this work.

Received: 29 April 2020; Accepted: 15 May 2020; Published: 18 May 2020

Abstract: Human and murine studies identified the lysosomal enzyme acid sphingomyelinase (ASM) as a target for antidepressant therapy and revealed its role in the pathophysiology of major depression. In this study, we generated a mouse model with overexpression of Asm (Asm-tg^{fb}) that is restricted to the forebrain to rule out any systemic effects of Asm overexpression on depressive-like symptoms. The increase in Asm activity was higher in male Asm-tg^{fb} mice than in female Asm-tg^{fb} mice due to the breeding strategy, which allows for the generation of wild-type littermates as appropriate controls. Asm overexpression in the forebrain of male mice resulted in a depressive-like phenotype, whereas in female mice, Asm overexpression resulted in a social anxiogenic-like phenotype. Ceramides in male Asm-tg^{fb} mice were elevated specifically in the dorsal hippocampus. mRNA expression analyses indicated that the increase in Asm activity affected other ceramide-generating pathways, which might help to balance ceramide levels in cortical brain regions. This forebrain-specific mouse model offers a novel tool for dissecting the molecular mechanisms that play a role in the pathophysiology of major depression.

Keywords: *Smpd1*; acid sphingomyelinase; forebrain; depressive-like behavior; anxiety-like behavior; ceramide

---

1. Introduction

Major depressive disorder (MDD) is a severe and chronic mood disorder with a lifetime prevalence of more than 10% [1]. Key symptoms of MDD are a depressed mood and loss of interest, anhedonia, feelings of worthlessness, weight loss, and insomnia. Although MDD is a very common disorder, its pathogenesis is still unclear. The acid sphingomyelinase (ASM)/ceramide system was recently implicated in the pathogenesis of MDD [2]. ASM (human; murine: Asm) is a lysosomal glycoprotein that catalyzes the hydrolysis of sphingomyelin into ceramide and phosphorylcholine [3]. Ceramide is generated by the hydrolysis of sphingomyelin through the activity of ASM, neutral sphingomyelinase

(NSM), or alkaline sphingomyelinase depending on the optimum pH of the enzyme [4]. Ceramide can also be generated by de novo synthesis [5], by the degradation of complex (gluco)sphingolipids [6] or through a salvage pathway involving reacylation of the degradation product sphingosine [7]. Several studies have reported altered sphingomyelin and ceramide metabolism in MDD, which increased ASM activity in peripheral blood mononuclear cells of patients experiencing a major depressive episode [8]. Similarly, plasma levels of several ceramide species, including Cer16:0, Cer18:0, Cer20:0, Cer24:1, and Cer26:1 but not Cer22:0 or Cer24:0, were increased in patients experiencing a major depressive episode during the past 2 years [9]. Higher plasma ceramide Cer16:0, Cer18:0, Cer20:0, Cer22:0, Cer24:0, and Cer24:1 levels were also observed in patients with MDD and bipolar disorder [10], and higher plasma levels of ceramide Cer16:0 and Cer18:0 and sphingomyelin SM18:1 were associated with the increased severity of depression symptoms in patients with coronary artery disease [11]. In contrast, plasma sphingomyelin SM26:1 [12], SM21:0 and SM21:1 [13] levels were decreased in MDD patients, and the SM23:1/SM16:0 ratio was negatively correlated with the severity of depressive symptoms in a Dutch family [14].

Similar deregulation of sphingolipid metabolism was found in rodent models of depression. For example, transgenic mice overexpressing Asm (Asm-tg) throughout the body showed an increased serum and hippocampal Asm activity and an increased hippocampal ceramide concentration, which was associated with a depressive- and anxiogenic-like phenotype in both social and nonsocial contexts [2,15,16]. Exposure to chronic unpredictable stress, a model that induces a depressive-like and anxiogenic-like phenotype [17], increased the levels of ceramide Cer16:0, Cer16:1, Cer18:1, Cer22:1, and Cer26:1 but not Cer18:0, Cer20:0, Cer20:1, Cer22:0, Cer24:0, Cer24:1, and Cer26:0 in the hippocampus and frontal cortex but not in the amygdala or cerebellum in mice. In contrast, the levels of sphingomyelin SM16:0, SM20:0, SM22:0, SM24:0, and SM26:0 but not SM18:0, SM18:1, SM24:1, and SM26:1 were reduced by chronic unpredictable stress [18]. Chronic administration of corticosterone, which is known to induce a depressive-like and anxiogenic-like phenotype [19], also increased ceramide Cer22:1 levels in the dorsal hippocampus and ceramide Cer20:0, Cer22:1, Cer24:1, Cer26:0 and Cer26:1 levels in the ventral hippocampus. Sphingomyelin SM16:0, SM18:0, SM18:1, SM20:0, SM22:0, SM24:0, SM24:1, SM26:0, and SM26:1 and ceramide Cer16:0, Cer16:1, Cer18:0, Cer18:1, Cer20:1, Cer22:0, and Cer24:0, however, were not altered by chronic corticosterone administration [20], suggesting that specific stressors might alter sphingolipid metabolism in a different way. The direct involvement of ceramide in the pathogenesis of depression was demonstrated in naïve mice, which developed a depressive-like phenotype after infusion of ceramide Cer16 but not Cer8 or Cer20 into the dorsal hippocampus [2,21]. Interestingly, Cer16 induced a predominantly nonsocial anxiogenic-like phenotype when infused into the basolateral amygdala, suggesting that ceramides alter depressive-like and anxiety-like behavior in a brain region- and ceramide species-specific way [21].

Understanding the role of the ASM/ceramide system in the pathogenesis of MDD might prove to be relevant for the development of an optimized treatment for MDD. The constitutive Asm-tg mouse model is an important tool for investigating the effects of Asm overexpression in the absence of a stressor-specific bias; however, one cannot exclude the effects of a systemic phenotype. Here, we report the generation and characterization of a conditional transgenic mouse model in which the expression of Asm is restricted to the forebrain (Asm-tg^{fb}). Restriction to the forebrain is possible via the Emx1-cre mouse strain [22], a widely used strain to generate conditional transgenic mouse models [23]. *Emx1* encodes a transcription factor and is expressed in the developing forebrain [24], specifically in the excitatory neurons and astrocytes [22]. Asm overexpression is therefore restricted, thereby excluding the influence of any systemic phenotype.

2. Materials and Methods

2.1. Animals

Male and female mice (12 weeks old) overexpressing Asm in the forebrain (Asm-tg^{fb}) were used in this study. These mice were generated by crossing female Asm-tg mice [2] with male Emx1IREScre homozygous mice, which possess the IRESCre recombinase-encoding sequence in the 3′ untranslated region of the Emx1 gene. IRESCre recombinase drives the expression of Cre recombinase starting on embryonic day 10.5, and this expression is restricted to the forebrain [22]. The Asm transgene is located on the X-chromosome. Therefore, the resulting female Asm-tg^{fb} mice were heterozygous, while males were hemizygous for the transgene. Male and female WT and Asm-tg^{fb} mice were individually housed for one week before the experiments started and remained so throughout the experiments. Mice were kept under standard laboratory conditions (12:12 light/dark cycle, lights on at 06:00 h, 22 °C, 60% humidity, with food and water *ad libitum*). Experiments were performed during the light phase between 09:00 and 14:00 in accordance with the recommendations in the Guide for the Care and Use of Laboratory Animals of the Government of Unterfranken and the Guidelines of the National Institutes of Health. All efforts were made to minimize animal suffering and to reduce the number of animals used.

2.2. Experimental Overview

In the first experiment, we assessed the behavior of Asm-tg^{fb} mice in comparison with that of WT mice. After one week of being housed individually, the social anxiety-like behavior of mice was tested in the social preference-avoidance test (SPAT). Four days later, the depressive-like behavior of mice was tested in the novelty-suppressed feeding (NSF) paradigm. Twenty-four hours later, mice were rapidly killed under CO_2 anesthesia, and the blood and brains were collected for further analysis. Blood was collected through cardiac puncture and centrifuged for 10 min at 4 °C and 2000 rpm. The serum was extracted and stored at −80 °C until it was assayed. Brains were removed, snap-frozen and stored at −80 °C. Several regions in the forebrain (i.e., frontal cortex, dorsal striatum, septum, amygdala, hypothalamus, dorsal hippocampus, and ventral hippocampus), midbrain (ventral mesencephalon) and hindbrain (cerebellum) were dissected from coronal brain slices as described in previous studies [15,21]. In one hemisphere (counterbalanced between mice), we analyzed Asm activity in all dissected brain regions. In the frontal cortex, ventral hippocampus and dorsal hippocampus of the second hemisphere, we quantified several sphingolipids, including the ceramide species Cer16:0, Cer18:0, Cer20:0, Cer22:0, Cer24:0, and Cer24:1, the sphingomyelin species SM16:0, SM18:0, SM20:0, SM22:0, SM24:0, and SM24:1, sphingosine and sphingosine-1-phosphate (S1P).

In a separate experiment, we collected brains from male Asm-tg^{fb} and WT mice, which were snap-frozen and stored at −80 °C. The frontal cortex and total hippocampus were dissected from coronal brain slices. We isolated RNA and performed quantitative real-time PCR (qPCR) analysis to investigate the expression of *Smpd1* mRNA encoding Asm and the expression of mRNAs encoding a variety of enzymes involved in sphingolipid metabolism, including neutral sphingomyelinase (*Smpd3*), glucosylceramidase (*Gba2*) and sphingosine-1-phosphate lyase (*Sgpl1*).

2.3. Social Preference-Avoidance Test (SPAT)

The social anxiety-like behavior of mice was tested in the SPAT as previously described [21]. Mice were placed in a novel arena (42 × 24 × 35 cm), and after a 30-s habituation period, an empty wire mesh cage (7 × 7 × 6 cm) was placed near one of the short walls. After 2.5 min, the empty cage was replaced by an identical cage containing an unfamiliar age-, weight- and sex-matched mouse for an additional 2.5 min. The test was recorded and analyzed using JWatcher (V 1.0, Macquarie University and UCLA). An increase in the investigation time directed towards the mouse versus the empty cage indicated social preference and, thus, a lack of social anxiety. A decrease in the investigation time directed towards the mouse indicated social avoidance and, thus, a social anxiogenic-like phenotype.

2.4. Novelty-Suppressed Feeding (NSF) Paradigm

The depressive-like behavior of mice was tested in the NSF paradigm as previously described [16,21]. Mice were food-deprived for 24 h prior to testing with unlimited access to fluids. Mice were placed in a novel arena ($50 \times 50 \times 50$ cm) with their head facing one of the corners. Immediately afterward, a single food pellet (ssniff Spezialdiäten GmbH, Soest, Germany) was placed in the center of the arena. The feeding latency, which was defined as biting the food pellet for longer than 3 s, was manually analyzed according to the videos. An increased feeding latency indicated a depressive-like phenotype.

2.5. Determination of Asm Activity In Vitro

Asm activity was determined in homogenates from several brain regions of the forebrain (i.e., frontal cortex, dorsal striatum, septum, amygdala, hypothalamus, dorsal hippocampus, and ventral hippocampus), midbrain (ventral mesencephalon) and hindbrain (cerebellum) and blood serum. For the preparation of brain homogenates, 10–20 mg pieces of tissue were homogenized in 0.5 mL sucrose lysis buffer (250 mM sucrose, 1 mM EDTA, and 0.2% Triton X—100) using a TissueLyser LT bead mill (Qiagen, Hilden, Germany). Raw lysates were centrifuged at $\geq 10.000 \times g$ at 4 °C for 10 min, and the supernatants were transferred to new tubes. The protein concentrations were determined using a bicinchoninic acid kit (Sigma, Darmstadt, Germany). For the determination of Asm activity, 1 µg of protein was incubated with 0.58 µM N—(4,4—difluoro—5,7—dimethyl—4—bora—3a,4a—diaza—s—indacene—3—dodecanoyl)—sphingosylphosphocholine (BODIPY®FL C_{12}—sphingomyelin; D—7711; Life Technologies, Darmstadt, Germany) in a 50 µL reaction buffer (50 mM sodium acetate pH 5.0, 0.3 M NaCl, and 0.2% NP—40) for 2 h at 37 °C; after incubation, 3 µL of the reaction mixture was spotted on a silica gel 60 plate (Macherey-Nagel; Düren, Germany), and the spots of ceramide and sphingomyelin were separated by thin-layer chromatography using 99% ethyl acetate/1% acetic acid (v/v) as a solvent [25]. The intensities of the BODIPY-conjugated ceramide and sphingomyelin fractions were determined using a Typhoon Trio scanner (GE Healthcare, München, Germany) and quantified with QuantityOne software (BioRad, München, Germany).

2.6. Sphingolipid Quantification by Liquid Chromatography Tandem-Mass Spectrometry (LC-MS/MS)

Tissue from the frontal cortex and ventral and dorsal hippocampus was subjected to lipid extraction using 1.5 mL methanol/chloroform (2:1, v/v) [26]. The extraction solvent contained d_7—sphingosine (d_7—Sph), d_7—sphingosine—1—phosphate (d_7—S1P), ceramide C17:0 (Cer17:0) and sphingomyelin C16:0—d_{31} (SM16:0—d_{31}) (all Avanti Polar Lipids, Alabaster, Alabama, USA) as internal standards. Sample analysis was carried out by liquid chromatography tandem-mass spectrometry (LC-MS/MS) using either a TQ 6490 mass spectrometer (for Sph and S1P) or a QTOF 6530 mass spectrometer (for Cer and SM species) (Agilent Technologies, Waldbronn, Germany) operating in the positive electrospray ionization mode (ESI+). The following selected reaction monitoring (SRM) transitions were used for quantification: m/z 300.3 $\rightarrow$ 282.3 for Sph, m/z 380.3 $\rightarrow$ 264.3 for S1P, m/z 307.3 $\rightarrow$ 289.3 for d_7—Sph and m/z 387.3 $\rightarrow$ 271.3 for d_7—S1P. The precursor ions of the Cer or SM species (which differed in their fatty acid chain lengths) were cleaved into the fragment ions corresponding to m/z 264.270 or m/z 184.074, respectively [27]. Quantification of the ceramide species Cer16:0, Cer18:0, Cer20:0, Cer22:0, Cer24:0, and Cer24:1, the sphingomyelin species SM16:0, SM18:0, SM20:0, SM22:0, SM24:0, and SM24:1, sphingosine and S1P was performed with MassHunter Software (Agilent Technologies, Waldbronn, Germany). The determined sphingolipid amounts were normalized to the actual protein content (determined by the Bradford assay) of the tissue homogenate used for lipid extraction. The used nomenclature of sphingolipids indicates the number of carbon atoms and double bonds of the fatty acid side chain. All sphingolipid species analyzed contain a d18:1 sphingosine backbone. For example, Cer16:0 has a fatty acid side chain length of 16 carbon atoms and no double bond.

2.7. Extraction of RNA and Synthesis of cDNA

Total RNA was isolated from cortical and hippocampal tissue (<30 mg) using a TissueLyser LT bead mill (Qiagen, Hilden, Germany) and peqGOLD Trifast reagent (Peqlab, Erlangen, Germany) according to the manufacturers' instructions, which was followed by RNA purification performed with the Purelink RNA Kit from Thermo Scientific (Schwerte, Germany) according to the manufacturer's protocol. RNA qualities and concentrations were assessed using a Nanodrop ND-1000 UV-Vis spectrophotometer. A total of 500 ng of RNA was transformed into cDNA using the Quanta cDNA Kit (Gaithersburg, MD, USA) according to the manufacturer's protocol.

2.8. Quantitative PCR Analysis

Quantitative real-time PCR was performed using cDNA from cortical and hippocampal tissue using a LightCycler 480 real-time PCR system (Roche, Mannheim, Germany) in SYBR Green format. We analyzed the expression of the following genes, for which the primer sequences can be found in our earlier publication [28]: *Asah1, Asah2, Cerk, CerS1, CerS2, CerS3, CerS4, CerS5, CerS6, Galc, Gba, Gba2, Sgms1, Sgms2, Sgpl1, Smpd1, Smpd3, Sphk1,* and *Sphk2; Gapdh* was used as a reference gene. qPCR reactions contained 5 µL FastStart Essential DNA Green Master Mix (Roche, Mannheim, Germany), 0.5 µM of each primer (20 µM) and 2.5 µL diluted cDNA (corresponding to 12.5 ng RNA) in a total volume of 10 µL. The temperature profile used consisted of 95 °C for 5 min followed by 40 cycles of amplification (95 °C for 10 s, 60 °C for 20 s, and 72 °C for 30 s). The threshold cycles (Ct) were determined with the "second derivative maximum" method, and the relative mRNA expression levels were calculated with the $2^{-\Delta\Delta Ct}$ method [29] using LightCycler 480 software (release 1.5.0).

2.9. Statistical Analyses

Statistical analyses were performed using SPSS Statistics version 21. Statistical significance was determined using Student's *t*-test and two-way ANOVA, followed by Bonferroni post-hoc analysis when appropriate. Statistical significance was set at $p < 0.05$.

3. Results

3.1. Asm-tg^{fb} Mice Show an Increase in the Expression of Smpd1 mRNA Encoding Asm

In the first analysis, we assessed *Smpd1* mRNA levels in male Asm-tg^{fb} mice to confirm our breeding strategy. In both cortical and hippocampal tissues, Asm-tg^{fb} mice showed a significant increase in *Smpd1* mRNA expression in comparison with WT mice (Figure 1A, frontal cortex, t(5) = −16.7; $p < 0.001$; Figure 1B, hippocampus, t(6) = −6.9; $p < 0.001$).

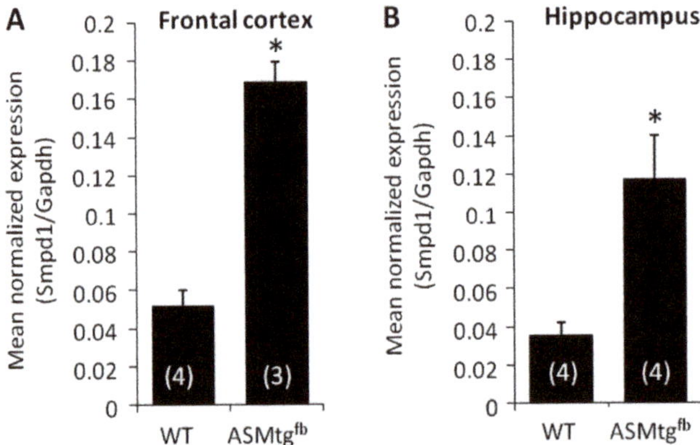

Figure 1. Asm-tg^{fb} mice show an increase in the mRNA expression of *Smpd1*, which encodes Asm. In both cortical and hippocampal tissues, male Asm-tg^{fb} mice showed a significant increase in *Smpd1* mRNA expression in comparison with WT mice. Data represent the means + SD, and numbers in parentheses indicate group sizes; * $p < 0.05$.

3.2. Asm-tg^{fb} Mice Show an Increase in Asm Activity in Forebrain-Related Brain Regions

To analyze whether increased *Smpd1* mRNA expression results in increased enzyme activity levels, we measured Asm activity in several regions of the forebrain (i.e., frontal cortex, dorsal striatum, septum, amygdala, hypothalamus, dorsal hippocampus, and ventral hippocampus), midbrain (ventral mesencephalon) and hindbrain (cerebellum) and the serum of male and female Asm-tg^{fb} mice. When compared with WT controls, both male and female Asm-tg^{fb} mice showed increased Asm activity in the dorsal striatum, dorsal hippocampus, ventral hippocampus, and amygdala. Male Asm-tg^{fb} mice also showed increased Asm activity in the frontal cortex, septum, and ventral mesencephalon. Female Asm-tg^{fb} mice showed increased Asm activity in the hypothalamus. Neither female nor male Asm-tg^{fb} mice showed increased Asm activity in the cerebellum or serum, confirming the regional specificity of ASM overexpression in the forebrain.

Statistical results of increased Asm activity in male and female Asm-tg^{fb} mice: Dorsal striatum, Figure 2B, genotype effect $F(1,34) = 34.2$, $p < 0.001$; dorsal hippocampus, Figure 2C, genotype effect $F(1,34) = 102.7$, $p < 0.001$, sex × genotype effect $F(1,34) = 16.7$, $p < 0.001$; ventral hippocampus, Figure 2D, genotype effect $F(1,34) = 72.1$, $p < 0.001$, sex × genotype effect $F(1,34) = 10.2$, $p = 0.003$; amygdala, Figure 2F, genotype effect $F(1,32) = 36.0$, $p < 0.001$, sex × genotype effect $F(1,32) = 5.71$, $p = 0.02$. Statistical results of increased Asm activity only in male Asm-tg^{fb} mice: Frontal cortex, Figure 2A, genotype effect $F(1,34) = 35.2$; $p < 0.001$, sex × genotype effect $F(1,34) = 11.9$; p = 0.002); septum, Figure 2E, genotype effect $F(1,33) = 13.3$, $p = 0.01$, sex × genotype effect $F(1,33) = 5.2$, $p = 0.03$; ventral mesencephalon, Figure 2H, genotype effect $F(1,30) = 4.67$, p = 0.04. Statistical results of increased Asm activity only in female Asm-tg^{fb} mice: Hypothalamus, Figure 2G; genotype effect $F(1,34) = 4.84$; $p = 0.04$. No increase in Asm activity: Cerebellum, Figure 2I; genotype effect $F(1,33) = 0.11$; $p = 0.74$; serum, Figure 2J; genotype effect $F(1,34) = 1.25$; $p = 0.27$.

3.3. Male Asm-tg^{fb} Mice Show Increased Depressive-Like Behavior

To investigate whether the increase in Asm activity induced a depressive-like phenotype, male and female Asm-tg^{fb} mice were tested in the NSF paradigm. Male but not female Asm-tg^{fb} mice showed an increase in the feeding latency after a fasting period of 24 h, reflecting an increase in depressive-like behavior (Figure 3A; sex × genotype effect $F(1,34) = 5.37$; $p = 0.03$).

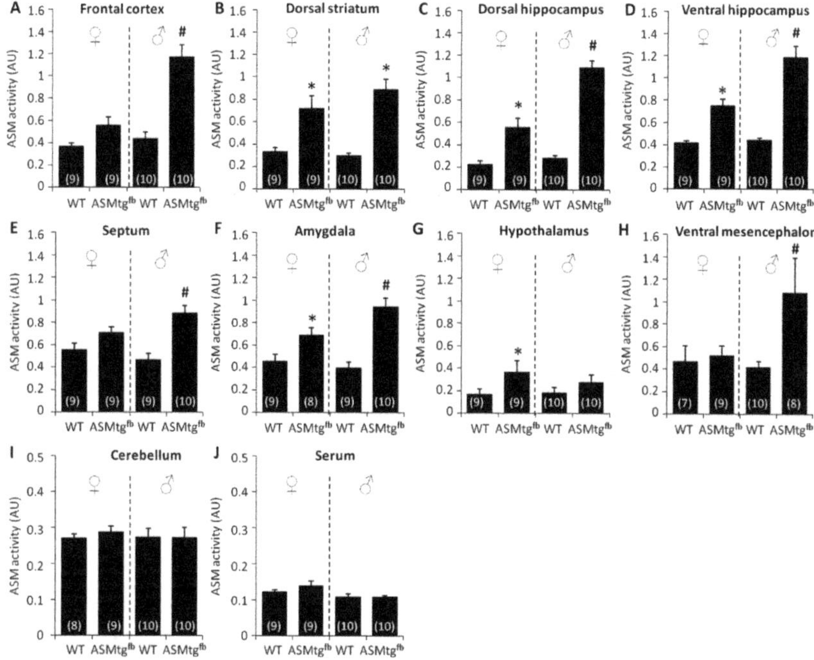

Figure 2. Brain Asm activity in WT and Asm-tg^{fb} mice. Asm activity was analyzed in nine different brain regions and blood serum for both males and females. In forebrain regions, Asm-tg^{fb} mice showed a significant increase in Asm activity levels compared with WT mice. Data represent the means + SEM, and numbers in parentheses indicate group sizes. * $p < 0.05$ versus same-sex WT; # $p < 0.05$ versus ♂ WT and ♀ Asm-tg^{fb}.

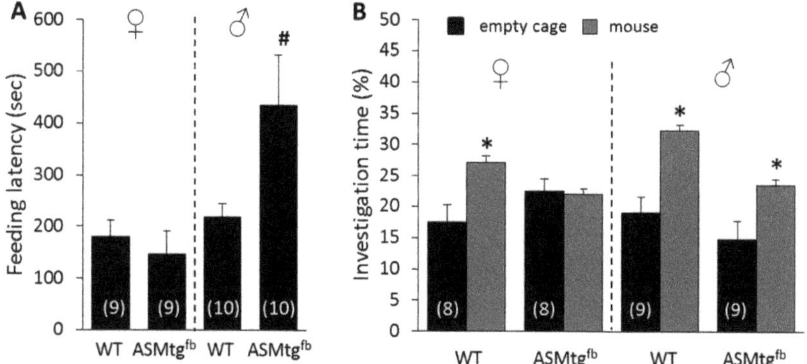

Figure 3. Asm overexpression alters depressive-like and social anxiety-like behavior in a sex-specific way. (**A**). Feeding latency, as an indicator of depressive-like behavior, was assessed in the novelty-suppressed feeding paradigm. Male Asm-tg^{fb} mice showed increased depressive-like behavior compared with WT mice. (**B**). The time of investigation of an unknown mouse compared with that of an empty cage, as an indicator of social anxiety-like behavior, was assessed in the social preference-avoidance test. Female Asm-tg^{fb} mice showed increased social anxiety-like behavior compared with WT mice. Data represent the means + SEM, and numbers in parentheses indicate group sizes. * $p < 0.05$ versus empty cage; # $p < 0.05$ versus ♂ WT and ♀ Asm-tg^{fb}.

3.4. Female Asm-tg^fb Mice Show Increased Social Anxiety-Like Behavior

To investigate whether the increase in Asm activity induced a social anxiogenic-like phenotype, male and female Asm-tg^fb mice were tested in the SPAT. Whereas WT females showed increased investigation of the mouse versus the empty cage during SPAT, which reflected a social preference and a lack of social anxiety, Asm-tg^fb females showed decreased investigation of the mouse, reflecting a social anxiogenic-like phenotype. In males, however, there was no effect of genotype (Figure 3B; group × stimulus effect $F_{(3,60)} = 2.86$; $p = 0.04$). Although male Asm-tg^fb mice showed decreased investigation of the mouse when compared with male WT mice, this did not reach statistical significance ($p = 0.10$).

3.5. Male Asm-tg^fb Mice Show Changes in Ceramide Levels Only in the Hippocampus

To investigate whether the increase in ASM activity affects sphingolipid levels in brain areas relevant for MDD, tissue from the dorsal and ventral hippocampus and frontal cortex was used for lipidomic analysis. In the dorsal hippocampus, the percentage of Cer24:0 compared to total ceramides varied in a sex-specific manner, with only male ASM-tg^fb mice displaying higher Cer24:0 levels than WT males (sex × genotype effect $F_{(1,34)} = 4.5$; $p = 0.04$). In the ventral hippocampus, male ASM-tg^fb mice showed a decreased percentage of Cer18:0 versus total ceramides compared to WT males (sex × genotype effect $F_{(1,34)} = 4.6$; $p = 0.04$). Other sphingolipids were not changed in hippocampal tissue. In the frontal cortex, no effects of ASM overexpression on sphingolipids were detected.

3.6. Asm-tg^fb Mice Show Changes in the mRNA Expression of Other Sphingolipid-Metabolizing Enzymes

To assess the cause of the relatively slight changes in ceramide levels despite the significant increase in Asm activity, we analyzed the mRNA expression of a variety of enzymes involved in sphingolipid metabolism. Interestingly, in cortical tissue of Asm-tg^fb mice, mRNA expression of neutral sphingomyelinase (Figure 4A; *Smpd3*; $t(5) = 2.6$; $p = 0.049$) and glucosylceramidase 2 (Figure 4B; *Gba2*; $t(5) = 2.9$; $p = 0.04$) were significantly decreased compared with that in WT mice. In the hippocampus, mRNA expression of sphingosine—1—phosphate lyase (*Sgpl1*) was significantly increased in Asm-tg^fb mice compared with that in WT mice (Figure 4C; $t(6) = -3.6$; $p = 0.01$). No changes were found in the expression of *Asah1*, *Asah2*, *Cerk*, *CerS1*, *CerS2*, *CerS3*, *CerS4*, *CerS5*, *CerS6*, *Galc*, *Gba*, *Sgms1*, *Sgms2*, *Sphk1*, and *Sphk2*.

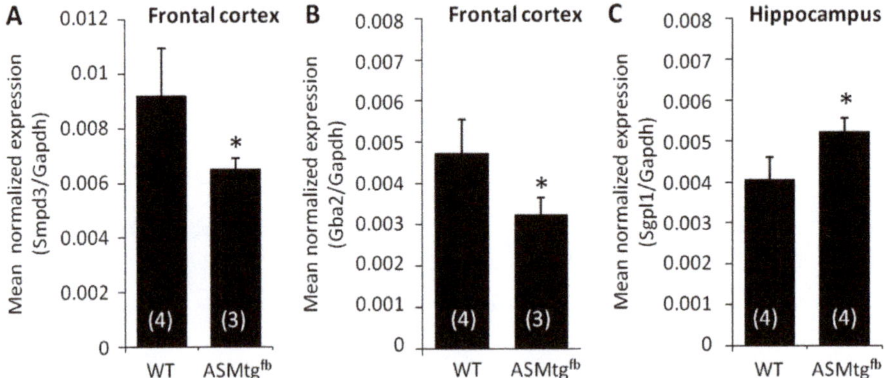

Figure 4. Asm-tg^fb mice show changes in mRNA expression of sphingolipid-metabolizing enzymes. In cortical tissue, male Asm-tg^fb mice showed a significant decrease in (**A**) *Smpd3* and (**B**) *Gba2* mRNA expression in comparison with WT mice. In hippocampal tissue, male Asm-tg^fb mice showed a significant increase in (**C**) *Sgpl1* mRNA expression in comparison with WT mice. Data represent the means + SD, and numbers in parentheses indicate group sizes; * $p < 0.05$.

4. Discussion

Our study characterizes a mouse model with increased Asm activity specifically in the forebrain. This increased Asm activity resulted in a depressive-like phenotype in males and a social anxiogenic-like phenotype in females. Compared with the Asm-tg model, which overexpresses Asm in the whole body [2,16], this conditional transgenic mouse model excludes the influence of a systemic phenotype.

The significant increase in *Smpd1* mRNA encoding Asm in cortical and hippocampal brain areas of Asm-tg^{fb} mice confirms the success of our breeding strategy. Similarly, ASM enzymatic activity was significantly increased in the forebrain as well as in areas with forebrain projections, including the frontal cortex, dorsal striatum, dorsal and ventral hippocampus, septum, amygdala, and hypothalamus. As expected, no increase in Asm activity in Asm-tg^{fb} mice was detected in the cerebellum, where no Cre is expressed in Emx1-cre strain mice [22], or in blood serum.

In most forebrain-related areas, the Asm activity levels of female Asm-tg^{fb} mice were lower than those of male Asm-tg^{fb} mice. In the frontal cortex and the septum, the increase in ASM activity levels in female ASM-tg^{fb} mice did not reach a significant level compared with that in WT female mice. Due to the location of the ASM transgene allele in the X chromosome, female ASM-tg^{fb} mice were heterozygous, while male ASM-tg^{fb} mice were hemizygous for the ASM transgene. In general, this results in the silencing of the respective transgene in female mice. Thus, the apparent sex differences in ASM activity levels might reflect the genetic situation more than sex differences per se. On the other hand, an exception is seen in the hypothalamus, where female Asm-tg^{fb} mice showed significantly higher Asm activity levels than male Asm-tg^{fb} mice. Given that the hypothalamus is a brain area highly relevant for several types of social behavior, including social anxiety [30], and that female but not male Asm-tg^{fb} mice showed a social anxiogenic-like phenotype, the increased Asm activity within the hypothalamus may contribute to social anxiety. Although female Asm-tg^{fb} mice showed a significant increase in Asm activity in the hippocampus, which is a brain area that was shown to be highly relevant for the pathology of MDD, they did not show a depressive-like phenotype. In contrast, male Asm-tg^{fb} mice showed clear changes in their behavior and displayed significant depressive-like behavior in the NSF test. Possibly, a very high threshold level of Asm activity in the hippocampus, as seen in male but not in female Asm-tg^{fb} mice, might be necessary to elicit changes in ceramide levels and depressive-like behavior. However, interestingly, male Asm-tg^{fb} mice showed an increase in Asm activity in the ventral mesencephalon, which might result from the close connections between forebrain regions and the mesencephalon. This might suggest the important role of the frontal cortex, septum, and ventral mesencephalon in the pathophysiology of MDD, given that male but not female Asm-tg^{fb} mice showed increased Asm activity within these brain areas. This points to a sex-specific effect, whereby increased Asm activity affects different circuits in female versus male Asm-tg^{fb} mice. The projections from the frontal cortex to the mesencephalon affecting the reward system could be essential for the control of emotional behavior in males and might be regulated sex-specifically. This could result in distinct subtypes of MDD for both sexes and might explain the different prevalence rates of MDD found in both sexes in human studies.

As ceramide is generated through the activity of Asm, a change in Asm activity levels is expected to alter ceramide levels, especially those in brain areas that are most relevant for depressive- and anxiety-like behavior, such as the hippocampus and frontal cortex. When looking at the effects of Asm overexpression on ceramides in the hippocampus, we found a sex- and brain region-specific effect. Male ASM-tg^{fb} mice displayed increased Cer24:0 levels in the dorsal hippocampus and decreased Cer18:0 levels in the ventral hippocampus compared with WT mice. This reflects the important role of the hippocampus in depressive- and social anxiety-like behavior. In particular, changes in ceramides in the dorsal hippocampus seem to be responsible for depressive-like behavior, which was also suggested by our earlier study [21]. In the hippocampus, the mRNA expression of sphingosine—1—phosphate lyase (*Sgpl1*) was significantly increased in Asm-tg^{fb} mice compared with that in WT mice. The enzyme S1P—lyase cleaves S1P to generate phosphoethanolamine and hexadecenal and plays an essential role in sphingolipid metabolism because this reaction cannot be reverted [31]. An increase in S1P—lyase would be associated with higher rates of irreversible cleavage of S1P. S1P is considered to be toxic in neurons since it induces stress in the endoplasmic reticulum and increases intracellular calcium currents [32–34]. Thus, neuronal cells might increase the expression of the *Sgpl1* gene, encoding S1P—lyase, to eliminate toxic S1P as a rescue mechanism. The increased ceramide levels, generated by increased Asm activity levels in Asm-tg[fb] mice, could result in increased production of S1P, which is irreversibly cleaved by S1P—lyase and eliminated from the rheostat.

Lipidomic analyses revealed no significant changes in ceramides in the frontal cortex in male Asm-tg[fb] mice showing a depressive-like phenotype. When we analyzed the mRNA expression of a variety of sphingolipid-metabolizing enzymes in the frontal cortex more closely, we found a significant decrease in the mRNA expression of neutral sphingomyelinase (*Smpd3*) and glucosylceramidase (*Gba2*) in Asm-tg[fb] male mice in comparison with that in WT male mice. Neutral sphingomyelinase converts sphingomyelin into ceramide, but this mainly occurs at a neutral pH and at the plasma membrane; in contrast, ASM mainly generates ceramide at an acidic pH and in the lysosome [35]. Glucosylceramidase 2 converts complex glucosylceramides into ceramide and is located at or close to the cell surface [36]. Thus, given that the mRNA expression levels reflect the enzymatic activity, a decrease in *Smpd3* and *Gba2* expression would result in a decrease in ceramide levels. The overexpression of Asm in our mouse model, which should result in increased ceramide levels, might be counterbalanced by decreases in other enzymes that generate ceramide, which might explain why no changes in ceramide levels in the frontal cortex were observed in our analyses. The question remains whether the subcellular determination of ceramide localization could determine the differences in ceramide distribution in the frontal cortex of Asm-tg[fb] mice.

Our results might provide novel insights into the complex regulation of the sphingolipid rheostat. Further studies should apply methods to determine the subcellular localization of ceramides to determine their specific roles. Sphingolipid metabolism is highly dynamic and well balanced [37]. In our mouse model, Asm was consistently overexpressed in the forebrain starting at an early developmental stage. The impact of ASM overexpression on ceramide species in different brain areas seems to reflect the complex mechanisms of sphingolipid metabolism.

Author Contributions: Conceptualization, C.R., A.F., J.K., and E.G.; methodology, M.R., A.F., E.G., C.R., I.Z., B.K., and F.S.; formal analysis, C.R., I.Z., A.F., J.K.; investigation, C.R., I.Z., F.S.; resources, J.K., E.G., A.F., and M.R.; data curation, C.R., I.Z., B.K., and F.S.; writing—original draft preparation, C.R. and I.Z.; writing—review and editing, A.F., E.G., and J.K.; visualization, I.Z. and C.R.; supervision, C.R., J.K., and A.F.; funding Acquisition, J.K., E.G., and A.F. All authors have read and agreed to the published version of the manuscript.

Funding: This research was funded by Forschungsstiftung Medizin at the University Hospital Erlangen and the German Research Foundation DFG (SFB779 A06, 270949263/GRK2162, GU 335/29-3, KO 947/13-3 to A.F., J.K. and E.G.). The APC was partly funded by the funding program Open Access Publishing of the Friedrich-Alexander University Erlangen-Nürnberg (FAU).

Acknowledgments: We thank Sabine Müller, Juliana Monti, Andrea Leicht, and Katrin Ebert for excellent technical assistance. We also thank Daniel Herrmann and Monika Haseloff for their help with the mass spectrometric analyses.

Conflicts of Interest: The authors declare no conflict of interest.

References

1. Belmaker, R.H.; Agam, G. Major depressive disorder. *N. Engl. J. Med.* **2008**, *358*, 55–68. [CrossRef] [PubMed]
2. Gulbins, E.; Palmada, M.; Reichel, M.; Lüth, A.; Böhmer, C.; Amato, D.; Müller, C.P.; Tischbirek, C.H.; Groemer, T.W.; Tabatabai, G.; et al. Acid sphingomyelinase-ceramide system mediates effects of antidepressant drugs. *Nat. Med.* **2013**, *19*, 934–938. [CrossRef]
3. Schneider, P.B.; Kennedy, E.P. Sphingomyelinase in normal human spleens and in spleens from subjects with Niemann-Pick disease. *J. Lipid Res.* **1967**, *8*, 202–209. [PubMed]
4. Goni, F.M.; Alonso, A. Sphingomyelinases: Enzymology and membrane activity. *FEBS Lett.* **2002**, *531*, 38–46. [CrossRef]

5. Lahiri, S.; Futerman, A.H. The metabolism and function of sphingolipids and glycosphingolipids. *Cell. Mol. Life Sci.* **2007**, *64*, 2270–2284. [CrossRef]

6. Ishibashi, Y.; Nakasone, T.; Kiyohara, M.; Horibata, Y.; Sakaguchi, K.; Hijikata, A.; Ichinose, S.; Omori, A.; Yasui, Y.; Imamura, A.; et al. A novel endoglycoceramidase hydrolyzes oligogalactosylceramides to produce galactooligosaccharides and ceramides. *J. Biol. Chem.* **2007**, *282*, 11386–11396. [CrossRef]

7. Okino, N.; He, X.; Gatt, S.; Sandhoff, K.; Ito, M.; Schuchman, E.H. The reverse activity of human acid ceramidase. *J. Biol. Chem.* **2003**, *278*, 29948–29953. [CrossRef]

8. Kornhuber, J.; Medlin, A.; Bleich, S.; Jendrossek, V.; Henkel, A.W.; Wiltfang, J.; Gulbins, E. High activity of acid sphingomyelinase in major depression. *J. Neural Transm.* **2005**, *112*, 1583–1590. [CrossRef]

9. Gracia-Garcia, P.; Rao, V.; Haughey, N.J.; Bandaru, V.V.; Smith, G.; Rosenberg, P.B.; Lobo, A.; Lyketsos, C.G.; Mielke, M.M. Elevated plasma ceramides in depression. *J. Neuropsychiatry Clin. Neurosci.* **2011**, *23*, 215–218. [CrossRef]

10. Brunkhorst-Kanaan, N.; Klatt-Schreiner, K.; Hackel, J.; Schröter, K.; Trautmann, S.; Hahnefeld, L.; Wicker, S.; Reif, A.; Thomas, D.; Geisslinger, G.; et al. Targeted lipidomics reveal derangement of ceramides in major depression and bipolar disorder. *Metabolism* **2019**, *95*, 65–76. [CrossRef]

11. Dinoff, A.; Saleem, M.; Herrmann, N.; Mielke, M.M.; Oh, P.I.; Venkata, S.L.V.; Haughey, N.J.; Lanctot, K.L. Plasma sphingolipids and depressive symptoms in coronary artery disease. *Brain Behav.* **2017**, *7*, e00836. [CrossRef] [PubMed]

12. Moaddel, R.; Shardell, M.; Khadeer, M.; Lovett, J.; Kadriu, B.; Ravichandran, S.; Morris, P.J.; Yuan, P.; Thomas, C.J.; Gould, T.D.; et al. Plasma metabolomic profiling of a ketamine and placebo crossover trial of major depressive disorder and healthy control subjects. *Psychopharmacology* **2018**, *235*, 3017–3030. [CrossRef] [PubMed]

13. Liu, X.; Li, J.; Zheng, P.; Zhao, X.; Zhou, C.; Hu, C.; Hou, X.; Wang, H.; Xie, P.; Xu, G. Plasma lipidomics reveals potential lipid markers of major depressive disorder. *Anal. Bioanal. Chem.* **2016**, *408*, 6497–6507. [CrossRef] [PubMed]

14. Demirkan, A.; Isaacs, A.; Ugocsai, P.; Liebisch, G.; Struchalin, M.; Rudan, I.; Wilson, J.F.; Pramstaller, P.P.; Gyllensten, U.; Campbell, H.; et al. Plasma phosphatidylcholine and sphingomyelin concentrations are associated with depression and anxiety symptoms in a Dutch family-based lipidomics study. *J. Psychiatr. Res.* **2013**, *47*, 357–362. [CrossRef]

15. Müller, C.P.; Kalinichenko, L.S.; Tiesel, J.; Witt, M.; Stockl, T.; Sprenger, E.; Fuchser, J.; Beckmann, J.; Praetner, M.; Huber, S.E.; et al. Paradoxical antidepressant effects of alcohol are related to acid sphingomyelinase and its control of sphingolipid homeostasis. *Acta Neuropathol.* **2017**, *133*, 463–483. [CrossRef]

16. Zoicas, I.; Reichel, M.; Gulbins, E.; Kornhuber, J. Role of Acid Sphingomyelinase in the Regulation of Social Behavior and Memory. *PLoS ONE* **2016**, *11*, e0162498. [CrossRef]

17. Willner, P. The chronic mild stress (CMS) model of depression: History, evaluation and usage. *Neurobiol. Stress* **2017**, *6*, 78–93. [CrossRef]

18. Oliveira, T.G.; Chan, R.B.; Bravo, F.V.; Miranda, A.; Silva, R.R.; Zhou, B.; Marques, F.; Pinto, V.; Cerqueira, J.J.; Di Paolo, G.; et al. The impact of chronic stress on the rat brain lipidome. *Mol. Psychiatry* **2016**, *21*, 80–88. [CrossRef]

19. Gregus, A.; Wintink, A.J.; Davis, A.C.; Kalynchuk, L.E. Effect of repeated corticosterone injections and restraint stress on anxiety and depression-like behavior in male rats. *Behav. Brain Res.* **2005**, *156*, 105–114. [CrossRef]

20. Miranda, A.M.; Bravo, F.V.; Chan, R.B.; Sousa, N.; Di Paolo, G.; Oliveira, T.G. Differential lipid composition and regulation along the hippocampal longitudinal axis. *Transl. Psychiatry* **2019**, *9*, 144. [CrossRef]

21. Zoicas, I.; Huber, S.E.; Kalinichenko, L.S.; Gulbins, E.; Müller, C.P.; Kornhuber, J. Ceramides affect alcohol consumption and depressive-like and anxiety-like behavior in a brain region- and ceramide species-specific way in male mice. *Addict. Biol.* **2019**, e12847. [CrossRef] [PubMed]

22. Gorski, J.A.; Talley, T.; Qiu, M.; Puelles, L.; Rubenstein, J.L.; Jones, K.R. Cortical excitatory neurons and glia, but not GABAergic neurons, are produced in the Emx1-expressing lineage. *J. Neurosci.* **2002**, *22*, 6309–6314. [CrossRef] [PubMed]

23. Altmüller, F.; Pothula, S.; Annamneedi, A.; Nakhaei-Rad, S.; Montenegro-Venegas, C.; Pina-Fernandez, E.; Marini, C.; Santos, M.; Schanze, D.; Montag, D.; et al. Aberrant neuronal activity-induced signaling and gene expression in a mouse model of RASopathy. *PLoS Genet.* **2017**, *13*, e1006684. [CrossRef] [PubMed]

24. Simeone, A.; Gulisano, M.; Acampora, D.; Stornaiuolo, A.; Rambaldi, M.; Boncinelli, E. Two vertebrate homeobox genes related to the Drosophila empty spiracles gene are expressed in the embryonic cerebral cortex. *EMBO J.* **1992**, *11*, 2541–2550. [CrossRef] [PubMed]

25. Mühle, C.; Kornhuber, J. Assay to measure sphingomyelinase and ceramidase activities efficiently and safely. *J. Chromatogr. A* **2017**, *1481*, 137–144. [CrossRef] [PubMed]

26. Gulbins, A.; Schumacher, F.; Becker, K.A.; Wilker, B.; Soddemann, M.; Boldrin, F.; Müller, C.P.; Edwards, M.J.; Goodman, M.; Caldwell, C.C.; et al. Antidepressants act by inducing autophagy controlled by sphingomyelin-ceramide. *Mol. Psychiatry* **2018**, *23*, 2324–2346. [CrossRef]

27. Kachler, K.; Bailer, M.; Heim, L.; Schumacher, F.; Reichel, M.; Holzinger, C.D.; Trump, S.; Mittler, S.; Monti, J.; Trufa, D.I.; et al. Enhanced Acid Sphingomyelinase Activity Drives Immune Evasion and Tumor Growth in Non-Small Cell Lung Carcinoma. *Cancer Res.* **2017**, *77*, 5963–5976. [CrossRef]

28. Reichel, M.; Rhein, C.; Hofmann, L.M.; Monti, J.; Japtok, L.; Langgartner, D.; Fuchsl, A.M.; Kleuser, B.; Gulbins, E.; Hellerbrand, C.; et al. Chronic Psychosocial Stress in Mice Is Associated With Increased Acid Sphingomyelinase Activity in Liver and Serum and With Hepatic C16:0-Ceramide Accumulation. *Front. Psychiatry* **2018**, *9*, 496. [CrossRef]

29. Schmittgen, T.D.; Livak, K.J. Analyzing real-time PCR data by the comparative C(T) method. *Nat. Protoc.* **2008**, *3*, 1101–1108. [CrossRef]

30. Choleris, E.; Devidze, N.; Kavaliers, M.; Pfaff, D.W. Steroidal/neuropeptide interactions in hypothalamus and amygdala related to social anxiety. *Prog. Brain Res.* **2008**, *170*, 291–303. [CrossRef]

31. Serra, M.; Saba, J.D. Sphingosine 1-phosphate lyase, a key regulator of sphingosine 1-phosphate signaling and function. *Adv. Enzym. Regul.* **2010**, *50*, 349–362. [CrossRef]

32. Hagen, N.; Hans, M.; Hartmann, D.; Swandulla, D.; van Echten-Deckert, G. Sphingosine-1-phosphate links glycosphingolipid metabolism to neurodegeneration via a calpain-mediated mechanism. *Cell Death Differ.* **2011**, *18*, 1356–1365. [CrossRef] [PubMed]

33. Hagen, N.; Van Veldhoven, P.P.; Proia, R.L.; Park, H.; Merrill, A.H., Jr.; van Echten-Deckert, G. Subcellular origin of sphingosine 1-phosphate is essential for its toxic effect in lyase-deficient neurons. *J Biol Chem* **2009**, *284*, 11346–11353. [CrossRef] [PubMed]

34. Mitroi, D.N.; Deutschmann, A.U.; Raucamp, M.; Karunakaran, I.; Glebov, K.; Hans, M.; Walter, J.; Saba, J.; Gräler, M.; Ehninger, D.; et al. Sphingosine 1-phosphate lyase ablation disrupts presynaptic architecture and function via an ubiquitin-proteasome mediated mechanism. *Sci. Rep.* **2016**, *6*, 37064. [CrossRef] [PubMed]

35. Clarke, C.J.; Wu, B.X.; Hannun, Y.A. The neutral sphingomyelinase family: Identifying biochemical connections. *Adv. Enzym. Regul.* **2011**, *51*, 51–58. [CrossRef]

36. Boot, R.G.; Verhoek, M.; Donker-Koopman, W.; Strijland, A.; van Marle, J.; Overkleeft, H.S.; Wennekes, T.; Aerts, J.M. Identification of the non-lysosomal glucosylceramidase as beta-glucosidase 2. *J. Biol. Chem.* **2007**, *282*, 1305–1312. [CrossRef]

37. Hagen-Euteneuer, N.; Lutjohann, D.; Park, H.; Merrill, A.H., Jr.; van Echten-Deckert, G. Sphingosine 1-phosphate (S1P) lyase deficiency increases sphingolipid formation via recycling at the expense of de novo biosynthesis in neurons. *J. Biol. Chem.* **2012**, *287*, 9128–9136. [CrossRef]

Review

Elusive Roles of the Different Ceramidases in Human Health, Pathophysiology, and Tissue Regeneration

Carolina Duarte [1],*, **Juliet Akkaoui [1]**, **Chiaki Yamada [1]**, **Anny Ho [1]**, **Cungui Mao [2,3]** and **Alexandru Movila [1,4,***

[1] Department of Periodontology, College of Dental Medicine, Nova Southeastern University, Fort Lauderdale, FL 33324, USA; ja1617@nova.edu (J.A.); cyamada@nova.edu (C.Y.); ah2197@nova.edu (A.H.)

[2] Department of Medicine, The State University of New York at Stony Brook, Stony Brook, NY 11794, USA; Cungui.Mao@stonybrookmedicine.edu

[3] Cancer Center, The State University of New York at Stony Brook, Stony Brook, NY 11794, USA

[4] Institute for Neuro-Immune Medicine, Nova Southeastern University, Fort Lauderdale, FL 33324, USA

* Correspondence: cduartep@nova.edu (C.D.); amovila@nova.edu (A.M.); Tel.: +1-954-262-7306 (A.M.)

Received: 18 April 2020; Accepted: 27 May 2020; Published: 2 June 2020

Abstract: Ceramide and sphingosine are important interconvertible sphingolipid metabolites which govern various signaling pathways related to different aspects of cell survival and senescence. The conversion of ceramide into sphingosine is mediated by ceramidases. Altogether, five human ceramidases—named acid ceramidase, neutral ceramidase, alkaline ceramidase 1, alkaline ceramidase 2, and alkaline ceramidase 3—have been identified as having maximal activities in acidic, neutral, and alkaline environments, respectively. All five ceramidases have received increased attention for their implications in various diseases, including cancer, Alzheimer's disease, and Farber disease. Furthermore, the potential anti-inflammatory and anti-apoptotic effects of ceramidases in host cells exposed to pathogenic bacteria and viruses have also been demonstrated. While ceramidases have been a subject of study in recent decades, our knowledge of their pathophysiology remains limited. Thus, this review provides a critical evaluation and interpretive analysis of existing literature on the role of acid, neutral, and alkaline ceramidases in relation to human health and various diseases, including cancer, neurodegenerative diseases, and infectious diseases. In addition, the essential impact of ceramidases on tissue regeneration, as well as their usefulness in enzyme replacement therapy, is also discussed.

Keywords: ceramides; ceramidases; inflammation; neurodegenerative diseases; infectious diseases

---

1. Introduction

Ceramides are bioactive sphingolipids responsible for cell apoptosis, senescence, and autophagy [1]. They are the precursors of other bioactive sphingolipids, including sphingosine (SPH), sphingosine-1-phosphate (S1P), and ceramide-1-phosphate, which play specific roles in signal transduction pathways (Figure 1) [1].

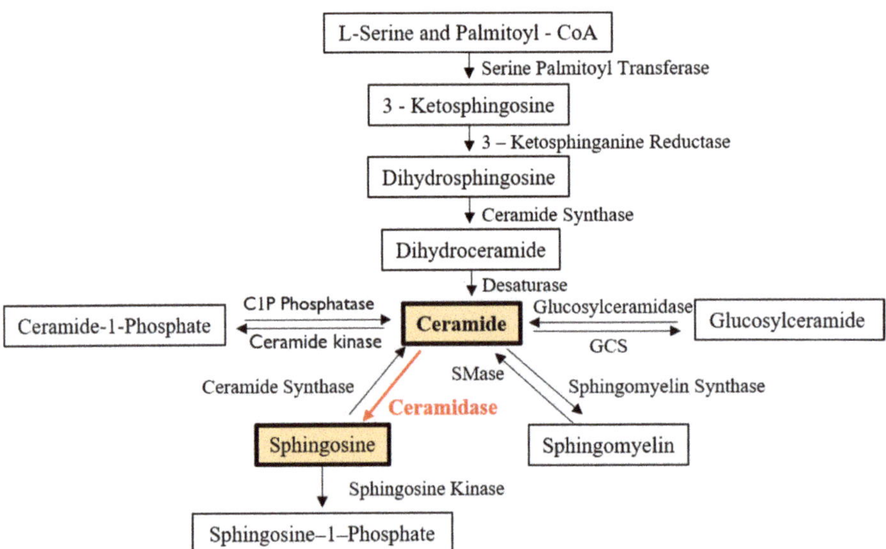

Figure 1. Role of ceramidases in ceramide metabolism. Ceramide in mammalian cells may be generated: (1) via the de novo synthesis pathway, which begins with the condensation of L-serine and Palmitoyl-CoA); (2) by the hydrolysis of sphingomyelin and glucosylceramide; or (3) from the dephosphorylation of ceramide-1-phosphate. Ceramidase is an enzyme that cleaves fatty acids from ceramide, producing sphingosine. Sphingosine may then be phosphorylated by a sphingosine kinase to form sphingosine-1-phosphate. SMase—sphingomyelinase; GCS—glucosylceramide synthase.

The de novo anabolic pathway for the biosynthesis of ceramide begins with the condensation of the amino acid, L-serine, and palmitoyl-CoA, producing 3-ketosphingonine. Then, 3-ketosphingonine is quickly converted into dihydrosphingosine (dhSPH) by 3-ketosphinganine reductase. The subsequent acylation of dhSPH by (dihydro)ceramide synthases gives rise to dihydroceramides. Finally, the removal of two hydrogens from a fatty acid chain of the dihydroceramides by the enzyme desaturase results in the formation of ceramides (Figure 1) [2]. Ceramides can be further hydrolyzed into sphingosine (SPH) and free fatty acids by ceramidases (Figure 1). SPH is the most common sphingolipid base molecule in mammalian cells and is the precursor of S1P [1].

The bioactive lipid mediator S1P is involved in cell proliferation, differentiation, and survival, whilst ceramides and SPH mediate cell death [1,2]. Notably, SPH is exclusively generated from the catabolism of ceramides by ceramidases [2]. Ceramidases control the balance between S1P and ceramides/SPH concentration, which leads to either cell survival or cell death [1,3]. Hence, a ceramidase-based enzyme replacement therapy that simultaneously achieves ceramide reduction and SPH elevation has been recently examined [4]. This therapeutic approach intends to reduce the negative pathophysiological impact of cell death mediated by ceramides [4]. To date, five human ceramidases have been identified and classified according to their optimal pH for catalytic activity: one acid ceramidase (ACDase) encoded by the gene ASAH1, one neutral ceramidase (NCDase) encoded by ASAH2, and three alkaline ceramidases (ALKCDase) encoded by the genes ACER1, ACER2 and ACER3 [3]. ACDase is the most widely studied ceramidase, and its effects on pathophysiology are variable. While NCDase and ALKCDases 1–3 have been the subject of studies in recent decades, our knowledge about their roles in pathophysiology remains limited. Thus, the aim of this review was to summarize the most recent knowledge of the biology of ceramidases and their role in the pathology of various common human diseases, including cancer, diabetes, and neurodegenerative diseases

(Figure 2). In addition, the roles of the ceramidases in infectious diseases, tissue regeneration, and healing were also addressed.

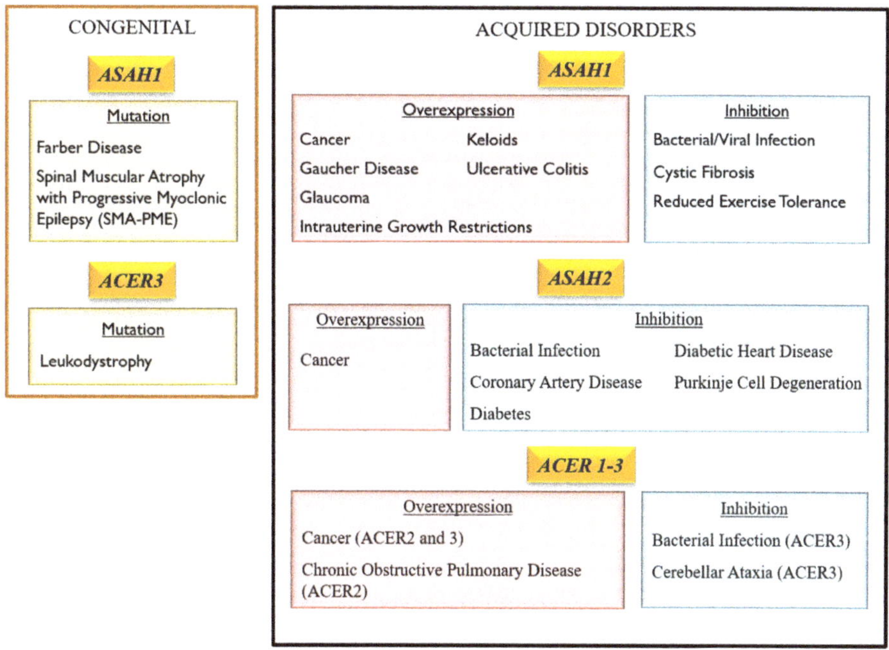

Figure 2. Pathological consequences of ceramidase dysregulation in mammalian cells which occur upon a loss or a gain of function.

2. Characteristics and Regulatory Pathways of Ceramidases

2.1. General Characteristics

2.1.1. Acid Ceramidase

ACDase (ASAH1) is synthesized from a 53–55 kDa polypeptide precursor which is proteolytically processed into the enzyme's 13 kDa α-subunit and 30 kDa β-subunit inside the lysosomes [5,6]. ACDase is a lipid hydrolase found in the lysosomal compartment of cells which catalyzes the hydrolysis of $C_{6:0}$–$C_{18:0}$ ceramides to SPH [6]. It is ubiquitously expressed in human tissues, with a particularly high expression in the heart and kidneys and is known for its role in senescence and apoptosis [7].

2.1.2. Neutral Ceramidase

NCDase (ASAH2) is synthesized as 118 kDa and 142 kDa isoforms in humans and as a 96 kDa molecule in mice [6]. NCDase is a transmembrane glycoprotein that is highly expressed in the human intestinal system and uses various ceramides and dihydroceramides as a substrate, with a reported preference for $C_{16:0}$ and $C_{18:0}$ ceramides [8,9]. It is localized to different cellular compartments, including the plasma membrane of cells; regulates the conversion of ceramide into SPH and S1P; and is important for the metabolism of dietary sphingolipids [10].

2.1.3. Alkaline Ceramidase

ALKCDases 1–3 (*ACER1, ACER2, ACER3*) are the smallest proteins among the ceramidases, with molecular weights of 31–31.6 kDa [6,11]. ALKCDases are predominantly located in the Golgi complex and endoplasmic reticulum and play a role in cell differentiation [11]. ACER1 hydrolyses $C_{20:0}$-$C_{24:0}$ ceramides [9]. It is predominantly expressed by skin cells and is involved in their differentiation, as well as in the viability of hair follicle stem cells [6,12–14]. ACER2 hydrolyses ceramides and dihydroceramides $C_{18:1}$ and $C_{20:1}$ [9]. It is upregulated during DNA damage and induces programmed cell death through an SPH-dependent pathway [15]. ACER3 hydrolyses ceramides, dihydroceramides, and phytoceramides with long unsaturated acyl chains [9]. It has been described as a seven-transmembrane protein, much like the adipocyte receptor (ADIPOR), and is associated with cytokine upregulation [16].

This brief overview of the general characteristics of ceramidases indicates that they have been classified according to their optimal pH. However, ceramidases also differ in molecular weight and expression patterns. Importantly, all three groups of ceramidases have a specific ceramide affinity and reported cellular functions. It is important to highlight that the ALKCDases, although classified together, differ in ceramide affinity and function. Moreover, ALKCDase-3, like NCDase, is a transmembrane protein and not a soluble enzyme. Thus, further consideration should be given to the classification of ceramidases and, particularly, the ALKCDases.

2.2. Regulatory Pathways

2.2.1. Acid Ceramidase

The activation of ACDase induces a pro-survival state, while its inhibition leads to cell death through a variety of apoptotic pathways mediated by caspases (CASP), poly (adp-ribose) polymerase (PARP), or cathepsins (CTS) [10,17–25]. Cathepsin B and Cathepsin D are activated during ceramide-induced apoptosis but are inhibited by ACDase activity [19,25]. Interestingly, the downregulation of Cathepsin B by ACDase increases ACDase's own activation, triggering a feedback mechanism through which ACDase prolongs its own activation through Cathepsin B inhibition [10]. Additionally, ACDase activity can be regulated by Ceramide Synthase 6 (CerS6) [26]. CerS6 increases the levels of $C_{16:0}$, which, in turn, activate ACDase through JNK-AP1-dependent mechanisms. However, this same mechanism mediates the inhibition of the gene expression of NCDase and ALKCDases in colorectal adenocarcinoma [26].

An age-dependent inhibition of ACDase leads to ceramide accumulation, an increase in oxidative stress, and the death of retinal cells and erythrocytes [27,28]. By contrast, it was reported that kidney cells collected from aged mice show an elevated expression of *Asah1* mRNA compared to that of young mice [29]. Thus, this published evidence suggests a tissue-specific ACDase activity in relation to cellular senescence and aging.

2.2.2. Neutral Ceramidase

The activity and gene expression of NCDase have been linked to cell-cycle arrest and growth regulation [30]. Biochemically, NCDase is a lipid amidase with a mechanistic similarity to a bacterial NCDase [8]. NCDase activates nitric oxide (NO), the WNT/β-catenin pathway, caspase apoptotic pathways, and autophagosomal activity in vivo and is associated with mitochondrial integrity [31–34]. Its gene expression and activity are regulated by c-Jun/AP-1 signaling, NO, all-trans retinoic acid, and ultra-violet radiation [35,36].

2.2.3. Alkaline Ceramidase

The ALKCDases 1–3 are regulated through markedly different mechanisms. ACER 1 is upregulated by extracellular calcium, through which it contributes to the regulation of cell differentiation and growth arrest [37]. Meanwhile, ACER2 is induced by p53 and activates p38 MAPK and AP-1 signaling to mediate DNA damage response, autophagy, and apoptosis [15,38,39]. ACER3 is associated with the AKT/BAX pathway and activates the S1P phosphorylation of AKT through S1PR2 and PI3K in cancer cells [40,41].

3. Association of Ceramidase Gene Mutations with Human Inheritable Diseases

3.1. Acid Ceramidase

3.1.1. Farber Lipogranulomatosis (FRBRL)

FRBRL is an autosomal recessive lysosomal disorder with a broad spectrum of phenotypes caused by 16 identified mutations of *Asah1* [5]. It is characterized by a substantial neurologic deficit, subcutaneous nodules, progressive arthritis with joint deformities, laryngeal hoarseness, and an accumulation of storage-laden CD68[+] macroglia/macrophages in white matter, periventricular zones, and meninges of the brain [42]. Animal models of ACDase deletion present hematopoietic organ hypertrophy, characterized by a foamy macrophage infiltration and increased myeloid progenitor colonies [43]. These myeloid progenitor colonies are comprised of cells that can develop normally when treated with ACDase [43]. Additionally, ACDase deletion causes an impaired airway resistance, elastance, and compliance; reduced blood oxygenation; lung edema; and increased immune cell infiltration of the lungs by foamy macrophages and neutrophils [44]. Furthermore, an increased vascular permeability of the lungs, heart, thymus, liver and spleen, as well as neurologic problems, including decreased deambulation, anxiety, and impaired motor coordination, are also observed [42]. These neurologic problems are caused by abnormal sphingolipid profiles in the brain and CD68+ microglia [42].

Changes in the *Asah1* gene expression in FRBRL patients result in the upregulation of the inflammatory cytokines interleukin 4 (IL-4), IL-6, tumor necrosis factor alpha (TNFα), and macrophage colony stimulating factor (M-CSF) in addition to the angiogenic marker, vascular endothelial growth factor (VEGF) [42,45]. Likewise, the expressions of the chemo-attractants, monocyte chemotactic protein-1 (MCP-1), and interferon gamma-induced protein 10 (IP10), are inversely correlated with the level of ACDase activity [45]. These mediators of inflammation, angiogenesis, and insulin resistance may be associated with the immune cell infiltration found in the organ tissues of FRBRL animal models [45]. MCP-1 deletion can partially rescue FRBRL phenotypes by improving organomegaly, blood cell counts, and liver and lung damage by inflammatory infiltrates, as well as the behavioral and neurologic aspects of the disease. However, hematopoiesis is not improved [46]. Similarly, the overexpression of MCP-1, IP-10, and IL-6 can be partially corrected by hematopoietic stem cell transplants [45]. Moreover, treatment with ACDase induces a dose-dependent decrease in hematopoietic organ weight, macrophage infiltration, and *MCP-1* expression, as well as increased expression of Collagen Type 2 *(Col2)*, aggrecan, and Sox-9 by chondrocytes [4].

3.1.2. Spinal Muscular Atrophy with Progressive Myoclonic Epilepsy (SMA-PME)

SMA-PME is a rare autosomal recessive disorder that is frequently associated with FRBRL and is caused by two identified mutations of *Asah1* [5]. This disorder is characterized by motor neuron disease and progressive myoclonic epilepsy, with a variable occurrence of sensorineural hearing loss, action tremor, cognitive dysfunction, and cerebral/cerebellar atrophy. Patients with SMA-PME present a 70–95% reduction in ACDase activity, a low ACDase/β-galactosidase ratio, and increased creatine kinase levels [47]. Additionally, the muscle atrophy associated with SMA can be accompanied by cyclooxygenase deficiency [48].

3.1.3. Intrauterine Growth Restrictions (IUGR)

The consequence of ACDase gene overexpression during gestation and its therapeutic effect on associated genetic disorders has also been described. IUGR can result from the TGFβ/ALK5-mediated overexpression of *Asah1* mRNA and increased ACDase activity, which upregulates SPH but not S1P concentrations during pregnancy [49]. S1P is not upregulated at the same rate as SPH in IUGR due to the inactivation of SPH kinase 1 through the ALK1-SMAD1/5 pathway [49].

This suggests that ACDase may induce embryonic cell death through SPH rather than affect embryonic cell proliferation and differentiation through S1P in IUGR.

3.1.4. Krabbe Disease

Globoid cell leukodystrophy, or Krabbe disease, is a congenital disorder caused by mutations in the galactosylceramidase gene, *GALC*, and is characterized by psychomotor regression, muscular hypertonia, muscular spasticity, truncal hypotonia, irritability, seizures, and nystagmus [50]. This disorder is caused by an accumulation of psychosine, a by-product of the deacylation of GALC by ACDase [51]. Therefore, the inhibition of ACDase activity, as observed in FRBRL or after treatment with ACDase inhibitors, can rescue the Krabbe Disease phenotype by preventing psychosine accumulation [51].

3.2. Neutral ceramidase

There are no reports demonstrating the association of the point genetic mutations of ASAH2 with inheritable diseases in humans. It is important to mention that *Asah2*$^{-/-}$ mice are viable and appear without severe defects [52].

3.3. Alkaline Ceramidase

Progressive Leukodystrophy

Progressive leukodystrophy is a group of disorders that affect the white matter of the brain and can occur as a consequence of ACER3 deficiency [53]. This condition is caused by a loss of function mutation in p.E33G which inactivates the catalytic activity of ACER3 and leads to an accumulation of sphingolipids in the blood [53]. The clinical phenotype associated with ACER3 mutations is caused by incorrect central nervous system myelination due to abnormal levels of ceramides in the brain [16]. While the study reporting the loss of function mutation in p.E33G did not report a sphingolipid accumulation or pattern in the brain, it is reasonable to assume that sphingolipid accumulation due to ACER3 inactivation results in abnormal sphingolipid patterns in the brain.

4. Role of Ceramidase Activity in Human Non-heritable Diseases

4.1. Role of Ceramidases in Cancer Pathology

The overexpression of ceramidases have been identified in various cancer cell types, and growing evidence suggests that they can be considered molecular markers and/or therapeutic targets for cancer [54] (Figure 3).

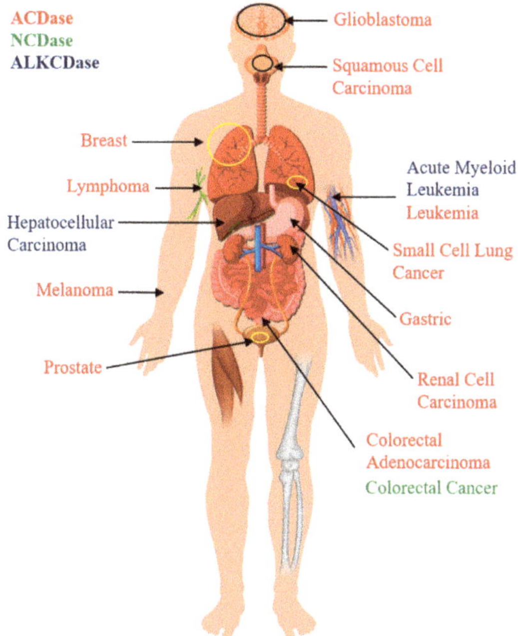

Figure 3. Overexpression of acid (ACDase), neutral (NCDase), and alkaline (ALKCDase) ceramidases in specific types of cancer.

4.1.1. Acid Ceramidase

ASAH1 has been identified in cancer cells and is associated with radiotherapy/chemotherapy-resistant tumors [22,55–57], metastatic cell lines [58], and estrogen/progesterone/androgen receptor-positive cells [59,60]. While ACDase gene overexpression has been identified in low-survival-rate colorectal adenocarcinoma and glioblastoma [57,61], it has also been observed in node-negative melanoma and breast cancer [59,62], which makes it a questionable marker for the aggressiveness or invasiveness of the disease. The *ASAH1* mRNA expression in cancer cells can be increased by radiotherapy, thereby generating resistance [56]. Likewise, the overexpression of *ASAH1* can be driven by the oncogene microphthalmia-associated transcription factor (MITF) [63]. ACDase activity is increased by the androgen receptor activation by dihydrotestosterone in prostate cancer, leading to decreased $C_{16:0}$ levels and reduced cell apoptosis [60]. Incidentally, the ACDase activity is significantly more upregulated than the *ASAH1* expression in melanoma cells [62]. This may suggest that gene expression alone should not be the determining factor in the use of ACDase as a marker for cancer; ACDase activity should also be assessed.

Multiple molecular mechanisms by which ACDase activation regulates cancer development and progression have been identified. For instance, drug resistance in leukemia is mediated by the ACDase activation of the drug transporter molecule ATP-Binding Cassette, Subfamily B, Member 1 (ABCB1), through nuclear factor kappa B (NF-κB) [64], whilst leukemic cancer cell survival is increased by the ACDase-mediated upregulation of the myeloid cell leukemia sequence 1 (MCL-1) [10]. Furthermore, cancer cell necrosis is mediated by ACDase gene overexpression in polynuclear giant cancer cells that undergo asymmetric cell division [65]. ACDase also regulates cancer cell motility through the activation of the ITGαVβ5/FAK signaling cascade [63]. Additionally, the significant roles of ACDase in angiogenesis, chronic inflammation, and tumorigenesis may contribute to cancer development and progression [66,67]. ACDase affects multiple factors in cancer pathogenicity, which adds to the complexity of the enzyme in the diagnosis and treatment of the disease.

A variety of ACDase inhibitors have been developed and successfully tested in different cancer cell types. ACDase deletion blocks the cell cycle at G1/S, promotes senescence through the β-Galactosidase/MITF pathway, induces apoptosis, reduces tumorigenesis, increases growth arrest, and decreases malignancy [68]. It was demonstrated that the activity of ACDase was significantly inhibited by Carmofur [55,58], LCL521 [19,65,69,70], Ceranib2 [24,71–73], N-oleocylethanolamine (NOE) [22,57], ARN14988 [74], LCL204 [10,64], Monascus Purperus (MP) [18], Hesperetin (Hst) [17], Hesperetine-7-O-acetate (HTA) [17], Silibinin [20], Curcumin [23], and Sanguinarine [21,75], leading to an increased accumulation of intracellular ceramide and apoptosis in various types of cancer cells, including glioblastoma; squamous cell carcinoma; acute myeloid leukemia; colorectal adenocarcinoma; and breast, prostate, lung, gastric, and kidney cancer. Furthermore, Carmofur, NOE, LCL521, and Ceranib2 have been used in combination with chemotherapeutic drugs or photodynamic therapy to either overcome cancer cell resistance to treatment, increase cell sensitivity to specific drugs, or increase the overall effectiveness of cancer cell apoptosis [22,55,58,70,72,73]. Ceranib2 treatment leads to an abnormal cell and mitochondria morphology and decreases the ability of cells to cluster [24,74]. It activates PARP and CASP3/7/8/9-mediated cell apoptosis; increases the expression of the pro-apoptotic markers BID, BCL2-Associated Agonist Of Cell Death (*BAD*), and BCL2-Associated X Protein (BAX); and decreases the expression of anti-apoptotic protein B-Cell Cll/Lymphoma 2 (BCL-2) [72,73]. Furthermore, MP, Hst, HTA, Curcumin, and Sanguinarine activate the apoptotic pathways dependent on Casp3/9 or reactive oxygen species (ROS) [17,18,21,23,75]. Sanguinarine induces peroxide-dependent ceramide generation and the inhibition of the AKT activation pathway [75]. NOE and LCL204 induce PARP- and Casp3-mediated apoptosis [10,22], whereas LCL521 increases $C_{16:0}$ levels, autophagosome accumulation, ER stress, and Cathepsin B- or Cathepsin D-mediated apoptosis [19]. Altogether, ACDase inhibitors are effective promoters of cancer cell death through different apoptotic pathways and have been shown to affect not only apoptosis but also cancer treatment resistance and cancer cell adhesion.

4.1.2. Neutral Ceramidase

An elevated gene expression of NCDase has been identified in both the plasma membrane and Golgi apparatus of colorectal cancer (CRC) cells, where its overexpression inhibits ceramide C6-mediated cell death [8]. Meanwhile, its deletion induces caspase and autophagosome-mediated apoptosis in the presence of C6 [32]. NCDase regulates CRC cell proliferation through the WNT/β-catenin pathway and by increasing the accumulation of SPH and S1P [31,32]. NCDase inhibition may affect cell-to-cell adhesion by reducing the β-catenin levels through AKT phosphorylation and, subsequently, GSK3β activation [31]. It also significantly reduces Azoxymethane-induced colon carcinogenesis by inhibiting aberrant crypt foci formation and transformation [32]. NCDase inhibition does not affect non-cancerous cell function, which makes it a suitable target for colon cancer therapy [32].

We can conclude that NCDase inhibition, like that of ACDase, activates apoptosis and affects adhesion in cancer cells. In addition, it may be a contributing factor in cancerous transformation.

4.1.3. Alkaline Ceramidase

ALKCDases can also affect cancer development and treatment. ACER2 is upregulated by the tumor suppressor gene p53 [38,39]. It was demonstrated that a moderate upregulation of ACER2 increases the levels of SPH and S1P and inhibits cell cycle arrest and senescence. However, when overexpressed, ACER2 mediates programmed cell death, autophagy, and apoptosis through ROS [15,38,39]. ACER2 also contributes to the effects of ionizing radiation treatment [39]. It also increases the phosphorylation of Ezrin-radixin-moesin through intracellular S1P production, hereby inactivating this group of proteins that regulate cell shape and motility and have been associated with cancer progression and metastasis [76]. ACER3 is expressed in low-survival hepatocellular carcinomas and acute myeloid leukemia [40,41]. It induces the S1P phosphorylation of AKT through the S1P receptor 2 and PI3K and inhibits the AKT/BAX apoptotic pathway in cancer cells [40,41]. Therefore, the inhibition of ACER3 reduces cell growth and increases cancer cell apoptosis.

These published observations indicate that ALKCDases are also associated with the regulation of cancer cell apoptosis. However, the observations of ACER2 overexpression reflect molecular effects contrary to those expected of ceramidases. Nonetheless, ALKCDases are associated with drug resistance and cancer metastasis, like the previously described ACDase.

4.2. Role of Ceramidases in the Onset of Age-Related Neurodegenerative Diseases

Ceramidases are involved in myelin and fatty acid metabolism and are associated with changes in the brain during aging [77]. For instance, ACER3 is upregulated with age and leads to a decrease in the brain levels of $C_{18:0}$ and $C_{18:1}$ ceramide, and its deletion results in purkinje cell degeneration and impaired motor coordination and balance in mice [78]. It has been reported that the overexpression of ACDase has implications for the onset and progression of neurodegenerative diseases, including Alzheimer's disease (AD) and Gaucher disease. Furthermore, treatment with ACDase inhibitors can control AD and Gaucher Disease, as well as Type IV Mucolipidosis.

Acid Ceramidase

AD is a multifactorial, highly heterogeneous, and complex disorder that affects the memory and cognitive functions of patients to the extent that they are completely dependent upon nursing care. It is now estimated that nearly 35.6 million patients are affected by AD worldwide and that about 4.6 million new cases are added each year, causing enormous societal and economic burdens, with the estimated cost reaching $1 trillion/year [79]. AD is caused by an accumulation of derivates from the amyloid precursor protein (APP), which can be modulated by the ATP-binding cassette transporter-2 (ABCA2). ABCA2 is a phospholipid transporter which increases the transcription of APP by activating the ACDase-mediated production of SPH [80]. Furthermore, ACDase inhibition by Ceranib 1 decreases SPH concentration and, subsequently, APP production in ABCA2-overexpressing cells [80].

Gaucher disease is a disorder caused by a loss of function mutations in the glucocerebrosidase (GCase)-encoding gene, Gba1. In a GCase deficiency, the breakdown of glucocylceramide (GlcCer) into ceramide and glucose by GCase is replaced by the ACDase deacylation of GlcCer into glucocylSPH (Glc-Sph), a cytotoxic compound [80]. The inhibition of ACDase by Carmofur corrects the lipid abnormalities in the GCase deficiency by reducing the accumulation of Glc-Sph [81,82]. GBA1 mutations are also a risk factor for Parkinson's disease, a neurodegenerative disorder characterized by Lewy body inclusions containing α-synuclein. Treatment with ACDase inhibitors decreases the accumulation of α-synuclein in cases of GBA1 mutation [81]. Similar lipid patterns are observed in the optic nerves of glaucoma patients, where Asah1 and Asah2 genes are overexpressed, but non-lysosomal GCase-GBA2 is inhibited, resulting in a lower total lipid content and significantly higher concentrations of Glc-Sph [83].

Type IV Mucolipidosis is a neurodegenerative disease caused by a loss-of-function mutation of human transient receptor potential-mucolipin-1 (TRPML-1). Treatment with the ACDase inhibitor, carmofur, induces the activity of TRPML-1 tunnels by increasing the SPH concentration in kidney cells and acting as a mediator of lysosome fusion and trafficking in multivesicular bodies, which can potentially compensate for the loss of function of TRPML-1 [84].

4.3. Role of Ceramidases in Cardio-Pulmonary Disease

Elevated levels of ceramide are known to be correlated with adverse cardiac events, whereas SPH has been shown to increase intracellular NO levels and maintain the mitochondrial integrity of the cardiovascular system [33]. Conversely, increased blood S1P levels are associated with the pathogenesis of inflammatory and cardiovascular diseases [85]. Hence, an association between ceramidase and cardiopulmonary events is expected.

4.3.1. Acid Ceramidase

The inhibition of ACDase activity is associated with cystic fibrosis (CF), which is caused by a dysregulation of the epithelial fluid transport in the lungs, resulting in a sticky dry mucous accumulation [86]. In CF, β1-Integrins are ectopically expressed in the luminal pole of epithelial cells and downregulate ACDase, leading to an increased ceramide accumulation. However, treatment with recombinant ACDase internalizes the β-Integrins and regulates ceramide accumulation, rescuing the CF phenotype [86].

4.3.2. Neutral Ceramidase

NCDase is inhibited in coronary artery disease vessels. NCDase and ADIPOR mediate the NO-dependent flow-induced dilation (FID) through S1P. Meanwhile, NCDase inhibition induces the damaging peroxide-dependent FID [33]. In addition, the inhibition of NCDase also leads to mitochondrial dysfunction in diabetic hearts through a lactocylceramide accumulation [87].

4.3.3. Alkaline Ceramidase

A high expression of ALKCDase genes, particularly ACER2, has been observed in cardiac tissue during hypoxia, where it plays a protective role [88]. However, an overexpression of ACER2 has been associated with chronic obstructive pulmonary disease (COPD) [89]. ACER2 inhibition contributes to a reduction in the circulating S1P and its analogue, dhS1P, as well as their precursors, SPH and dhSPH, in hematopoietic cells and reduces the concentration of dhS1P in the lungs [85,88].

These data indicate that ACDase and ALKCDase are increased in CF and COPD, respectively, whereas NCDase is decreased in coronary artery disease.

4.4. Role of Ceramidases in Metabolic Disorders

4.4.1. Acid Ceramidase

Multiple factors are involved in the onset and progression of metabolic disease, including the activities of ceramidases. Genetic variations of *ASAH1* have been associated with exercise tolerance and skeletal/cardiac muscle adaptation to exercise, which can condition adherence to physical activity regimens necessary for a healthy lifestyle, thereby increasing the individual risk of metabolic diseases [90]. After onset, metabolic disorders affect the physiology of the cardiovascular system, kidneys, and liver. Hyperglycemia inhibits the Unc51-Like Autophagy-Activating Kinase 1 (ULK1) phosphorylation in aortic endothelial cells, which leads to a dysregulation of autophagy and atherogenesis. However, ACDase activity can increase the phosphorylation of ULK1 and restore its function even in nutrient-rich conditions, thus preventing atherogenesis [91]. In addition, obesity-induced kidney damage is caused by hyperglycemic conditions that stimulate the NLR Family Pyrin Domain-Containing 3 (NLRP3) inflammasomes to release IL-1β in podocytes, but the treatment of podocytes with ACDase decreases the NLPR3-induced cytokine release through extracellular vesicles [92]. ACDase reduces the activity of Pannexin-1 (Panx1), a transmembrane channel glycoprotein that activates NLRP3 through S1P accumulation [93]. Animal models of ACDase deficiency show significant damage to the liver and change to lipid profiles and metabolism, including hepatomegaly with higher serum levels of aspartate, aminotransferase, alanine aminotransferase, and alkaline phosphatase and decreased levels of free fatty acids, triglycerides, and cholesterol [25]. The inducible liver-specific overexpression of ACDase in the Alb-AC transgenic mice, results in significantly reduced $C_{16:0}$ ceramide in the liver and improved total body glucose homeostasis and insulin sensitivity under a high-fat diet [94]. However, aberrant ACDase overexpression in very low-density lipoprotein (VLDL) deficiency may result in non-alcoholic fatty liver disease, which can be normalized by supplementation with Vitamin E [95]. Adipocyte-specific ACDase overexpression improves glucose metabolism by white adipose tissue, reverses insulin resistance, reduces lipid accumulation in the liver, and reduces adipose inflammation and fibrosis [94]. This could be due to an ACDase-mediated activation of the adiponectin receptor that triggers an AMP-dependent kinase pathway, which subsequently inhibits adipogenesis and induces fatty acid oxidation [96].

4.4.2. Neutral Ceramidase

Palmitate is a precursor of palmitoyl-CoA, a thioester used in the de novo biosynthesis of ceramide that is associated with pancreatic β-cell apoptosis and insulin resistance. Palmitate inhibits NCDase gene expression and activity in pancreatic β cells, which, in turn, exacerbates apoptosis through ceramide accumulation [97]. Pancreatic β cells secrete NCDase via exosomes that reduce palmitate-induced ROS and act as a protective mechanism against free fatty acid-induced apoptosis [98,99]. An overexpression of NCDase inhibits palmitate-induced apoptosis and may be a therapeutic target for type 2 diabetes mellitus and lipotoxicity [97]. Furthermore, Asah2 is one of the four genes related to sphingolipid metabolism that are deregulated in animal models of type 3 maturity-onset diabetes of the young [100]. This pathology is characterized by increased ceramide and SPH levels as well as hypochromic microcytic anemia, with abnormally-shaped and osmotically fragile red blood cells characterized by an accumulation of SPH [100].

4.4.3. Alkaline Ceramidase

Non-alcoholic fatty liver disease is associated with an increased expression of ACER3, which reduces the accumulation of $C_{18:1}$-ceramide in the liver [101]. Acer3 deletion reduces inflammation, fibrosis, oxidative stress, and apoptosis of hepatocytes through a palmitic acid-induced increase in $C_{18:1}$-ceramide [101].

Altogether, we can conclude that the holistic beneficial effects of ACDase in metabolic disease have been demonstrated. ACDase activity controls atherogenesis, kidney damage, and liver damage, while improving glucose and lipid metabolism. In addition, NCDase also appears to improve metabolic conditions via a protective effect on pancreatic β cells, while ALKCDase3 mediates liver damage.

5. Role of Ceramidase Activity in Infectious Diseases

5.1. Role of Ceramidase Activity in Bacterial Infection

5.1.1. Acid Ceramidase

Ceramidases have been identified as contributors to bacterial infection and mediators of the immune response and inflammation. The α-toxin released by *Staphylococcus aureus* inhibits ACDase gene expression, causing decreased levels of SPH that contribute to bacterial infection susceptibility [102]. Moreover, this mechanism further increases the risk of infection by *S. aureus* and *Pseudomonas aeruginosa* in already ACDase- and SPH-deficient CF patients [86,102]. *Porphyromonas gingivalis*, an etiological factor for periodontitis, downregulates ACDase in periodontal tissues, thereby increasing its own apoptotic potential and inhibiting the host's inflammatory response [103]. The inhibition of ACDase by bacteria increases host cell apoptosis and reduces the production of inflammatory cytokines, such as TNF-α, IL-1β, IL-6, and IL-17A, which delay the immune response [67,103]. Conversely, ACDase overexpression upregulates the inflammatory cytokines involved in the recruitment of neutrophils and macrophages, as demonstrated in ulcerative colitis, where ACDase mediates the associated histopathological characteristics of the disease [67].

5.1.2. Neutral Ceramidase

Ceramide accumulation is increased after burn injuries and may be associated with bacterial infections that frequently lead to death. NCDase treatment protects against *Pseudomonas aeruginosa* infection after burn injuries by controlling ceramide accumulation and inducing the accumulation of SPH, which directly kills bacteria [104].

5.1.3. Alkaline Ceramidase

Bacterial lipopolysaccharides may also downregulate the expression and activity of *Acer3* and increase $C_{18:1}$ ceramide accumulation in mice [105]. A loss of *Acer3* expression leads to the production of pro-inflammatory IL-1β, IL-6, IL-23α, and TNF-α cytokines from peritoneal macrophages, bone mononuclear cells, and colonic epithelial cells isolated from Acer3$^{-/-}$ mice [105]. Overall, the bacterial species inhibit ceramidase activity to reduce the concentration of SPH in the host cells, which results in a reduced immune response.

5.2. Role of Ceramidase Activity in Viral Infection

Viruses have the potential to spread among individuals, resulting in epidemics that cause loss of human life and heavy burdens to healthcare systems [106]. The influenza, Ebola, and Zika epidemics are recent examples of the effects of broad viral infection and of the mechanisms by which viruses can be studied and controlled [106–108]. A recent mutation of the coronavirus, named SARS-CoV-2, has caused a pandemic of unprecedented magnitude. This virus has a lower mortality rate but is exponentially more contagious than the closely related SARS-CoV and MERS-CoV [109]. However, our knowledge of the potential role of host ceramidases in viral pathology remains elusive. It was reported that the overall inhibition of ceramidase activity in host peripheral blood lymphocytes using Ceranib 1 and Ceranib 2 significantly reduces the replication of the rhinovirus and measles virus, respectively [110,111]. Furthermore, the inhibition of ACDase activity in macrophages significantly increases the propagation of herpes simplex virus-1, which, in turn, elevates the mortality rate in Asah1$^{-/-}$ mice [112].

Collectively, these published observations indicate that ceramidases may have an important antiviral effector role that should further studied.

6. Role of Ceramidases in Tissue Regeneration and Healing

Ceramidases are expressed in epithelial cells and fibroblasts and may be involved in their response through S1P [12,103,113]. However, only a limited number of studies have demonstrated the effects of these enzymes in tissue regeneration and healing.

6.1. Acid Ceramidase

ACDase activity contributes to physiological processes involving collagen turnover. ASAH1 is associated with familial keloid healing and is overexpressed in keloid scar tissue and hypertrophic scars caused by excessive collagen deposition during epidermal healing [114]. In the liver, hepatic stellate cells (HSC) are activated during normal wound healing but can, after multiple activations, cause hepatic fibrosis. However, the inhibition of ACDase by tricyclic antidepressants leads to ceramide accumulation, which inactivates HSCs and prevents hepatic fibrosis [115]. In vivo studies have also demonstrated a positive effect of ACDase in chondrocyte differentiation. In cartilage replacement therapy, pre-treatment with ACDase induces chondrocyte proliferation, the production of glycosaminoglycan, the expression of *COL2*, the adhesion of chondrocytes to a scaffold, a reduced resorption after implantation, and an improved differentiation to cartilage [116]. Furthermore, a variation of FRBRL characterized by peripheral osteolysis not associated with MMP-2 and MMP-14 was found, suggesting the involvement of ASAH1 in bone remodeling [117].

6.2. Neutral Ceramidase

Various studies have focused on the use of exosomes for tissue repair and regeneration [118,119]. The results of a recent study indicated that hepatocyte exosomes show significant NCDase activity and promote hepatocyte proliferation in vitro and liver regeneration in vivo [118]. This suggests a role of NCDase in tissue regeneration. The ceramidase has also been identified as an antagonist of cell necrosis caused by 2DG/AA-dependent ceramide accumulation and mitochondrial damage [34]. Furthermore, NCDase increases autophagy and protects cells from ER stress-mediated cell death [34].

6.3. Alkaline Ceramidase

ACER1 inhibition leads to abnormal hair, alopecia, hyperproliferation, inflammation, an abnormal differentiation of the epidermis, sebaceous gland abnormalities, and infundibulum expansion, as well as an increased trans-epidermal water loss and hypermetabolism with an associated reduction in fat content during aging [12]. Its inhibition gradually depletes the number of hair follicle stems and causes alopecia through decreased hair follicle activity [14]. The specific mechanisms through which these effects of ALKCDase occur are still not detailed in the literature. However, its expression has been associated with keratinocyte growth arrest and differentiation [13]. Altogether, these data suggest that ACDase is involved in collagen matrix metabolism, whereas NCDase and ALKCDase appear to affect tissue regeneration and healing through their anti-apoptotic effects.

7. Conclusions

Ceramidases (acid, neutral, alkaline) are key enzymes that maintain the intracellular homeostasis of ceramide/SPH and are critical regulators of signals that tilt the balance between cell survival and death. Various studies have demonstrated the involvement and potential therapeutic role of these enzymes in a diverse set of common human diseases, including bacterial-induced infectious diseases, neurodegenerative diseases, cancer, diabetes, and others (Figure 4). Therefore, the clinical applicability of studies examining the versatility of the effects of ceramidases in health and disease deserves further examination.

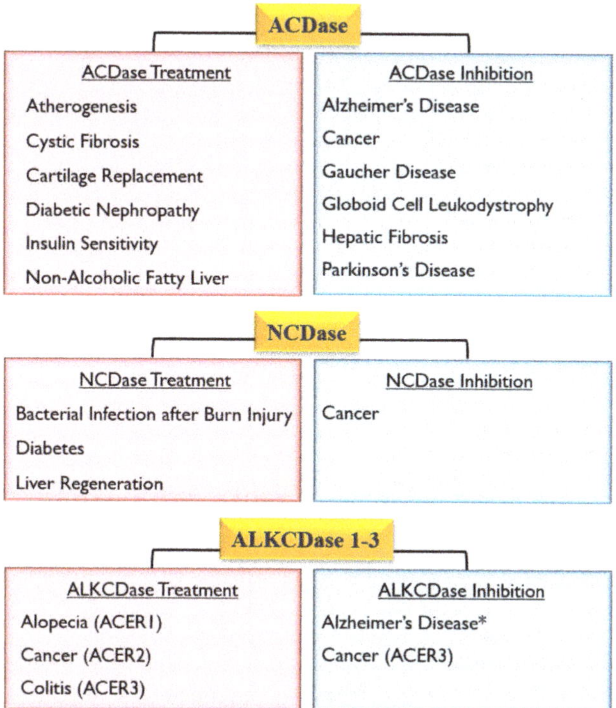

Figure 4. Potential therapeutic targeting of acid (ACDase), neutral (NCDase), and alkaline (ALKCDase) ceramidases. *Unspecified class of ALKCDase [80].

Funding: This research was funded by NIH, grant number AG064003, DE027153, and DE028699 (A.M), and NIH Research supplements to Promote Diversity in Health-Related Research (C.D., J.A.).

Conflicts of Interest: The authors declare no conflict of interest.

References

1. Bartke, N.; Hannun, Y.A. Bioactive sphingolipids: Metabolism and function. *J. Lipid Res.* **2008**, *50*, S91–S96. [CrossRef] [PubMed]
2. Książek, M.; Chacińska, M.; Chabowski, A.; Baranowski, M. Sources, metabolism, and regulation of circulating sphingosine-1-phosphate. *J. Lipid Res.* **2015**, *56*, 1271–1281. [CrossRef] [PubMed]
3. Coant, N.; Sakamoto, W.; Mao, C.; Hannun, Y.A. Ceramidases, roles in sphingolipid metabolism and in health and disease. *Adv. Boil. Regul.* **2016**, *63*, 122–131. [CrossRef] [PubMed]
4. He, X.; Dworski, S.; Zhu, C.; DeAngelis, V.; Solyom, A.; Medin, J.A.; Simonaro, C.M.; Schuchman, E.H. Enzyme replacement therapy for Farber disease: Proof-of-concept studies in cells and mice. *BBA Clin.* **2017**, *7*, 85–96. [CrossRef] [PubMed]
5. Gebai, A.; Gorelik, A.; Li, Z.; Illes, K.; Nagar, B. Structural basis for the activation of acid ceramidase. *Nat. Commun.* **2018**, *9*, 1621. [CrossRef]
6. Mao, C.; Obeid, L.M. Ceramidases: Regulators of cellular responses mediated by ceramide, sphingosine, and sphingosine-1-phosphate. *Biochim. et Biophys. Acta (BBA) - Bioenerg.* **2008**, *1781*, 424–434. [CrossRef]
7. Houben, E.; Holleran, W.M.; Yaginuma, T.; Mao, C.; Obeid, L.M.; Rogiers, V.; Takagi, Y.; Elias, P.M.; Uchida, Y. Differentiation-associated expression of ceramidase isoforms in cultured keratinocytes and epidermis. *J. Lipid Res.* **2006**, *47*, 1063–1070. [CrossRef]
8. Sakamoto, W.; Coant, N.; Canals, D.; Obeid, L.M.; Hannun, Y.A. Functions of neutral ceramidase in the Golgi apparatus. *J. Lipid Res.* **2018**, *59*, 2116–2125. [CrossRef]
9. Casasampere, M.; Camacho, L.; Cingolani, F.; Casas, J.; Egido-Gabás, M.; Abad, J.L.; Bedia, C.; Xu, R.; Wang, K.; Canals, D.; et al. Activity of neutral and alkaline ceramidases on fluorogenicN-acylated coumarin-containing aminodiols. *J. Lipid Res.* **2015**, *56*, 2019–2028. [CrossRef]
10. Tan, S.-F.; Liu, X.; Fox, T.E.; Barth, B.M.; Sharma, A.; Turner, S.D.; Awwad, A.; Dewey, A.; Doi, K.; Spitzer, B.; et al. Acid ceramidase is upregulated in AML and represents a novel therapeutic target. *Oncotarget* **2016**, *7*, 83208–83222. [CrossRef]
11. Hu, W.; Xu, R.; Sun, W.; Szulc, Z.M.; Bielawski, J.; Obeid, L.M.; Mao, C. Alkaline Ceramidase 3 (ACER3) Hydrolyzes Unsaturated Long-chain Ceramides, and Its Down-regulation Inhibits Both Cell Proliferation and Apoptosis*. *J. Boil. Chem.* **2010**, *285*, 7964–7976. [CrossRef] [PubMed]
12. Liakath-Ali, K.; Vancollie, V.; Lelliott, C.J.; Speak, A.O.; Lafont, D.; Protheroe, H.J.; Ingvorsen, C.; Galli, A.; Green, A.; Gleeson, D.; et al. Alkaline ceramidase 1 is essential for mammalian skin homeostasis and regulating whole-body energy expenditure. *J. Pathol.* **2016**, *239*, 374–383. [CrossRef] [PubMed]
13. Bermudez, L.E. Use of Liposome Preparation to Treat Mycobacterial Infections. *Immunobiol.* **1994**, *191*, 578–583. [CrossRef]
14. Lin, C.-L.; Xu, R.; Yi, J.K.; Li, F.; Chen, J.; Jones, E.C.; Slutsky, J.B.; Huang, L.; Rigas, B.; Cao, J.; et al. Alkaline Ceramidase 1 Protects Mice from Premature Hair Loss by Maintaining the Homeostasis of Hair Follicle Stem Cells. *Stem Cell Rep.* **2017**, *9*, 1488–1500. [CrossRef] [PubMed]
15. Xu, R.; Wang, K.; Mileva, I.; Hannun, Y.A.; Obeid, L.M.; Mao, C. Alkaline ceramidase 2 and its bioactive product sphingosine are novel regulators of the DNA damage response. *Oncotarget* **2016**, *7*, 18440–18457. [CrossRef]
16. Vasiliauskaite-Brooks, I.; Healey, R.D.; Rochaix, P.; Saint-Paul, J.; Sounier, R.; Grison, C.; Waltrich-Augusto, T.; Fortier, M.; Hoh, F.; Saied, E.M.; et al. Structure of a human intramembrane ceramidase explains enzymatic dysfunction found in leukodystrophy. *Nat. Commun.* **2018**, *9*, 5437. [CrossRef]
17. Boojar, M.M.A.; Boojar, M.M.A.; Golmohammad, S.; Bahrehbar, I. Data on cell survival, apoptosis, ceramide metabolism and oxidative stress in A-494 renal cell carcinoma cell line treated with hesperetin and hesperetin-7-O-acetate. *Data Brief.* **2018**, *20*, 596–601. [CrossRef]
18. Kurokawa, H.; Ito, H.; Matsui, H. Monascus purpureus induced apoptosis on gastric cancer cell by scavenging mitochondrial reactive oxygen species. *J. Clin. Biochem. Nutr.* **2017**, *61*, 189–195. [CrossRef]

19. Liu, F.; Li, X.; Lu, C.; Bai, A.; Bielawski, J.; Bielawska, A.; Marshall, B.; Schoenlein, P.V.; Lebedyeva, I.O.; Liu, K. Ceramide activates lysosomal cathepsin B and cathepsin D to attenuate autophagy and induces ER stress to suppress myeloid-derived suppressor cells. *Oncotarget* **2016**, *7*, 83907–83925. [CrossRef]

20. Boojar, M.M.A.; Hassanipour, M.; Mehr, S.E.; Boojar, M.M.A.; Dehpour, A.R. New Aspects of Silibinin Stereoisomers and their 3-O-galloyl Derivatives on Cytotoxicity and Ceramide Metabolism in Hep G2 hepatocarcinoma Cell Line. *Iran. J. Pharm. Res. IJPR* **2016**, *15*, 421–433.

21. Rahman, A.; Pallichankandy, S.; Thayyullathil, F.; Galadari, S. Critical role of H2O2 in mediating sanguinarine-induced apoptosis in prostate cancer cells via facilitating ceramide generation, ERK1/2 phosphorylation, and Par-4 cleavage. *Free. Radic. Boil. Med.* **2019**, *134*, 527–544. [CrossRef] [PubMed]

22. Roh, J.-L.; Park, J.Y.; Kim, E.H.; Jang, H.J.; Information, P.E.K.F.C. Targeting acid ceramidase sensitises head and neck cancer to cisplatin. *Eur. J. Cancer* **2016**, *52*, 163–172. [CrossRef] [PubMed]

23. Thangavel, S.; Yoshitomi, T.; Sakharkar, M.K.; Nagasaki, Y. Redox nanoparticles inhibit curcumin oxidative degradation and enhance its therapeutic effect on prostate cancer. *J. Control. Release* **2015**, *209*, 110–119. [CrossRef] [PubMed]

24. Ozer, M.Y.; Oztopcu-Vatan, P.; Kus, G. The investigation of ceranib-2 on apoptosis and drug interaction with carboplatin in human non small cell lung cancer cells in vitro. *Cytotechnology* **2017**, *70*, 387–396. [CrossRef]

25. Yu, F.P.S.; Molino, S.; Sikora, J.; Rasmussen, S.; Rybova, J.; Tate, E.; Geurts, A.M.; Turner, P.V.; McKillop, W.M.; Medin, J.A. Hepatic pathology and altered gene transcription in a murine model of acid ceramidase deficiency. *Lab. Investig.* **2019**, *99*, 1572–1592. [CrossRef]

26. Tirodkar, T.S.; Lu, P.; Bai, A.; Scheffel, M.J.; Gencer, S.; Garrett-Mayer, E.; Bielawska, A.; Ogretmen, B.; Voelkel-Johnson, C. Expression of Ceramide Synthase 6 Transcriptionally Activates Acid Ceramidase in a c-Jun N-terminal Kinase (JNK)-dependent Manner. *J. Boil. Chem.* **2015**, *290*, 13157–13167. [CrossRef]

27. Signoretto, E.; Zierle, J.; Bhuyan, A.A.M.; Castagna, M.; Lang, F. Ceranib-2-induced suicidal erythrocyte death. *Cell Biochem. Funct.* **2016**, *34*, 359–366. [CrossRef]

28. Sugano, E.; Edwards, G.; Saha, S.; Wilmott, L.A.; Grambergs, R.C.; Mondal, K.; Qi, H.; Stiles, M.; Tomita, H.; Mandal, N. Overexpression of acid ceramidase (ASAH1) protects retinal cells (ARPE19) from oxidative stress. *J. Lipid Res.* **2018**, *60*, 30–43. [CrossRef]

29. Braun, F.; Rinschen, M.M.; Bartels, V.; Frommolt, P.; Habermann, B.; Hoeijmakers, J.H.; Schumacher, B.; Dollé, M.E.; Müller, R.-U.; Benzing, T.; et al. Altered lipid metabolism in the aging kidney identified by three layered omic analysis. *Aging* **2016**, *8*, 441–454. [CrossRef]

30. Airola, M.V.; Allen, W.J.; Pulkoski-Gross, M.J.; Obeid, L.M.; Rizzo, R.C.; Hannun, Y.A. Structural basis for ceramide recognition and hydrolysis by human Neutral Ceramidase. *Structure* **2015**, *23*, 1482–1491. [CrossRef]

31. Coant, N.; García-Barros, M.; Zhang, Q.; Obeid, L.M.; Hannun, Y.A. AKT as a key target for growth promoting functions of neutral ceramidase in colon cancer cells. *Oncogene* **2018**, *37*, 3852–3863. [CrossRef] [PubMed]

32. García-Barros, M.; Coant, N.; Kawamori, T.; Wada, M.; Snider, A.J.; Truman, J.-P.; Wu, B.X.; Furuya, H.; Clarke, C.J.; Bialkowska, A.B.; et al. Role of neutral ceramidase in colon cancer. *FASEB J.* **2016**, *30*, 4159–4171. [CrossRef] [PubMed]

33. Schulz, M.E.; Katunaric, B.; Hockenberry, J.C.; Gutterman, D.D.; Freed, J.K. Manipulation of the Sphingolipid Rheostat Influences the Mediator of Flow-Induced Dilation in the Human Microvasculature. *J. Am. Hear. Assoc.* **2019**, *8*, e013153. [CrossRef] [PubMed]

34. Sundaram, K.; Mather, A.R.; Marimuthu, S.; Shah, P.P.; Snider, A.J.; Obeid, L.M.; Hannun, Y.A.; Beverly, L.J.; Siskind, L.J. Loss of neutral ceramidase protects cells from nutrient- and energy -deprivation-induced cell death. *Biochem. J.* **2016**, *473*, 743–755. [CrossRef]

35. Coant, N.; Hannun, Y.A. Neutral ceramidase: Advances in mechanisms, cell regulation, and roles in cancer. *Adv. Boil. Regul.* **2019**, *71*, 141–146. [CrossRef] [PubMed]

36. Ito, M.; Okino, N.; Tani, M. New insight into the structure, reaction mechanism, and biological functions of neutral ceramidase. *Biochim. et Biophys. Acta (BBA) - Mol. Cell Boil. Lipids* **2014**, *1841*, 682–691. [CrossRef]

37. Sun, W.; Xu, R.; Hu, W.; Jin, J.; Crellin, H.A.; Bielawski, J.; Szulc, Z.M.; Thiers, B.H.; Obeid, L.M.; Mao, C. Upregulation of the Human Alkaline Ceramidase 1 and Acid Ceramidase Mediates Calcium-Induced Differentiation of Epidermal Keratinocytes. *J. Investig. Dermatol.* **2008**, *128*, 389–397. [CrossRef]

38. Wang, Y.; Zhang, C.; Jin, Y.; Wang; He, Q.; Liu, Z.; Ai, Q.; Lei, Y.; Li, Y.; Song, F.; et al. Alkaline ceramidase 2 is a novel direct target of p53 and induces autophagy and apoptosis through ROS generation. *Sci. Rep.* **2017**, *7*, 44573. [CrossRef]

39. Xu, R.; Garcia-Barros, M.; Wen, S.; Li, F.; Lin, C.-L.; Hannun, Y.A.; Obeid, L.M.; Mao, C. Tumor suppressor p53 links ceramide metabolism to DNA damage response through alkaline ceramidase 2. *Cell Death Differ.* **2017**, *25*, 841–856. [CrossRef]

40. Chen, C.; Yin, Y.; Li, C.; Chen, J.; Xie, J.; Lu, Z.; Li, M.; Wang, Y.; Zhang, C.C. ACER3 supports development of acute myeloid leukemia. *Biochem. Biophys. Res. Commun.* **2016**, *478*, 33–38. [CrossRef]

41. Yin, Y.; Xu, M.; Gao, J.; Li, M. Alkaline ceramidase 3 promotes growth of hepatocellular carcinoma cells via regulating S1P/S1PR2/PI3K/AKT signaling. *Pathol. Res. Pract.* **2018**, *214*, 1381–1387. [CrossRef] [PubMed]

42. Sikora, J.; Dworski, S.; Jones, E.E.; Kamani, M.A.; Micsenyi, M.C.; Sawada, T.; Le Faouder, P.; Bertrand-Michel, J.; Dupuy, A.; Dunn, C.K.; et al. Acid Ceramidase Deficiency in Mice Results in a Broad Range of Central Nervous System Abnormalities. *Am. J. Pathol.* **2017**, *187*, 864–883. [CrossRef] [PubMed]

43. Dworski, S.; Berger, A.; Furlonger, C.; Moreau, J.M.; Yoshimitsu, M.; Trentadue, J.; Au, B.C.; Paige, C.J.; A Medin, J. Markedly perturbed hematopoiesis in acid ceramidase deficient mice. *Haematologica* **2015**, *100*, e162–e165. [CrossRef] [PubMed]

44. Yu, F.P.S.; Islam, D.; Sikora, J.; Dworski, S.; Gurka', J.; Lopez-Vasquez, L.; Liu, M.; Kuebler, W.M.; Levade, T.; Zhang, H.; et al. Chronic lung injury and impaired pulmonary function in a mouse model of acid ceramidase deficiency. *Am. J. Physiol. Cell. Mol. Physiol.* **2018**, *314*, L406–L420. [CrossRef]

45. Dworski, S.; Lu, P.; Khan, A.; Maranda, B.; Mitchell, J.; Parini, R.; Di Rocco, M.; Hugle, B.; Yoshimitsu, M.; Magnusson, B.; et al. Acid Ceramidase Deficiency is characterized by a unique plasma cytokine and ceramide profile that is altered by therapy. *Biochim. et Biophys. Acta (BBA) - Mol. Basis Dis.* **2016**, *1863*, 386–394. [CrossRef]

46. Yu, F.P.S.; Dworski, S.; Medin, J.A. Deletion of MCP-1 Impedes Pathogenesis of Acid Ceramidase Deficiency. *Sci. Rep.* **2018**, *8*, 1808. [CrossRef]

47. Gan, J.J.; Garcia, V.; Tian, J.; Tagliati, M.; Parisi, J.E.; Chung, J.M.; Lewis, R.; Baloh, R.; Levade, T.; Pierson, T.M. Acid ceramidase deficiency associated with spinal muscular atrophy with progressive myoclonic epilepsy. *Neuromuscul. Disord.* **2015**, *25*, 959–963. [CrossRef]

48. Rubboli, G.; Veggiotti, P.; Pini, A.; Berardinelli, A.; Cantalupo, G.; Bertini, E.; Tiziano, D.F.; D'Amico, A.; Piazza, E.; Abiusi, E.; et al. Spinal muscular atrophy associated with progressive myoclonic epilepsy: A rare condition caused by mutations in ASAH1. *Epilepsia* **2015**, *56*, 692–698. [CrossRef]

49. Chauvin, S.; Yinon, Y.; Xu, J.; Ermini, L.; Sallais, J.; Tagliaferro, A.; Todros, T.; Post, M.; Caniggia, I. Aberrant TGFbeta Signalling Contributes to Dysregulation of Sphingolipid Metabolism in Intrauterine Growth Restriction. *J. Clin. Endocrinol. Metab.* **2015**, *100*, E986–E996. [CrossRef]

50. Tappino, B.; Biancheri, R.; Mort, M.; Regis, S.; Corsolini, F.; Rossi, A.; Stroppiano, M.; Lualdi, S.; Fiumara, A.; Bembi, B.; et al. Identification and characterization of 15 novel GALC gene mutations causing Krabbe disease. *Hum. Mutat.* **2010**, *31*, E1894–E1915. [CrossRef]

51. Li, Y.; Xu, Y.; Benitez, B.A.; Nagree, M.S.; Dearborn, J.T.; Jiang, X.; Guzman, M.A.; Woloszynek, J.C.; Giaramita, A.; Yip, B.K.; et al. Genetic ablation of acid ceramidase in Krabbe disease confirms the psychosine hypothesis and identifies a new therapeutic target. *Proc. Natl. Acad. Sci. USA* **2019**, *116*, 20097–20103. [CrossRef] [PubMed]

52. Kono, M.; Dreier, J.L.; Ellis, J.M.; Allende, M.L.; Kalkofen, D.N.; Sanders, K.M.; Bielawski, J.; Bielawska, A.; Hannun, Y.A.; Proia, R.L. Neutral Ceramidase Encoded by theAsah2Gene Is Essential for the Intestinal Degradation of Sphingolipids. *J. Boil. Chem.* **2005**, *281*, 7324–7331. [CrossRef] [PubMed]

53. Edvardson, S.; Yi, J.K.; Jalas, C.; Xu, R.; Webb, B.D.; Snider, J.; Fedick, A.; Kleinman, E.; Treff, N.R.; Mao, C.; et al. Deficiency of the alkaline ceramidase ACER3 manifests in early childhood by progressive leukodystrophy. *J. Med. Genet.* **2016**, *53*, 389–396. [CrossRef] [PubMed]

54. Parveen, F.; Bender, D.; Law, S.-H.; Mishra, V.K.; Chen, C.-C.; Ke, L. Role of Ceramidases in Sphingolipid Metabolism and Human Diseases. *Cells* **2019**, *8*, 1573. [CrossRef]

55. Doan, N.; Nguyen, H.S.; Montoure, A.; Al-Gizawiy, M.M.; Mueller, W.M.; Kurpad, S.; Rand, S.D.; Connelly, J.M.; Chitambar, C.R.; Schmainda, K.M.; et al. Acid ceramidase is a novel drug target for pediatric brain tumors. *Oncotarget* **2017**, *8*, 24753–24761. [CrossRef] [PubMed]

56. Doan, N.; Nguyen, H.S.; Al-Gizawiy, M.M.; Mueller, W.M.; Sabbadini, R.A.; Rand, S.D.; Connelly, J.M.; Chitambar, C.R.; Schmainda, K.; Mirza, S.P. Acid ceramidase confers radioresistance to glioblastoma cells. *Oncol. Rep.* **2017**, *38*, 1932–1940. [CrossRef]

57. Doan, N.B.; Alhajala, H.; Al-Gizawiy, M.M.; Mueller, W.M.; Rand, S.D.; Connelly, J.M.; Cochran, E.J.; Chitambar, C.R.; Clark, P.; Kuo, J.; et al. Acid ceramidase and its inhibitors: A de novo drug target and a new class of drugs for killing glioblastoma cancer stem cells with high efficiency. *Oncotarget* **2017**, *8*, 112662–112674. [CrossRef]

58. Klobucar, M.; Grbcic, P.; Pavelic, S.K.; Jonjic, N.; Visentin, S.; Sedic, M. Acid ceramidase inhibition sensitizes human colon cancer cells to oxaliplatin through downregulation of transglutaminase 2 and beta1 integrin/FAK-mediated signalling. *Biochem. Biophys. Res. Commun.* **2018**, *503*, 843–848. [CrossRef]

59. Sänger, N.; Ruckhäberle, E.; Győrffy, B.; Engels, K.; Heinrich, T.; Fehm, T.; Graf, A.; Holtrich, U.; Becker, S.; Karn, T. Acid ceramidase is associated with an improved prognosis in both DCIS and invasive breast cancer. *Mol. Oncol.* **2014**, *9*, 58–67. [CrossRef]

60. Mizutani, N.; Inoue, M.; Omori, Y.; Ito, H.; Tamiya-Koizumi, K.; Takagi, A.; Kojima, T.; Nakamura, M.; Iwaki, S.; Nakatochi, M.; et al. Increased acid ceramidase expression depends on upregulation of androgen-dependent deubiquitinases, USP2, in a human prostate cancer cell line, LNCaP. *J. Biochem.* **2015**, *158*, 309–319. [CrossRef]

61. Bowden, D.; Sutton, P.A.; Wall, M.; Jithesh, P.V.; Jenkins, R.; Palmer, D.; Goldring, C.; Parsons, J.L.; Park, B.K.; Kitteringham, N.; et al. Proteomic profiling of rectal cancer reveals acid ceramidase is implicated in radiation response. *J. Proteom.* **2018**, *179*, 53–60. [CrossRef] [PubMed]

62. Realini, N.; Palese, F.; Pizzirani, D.; Pontis, S.; Basit, A.; Bach, A.; Ganesan, A.; Piomelli, D. Acid Ceramidase in Melanoma: Expression, Localization, and Effects of Pharmacological Inhibition. *J. Biol. Chem.* **2016**, *291*, 2422–2434. [CrossRef] [PubMed]

63. Leclerc, J.; Garandeau, D.; Pandiani, C.; Gaudel, C.; Bille, K.; Nottet, N.; Garcia, V.; Colosetti, P.; Pagnotta, S.; Bahadoran, P.; et al. Lysosomal acid ceramidase ASAH1 controls the transition between invasive and proliferative phenotype in melanoma cells. *Oncogene* **2018**, *38*, 1282–1295. [CrossRef] [PubMed]

64. Tan, S.-F.; Dunton, W.; Liu, X.; Fox, T.E.; Morad, S.A.F.; Desai, D.; Doi, K.; Conaway, M.R.; Amin, S.; Claxton, D.F.; et al. Acid ceramidase promotes drug resistance in acute myeloid leukemia through NF-κB-dependent P-glycoprotein upregulation. *J. Lipid Res.* **2019**, *60*, 1078–1086. [CrossRef] [PubMed]

65. White-Gilbertson, S.; Lu, P.; Norris, J.S.; Voelkel-Johnson, C. Genetic and pharmacological inhibition of acid ceramidase prevents asymmetric cell division by neosis. *J. Lipid Res.* **2019**, *60*, 1225–1235. [CrossRef]

66. Cho, S.M.; Lee, H.K.; Liu, Q.; Wang, M.-W.; Kwon, H.J. A Guanidine-Based Synthetic Compound Suppresses Angiogenesis via Inhibition of Acid Ceramidase. *ACS Chem. Boil.* **2018**, *14*, 11–19. [CrossRef]

67. Espaillat, M.P.; Snider, A.J.; Qiu, Z.; Channer, B.; Coant, N.; Schuchman, E.H.; Kew, R.R.; Sheridan, B.S.; A Hannun, Y.; Obeid, L.M. Loss of acid ceramidase in myeloid cells suppresses intestinal neutrophil recruitment. *FASEB J.* **2017**, *32*, 2339–2353. [CrossRef]

68. Lai, M.; Realini, N.; La Ferla, M.; Passalacqua, I.; Matteoli, G.; Ganesan, A.; Pistello, M.; Mazzanti, C.M.; Piomelli, D. Complete Acid Ceramidase ablation prevents cancer-initiating cell formation in melanoma cells. *Sci. Rep.* **2017**, *7*, 7411. [CrossRef]

69. Bai, A.; Bielawska, A.; Rahmaniyan, M.; Kraveka, J.M.; Bielawski, J.; Hannun, Y.A. Dose dependent actions of LCL521 on acid ceramidase and key sphingolipid metabolites. *Bioorganic Med. Chem.* **2018**, *26*, 6067–6075. [CrossRef]

70. Bai, A.; Mao, C.; Jenkins, R.W.; Szulc, Z.M.; Bielawska, A.; Hannun, Y.A. Anticancer actions of lysosomally targeted inhibitor, LCL521, of acid ceramidase. *PLoS ONE* **2017**, *12*, e0177805. [CrossRef]

71. Baspinar, M.; Ozyurt, R.; Kus, G.; Kutlay, O.; Ozkurt, M.; Erkasap, N.; Kabadere, S.; Yasar, N.F.; Erkasap, S. Effects of ceranib-2 on cell survival and TNF-alpha in colon cancer cell line. *Bratisl. Med. J.* **2017**, *118*, 391–393. [CrossRef]

72. Vethakanraj, H.S.; Sesurajan, B.P.; Padmanaban, V.P.; Jayaprakasam, M.; Murali, S.; Sekar, A.K. Anticancer effect of acid ceramidase inhibitor ceranib-2 in human breast cancer cell lines MCF-7, MDA MB-231 by the activation of SAPK/JNK, p38 MAPK apoptotic pathways, inhibition of the Akt pathway, downregulation of ERα. *Anti-Cancer Drugs* **2018**, *29*, 50–60. [CrossRef] [PubMed]

73. Vethakanraj, H.S.; Babu, T.A.; Sudarsanan, G.B.; Duraisamy, P.K.; Kumar, S.A. Targeting ceramide metabolic pathway induces apoptosis in human breast cancer cell lines. *Biochem. Biophys. Res. Commun.* **2015**, *464*, 833–839. [CrossRef] [PubMed]

74. Vejselova, D.; Kutlu, H.M.; Kuş, G. Examining impacts of ceranib-2 on the proliferation, morphology and ultrastructure of human breast cancer cells. *Cytotechnology* **2016**, *68*, 2721–2728. [CrossRef] [PubMed]

75. Rahman, A.; Thayyullathil, F.; Pallichankandy, S.; Galadari, S. Hydrogen peroxide/ceramide/Akt signaling axis play a critical role in the antileukemic potential of sanguinarine. *Free. Radic. Boil. Med.* **2016**, *96*, 273–289. [CrossRef]

76. Adada, M.; Canals, D.; Jeong, N.; Kelkar, A.D.; Hernandez-Corbacho, M.; Pulkoski-Gross, M.J.; Donaldson, J.C.; Hannun, Y.A.; Obeid, L.M. Intracellular sphingosine kinase 2-derived sphingosine-1-phosphate mediates epidermal growth factor-induced ezrin-radixin-moesin phosphorylation and cancer cell invasion. *FASEB J.* **2015**, *29*, 4654–4669. [CrossRef]

77. Klosinski, L.P.; Yao, J.; Yin, F.; Fonteh, A.N.; Harrington, M.G.; Christensen, T.A.; Trushina, E.; Brinton, R.D. White Matter Lipids as a Ketogenic Fuel Supply in Aging Female Brain: Implications for Alzheimer's Disease. *EBioMedicine* **2015**, *2*, 1888–1904. [CrossRef]

78. Wang, K.; Xu, R.; Schrandt, J.; Shah, P.; Gong, Y.Z.; Preston, C.; Wang, L.; Yi, J.K.; Lin, C.-L.; Sun, W.; et al. Alkaline Ceramidase 3 Deficiency Results in Purkinje Cell Degeneration and Cerebellar Ataxia Due to Dyshomeostasis of Sphingolipids in the Brain. *PLoS Genet.* **2015**, *11*, e1005591. [CrossRef]

79. Wimo, A.; Winblad, B.; Jönsson, L. The worldwide societal costs of dementia: Estimates for 2009. *Alzheimer's Dement.* **2010**, *6*, 98–103. [CrossRef]

80. Davis, W. The ATP-Binding Cassette Transporter-2 (ABCA2) Overexpression Modulates Sphingosine Levels and Transcription of the Amyloid Precursor Protein (APP) Gene. *Curr. Alzheimer Res.* **2015**, *12*, 847–859. [CrossRef]

81. Kim, M.J.; Jeon, S.; Burbulla, L.F.; Krainc, D. Acid ceramidase inhibition ameliorates α-synuclein accumulation upon loss of GBA1 function. *Hum. Mol. Genet.* **2018**, *27*, 1972–1988. [CrossRef] [PubMed]

82. Ferraz, M.J.; Marques, A.R.A.; Appelman, M.D.; Verhoek, M.; Strijland, A.; Mirzaian, M.; Scheij, S.; Ouairy, C.M.; Lahav, D.; Wisse, P.; et al. Lysosomal glycosphingolipid catabolism by acid ceramidase: Formation of glycosphingoid bases during deficiency of glycosidases. *FEBS Lett.* **2016**, *590*, 716–725. [CrossRef] [PubMed]

83. Chauhan, M.Z.; Valencia, A.-K.; Piqueras, M.C.; Enriquez-Algeciras, M.; Bhattacharya, S.K. Optic Nerve Lipidomics Reveal Impaired Glucosylsphingosine Lipids Pathway in Glaucoma. *Investig. Opthalmology Vis. Sci.* **2019**, *60*, 1789–1798. [CrossRef] [PubMed]

84. Li, G.; Huang, D.; Hong, J.; Bhat, O.M.; Yuan, X.; Li, P.-L. Control of lysosomal TRPML1 channel activity and exosome release by acid ceramidase in mouse podocytes. *Am. J. Physiol. Physiol.* **2019**, *317*, C481–C491. [CrossRef] [PubMed]

85. Li, F.; Xu, R.; Low, B.E.; Lin, C.-L.; García-Barros, M.; Schrandt, J.; Mileva, I.; Snider, A.; Luo, C.K.; Jiang, X.-C.; et al. Alkaline ceramidase 2 is essential for the homeostasis of plasma sphingoid bases and their phosphates. *FASEB J.* **2018**, *32*, 3058–3069. [CrossRef]

86. Grassme, H.; Henry, B.; Ziobro, R.; Becker, K.A.; Riethmuller, J.; Gardner, A.; Seitz, A.P.; Steinmann, J.; Lang, S.; Ward, C.; et al. beta1-Integrin Accumulates in Cystic Fibrosis Luminal Airway Epithelial Membranes and Decreases Sphingosine, Promoting Bacterial Infections. *Cell Host Microbe.* **2017**, *21*, 707–718.e8. [CrossRef]

87. Novgorodov, S.A.; Riley, C.L.; Yu, J.; Keffler, J.A.; Clarke, C.J.; Van Laer, A.O.; Baicu, C.F.; Zile, M.R.; Gudz, T.I. Lactosylceramide contributes to mitochondrial dysfunction in diabetes. *J. Lipid Res.* **2016**, *57*, 546–562. [CrossRef]

88. Nedvedova, I.; Kolar, D.; Neckar, J.; Kalous, M.; Pravenec, M.; Šilhavý, J.; Korenkova, V.; Kolar, F.; Zurmanova, J. Cardioprotective Regimen of Adaptation to Chronic Hypoxia Diversely Alters Myocardial Gene Expression in SHR and SHR-mtBN Conplastic Rat Strains. *Front. Endocrinol.* **2019**, *9*. [CrossRef]

89. Jeong, I.; Lim, J.-H.; Oh, D.K.; Kim, W.J.; Oh, Y.-M. Gene expression profile of human lung in a relatively early stage of COPD with emphysema. *Int. J. Chronic Obstr. Pulm. Dis.* **2018**, *13*, 2643–2655. [CrossRef]

90. Lewis, L.S.; Huffman, K.M.; Smith, I.J.; Donahue, M.P.; Slentz, C.A.; Houmard, J.A.; Hubal, M.J.; Hoffman, E.P.; Hauser, E.R.; Siegler, I.C.; et al. Genetic Variation in Acid Ceramidase Predicts Non-completion of an Exercise Intervention. *Front. Physiol.* **2018**, *9*. [CrossRef]

91. Weikel, K.A.; Cacicedo, J.M.; Ruderman, N.B.; Ido, Y. Glucose and palmitate uncouple AMPK from autophagy in human aortic endothelial cells. *Am. J. Physiol. Cell Physiol.* **2015**, *308*, C249–C263. [CrossRef] [PubMed]

92. Hong, J.; Bhat, O.M.; Li, G.; Dempsey, S.K.; Zhang, Q.; Ritter, J.K.; Li, W.; Li, P.-L. Lysosomal regulation of extracellular vesicle excretion during d-ribose-induced NLRP3 inflammasome activation in podocytes. *Biochim. et Biophys. Acta (BBA) - Bioenerg.* **2019**, *1866*, 849–860. [CrossRef] [PubMed]

93. Li, G.; Zhang, Q.; Hong, J.; Ritter, J.K.; Li, P.-L. Inhibition of pannexin-1 channel activity by adiponectin in podocytes: Role of acid ceramidase activation. *Biochim. et Biophys. Acta (BBA) - Mol. Cell Boil. Lipids* **2018**, *1863*, 1246–1256. [CrossRef] [PubMed]

94. Xia, J.Y.; Holland, W.L.; Kusminski, C.M.; Sun, K.; Sharma, A.; Pearson, M.J.; Sifuentes, A.J.; McDonald, J.G.; Gordillo, R.; Scherer, P.E.; et al. Targeted Induction of Ceramide Degradation Leads to Improved Systemic Metabolism and Reduced Hepatic Steatosis. *Cell Metab.* **2015**, *22*, 266–278. [CrossRef] [PubMed]

95. Presa, N.; Clugston, R.D.; Lingrell, S.; Kelly, S.E.; Merrill, A.H.; Jana, S.; Kassiri, Z.; Gomez-Muñoz, A.; Vance, D.E.; Jacobs, R.; et al. Vitamin E alleviates non-alcoholic fatty liver disease in phosphatidylethanolamine N-methyltransferase deficient mice. *Biochim. et Biophys. Acta (BBA) - Mol. Basis Dis.* **2019**, *1865*, 14–25. [CrossRef]

96. Choi, S.R.; Lim, J.H.; Kim, M.Y.; Kim, E.N.; Kim, Y.; Choi, B.S.; Kim, Y.-S.; Kim, H.W.; Lim, K.-M.; Kim, M.J.; et al. Adiponectin receptor agonist AdipoRon decreased ceramide, and lipotoxicity, and ameliorated diabetic nephropathy. *Metabolism* **2018**, *85*, 348–360. [CrossRef]

97. Luo, F.; Feng, Y.; Ma, H.; Liu, C.; Chen, G.; Wei, X.; Mao, X.; Li, X.; Xu, Y.; Tang, S.; et al. Neutral ceramidase activity inhibition is involved in palmitate-induced apoptosis in INS-1 cells. *Endocr. J.* **2017**, *64*, 767–776. [CrossRef]

98. Aguirre, C.; Castillo, V.; Llanos, M. Oral Administration of the Endocannabinoid Anandamide during Lactation: Effects on Hypothalamic Cannabinoid Type 1 Receptor and Food Intake in Adult Mice. *J. Nutr. Metab.* **2017**, *2017*, 1–5. [CrossRef]

99. Zhu, Q.; Zhu, R.; Jin, J. Neutral ceramidase-enriched exosomes prevent palmitic acid-induced insulin resistance in H4IIEC3 hepatocytes. *FEBS Open Bio* **2016**, *6*, 1078–1084. [CrossRef]

100. Lipinski, K.V.W.; Weske, S.; Keul, P.; Peters, S.; Baba, H.A.; Heusch, G.; Gräler, M.H.; Levkau, B. Hepatocyte nuclear factor 1A deficiency causes hemolytic anemia in mice by altering erythrocyte sphingolipid homeostasis. *Blood* **2017**, *130*, 2786–2798. [CrossRef]

101. Wang, K.; Li, C.; Lin, X.; Sun, H.; Xu, R.; Li, Q.; Wei, Y.; Li, Y.; Qian, J.; Liu, C.; et al. Targeting alkaline ceramidase 3 alleviates the severity of nonalcoholic steatohepatitis by reducing oxidative stress. *Cell Death Dis.* **2020**, *11*, 28. [CrossRef]

102. Keitsch, S.; Riethmüller, J.; Soddemann, M.; Sehl, C.; Wilker, B.; Edwards, M.J.; Caldwell, C.C.; Fraunholz, M.; Gulbins, E.; Becker, K.A.; et al. Pulmonary infection of cystic fibrosis mice with Staphylococcus aureus requires expression of α-toxin. *Boil. Chem.* **2018**, *399*, 1203–1213. [CrossRef]

103. Azuma, M.M.; Balani, P.; Boisvert, H.; Gil, M.; Egashira, K.; Yamaguchi, T.; Hasturk, H.; Duncan, M.; Kawai, T.; Movila, A. Endogenous acid ceramidase protects epithelial cells from Porphyromonas gingivalis-induced inflammation in vitro. *Biochem. Biophys. Res. Commun.* **2017**, *495*, 2383–2389. [CrossRef] [PubMed]

104. Rice, T.C.; Seitz, A.P.; Edwards, M.J.; Gulbins, E.; Caldwell, C.C. Frontline Science: Sphingosine rescues burn-injured mice from pulmonary Pseudomonas aeruginosa infection. *J. Leukoc. Biol.* **2016**, *100*, 1233–1237. [CrossRef] [PubMed]

105. Wang, K.; Xu, R.; Snider, A.J.; Schrandt, J.; Li, Y.; Bialkowska, A.B.; Li, M.; Zhou, J.; A Hannun, Y.; Obeid, L.M.; et al. Alkaline ceramidase 3 deficiency aggravates colitis and colitis-associated tumorigenesis in mice by hyperactivating the innate immune system. *Cell Death Dis.* **2016**, *7*, e2124. [CrossRef] [PubMed]

106. Holmberg, M. The ghost of pandemics past: Revisiting two centuries of influenza in Sweden. *Med. Humanit.* **2016**, *43*, 141–147. [CrossRef] [PubMed]

107. Holmes, E.C.; Dudas, G.; Rambaut, A.; Andersen, K.G. The evolution of Ebola virus: Insights from the 2013–2016 epidemic. *Nature* **2016**, *538*, 193–200. [CrossRef]

108. Weaver, S.C.; Costa, F.; Blanco, M.A.G.; Ko, A.; Ribeiro, G.S.; Saade, G.; Shi, P.-Y.; Vasilakis, N. Zika virus: History, emergence, biology, and prospects for control. *Antivir. Res.* **2016**, *130*, 69–80. [CrossRef]

109. Prompetchara, E.; Ketloy, C.; Palaga, T. Immune responses in COVID-19 and potential vaccines: Lessons learned from SARS and MERS epidemic. *Asian Pac. J. Allergy Immunol.* **2020**, *38*, 1–9. [CrossRef]

110. Nguyen, A.; Guedán, A.; Mousnier, A.; Swieboda, D.; Zhang, Q.; Horkai, D.; Le Novère, N.; Solari, R.; Wakelam, M. Host lipidome analysis during rhinovirus replication in HBECs identifies potential therapeutic targets. *J. Lipid Res.* **2018**, *59*, 1671–1684. [CrossRef]

111. Grafen, A.; Schumacher, F.; Chithelen, J.; Kleuser, B.; Beyersdorf, N.; Schneider-Schaulies, J. Use of Acid Ceramidase and Sphingosine Kinase Inhibitors as Antiviral Compounds Against Measles Virus Infection of Lymphocytes in vitro. *Front. Cell Dev. Boil.* **2019**, *7*, 218. [CrossRef] [PubMed]

112. Lang, J.; Bohn, P.; Bhat, H.; Jastrow, H.; Walkenfort, B.; Cansiz, F.; Fink, J.; Bauer, M.; Olszewski, D.; Ramos-Nascimento, A.; et al. Acid ceramidase of macrophages traps herpes simplex virus in multivesicular bodies and protects from severe disease. *Nat. Commun.* **2020**, *11*, 1–15. [CrossRef] [PubMed]

113. Plöhn, S.; Edelmann, B.; Japtok, L.; He, X.; Hose, M.; Hansen, W.; Schuchman, E.H.; Eckstein, A.; Berchner-Pfannschmidt, U. CD40 Enhances Sphingolipids in Orbital Fibroblasts: Potential Role of Sphingosine-1-Phosphate in Inflammatory T-Cell Migration in Graves' Orbitopathy. *Investig. Opthalmology Vis. Sci.* **2018**, *59*, 5391–5397. [CrossRef] [PubMed]

114. Santos-Cortez, R.L.P.; Hu, Y.; Sun, F.; Benahmed-Miniuk, F.; Tao, J.; Kanaujiya, J.; Ademola, S.; Fadiora, S.; Odesina, V.; A Nickerson, D.; et al. Identification of ASAH1 as a susceptibility gene for familial keloids. *Eur. J. Hum. Genet.* **2017**, *25*, 1155–1161. [CrossRef]

115. Chen, J.Y.; Newcomb, B.; Zhou, C.; Pondick, J.V.; Ghoshal, S.; York, S.R.; Motola, D.L.; Coant, N.; Yi, J.K.; Mao, C.; et al. Tricyclic Antidepressants Promote Ceramide Accumulation to Regulate Collagen Production in Human Hepatic Stellate Cells. *Sci. Rep.* **2017**, *7*, 44867. [CrossRef]

116. Frohbergh, M.E.; Guevara, J.; Grelsamer, R.P.; Barbe, M.F.; He, X.; Simonaro, C.M.; Schuchman, E.H. Acid ceramidase treatment enhances the outcome of autologous chondrocyte implantation in a rat osteochondral defect model. *Osteoarthr. Cartil.* **2015**, *24*, 752–762. [CrossRef]

117. Bonafe, L.; Kariminejad, A.; Li, J.; Royer-Bertrand, B.; Garcia, V.; Mahdavi, S.; Bozorgmehr, B.; Lachman, R.L.; Mittaz-Crettol, L.; Campos-Xavier, B.; et al. Brief Report: Peripheral Osteolysis in Adults Linked to ASAH1 (Acid Ceramidase) Mutations: A New Presentation of Farber's Disease. *Arthritis Rheumatol.* **2016**, *68*, 2323–2327. [CrossRef]

118. Nojima, H.; Freeman, C.M.; Schuster, R.M.; Japtok, L.; Kleuser, B.; Edwards, M.J.; Gulbins, E.; Lentsch, A.B. Hepatocyte exosomes mediate liver repair and regeneration via sphingosine-1-phosphate. *J. Hepatol.* **2015**, *64*, 60–68. [CrossRef]

119. Pethő, A.; Chen, Y.; George, A. Exosomes in Extracellular Matrix Bone Biology. *Curr. Osteoporos. Rep.* **2018**, *16*, 58–64. [CrossRef]

Review

Sphingosine-1-Phosphate Metabolism in the Regulation of Obesity/Type 2 Diabetes

Jeanne Guitton [1], Cécile L. Bandet [2,3], Mohamed L. Mariko [1], Sophie Tan-Chen [2,3], Olivier Bourron [2,3,4], Yacir Benomar [1], Eric Hajduch [2,3] and Hervé Le Stunff [1,*]

[1] Institut des Neurosciences Paris-Saclay, Université Paris Saclay, CNRS UMR 9197, F-91190 Orsay, France; Jeanne.guitton@u-psud.fr (J.G.); mohamed.mariko@universite-paris-saclay.fr (M.L.M.); yacir.benomar@universite-paris-saclay.fr (Y.B.)

[2] Centre de Recherche des Cordeliers, INSERM, Sorbonne Université, Université de Paris, F-75006 Paris, France; cecile.bandet@gmail.com (C.L.B.); sophie.tan.crc@gmail.com (S.T.-C.); olivier.bourron@aphp.fr (O.B.); eric.hajduch@crc.jussieu.fr (E.H.)

[3] Institut Hospitalo-Universitaire ICAN, F-75013 Paris, France

[4] Assistance Publique-Hôpitaux de Paris, Département de Diabétologie et Maladies Métaboliques, Hôpital Pitié-Salpêtrière, F-75013 Paris, France

* Correspondence: herve.le-stunff@universite-paris-saclay.fr

Received: 9 June 2020; Accepted: 7 July 2020; Published: 13 July 2020

Abstract: Obesity is a pathophysiological condition where excess free fatty acids (FFA) target and promote the dysfunctioning of insulin sensitive tissues and of pancreatic β cells. This leads to the dysregulation of glucose homeostasis, which culminates in the onset of type 2 diabetes (T2D). FFA, which accumulate in these tissues, are metabolized as lipid derivatives such as ceramide, and the ectopic accumulation of the latter has been shown to lead to lipotoxicity. Ceramide is an active lipid that inhibits the insulin signaling pathway as well as inducing pancreatic β cell death. In mammals, ceramide is a key lipid intermediate for sphingolipid metabolism as is sphingosine-1-phosphate (S1P). S1P levels have also been associated with the development of obesity and T2D. In this review, the current knowledge on S1P metabolism in regulating insulin signaling in pancreatic β cell fate and in the regulation of feeding by the hypothalamus in the context of obesity and T2D is summarized. It demonstrates that S1P can display opposite effects on insulin sensitive tissues and pancreatic β cells, which depends on its origin or its degradation pathway.

Keywords: Sphingosine-1-phosphate; obesity; type 2 diabetes; insulin resistance; pancreatic β cell fate; hypothalamus

---

1. Introduction

Obesity is a major public health problem, which results in over nutrition that leads to a net-positive energy balance characterized by the storage of excess fat in the subcutaneous and visceral adipose tissues, as well as in ectopic tissues, including skeletal muscles, liver, and pancreatic β cells [1]. In physiological conditions, ingested lipids are usually used as an energy source by most organisms and can be substituted by carbohydrates for ATP production, based on acute changes in nutrient availability and energy requirements [2]. However, in pathophysiological conditions, adipose tissue lipid metabolism becomes dysfunctional, which leads to increased delivery of fatty acids to other peripheral tissues [3]. Increased free fatty acids (FFA) produced from adipose tissue as well as secretion of hormones, cytokines, and pro-inflammatory markers, which are directly linked to obesity, induce reduced glucose uptake in muscle cells and increased hepatic glucose production. These metabolic dysfunctions lead to a glucose overflow in the circulation, which culminates in glucose intolerance and the installation of type 2 diabetes (T2D) [4].

T2D is a serious metabolic condition due to the insufficient secretion of insulin by pancreatic β cells, and due to an inefficient response from the body to secreted insulin. Diabetes is one of the fastest growing global health emergencies of the 21st century. In 2019, the world prevalence of diabetes was estimated as 463 million people, and this number is projected to reach 578 million by 2030, and 700 million by 2045. T2D is also the most common form of diabetes and accounts for 90% of the disease worldwide [5]. T2D is most commonly observed in older adults but is increasingly seen in children and younger adults due to the rise of obesity, physical inactivity, and inappropriate diet.

High levels of circulating FFA are known to induce not only insulin resistance, but also defects in the insulin secretory capacity of β cells, as well as in insulin gene expression [6,7]. The nature of FFA, that is, its degree of saturation and carbon chain length, is one of the critical factors involved in the induction of lipotoxicity, such as inhibition of insulin secretion, β cell apoptosis, and insulin resistance [8,9]. Non-adipose tissue accumulated FFAs are metabolized into lipid derivatives such as ceramides, which, in turn, lead to lipotoxicity in these tissues [10]. In mammalian cells, ceramides are key lipids of sphingolipid metabolism and are widely distributed in cell membranes where they play a crucial structural role. It also has important functions in intracellular signaling, regulation of growth, proliferation, cell migration, apoptosis, and differentiation [11–13]. Ceramides consist of a sphingoid long chain base to which a fatty acid is attached via an amide bond. In the context of fatty acid overload, ceramide is mainly produced *de novo* in the endoplasmic reticulum (ER), through different enzymatic reactions [14,15]. It is now clearly established that *de novo* synthetized ceramides are among the most active lipid second messengers, which inhibits the function of some key proteins of the insulin signaling pathway [16,17] and stimulates pancreatic β cell death [18]. Apart from its structural and signaling functions, ceramide is a central lipid intermediate in sphingolipid metabolism. It is a precursor for other bioactive sphingolipids, from complex glycosphingolipids or sphingomyelin to more "simple" lipids such as ceramide-1-phosphate, sphingosine, and sphingosine-1-phosphate (S1P) [12].

The most important site of S1P production is the plasma membrane where sphingomyelin is metabolized into ceramide by sphingomyelinase, then S1P is produced by cooperating two enzyme families, namely ceramidases, and sphingosine kinases (SphK) (Figure 1) [19]. Contrary to most sphingolipids, S1P does not possess a structural function, but is a potent signal mediator that modulates multiple cellular functions important for health and diseases [14]. The multimodal actions of S1P can be explained by the fact that the sphingolipid, on the one hand, directly modulates intracellular functions, and, on the other hand, acts as a ligand of G protein-coupled receptors (GPCR) after secretion into the extracellular environment, transported by ApoM-containing high density lipoproteins (HDL) or albumin, to exert either autocrine and/or paracrine functions [19].

In many cellular and animal models, both ceramide and S1P display opposite effects. This is well documented in cancer cells where ceramide stimulates apoptosis, whereas S1P promotes cell survival [11]. While the role of ceramide in the development of muscle insulin resistance is now well established [16], the relationship of S1P with insulin resistance and T2D still remains controversial in some tissues. Elevation of tissue and plasma S1P levels has been recognized as a critical feature of both human and rodent obesity [20], which suggests that S1P metabolism could be involved in the onset of T2D, or that its regulation is an adaptive process in the presence of high levels of circulating lipids. Thus, this review describes the role played by S1P metabolism in the development of obesity/T2D by analysing the enzymes regulating both its tissue and circulating levels in insulin resistance of peripheral tissues and pancreatic β cell fate.

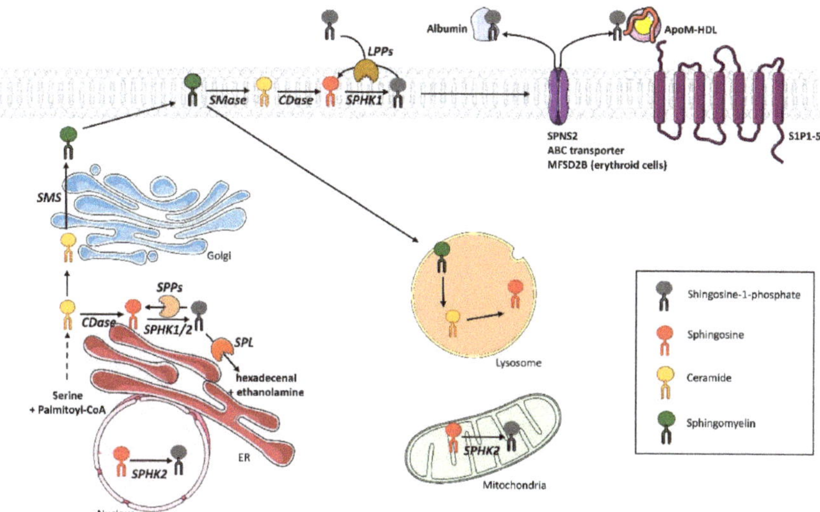

Figure 1. Sphingosine-1-phosphate metabolism in mammals. Sphingolipid *de novo* synthesis is initiated in the endoplasmic reticulum (ER), starting by the condensation of serine and palmitoyl-coA followed by a cascade of enzymatic reactions to produce ceramide. In the ER, ceramide is deacylated by neutral CDase into sphingosine. Sphingosine is phosphorylated to produce S1P by SphK1/2. Produced S1P can be either dephosphorylated back to sphingosine by ER resident SPPs, or irreversibly transformed into hexadecenal and phosphoethanolamine by S1P lyase. Ceramide is transported to the Golgi apparatus to be transformed into SM, which will reach the plasma membrane. In the plasma membrane, SM can be transformed into ceramide through the action of SMases. Ceramide will then be deacylated by acidic CDase to give sphingosine that will be phosphorylated into S1P by SphK1. Produced S1P can be dephosphorylated by ecto-LPPs. S1P can also be secreted through ABC, SPNS2, and MFSD2B transporters in extracellular space to activate S1P receptors. Extracellular S1P can also be transported by either albumin or ApoM/HDL. The latter can activate S1P receptors. SM can be endocytosed to be recycled into ceramide and sphingosine inside lysosomes. SphK2 can catalyze S1P production in the mitochondria and the nucleus. ABC: ATP-binding cassette. CDase: ceramidase. ER: endoplasmic reticulum. HDL: high density lipoproteins. MFSD2B: Major Facilitator Superfamily Domain Containing 2B. S1P: sphingosine-1-phosphate. SM: sphingomyelin. SMase: sphingomyelinase. SphK: sphingosine kinase. S1P1-5: S1P receptor 1 to 5. SPNS2: Spinster homolog 2. SPP: Sphingosine-1-phosphate phosphohydrolase.

2. S1P Metabolism in Mammals

2.1. S1P Synthesis

S1P is produced by deacylation of ceramide by ceramidases to give sphingosine. Subsequently, sphingosine kinases (SphK) are responsible for the phosphorylation of sphingosine, which results in the formation of S1P (Figure 1). As to the anabolic pathway of S1P, two isoforms of sphingosine kinases (SphK) have been discovered, called SphK1 and SphK2. Both are widely expressed [21]. Compared to SphK1, SphK2 possesses 240 additional amino acids in its N-terminal region corresponding to a nuclear export sequence [22]. Although they have similar sequences, these enzymes differ in their intracellular localization, regulation, level of tissue expression, and, therefore, in their functions [23], especially in sphingolipid metabolism and, thus, the level of ceramide [24].

While SphK1 resides in the cytosol, SphK2 is localized in the nucleus, the inner mitochondrial membrane, and the endoplasmic reticulum (ER) (Figure 1). Under basal conditions, SphK1 is mostly present in the cytoplasm. SphK1 catalytic activity increases from 1.5 to 4-fold as it translocates to the plasma membrane upon stimulation. Both translocation and activity are regulated not only by the phosphorylation of SphK1 Ser^{225} residue by extracellular signal-regulated kinases (ERK1/2) [25], but also by anionic lipids (phosphatidylserine and phosphatidic acid) and Ca^{2+}/calmodulin [22].

SphK2 can also be phosphorylated by ERK1/2, but the exact phosphorylation site remains unclear, as Ser351 and/or Thr578 residues may be involved [25]. As SphK2 is localized in the nucleus, it can directly interact and form a complex with H3 histone and histone deacetylases 1 and 2 (HDAC1/2) in the promotor of transcriptional regulator c-fos and dependent kinase inhibitor p21 genes, where it enhances local histone H3 acetylation and transcription [26]. Synthetized S1P by SphK2 binds to and inhibits both HDAC1 and HDAC2, which suggests that nucleus-generated S1P via SphK2 influences the dynamic balance of histone acetylation and, thus, the epigenetic modulation of specific target genes [27]. In addition, when produced in the mitochondria by SphK2, S1P regulates prohibitin 2 (PHB2) function, which is a highly conserved protein that regulates mitochondrial homeostasis [28].

According to Maceyka et al., SphK1 and SphK2 display opposite functions in sphingolipid metabolism in the regulation of ceramide biosynthesis. Indeed, in HEK293 cells, specific down-regulation of SphK2 reduced conversion of sphingosine into ceramide in the recycling pathway and, conversely, down-regulation of SphK1 increased it [24]. This difference could be linked to a potent dialogue between SphK2 and the S1P phosphatase 1 (SPP1) that favors the conversion of S1P into ceramide [29] (see below Section 2.2).

2.2. S1P Recycling and Degradation

S1P can be quickly and irreversibly degraded by the endoplasmic reticulum resident enzyme S1P lyase (SPL), which cleaves the C2-C3 bond of S1P to generate two products: hexadecenal (palmitaldehyde) and phosphoethanolamine [30] (Figure 1). Both products can then be transferred as glycerol substrates and phospholipid substrates in the glycerophospholipid pathway [31]. Phosphoethanolamine will be used for the synthesis of phosphatidylethanolamine and hexadecenal will be used for reloading the palmitoyl-CoA pool [31].

Alternatively, S1P can also be reversibly dephosphorylated by several phosphohydrolases to regenerate sphingosine. The first lipid phosphohydrolases involved are lipid phosphate phosphohydrolases (LPPs) (Figure 1). They belong to the superfamily of lipid phosphatases that includes three isoforms characterized in mammals: LPP1, LPP2, and LPP3. LPPs are membrane-associated, magnesium-independent and N-ethylmaleimide-insensitive enzymes [29]. Their active sites are located on the outer surface of plasma membranes or at the lumenal surface of internal membranes (Golgi and endosomes) [32]. LPP2 resides intracellularly, whereas LPP1 and LPP3 are mainly localized at the plasma membrane and function as ecto-enzymes, while degrading lipid phosphate substrates such as S1P as well as lysophosphatidic acid in the extracellular space [33].

S1P can also be dephosphorylated by two specific S1P phosphohydrolases called SPP1 and SPP2 (Figure 1). These two mammalian isoforms are differentially expressed-sphingoid base-specific phosphatases localized in the ER. SPP1 regulates the salvage of sphingosine for the synthesis of ceramide in the ER (rescue pathway) [33], and it has been shown that SPP1 overexpression induces ceramide accumulation in the ER, which suggests that dephosphorylation of S1P is a limiting step for the recycling pathway [33]. A regulatory role in the recycling pathway for SPP2 has not yet been demonstrated, but its expression was increased during the inflammatory response [33]. In addition, it was reported that both SPP1 and SPP2 were also involved in ER stress-induced-autophagy [34] and proliferation [35].

2.3. S1P Transport

Contrary to most sphingolipids, S1P does not possess any structural function, but is a potent signal mediator that affects multiple cellular functions important for health and diseases. The multitude of different S1P-mediated actions is linked to its capacity to be secreted by various cells and tissues. To exert its extracellular functions, intracellularly generated S1P is transported across the plasma membrane. Since S1P is too hydrophilic to simply diffuse through the membrane, it is exported by specific ATP-binding cassette (ABC) transporters or the spinster homolog 2 (SPNS2) transporter, which is a member of non-ATP-dependent organic ion transporter family [36]. In the erythrocyte, S1P was recently shown to be secreted through the protein MFSD2B [37]. Once outside the cell, S1P can either bind to albumin [38], or ApoM [39]. Approximately 35% of plasma S1P is bound to albumin and 65% to ApoM, which is found on a small percentage (~5%) of high density lipoprotein (HDL) particles [40]. S1P has a four-times longer half-life when bound to ApoM/HDL than to albumin, as seen when tested in vivo (15 min) and in vitro (30 min) under albumin binding conditions [41,42]). This suggests that the binding of S1P to HDL prevents its degradation. ApoM/HDL-bound S1P has been proposed as a primary contributor to the vasoprotective properties of HDLs [43], and S1P has also been shown to be a key component in the anti-atherogenic properties of HDL [44]. However, S1P-bound albumin has been suggested to represent a reservoir for free S1P [39].

2.4. S1P Receptors

As an extracellular second messenger, S1P is a high-affinity ligand (Kd from 2 to 63 nM) of a family of five GPCRs, termed S1P1-5 [45]. Receptor-bound S1P induces a wide range of physiological responses such as proliferation, migration, inhibition of apoptosis, formation of actin stress fibers, stimulation of adherent junctions, and enhanced extracellular matrix assembly [46]. S1P1–3 are ubiquitously expressed throughout tissues, whereas S1P4 is predominantly expressed in the immune system, and S1P5 is expressed in the central nervous system and the spleen [27]. S1P receptor activation on different cell types depends on specific G protein coupling. S1P1 couples exclusively with the inhibitory G protein alpha subunit (Gαi), whereas S1P2 and S1P3 bind to Gαi, Gαq, and Gα13, while S1P4 and S1P5 couple to both Gαi and Gα13 [47]. Following ligand binding and subsequent activation, the α subunit of the heterotrimeric G protein is released and interacts with various downstream effectors (see review [48] for more information).

3. S1P Metabolism and Insulin Action: Muscle, Liver, and Adipose Tissue

Since the early 2000s, several studies have looked for the potential role of S1P in mediating insulin action in insulin-sensitive tissues such as liver, skeletal muscle, and adipose tissue.

3.1. Liver

Liver is a major organ for glucose and lipid metabolism, and it has been known for many years that lipid accumulation is linked to the development of insulin resistance and constitutes the first stage of non-alcoholic fatty liver diseases (NAFLD) [49]. Several studies have shown that the SphK1/S1P axis can control the insulin response in the liver. One such study highlighted that hepatic SphK1 expression increased in animals under lipid overload induced by a high-fat diet (HFD) [40]. This increased expression of SphK1 was also found in the liver of human patients displaying NAFLD [40].

Several other studies have also highlighted a positive action of the SphK1/S1P axis on glucose metabolism in hepatocytes. A pioneer study performed in human hepatocytes showed that tumor necrosis factor α (TNFα), which is a cytokine involved in inflammation [50], was unable to activate the NFκB pathway-induced apoptosis, but rather activated the pro-survival SphK1 pathway [51]. The authors also emphasized that TNFα protected hepatocytes from apoptosis by activating SphK1 upstream of the PI3K/Akt pathway [51]. An increase in Akt phosphorylation was observed after 5 min of treatment with exogenous S1P (without insulin), which suggests that a relationship between the SphK/S1P axis and the PI3K/Akt pathway exists in hepatocytes [51]. Similar results were observed by treating primary rat hepatocytes with exogenous S1P [52]. In addition, treatment of a human liver cell line (LO2 cells) with S1P induced an increase in glucose uptake [53]. SphK1 overexpression also induced glucose uptake in the absence of insulin in hepatocellular carcinoma [54]. Conversely, inhibition of SphK1 reduced glucose uptake in the presence or absence of insulin in the same cell line [54]. The authors extended their observation in vivo by injecting an adenoviral vector containing the human SphK1 cDNA in diabetic KK/Ay mice [54]. Under these conditions, transfected diabetic KK/Ay mice displayed a decrease in basal glycemia and a better glucose tolerance compared to control animals [54]. In parallel, they observed a decrease in total cholesterol, triglycerides, and low density lipoproteins as well as an increase in circulating HDL in SphK1-transfected animals compared to control animals [54].

All of these parameters demonstrated that SphK1 overexpression in diabetic animals improved glucose homeostasis at the systemic level. The authors also assessed the hepatic insulin response in these animals, and they showed that both Akt and GSK3 phosphorylation levels were increased in animals overexpressing SphK1 when compared to control animals (Figure 2) [54].

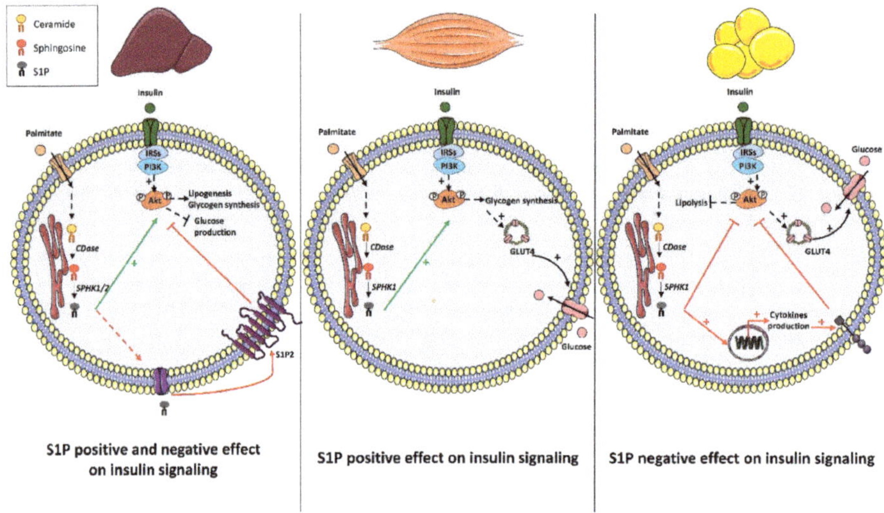

Figure 2. Role of S1P metabolism on insulin in peripheral tissues in response to palmitate. In hepatocytes, palmitate increases intracellular S1P content through SphK1/2 activities. According to studies, produced S1P seems to exert a direct positive action on insulin signaling, or a negative action by stimulating its S1P2 receptors. In muscle cells, palmitate increased intracellular S1P through SphK1 activity, which favors Akt activation, glucose uptake, and glycogen synthesis in response to insulin. In adipocytes, palmitate increases intracellular S1P to inhibit Akt activation in response to insulin. Produced S1P also favors expression of pro-inflammatory cytokines that will contribute to inhibit Akt activity. CDase: ceramidase. GLUT4: glucose transporter 4. IRS: insulin receptor substrate. PI3K: phosphatidylinositol-3-kinase. SphK: sphingosine kinase. S1P2: S1P receptor 2.

These results were confirmed in another study showing that increased mouse hepatic S1P following the overexpression of acid sphingomyelinase (ASM) favored hepatic Akt phosphorylation as well as improved glucose tolerance [55]. When SphK1 expression was reduced in these animals, increased Akt phosphorylation was no longer observed (Figure 2) [55]. In the same study, these observations were confirmed in vitro by treating isolated primary hepatocytes with exogenous S1P, which induced an increase in Akt phosphorylation in the absence of insulin [55].

Contrary to SphK1 expression, incubation of primary mouse hepatocytes with palmitate did not induce any increase in SphK2 expression [56]. Nonetheless, a hepatic role of SphK2-produced S1P in regulating glucose metabolism was investigated by Lee et al. [56]. Endoplasmic reticulum (ER) stress is known to participate in the development of insulin resistance in liver, mainly by promoting the accumulation of lipids in the liver, by directly blocking insulin signaling, and by modifying the expression of key enzymes of gluconeogenesis or lipolysis [49]. Lee et al. found that ER stress transcriptionally up-regulated SphK2 in liver [56]. Overexpression of SphK2 in the AML12 hepatocyte cell line induced an increase in S1P concentration, which was associated with increased Akt phosphorylation in the absence of insulin [57]. In addition, SphK2 overexpression induced a decrease of some sphingolipid species (C16-ceramide, C18-ceramide, C18:1-ceramide, C16-sphingomyelin, C18-sphingomyelin) as well as a decrease in cholesterol and hepatic triglyceride concentration [57]. As observed previously in SphK1-overexpressing animals, hepatic SphK2 overexpression induced an improvement in insulin sensitivity, an increase in hepatic Akt phosphorylation, and, therefore, an improvement in glucose tolerance of these animals when fed an HFD (Figure 2) [57].

Although these studies demonstrate that the hepatic SphKs/S1P axis positively regulates liver insulin response and carbohydrate metabolism under lipotoxic conditions, some other studies showed the opposite and gave a deleterious role to this axis of hepatic insulin signaling. One study reported that S1P inhibited insulin signaling in the liver both in vitro and in vivo [58]. As already described above, Fayyaz et al. showed that, after palmitate treatment, concentrations of intra- and extracellular S1P were increased in primary rat hepatocytes [58]. However, they also observed generated S1P counteracted insulin signaling [58]. The negative role of S1P on insulin signaling in rat or human hepatocytes with exogenous S1P was counteracted in the presence of JTE-013, which is an S1P2 antagonist. This suggests that S1P inhibited the insulin signal through the activation of S1P2 receptor [58]. These observations were extended in vivo Diabetic New Zealand obese (NZO) mice were treated with JTE-013 for seven days before being sacrificed. Both an increase in liver Akt phosphorylation and a decrease in basal glycaemia were observed [58]. Overall, this study demonstrates that palmitate-produced S1P stimulates S1P2 to impair hepatocyte insulin signaling (Figure 2).

In addition, other studies have also highlighted a relationship between liver S1P levels and hepatic lipid accumulation. Mouse hepatic overexpression of ASM has been shown to increase hepatic triglyceride content, which was blunted by SphK1 deletion [55]. This suggests a potent role of SphK1 in steatosis. SphK1 knock-out (KO) mice fed an HFD for 24 weeks displayed an increase in circulating triglycerides compared to wild-type (WT) animals fed the same diet [59]. By contrast, mice displaying a liver-specific overexpression of SphK1 via the use of an adeno-associated-viral (AAV) 8, whose tropism is specific of the liver [60], exhibited reduced hepatic triglyceride levels (steatosis) without affecting glucose metabolism on a low-fat diet [60]. However, no impact of increasing SphK activity on hepatic lipid content or glucose metabolism was observed in HFD fed mice [60]. The discrepancies between these studies could arise from animal models, which use enzyme overexpression (i.e., ASM and SphK1). Lastly, a study showed that SphK1 expression increased in hepatic steatosis and that SphK1 KO mice were protected against hepatic steatosis induced by HFD [56], which supports the idea that endogenous S1P/SphK1 axis could be a major promoter of lipid accumulation in liver (steatosis) [61]. In contrast, SphK2 KO mice fed with HFD showed an increase in hepatic lipid accumulation, which supports the idea that this isoform protected mice from steatosis [62].

Overall, the effect of the SphK/S1P axis on liver glucose metabolism remains not completely solved. Most of the genetic approaches used, to either overexpress or invalidate SphK1 (and SphK2), showed a positive action of the SphK/S1P axis on hepatic insulin response. However, these studies were carried out at the level of the whole organism and, thus, were not liver-specific. It is, therefore, possible that, in addition to hepatic S1P, circulating S1P coming from other tissues could also affect hepatic homeostasis. It has already been shown that hepatic S1P could be secreted to regulate macrophage chemotaxis [63]. The divergent effect of hepatic S1P could also be related to the specific activation of S1P receptors [58] and will require more exploration as to their role in liver homeostasis during obesity. Moreover, it also remains to determine how S1P signals could move from the beneficial effect through insulin signaling to the dysregulation of lipid homeostasis (steatosis). Only one clinical study has shown SphK1 expression increase in liver biopsies from patients with steatosis compared to healthy lean people, which supports the notion that SphK1/S1P axis could play a role in the onset of these diseases [64]. However, whether the localization of increased SphK1 in the human liver is specific just to hepatocyte, as well as its role, still remain unknown. Therefore, future work and analysis will be required to translate data obtained in cell/mouse to those in humans.

3.2. Muscle

Muscles constitute 40% of the body weight and are responsible for 40–75% of the glucose uptake in response to insulin in the postprandial period [65]. They are, therefore, major tissues toward the regulation of carbohydrate homeostasis within the body. Compared to liver, few studies have looked for the role of S1P on glucose metabolism in muscle, but, unlike liver, it seems that they all demonstrate a positive action of this lipid.

Saturated fatty acid (palmitate) induced an increase in SphK1 expression as well as an increase in S1P concentrations in a muscle cells line (C2C12 myotubes) [66], and in mouse primary myotubes [67]. It is important to note that no increase in SphK1 expression was observed in response to unsaturated fatty acids such as oleate [66]. These data were confirmed in vivo where a 2.5-fold increase in SphK1 expression was observed in skeletal muscles of mice fed an HFD compared to control mice (Figure 2) [66]. The addition of exogenous S1P on C2C12 myotubes increased basal Akt phosphorylation, which led to a concomitant increase in glucose uptake [68].

In vivo studies also reported a positive role of SphK1/S1P on insulin signaling in muscle. SphK1-overexpressing mice displayed increased SphK activity in skeletal muscle, and when fed a HFD, skeletal muscle and whole-body insulin sensitivity were improved in these mice compared with control mice fed the same diet [69]. In addition, animals overexpressing SphK1 fed an HFD for six weeks displayed better muscle Akt phosphorylation and were more glucose-tolerant and more sensitive to insulin than wild type animals (Figure 2). However, although skeletal muscles show an increase in SphK1 overexpression, it cannot be excluded that other untested tissues could also overexpress SphK1 and, thus, participate with the observed phenotype.

To complicate the picture, Bruce et al. showed that SphK1 overexpression induced a decrease in muscle ceramide concentration [69]. Considering the importance of this sphingolipid species in the development of insulin resistance [66,67], it remains difficult to ascertain if the observed phenotypes were linked to a decrease in ceramide content or rather from an independent action of S1P. Likewise, both Bruce et al. [70] and Kendall et al. [71] showed that administration of FTY720, which is an S1P analogue that downregulates all S1PR expressions except for S1P2 [72], to animals fed an HFD induced a better muscle insulin signaling as well as a better glucose tolerance compared to animals receiving vehicle only. However, FTY720, which has also been shown as a potent inhibitor of CerS [73], inhibited ceramide production in mice under HFD [70]. These data suggest that the insulin sensitizer effect of FTY720 was associated with a decrease of ceramide levels in muscle rather than an antagonist action on S1P receptors.

Altogether, even if all studies reported a positive role of the SphK1/S1P axis on muscle insulin signaling, and, consequently, on the systemic glucose metabolism, no specific muscle approach was performed. Thus, this possibly hid some cross-talk mediated by S1P between muscles and other S1P producing tissues such as liver, adipose tissue, or even immune cells. In addition, no study, so far, has investigated the role of SphK2-produced S1P in this tissue, nor shown the opposite roles of SphK1 and SphK2 [24]. However, it would be interesting to study the role of the latter on insulin signaling in muscle. It would also be important to explore the role of S1P catabolism and S1P signaling through its receptors in muscle homeostasis. To date, clinical data demonstrating the role of S1P metabolism in regulating muscle insulin resistance in man are lacking and will, therefore, require extensive study.

3.3. Adipose Tissue

Adipose tissue (AT), in addition to its storage functions, is an endocrine tissue that secretes several adipokines and chemokines [74]. AT also participates in the development of insulin resistance when it is in a state of inflammation known as "low-grade" [75]. In addition, when maximum AT storage capacities are reached, excess lipids are then stored in peripheral tissues, which causes insulin resistance or apoptosis in these various tissues [16]. Homeostasis of adipose tissue is, therefore, important for maintaining sensitivity to systemic insulin [76].

Expression of SphK1, but not SphK2, has been reported to be increased in subcutaneous adipose tissue from *ob/ob* mice compared to wild type mice [77]. Similar results were observed in epididymal adipose tissue and isolated mature adipocytes from mice fed an HFD compared to animals on a low-fat diet [78]. Similar profiles were also reported in human inflamed subcutaneous AT compared to less inflamed AT [58]. Concentrations of S1P are also increased in subcutaneous AT from obese patients compared to those from lean people [79].

One study reported a positive role of S1P on insulin signaling in AT. Administration of the S1P analogue FTY720 improved insulin sensitivity in animals fed an HFD [71]. Immune cell infiltration is known to play an important role in insulin resistance [80], and it was found that FTY720 decreased lymphocyte and macrophage infiltration in TA of this mice, likely through its lymphopenic properties [81]. This phenomenon contributes to improving insulin sensitivity in mice. However, another study demonstrated the opposite results. It showed that, in HFD-fed mice, SphK1 deficiency increased adipogenic markers such as adiponectin and the anti-inflammatory cytokine IL-10, but reduced adipose tissue macrophage recruitment as well as pro-inflammatory molecules TNFα and IL-6 (Figure 2). These changes were associated with a better insulin response in the AT and improved insulin sensitivity and glucose tolerance (Figure 2) [78]. Obesity was found to increase SphK1 expression in AT macrophages of both M1 and M2 phenotypes [82]. Elevated SphK1 expression in AT macrophages was associated with the reduction of endoplasmic reticulum stress related genes, which suggests that Sphk1 promotes AT macrophage survival [82].

Overall, these few studies indicate that SphK1/S1P axis leans towards a pro-inflammatory and negative action on AT insulin signaling. However, extracellular S1P through S1P receptors may have the opposite effect [71]. Therefore, analysis of the role of other S1P metabolic enzymes in adipose tissue homeostasis will be necessary to confirm this tendency. It will also be important to decipher whether differences in S1P function exist between AT distributions (visceral vs. subcutaneous) known to play a distinct role in obesity. Although SphK1 expression is increased in the adipose tissue from obese patients [78], no clinical study has, so far, described the functional role of S1P in human adipocyte insulin resistance.

4. S1P Metabolism and Pancreatic β Cell Fate

Pancreatic β cells secrete insulin in response to glucose and various hormones to maintain glycaemia and, therefore, regulate glucose homeostasis. However, obesity is associated with deleterious effects of elevated fatty acid levels on pancreatic β cell function and survival. Excessive fatty acids leads to the loss of β cell insulin secretory responsiveness and β cell death by apoptosis, which favors induction of chronic hyperglycemia [18]. Sphingolipids and, in particular, ceramide have been shown to play a central role in pancreatic β-cell apoptosis induced by palmitate [18]. More recently, S1P has also been implicated in mediating β-cell function and viability with a specific role for its metabolizing enzymes.

In 2005, a pioneering study characterized the SphK/S1P axis in rat pancreatic β cells and in INS-1 cells [83]. This study was followed by numerous others that focused on the SphK/S1P axis involvement in β-cell secretory function. Hasan et al. reported for the first time that SphK1 activity was important for insulin synthesis and secretion [84]. The knock-down of SphK1 expression in pancreatic β INS-1 cells resulted in both lowered glucose-stimulated insulin secretion (GSIS) and insulin content associated with decreased insulin gene expression. Conversely, SphK1 overexpression restored both insulin synthesis and secretion [84]. In contrast, pancreatic β MIN6 cells exposed to high glucose concentrations displayed an increase in S1P levels due to SphK2 activity, which is concomitant with higher insulin secretion. In addition, inhibition of S1P production through SphK2 KO in MIN6 cells resulted in the abolition of GSIS [85]. Overall, these data suggest that S1P synthesis through both SphK1 and SphK2 could be positively involved in regulating insulin secretion (Figure 3). However, this conclusion still needs in vivo and in vitro exploration of GSIS in mice KO for either SphK1 or SphK2.

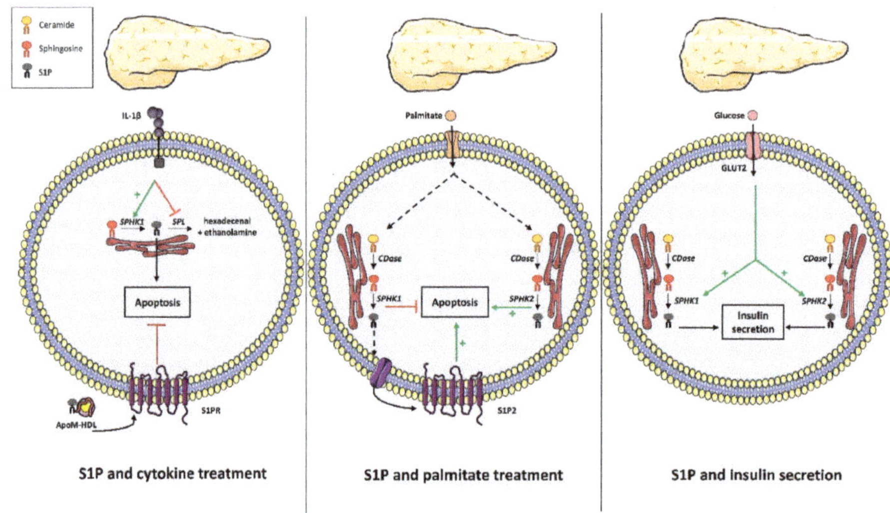

Figure 3. Role of S1P metabolism on pancreatic β cell fate. In pancreatic β cells, cytokines, such as IL1β increase SphK1 expression and repress SPL expression. This contributes to the increase of intracellular S1P content and apoptosis. In contrast, extracellular ApoM/HDL-bound S1P y antagonizes apoptosis induced by IL1β. In pancreatic β cells, palmitate increases the expression of both SphK1 and2. SphK1 activation represses palmitate-induced pancreatic β cell apoptosis, whereas SphK2 activation promotes apoptosis. SphK1-produced S1P can be secreted and stimulates S1P2 to promote apoptosis. High glucose levels could activate both SphK1 and 2, which contribute to the secretion of insulin. CDase: ceramidase. SPL: S1P lyase. SphK: sphingosine kinase. S1P2: S1P receptor 2.

The SphK/S1P axis was shown to be stimulated by cytokines in rat pancreatic β cells and INS-1 cells (Figure 3) [83], which suggests a potential role in the pathological response to cytokines observed during low-grade inflammation induced by obesity. Later on, Hahn et al. showed that cytokines decreased SPL expression in pancreatic β cells, whereas overexpression of SPL protected them against cytokine toxicity (Figure 3) [86], which comforts a pathological role of intracellular S1P metabolism of pancreatic β cells in diabetes. In contrast, Laychock et al. showed that exogenous S1P counteracted pancreatic β cell apoptosis induced by cytokines [87], which suggests a divergent role of cellular S1P from circulating S1P (Figure 3). Supporting this notion, Rütti et al. found that HDL, known to be enriched in S1P through its binding to apoM, also counteracted pancreatic β cell apoptosis induced by cytokines (Figure 3) [88].

Although the SphK/S1P axis appears to regulate β-cell induced-apoptosis induced by cytokines, the circulating levels are increased by obesity, whether it is implicated in β-cell apoptosis induced by free fatty acids still remains unknown. Palmitate increased not only ceramide but also S1P levels, through SphK1 up-regulation in pancreatic β INS-1 cells (Figure 3) [89]. Japtok et al. also demonstrated that palmitate increased S1P levels in pancreatic β MIN6 cells, which were released in the extracellular medium (Figure 3) [90]. Apoptosis was abrogated in INS-1 cells over-expressing SphK1 [89]. Similarly, either S1P supplementation or SphK1 overexpression in palmitate-treated INS-1 or MIN6 cells prevented cell death (Figure 3) [59]. Conversely, dominant negative expression of SphK1 in these cell lines enhanced palmitate-induced apoptosis [59]. The protective role of SphK1 was independent of S1P receptors, but was mediated by decreasing formation of pro-apoptotic ceramides induced by palmitate [89]. In addition, endoplasmic reticulum-targeted SphK1 also partially inhibited apoptosis induced by lipotoxicity, which suggests a specific localization for the anti-apoptotic action of S1P [89]. Nevertheless, JTE-013, which is an antagonist of S1P2, partially counteracted pancreatic β-cell apoptosis and the reduced proliferation induced by palmitate [89,90], which suggests that the

S1P produced could determine pancreatic β-cell fate under lipotoxicity by interacting with specific receptors (Figure 3). Overall, these studies reported a survival and protective role of both intracellular S1P and its enzyme SphK1 against palmitate-induced β-cell apoptosis [59,90].

In addition, one study discovered that HFD-fed SphK1 KO mice displayed a reduction in β cell size, number, and mass associated with increased β cell apoptosis compared to WT HFD-fed mice, which all favor the installation of glucose intolerance [59]. These data indicated that in vivo SphK1 deficiency predisposes mice to T2D-onset by promoting pancreatic β cell death under lipotoxic conditions [59]. SphK1 has been shown to interact with SKIP (SPHK1-interacting protein) and that SKIP overexpression in NIH 3T3 fibroblasts reduces SphK1 activity and interferes with its biological functions [91]. In another study, SKIP-deficient mice improved glucose tolerance by increasing insulin and GLP-1 secretion [92], which suggests that SKIP deficiency in mice allow SphK1 to better regulate glucose tolerance. However, it remains to be established whether SKIP is acting only at the level of intestinal L cells or on the pancreatic β cell since it is already known that islet-derived GLP-1 is necessary for glucose-stimulated insulin secretion [93]. Not surprisingly, a consensus on the role of the SphK1/S1P axis in β-cells has not been reached. Although the above studies demonstrated a beneficial role of SphK1 on glucose homeostasis and β cell function, another study showed that SphK1 KO mice were protected from HFD-induced glucose intolerance due to a reduced adipocyte pro-inflammatory response, which suggests a negative role of SphK1/S1P axis on regulating glucose homeostasis [78].

In contrast, a negative role of the SphK2/S1P axis was observed on β-cell fate. SphK2 expression KO reversed palmitate-induced cell death, whereas SphK2 overexpression promoted cell death under lipotoxic conditions in both INS-1 and MIN6 cells (Figure 3) [94]. In fact, lipotoxicity induced the shuttling of SphK2 from the nucleus to the cytoplasm, where it led to mitochondrial apoptosis [94]. SphK2 KO diabetic mice under HFD significantly improved their diabetic phenotypes [94], which suggests that, contrary to SphK1, SphK2 exerts a major role in promoting lipotoxicity-induced apoptosis of β cells [94]. Mice with a deletion of the S1P phosphohydrylolase SPP2 exhibited glucose intolerance due to a defect in the adaptation of pancreatic β cell mass, which supports the idea that the rise of endogenous S1P regulated by SphK2 and SPP2 can promote β cell lipotoxicity [35].

Overall, the opposed functions on β cell survival between both SphKs could be explained by expression differences observed in pathophysiological situations but more likely by differences in produced S1P subcellular localization. Nevertheless, it remains crucial to determine the potent role of S1P receptors in pancreatic β cell fate during obesity. To date, there are no clinical studies available describing a potential role of S1P in human islets in the context of obesity or T2D.

5. S1P and the Hypothalamic Regulation of Body Weight and Energy Homeostasis

The hypothalamus is a key brain area that plays a crucial role in regulating energy metabolism. It consists of several nuclei including the arcuate nucleus (ARC), ventromedial (VMH), dorsomedial (DMH), lateral (LH), and paraventricular (PVH) hypothalamus, which interact functionally to coordinate adaptive physiological responses controlling feeding behavior and energy expenditure. This process involves the integration of metabolic, endocrine, and neural signals from the periphery and autonomic circuitries that encode information about energy availability and energy reserve in the body [95–98].

Growing evidence suggests that hypothalamic lipid sensing plays a key role in controlling food intake, fat deposition, and energy balance [99,100], and that its dysregulation could lead to the development of obesity and T2D. Recent investigations reported that S1P is involved in the hypothalamic control of energy homeostasis [101]. Precisely, the intracerebroventricular (ICV) administration of S1P decreased food intake and increased energy expenditure [101]. Conversely, selective disruption of S1P1 in the mediobasal hypothalamus (MBH) induced the opposite effects [101]. At the molecular level, S1P exerted its effects by activating S1P1, which is highly expressed in key hypothalamic nuclei, ARC, VMH, and DMH, which controls feeding, particularly in the anorectic pro-opiomelanocortin (POMC) neurons of the ARC. Altogether, these findings identified S1P/S1P1/JAK2/STAT3 as a new regulatory pathway that plays a crucial role in the hypothalamic control of energy homeostasis and body weight gain. A positive correlation between plasma S1P and body fat percentage exists [20,102], as rodent models of obesity also exhibited an increased hypothalamic S1P/S1P1/STAT3 signaling [101,103]. From a therapeutic point of view, the ICV injection of S1P or the S1P1 agonist, SEW2871, induced anorexigenic effects, and prevented the development of obesity and associated metabolic dysfunctions [101].

In the context of an HFD-induced obesity, and, as it has already been observed in the AT (see section on adipose tissue), inflammatory processes also occur in the brain [104]. HFD triggers brain inflammation, notably in the hypothalamus, by activating microglia and astrocytes, which results in reactive gliosis, production of pro-inflammatory cytokines such as IL1β and TNF, and the development of neuronal inflammation [105]. This contributes to the deregulation of hypothalamic control of energy homeostasis, which promotes the development of obesity and associated metabolic disorders [106–108]. Emerging evidence suggests a pivotal role of S1P metabolism and S1P-mediated signaling in the development of neuro-inflammation. It was shown that S1P was able to induce astrocytes activation [109,110] and increase the inflammatory response of activated microglia, which results in reactive gliosis and the upregulation of pro-inflammatory cytokines [111,112]. Additionally, S1P modulated neuro-inflammation by regulating the infiltration of peripheral immune cells in the central nervous system [113,114]. Considering the emerging importance of S1P metabolism in neuro-inflammation, further studies will be required to determine the role of S1P signaling in the early onset of hypothalamic inflammation and gliosis in the context of diet-induced obesity.

The role of hypothalamic S1P in the regulation of obesity and dysregulation of glucose homeostasis is actually an emerging area. Thus, future studies will be important to determine the role of S1P metabolism and signaling at the level of the hypothalamus in the context of obesity and T2D. This will constitute an important step toward identifying new targets for therapeutic intervention in obesity and obesity-related metabolic disorders.

6. Conclusions and Perspectives

The elevation of S1P levels in tissues and plasma has been associated with obesity, which suggests that S1P metabolism could be negatively or positively linked to this pathology and to the onset of T2D. Many studies performed in the last decade suggest that S1P metabolism plays a positive role in insulin signaling in peripheral tissues, which points to an adaptive role of S1P to counteract the installation of insulin resistance in muscle, adipose tissue, and liver [54,55,57,69]. However, it should be noted that some studies argue for a causative role of S1P metabolism in insulin resistance in the liver and in adipose tissue [58,78]. S1P metabolism has also been linked to pancreatic β-cell fate during obesity or T2D with opposite roles of S1P produced by SphK1 or SphK2 on pancreatic β cell apoptosis [59,94]. Moreover, the specific role of the SphK1/S1P axis in glucose homeostasis will require further attention since studies reveal divergent phenotypes of SphK1 KO mice under HFD [59,78].

These discrepancies in the results may be linked to the fact that cellular S1P levels are fine-tuned by a concerted regulation of S1P synthetizing enzymes (SphK) and S1P degradation enzymes (SPL and SPPs). The other reason could come from the intrinsic nature of S1P, which is both an intracellular mediator and a circulating bioactive lipid. This supports the idea that S1P could act not only intracellularly but also as an endocrine or paracrine signal through its secretion to regulate the insulin response in distant organs as well as in the pancreatic β-cell fate. Plasma apoM/HDL-bound S1P has been shown to regulate brown adipose tissue activity in the context of obesity [115] and also pancreatic β-cell survival induced by cytokines [88].

In conclusion, more work is required to understand the role of the enzymes involved in S1P metabolism/signaling, especially of the catabolizing enzymes SPL and SPPs in the development of obesity and diabetes. It will also be important to determine the role of its transporters and its receptors. Due to the duality of actions of S1P (intracellular and extracellular), the development of tissue-specific disruption of S1P metabolism enzyme genes in mice would also help us understand the divergent roles of S1P observed in whole KO models used to date. This would be crucial before the modulation of S1P metabolism can be considered as a potential therapeutic target for treating obesity/T2D.

Author Contributions: Conceived the idea: H.L.S. Wrote the manuscript: J.G., C.L.B., M.L.M., S.T.-C., O.B., Y.B., E.H. and H.L.S. Figure preparation: S.T.-C. and J.G. Critically reviewed the manuscript and figures: C.L.B., S.T.-C., J.G., E.H. and H.L.S. All authors approved the final manuscript. All authors have read and agreed to the published version of the manuscript.

Funding: The authors would like to acknowledge the support by the Centre National de la Recherche Scientifique (CNRS), the Institut National de la Recherche Médicale (INSERM), the Sorbonne Université, and the Université Paris Saclay. H.L.S. is funded by the Société Francophone de Diabétologie (SFD). E.H. is funded by the Fondation de France. J.G. is funded by a scholarship from the French Research Ministry/University Paris Saclay.

Acknowledgments: We are grateful to Froogh Darakhshan-Hajduch (Anglais Pour Vous, Melun, France) for professional editing of this review.

Conflicts of Interest: The authors declare no conflict of interest.

References

1. Galgani, J.E.; Moro, C.; Ravussin, E. Metabolic flexibility and insulin resistance. *Am. J. Physiol. Endocrinol. Metab.* **2008**, *295*, E1009–E1017. [CrossRef] [PubMed]

2. Zacharewicz, E.; Hesselink, M.K.C.; Schrauwen, P. Exercise counteracts lipotoxicity by improving lipid turnover and lipid droplet quality. *J. Intern. Med.* **2018**, *284*, 505–518. [CrossRef] [PubMed]

3. Montgomery, M.K.; De Nardo, W.; Watt, M.J. Impact of Lipotoxicity on Tissue "Cross Talk" and Metabolic Regulation. *Physiol. Bethesda Md.* **2019**, *34*, 134–149. [CrossRef] [PubMed]

4. Torretta, E.; Barbacini, P.; Al-Daghri, N.M.; Gelfi, C. Sphingolipids in Obesity and Correlated Co-Morbidities: The Contribution of Gender, Age and Environment. *Int. J. Mol. Sci.* **2019**, *20*, 5901. [CrossRef] [PubMed]

5. Saeedi, P.; Petersohn, I.; Salpea, P.; Malanda, B.; Karuranga, S.; Unwin, N.; Colagiuri, S.; Guariguata, L.; Motala, A.A.; Ogurtsova, K.; et al. Global and regional diabetes prevalence estimates for 2019 and projections for 2030 and 2045: Results from the International Diabetes Federation Diabetes Atlas, 9th edition. *Diabetes Res. Clin. Pract.* **2019**, *157*, 107843. [CrossRef] [PubMed]

6. Oh, Y.S.; Bae, G.D.; Baek, D.J.; Park, E.-Y.; Jun, H.-S. Fatty Acid-Induced Lipotoxicity in Pancreatic Beta-Cells during Development of Type 2 Diabetes. *Front. Endocrinol.* **2018**, *9*, 384. [CrossRef] [PubMed]

7. Bachmann, O.P.; Dahl, D.B.; Brechtel, K.; Machann, J.; Haap, M.; Maier, T.; Loviscach, M.; Stumvoll, M.; Claussen, C.D.; Schick, F.; et al. Effects of intravenous and dietary lipid challenge on intramyocellular lipid content and the relation with insulin sensitivity in humans. *Diabetes* **2001**, *50*, 2579–2584. [CrossRef]

8. Palomer, X.; Pizarro-Delgado, J.; Barroso, E.; Vázquez-Carrera, M. Palmitic and Oleic Acid: The Yin and Yang of Fatty Acids in Type 2 Diabetes Mellitus. *Trends Endocrinol. Metab.* **2018**, *29*, 178–190. [CrossRef]

9. Ralston, J.C.; Nguyen-Tu, M.-S.; Lyons, C.L.; Cooke, A.A.; Murphy, A.M.; Falvey, A.; Finucane, O.M.; McGillicuddy, F.C.; Rutter, G.A.; Roche, H.M. Dietary substitution of SFA with MUFA within high-fat diets attenuates hyperinsulinaemia and pancreatic islet dysfunction. *Br. J. Nutr.* **2020**, 1–9. [CrossRef] [PubMed]

10. Bandet, C.L.; Tan-Chen, S.; Bourron, O.; Stunff, H.L.; Hajduch, E. Sphingolipid Metabolism: New Insight into Ceramide-Induced Lipotoxicity in Muscle Cells. *Int. J. Mol. Sci.* **2019**, *20*, 479. [CrossRef]

11. Hannun, Y.A.; Obeid, L.M. Principles of bioactive lipid signalling: Lessons from sphingolipids. *Nat. Rev. Mol. Cell Biol.* **2008**, *9*, 139–150. [CrossRef] [PubMed]

12. Bartke, N.; Hannun, Y.A. Bioactive sphingolipids: Metabolism and function. *J. Lipid Res.* **2009**, *50*, S91–S96. [CrossRef] [PubMed]

13. Ségui, B.; Andrieu-Abadie, N.; Jaffrézou, J.-P.; Benoist, H.; Levade, T. Sphingolipids as modulators of cancer cell death: Potential therapeutic targets. *Biochim. Biophys. Acta* **2006**, *1758*, 2104–2120. [CrossRef] [PubMed]

14. Hannun, Y.A.; Obeid, L.M. Sphingolipids and their metabolism in physiology and disease. *Nat. Rev. Mol. Cell Biol.* **2018**, *19*, 175–191. [CrossRef] [PubMed]

15. Mullen, T.D.; Hannun, Y.A.; Obeid, L.M. Ceramide synthases at the centre of sphingolipid metabolism and biology. *Biochem. J.* **2012**, *441*, 789–802. [CrossRef]

16. Hage Hassan, R.; Bourron, O.; Hajduch, E. Defect of insulin signal in peripheral tissues: Important role of ceramide. *World J. Diabetes* **2014**, *5*, 244–257. [CrossRef]

17. Campana, M.; Bellini, L.; Rouch, C.; Rachdi, L.; Coant, N.; Butin, N.; Bandet, C.L.; Philippe, E.; Meneyrol, K.; Kassis, N.; et al. Inhibition of central de novo ceramide synthesis restores insulin signaling in hypothalamus and enhances β-cell function of obese Zucker rats. *Mol. Metab.* **2018**, *8*, 23–36. [CrossRef]

18. Bellini, L.; Campana, M.; Mahfouz, R.; Carlier, A.; Véret, J.; Magnan, C.; Hajduch, E.; Stunff, H.L. Targeting sphingolipid metabolism in the treatment of obesity/type 2 diabetes. *Expert Opin. Ther. Targets* **2015**, *19*, 1037–1050. [CrossRef]

19. Maceyka, M.; Harikumar, K.B.; Milstien, S.; Spiegel, S. Sphingosine-1-Phosphate Signaling and Its Role in Disease. *Trends Cell Biol.* **2012**, *22*, 50–60. [CrossRef]

20. Kowalski, G.M.; Carey, A.L.; Selathurai, A.; Kingwell, B.A.; Bruce, C.R. Plasma Sphingosine-1-Phosphate Is Elevated in Obesity. *PLoS ONE* **2013**, *8*. [CrossRef]

21. Liu, H.; Sugiura, M.; Nava, V.E.; Edsall, L.C.; Kono, K.; Poulton, S.; Milstien, S.; Kohama, T.; Spiegel, S. Molecular cloning and functional characterization of a novel mammalian sphingosine kinase type 2 isoform. *J. Biol. Chem.* **2000**, *275*, 19513–19520. [CrossRef] [PubMed]

22. Ng, M.L.; Wadham, C.; Sukocheva, O.A. The role of sphingolipid signalling in diabetes-associated pathologies (Review). *Int. J. Mol. Med.* **2017**, *39*, 243–252. [CrossRef] [PubMed]

23. Alemany, R.; van Koppen, C.J.; Danneberg, K.; Ter Braak, M.; Meyer zu Heringdorf, D. Regulation and functional roles of sphingosine kinases. *Naunyn. Schmiedebergs Arch. Pharmacol.* **2007**, *374*, 413–428. [CrossRef] [PubMed]

24. Maceyka, M.; Sankala, H.; Hait, N.C.; Le Stunff, H.; Liu, H.; Toman, R.; Collier, C.; Zhang, M.; Satin, L.S.; Merrill, A.H.; et al. SphK1 and SphK2, Sphingosine Kinase Isoenzymes with Opposing Functions in Sphingolipid Metabolism. *J. Biol. Chem.* **2005**, *280*, 37118–37129. [CrossRef]

25. *Lysophospholipid Receptors: Signaling and Biochemistry*; Chun, J., Ed.; Wiley: Hoboken, NJ, USA, 2013; ISBN 978-1-118-53130-3.

26. Hait, N.C.; Allegood, J.; Maceyka, M.; Strub, G.M.; Harikumar, K.B.; Singh, S.K.; Luo, C.; Marmorstein, R.; Kordula, T.; Milstien, S.; et al. Regulation of histone acetylation in the nucleus by sphingosine-1-phosphate. *Science* **2009**, *325*, 1254–1257. [CrossRef]

27. Kleuser, B. Divergent Role of Sphingosine 1-Phosphate in Liver Health and Disease. *Int. J. Mol. Sci.* **2018**, *19*, 722. [CrossRef]

28. Strub, G.M.; Paillard, M.; Liang, J.; Gomez, L.; Allegood, J.C.; Hait, N.C.; Maceyka, M.; Price, M.M.; Chen, Q.; Simpson, D.C.; et al. Sphingosine-1-phosphate produced by sphingosine kinase 2 in mitochondria interacts with prohibitin 2 to regulate complex IV assembly and respiration. *FASEB J. Off. Publ. Fed. Am. Soc. Exp. Biol.* **2011**, *25*, 600–612. [CrossRef]

29. Le Stunff, H.; Giussani, P.; Maceyka, M.; Lepine, S.; Milstien, S.; Spiegel, S. Recycling of Sphingosine Is Regulated by the Concerted Actions of Sphingosine-1-phosphate Phosphohydrolase 1 and Sphingosine Kinase 2. *J. Biol. Chem.* **2007**, *282*, 34372–34380. [CrossRef]

30. Aguilar, A.; Saba, J.D. Truth and consequences of sphingosine-1-phosphate lyase. *Adv. Biol. Regul.* **2012**, *52*, 17–30. [CrossRef]

31. Le Stunff, H. Sphingosine-1-phosphate and lipid phosphohydrolases. *Biochim. Biophys. Acta BBA—Mol. Cell Biol. Lipids* **2002**, *1582*, 8–17. [CrossRef]

32. Sigal, Y.J.; McDermott, M.I.; Morris, A.J. Integral membrane lipid phosphatases/phosphotransferases: Common structure and diverse functions. *Biochem. J.* **2005**, *387*, 281–293. [CrossRef] [PubMed]

33. *Sphingolipids as Signaling and Regulatory Molecules*; Chalfant, C.; Del Poeta, M. (Eds.) Advances in Experimental Medicine and Biology; Springer Science+Business Media: New York, NY, USA; Landes Bioscience: Austin, TX, USA, 2010; ISBN 978-1-4419-6740-4.

34. Lépine, S.; Allegood, J.C.; Park, M.; Dent, P.; Milstien, S.; Spiegel, S. Sphingosine-1-phosphate phosphohydrolase-1 regulates ER stress-induced autophagy. *Cell Death Differ.* **2011**, *18*, 350–361. [CrossRef] [PubMed]

35. Taguchi, Y.; Allende, M.L.; Mizukami, H.; Cook, E.K.; Gavrilova, O.; Tuymetova, G.; Clarke, B.A.; Chen, W.; Olivera, A.; Proia, R.L. Sphingosine-1-phosphate Phosphatase 2 Regulates Pancreatic Islet β-Cell Endoplasmic Reticulum Stress and Proliferation. *J. Biol. Chem.* **2016**, *291*, 12029–12038. [CrossRef]

36. Reitsema, V.; Bouma, H.; Willem Kok, J. Sphingosine-1-phosphate transport and its role in immunology. *AIMS Mol. Sci.* **2014**, *1*, 183–201. [CrossRef]

37. Kobayashi, N.; Kawasaki-Nishi, S.; Otsuka, M.; Hisano, Y.; Yamaguchi, A.; Nishi, T. MFSD2B is a sphingosine 1-phosphate transporter in erythroid cells. *Sci. Rep.* **2018**, *8*. [CrossRef]

38. Yatomi, Y. Plasma sphingosine 1-phosphate metabolism and analysis. *Glycobiol. Sphingobiology* **2008**, *1780*, 606–611. [CrossRef]

39. Christoffersen, C.; Obinata, H.; Kumaraswamy, S.B.; Galvani, S.; Ahnström, J.; Sevvana, M.; Egerer-Sieber, C.; Muller, Y.A.; Hla, T.; Nielsen, L.B.; et al. Endothelium-protective sphingosine-1-phosphate provided by HDL-associated apolipoprotein M. *Proc. Natl. Acad. Sci. USA* **2011**, *108*, 9613–9618. [CrossRef] [PubMed]

40. Blaho, V.A.; Hla, T. An update on the biology of sphingosine 1-phosphate receptors. *J. Lipid Res.* **2014**, *55*, 1596–1608. [CrossRef]

41. Venkataraman, K.; Lee, Y.-M.; Michaud, J.; Thangada, S.; Ai, Y.; Bonkovsky, H.L.; Parikh, N.S.; Habrukowich, C.; Hla, T. Vascular endothelium as a contributor of plasma sphingosine 1-phosphate. *Circ. Res.* **2008**, *102*, 669–676. [CrossRef]

42. Kimura, T.; Sato, K.; Kuwabara, A.; Tomura, H.; Ishiwara, M.; Kobayashi, I.; Ui, M.; Okajima, F. Sphingosine 1-phosphate may be a major component of plasma lipoproteins responsible for the cytoprotective actions in human umbilical vein endothelial cells. *J. Biol. Chem.* **2001**, *276*, 31780–31785. [CrossRef]

43. Tran-Dinh, A.; Diallo, D.; Delbosc, S.; Varela-Perez, L.M.; Dang, Q.; Lapergue, B.; Burillo, E.; Michel, J.; Levoye, A.; Martin-Ventura, J.; et al. HDL and endothelial protection: HDL and endothelial protection. *Br. J. Pharmacol.* **2013**, *169*, 493–511. [CrossRef]

44. Poti, F.; Simoni, M.; Nofer, J.-R. Atheroprotective role of high-density lipoprotein (HDL)-associated sphingosine-1-phosphate (S1P). *Cardiovasc. Res.* **2014**, *103*, 395–404. [CrossRef] [PubMed]

45. Cuvillier, O. Les récepteurs de la sphingosine 1-phosphate: De la biologie à la physiopathologie. *Médecine/Sciences* **2012**, *28*, 951–957. [CrossRef] [PubMed]

46. Park, S.-J.; Im, D.-S. Sphingosine 1-Phosphate Receptor Modulators and Drug Discovery. *Biomol. Ther.* **2017**, *25*, 80–90. [CrossRef] [PubMed]

47. Cannavo, A.; Liccardo, D.; Komici, K.; Corbi, G.; de Lucia, C.; Femminella, G.D.; Elia, A.; Bencivenga, L.; Ferrara, N.; Koch, W.J.; et al. Sphingosine Kinases and Sphingosine 1-Phosphate Receptors: Signaling and Actions in the Cardiovascular System. *Front. Pharmacol.* **2017**, *8*, 556. [CrossRef]

48. Kihara, Y.; Maceyka, M.; Spiegel, S.; Chun, J. Lysophospholipid receptor nomenclature review: IUPHAR Review 8. *Br. J. Pharmacol.* **2014**, *171*, 3575–3594. [CrossRef]

49. Flamment, M.; Hajduch, E.; Ferré, P.; Foufelle, F. New insights into ER stress-induced insulin resistance. *Trends Endocrinol. Metab.* **2012**, *23*, 381–390. [CrossRef]

50. Bradham, C.A.; Plümpe, J.; Manns, M.P.; Brenner, D.A.; Trautwein, C. Mechanisms of hepatic toxicity. I. TNF-induced liver injury. *Am. J. Physiol.* **1998**, *275*, G387–G392. [CrossRef]

51. Osawa, Y.; Banno, Y.; Nagaki, M.; Brenner, D.A.; Naiki, T.; Nozawa, Y.; Nakashima, S.; Moriwaki, H. TNF-α-Induced Sphingosine 1-Phosphate Inhibits Apoptosis Through a Phosphatidylinositol 3-Kinase/Akt Pathway in Human Hepatocytes. *J. Immunol.* **2001**, *167*, 173–180. [CrossRef]

52. Osawa, Y.; Uchinami, H.; Bielawski, J.; Schwabe, R.F.; Hannun, Y.A.; Brenner, D.A. Roles for C16-ceramide and Sphingosine 1-Phosphate in Regulating Hepatocyte Apoptosis in Response to Tumor Necrosis Factor-α. *J. Biol. Chem.* **2005**, *280*, 27879–27887. [CrossRef]

53. Fang, H.; Feng, Q.; Shi, Y.; Zhou, J.; Wang, Q.; Zhong, L. Hepatic insulin resistance induced by mitochondrial oxidative stress can be ameliorated by sphingosine 1-phosphate. *Mol. Cell. Endocrinol.* **2020**, *501*, 110660. [CrossRef] [PubMed]

54. Ma, M.M.; Chen, J.L.; Wang, G.G.; Wang, H.; Lu, Y.; Li, J.F.; Yi, J.; Yuan, Y.J.; Zhang, Q.W.; Mi, J.; et al. Sphingosine kinase 1 participates in insulin signalling and regulates glucose metabolism and homeostasis in KK/Ay diabetic mice. *Diabetologia* **2007**, *50*, 891–900. [CrossRef] [PubMed]

55. Osawa, Y.; Seki, E.; Kodama, Y.; Suetsugu, A.; Miura, K.; Adachi, M.; Ito, H.; Shiratori, Y.; Banno, Y.; Olefsky, J.M.; et al. Acid sphingomyelinase regulates glucose and lipid metabolism in hepatocytes through AKT activation and AMP-activated protein kinase suppression. *FASEB J.* **2011**, *25*, 1133–1144. [CrossRef] [PubMed]

56. Chen, J.; Wang, W.; Qi, Y.; Kaczorowski, D.; McCaughan, G.W.; Gamble, J.R.; Don, A.S.; Gao, X.; Vadas, M.A.; Xia, P. Deletion of sphingosine kinase 1 ameliorates hepatic steatosis in diet-induced obese mice: Role of PPARγ. *Biochim. Biophys. Acta BBA—Mol. Cell Biol. Lipids* **2016**, *1861*, 138–147. [CrossRef]

57. Lee, S.-Y.; Hong, I.-K.; Kim, B.-R.; Shim, S.-M.; Sung Lee, J.; Lee, H.-Y.; Soo Choi, C.; Kim, B.-K.; Park, T.-S. Activation of sphingosine kinase 2 by endoplasmic reticulum stress ameliorates hepatic steatosis and insulin resistance in mice: LEE, HONG, ET AL. *Hepatology* **2015**, *62*, 135–146. [CrossRef]

58. Fayyaz, S.; Henkel, J.; Japtok, L.; Krämer, S.; Damm, G.; Seehofer, D.; Püschel, G.P.; Kleuser, B. Involvement of sphingosine 1-phosphate in palmitate-induced insulin resistance of hepatocytes via the S1P2 receptor subtype. *Diabetologia* **2014**, *57*, 373–382. [CrossRef] [PubMed]

59. Qi, Y.; Chen, J.; Lay, A.; Don, A.; Vadas, M.; Xia, P. Loss of sphingosine kinase 1 predisposes to the onset of diabetes via promoting pancreatic β-cell death in diet-induced obese mice. *FASEB J.* **2013**, *27*, 4294–4304. [CrossRef]

60. Kowalski, G.M.; Kloehn, J.; Burch, M.L.; Selathurai, A.; Hamley, S.; Bayol, S.A.M.; Lamon, S.; Watt, M.J.; Lee-Young, R.S.; McConville, M.J.; et al. Overexpression of sphingosine kinase 1 in liver reduces triglyceride content in mice fed a low but not high-fat diet. *Biochim. Biophys. Acta* **2015**, *1851*, 210–219. [CrossRef] [PubMed]

61. Pacana, T.; Sanyal, A.J. Recent advances in understanding/management of non-alcoholic steatohepatitis. *F1000prime Rep.* **2015**, *7*, 28. [CrossRef] [PubMed]

62. Nagahashi, M.; Takabe, K.; Liu, R.; Peng, K.; Wang, X.; Wang, Y.; Hait, N.C.; Wang, X.; Allegood, J.C.; Yamada, A.; et al. Conjugated bile acid-activated S1P receptor 2 is a key regulator of sphingosine kinase 2 and hepatic gene expression. *Hepatol. Baltim. Md.* **2015**, *61*, 1216–1226. [CrossRef] [PubMed]

63. Liao, C.-Y.; Song, M.J.; Gao, Y.; Mauer, A.S.; Revzin, A.; Malhi, H. Hepatocyte-Derived Lipotoxic Extracellular Vesicle Sphingosine 1-Phosphate Induces Macrophage Chemotaxis. *Front. Immunol.* **2018**, *9*, 2980. [CrossRef] [PubMed]

64. Geng, T.; Sutter, A.; Harland, M.D.; Law, B.A.; Ross, J.S.; Lewin, D.; Palanisamy, A.; Russo, S.B.; Chavin, K.D.; Cowart, L.A. SphK1 mediates hepatic inflammation in a mouse model of NASH induced by high saturated fat feeding and initiates proinflammatory signaling in hepatocytes. *J. Lipid Res.* **2015**, *56*, 2359–2371. [CrossRef] [PubMed]

65. DeFronzo, R.A.; Ferrannini, E.; Sato, Y.; Felig, P.; Wahren, J. Synergistic interaction between exercise and insulin on peripheral glucose uptake. *J. Clin. Investig.* **1981**, *68*, 1468–1474. [CrossRef]

66. Hu, W.; Bielawski, J.; Samad, F.; Merrill, A.H.; Cowart, L.A. Palmitate increases sphingosine-1-phosphate in C2C12 myotubes via upregulation of sphingosine kinase message and activity. *J. Lipid Res.* **2009**, *50*, 1852–1862. [CrossRef]

67. Ross, J.S.; Hu, W.; Rosen, B.; Snider, A.J.; Obeid, L.M.; Cowart, L.A. Sphingosine Kinase 1 Is Regulated by Peroxisome Proliferator-activated Receptor α in Response to Free Fatty Acids and Is Essential for Skeletal Muscle Interleukin-6 Production and Signaling in Diet-induced Obesity. *J. Biol. Chem.* **2013**, *288*, 22193–22206. [CrossRef] [PubMed]

68. Rapizzi, E.; Taddei, M.L.; Fiaschi, T.; Donati, C.; Bruni, P.; Chiarugi, P. Sphingosine 1-phosphate increases glucose uptake through trans-activation of insulin receptor. *Cell. Mol. Life Sci.* **2009**, *66*, 3207–3218. [CrossRef] [PubMed]

69. Bruce, C.R.; Risis, S.; Babb, J.R.; Yang, C.; Kowalski, G.M.; Selathurai, A.; Lee-Young, R.S.; Weir, J.M.; Yoshioka, K.; Takuwa, Y.; et al. Overexpression of Sphingosine Kinase 1 Prevents Ceramide Accumulation and Ameliorates Muscle Insulin Resistance in High-Fat Diet-Fed Mice. *Diabetes* **2012**, *61*, 3148–3155. [CrossRef] [PubMed]

70. Bruce, C.R.; Risis, S.; Babb, J.R.; Yang, C.; Lee-Young, R.S.; Henstridge, D.C.; Febbraio, M.A. The sphingosine-1-phosphate analog FTY720 reduces muscle ceramide content and improves glucose tolerance in high fat-fed male mice. *Endocrinology* **2013**, *154*, 65–76. [CrossRef]

71. Kendall, M. FTY720, a sphingosine-1-phosphate receptor modulator, reverses high-fat diet-induced weight gain, insulin resistance and adipose tissue inflammation in C57BL/6 mice. *Diabetes Obes. Metab.* **2008**, *10*, 802–805. [CrossRef]

72. Brinkmann, V. Sphingosine 1-phosphate receptors in health and disease: Mechanistic insights from gene deletion studies and reverse pharmacology. *Pharmacol. Ther.* **2007**, *115*, 84–105. [CrossRef]

73. Berdyshev, E.V.; Gorshkova, I.; Skobeleva, A.; Bittman, R.; Lu, X.; Dudek, S.M.; Mirzapoiazova, T.; Garcia, J.G.N.; Natarajan, V. FTY720 Inhibits Ceramide Synthases and Up-regulates Dihydrosphingosine 1-Phosphate Formation in Human Lung Endothelial Cells. *J. Biol. Chem.* **2009**, *284*, 5467–5477. [CrossRef] [PubMed]

74. Zorena, K.; Jachimowicz-Duda, O.; Ślęzak, D.; Robakowska, M.; Mrugacz, M. Adipokines and Obesity. Potential Link to Metabolic Disorders and Chronic Complications. *Int. J. Mol. Sci.* **2020**, *21*, 3570. [CrossRef] [PubMed]

75. Calder, P.C.; Ahluwalia, N.; Brouns, F.; Buetler, T.; Clement, K.; Cunningham, K.; Esposito, K.; Jönsson, L.S.; Kolb, H.; Lansink, M.; et al. Dietary factors and low-grade inflammation in relation to overweight and obesity. *Br. J. Nutr.* **2011**, *106* (Suppl. 3), S5–S78. [CrossRef] [PubMed]

76. Burhans, M.S.; Hagman, D.K.; Kuzma, J.N.; Schmidt, K.A.; Kratz, M. Contribution of adipose tissue inflammation to the development of type 2 diabetes mellitus. *Compr. Physiol.* **2018**, *9*, 1–58. [CrossRef]

77. Hashimoto, T.; Igarashi, J.; Kosaka, H. Sphingosine kinase is induced in mouse 3T3-L1 cells and promotes adipogenesis. *J. Lipid Res.* **2009**, *50*, 602–610. [CrossRef]

78. Wang, J.; Badeanlou, L.; Bielawski, J.; Ciaraldi, T.P.; Samad, F. Sphingosine kinase 1 regulates adipose proinflammatory responses and insulin resistance. *Am. J. Physiol.—Endocrinol. Metab.* **2014**, *306*, E756–E768. [CrossRef]

79. Blachnio-Zabielska, A.U.; Koutsari, C.; Tchkonia, T.; Jensen, M.D. Sphingolipid content of human adipose tissue: Relationship to adiponectin and insulin resistance. *Obes. Silver Spring Md.* **2012**, *20*, 2341–2347. [CrossRef]

80. Lu, J.; Zhao, J.; Meng, H.; Zhang, X. Adipose Tissue-Resident Immune Cells in Obesity and Type 2 Diabetes. *Front. Immunol.* **2019**, *10*, 1173. [CrossRef]

81. Mandala, S.; Hajdu, R.; Bergstrom, J.; Quackenbush, E.; Xie, J.; Milligan, J.; Thornton, R.; Shei, G.-J.; Card, D.; Keohane, C.; et al. Alteration of lymphocyte trafficking by sphingosine-1-phosphate receptor agonists. *Science* **2002**, *296*, 346–349. [CrossRef] [PubMed]

82. Gabriel, T.L.; Mirzaian, M.; Hooibrink, B.; Ottenhoff, R.; van Roomen, C.; Aerts, J.M.F.G.; van Eijk, M. Induction of Sphk1 activity in obese adipose tissue macrophages promotes survival. *PLoS ONE* **2017**, *12*, e0182075. [CrossRef]

83. Mastrandrea, L.D.; Sessanna, S.M.; Laychock, S.G. Sphingosine kinase activity and sphingosine-1 phosphate production in rat pancreatic islets and INS-1 cells: Response to cytokines. *Diabetes* **2005**, *54*, 1429–1436. [CrossRef]

84. Hasan, N.M.; Longacre, M.J.; Stoker, S.; Kendrick, M.A.; Druckenbrod, N.R.; Laychock, S.G.; Mastrandrea, L.D.; MacDonald, M.J. Sphingosine Kinase 1 Knockdown Reduces Insulin Synthesis and Secretion in a Rat Insulinoma Cell Line. *Arch. Biochem. Biophys.* **2012**, *518*, 23–30. [CrossRef] [PubMed]

85. Stanford, J.C.; Morris, A.J.; Sunkara, M.; Popa, G.J.; Larson, K.L.; Özcan, S. Sphingosine 1-Phosphate (S1P) Regulates Glucose-stimulated Insulin Secretion in Pancreatic Beta Cells. *J. Biol. Chem.* **2012**, *287*, 13457–13464. [CrossRef] [PubMed]

86. Hahn, C.; Tyka, K.; Saba, J.D.; Lenzen, S.; Gurgul-Convey, E. Overexpression of sphingosine-1-phosphate lyase protects insulin-secreting cells against cytokine toxicity. *J. Biol. Chem.* **2017**, *292*, 20292–20304. [CrossRef]

87. Laychock, S.G.; Sessanna, S.M.; Lin, M.-H.; Mastrandrea, L.D. Sphingosine 1-phosphate affects cytokine-induced apoptosis in rat pancreatic islet beta-cells. *Endocrinology* **2006**, *147*, 4705–4712. [CrossRef] [PubMed]

88. Rütti, S.; Ehses, J.A.; Sibler, R.A.; Prazak, R.; Rohrer, L.; Georgopoulos, S.; Meier, D.T.; Niclauss, N.; Berney, T.; Donath, M.Y.; et al. Low- and high-density lipoproteins modulate function, apoptosis, and proliferation of primary human and murine pancreatic beta-cells. *Endocrinology* **2009**, *150*, 4521–4530. [CrossRef] [PubMed]

89. Véret, J.; Coant, N.; Gorshkova, I.A.; Giussani, P.; Fradet, M.; Riccitelli, E.; Skobeleva, A.; Goya, J.; Kassis, N.; Natarajan, V.; et al. Role of palmitate-induced sphingoid base-1-phosphate biosynthesis in INS-1 β-cell survival. *Biochim. Biophys. Acta BBA—Mol. Cell Biol. Lipids* **2013**, *1831*, 251–262. [CrossRef]

90. Japtok, L.; Schmitz, E.I.; Fayyaz, S.; Krämer, S.; Hsu, L.J.; Kleuser, B. Sphingosine 1-phosphate counteracts insulin signaling in pancreatic β-cells via the sphingosine 1-phosphate receptor subtype 2. *FASEB J. Off. Publ. Fed. Am. Soc. Exp. Biol.* **2015**, *29*, 3357–3369. [CrossRef]

91. Lacaná, E.; Maceyka, M.; Milstien, S.; Spiegel, S. Cloning and Characterization of a Protein Kinase A Anchoring Protein (AKAP)-related Protein That Interacts with and Regulates Sphingosine Kinase 1 Activity. *J. Biol. Chem.* **2002**, *277*, 32947–32953. [CrossRef]

92. Liu, Y.; Harashima, S.; Wang, Y.; Suzuki, K.; Tokumoto, S.; Usui, R.; Tatsuoka, H.; Tanaka, D.; Yabe, D.; Harada, N.; et al. Sphingosine kinase 1–interacting protein is a dual regulator of insulin and incretin secretion. *FASEB J.* **2019**, *33*, 6239–6253. [CrossRef]

93. De Souza, A.H.; Tang, J.; Yadev, A.K.; Saghafi, S.T.; Kibbe, C.R.; Linnemann, A.K.; Merrins, M.J.; Davis, D.B. Intra-islet GLP-1, but Not CCK, Is Necessary for β-Cell Function in Mouse and Human Islets. Available online: https://pubmed.ncbi.nlm.nih.gov/32071395/ (accessed on 20 May 2020).

94. Song, Z.; Wang, W.; Li, N.; Yan, S.; Rong, K.; Lan, T.; Xia, P. Sphingosine kinase 2 promotes lipotoxicity in pancreatic β-cells and the progression of diabetes. *FASEB J. Off. Publ. Fed. Am. Soc. Exp. Biol.* **2019**, *33*, 3636–3646. [CrossRef] [PubMed]

95. Morton, G.J.; Cummings, D.E.; Baskin, D.G.; Barsh, G.S.; Schwartz, M.W. Central nervous system control of food intake and body weight. *Nature* **2006**, *443*, 289–295. [CrossRef]

96. Jeong, J.H.; Lee, D.K.; Jo, Y.-H. Cholinergic neurons in the dorsomedial hypothalamus regulate food intake. *Mol. Metab.* **2017**, *6*, 306–312. [CrossRef] [PubMed]

97. Koch, M.; Horvath, T.L. Molecular and cellular regulation of hypothalamic melanocortin neurons controlling food intake and energy metabolism. *Mol. Psychiatry* **2014**, *19*, 752–761. [CrossRef] [PubMed]

98. Jeong, J.H.; Lee, D.K.; Liu, S.-M.; Chua, S.C.; Schwartz, G.J.; Jo, Y.-H. Activation of temperature-sensitive TRPV1-like receptors in ARC POMC neurons reduces food intake. *PLoS Biol.* **2018**, *16*, e2004399. [CrossRef]

99. Picard, A.; Moullé, V.S.; Le Foll, C.; Cansell, C.; Véret, J.; Coant, N.; Le Stunff, H.; Migrenne, S.; Luquet, S.; Cruciani-Guglielmacci, C.; et al. Physiological and pathophysiological implications of lipid sensing in the brain. *Diabetes Obes. Metab.* **2014**, *16* (Suppl. 1), 49–55. [CrossRef]

100. Le Stunff, H.; Coant, N.; Migrenne, S.; Magnan, C. Targeting lipid sensing in the central nervous system: New therapy against the development of obesity and type 2 diabetes. *Expert Opin. Ther. Targets* **2013**, *17*, 545–555. [CrossRef]

101. Silva, V.R.R.; Micheletti, T.O.; Pimentel, G.D.; Katashima, C.K.; Lenhare, L.; Morari, J.; Mendes, M.C.S.; Razolli, D.S.; Rocha, G.Z.; de Souza, C.T.; et al. Hypothalamic S1P/S1PR1 axis controls energy homeostasis. *Nat. Commun.* **2014**, *5*, 4859. [CrossRef]

102. Ito, S.; Iwaki, S.; Koike, K.; Yuda, Y.; Nagasaki, A.; Ohkawa, R.; Yatomi, Y.; Furumoto, T.; Tsutsui, H.; Sobel, B.E.; et al. Increased plasma sphingosine-1-phosphate in obese individuals and its capacity to increase the expression of plasminogen activator inhibitor-1 in adipocytes. *Coron. Artery Dis.* **2013**, *24*, 642–650. [CrossRef] [PubMed]

103. Samad, F.; Hester, K.D.; Yang, G.; Hannun, Y.A.; Bielawski, J. Altered adipose and plasma sphingolipid metabolism in obesity: A potential mechanism for cardiovascular and metabolic risk. *Diabetes* **2006**, *55*, 2579–2587. [CrossRef]

104. Dorfman, M.D.; Thaler, J.P. Hypothalamic inflammation and gliosis in obesity. *Curr. Opin. Endocrinol. Diabetes Obes.* **2015**, *22*, 325–330. [CrossRef]

105. Mendes, N.F.; Kim, Y.-B.; Velloso, L.A.; Araújo, E.P. Hypothalamic Microglial Activation in Obesity: A Mini-Review. *Front. Neurosci.* **2018**, *12*, 846. [CrossRef]

106. Gregor, M.F.; Hotamisligil, G.S. Inflammatory mechanisms in obesity. *Annu. Rev. Immunol.* **2011**, *29*, 415–445. [CrossRef] [PubMed]

107. Thaler, J.P.; Schwartz, M.W. Minireview: Inflammation and obesity pathogenesis: The hypothalamus heats up. *Endocrinology* **2010**, *151*, 4109–4115. [CrossRef] [PubMed]

108. Milanski, M.; Degasperi, G.; Coope, A.; Morari, J.; Denis, R.; Cintra, D.E.; Tsukumo, D.M.L.; Anhe, G.; Amaral, M.E.; Takahashi, H.K.; et al. Saturated fatty acids produce an inflammatory response predominantly through the activation of TLR4 signaling in hypothalamus: Implications for the pathogenesis of obesity. *J. Neurosci. Off. J. Soc. Neurosci.* **2009**, *29*, 359–370. [CrossRef] [PubMed]

109. Choi, J.W.; Gardell, S.E.; Herr, D.R.; Rivera, R.; Lee, C.-W.; Noguchi, K.; Teo, S.T.; Yung, Y.C.; Lu, M.; Kennedy, G.; et al. FTY720 (fingolimod) efficacy in an animal model of multiple sclerosis requires astrocyte sphingosine 1-phosphate receptor 1 (S1P1) modulation. *Proc. Natl. Acad. Sci. USA* **2011**, *108*, 751–756. [CrossRef]

110. Dusaban, S.S.; Chun, J.; Rosen, H.; Purcell, N.H.; Brown, J.H. Sphingosine 1-phosphate receptor 3 and RhoA signaling mediate inflammatory gene expression in astrocytes. *J. Neuroinflamm.* **2017**, *14*, 111. [CrossRef]

111. Assi, E.; Cazzato, D.; De Palma, C.; Perrotta, C.; Clementi, E.; Cervia, D. Sphingolipids and brain resident macrophages in neuroinflammation: An emerging aspect of nervous system pathology. *Clin. Dev. Immunol.* **2013**, *2013*, 309302. [CrossRef]

112. Lv, M.; Zhang, D.; Dai, D.; Zhang, W.; Zhang, L. Sphingosine kinase 1/sphingosine-1-phosphate regulates the expression of interleukin-17A in activated microglia in cerebral ischemia/reperfusion. *Inflamm. Res. Off. J. Eur. Histamine Res. Soc. Al* **2016**, *65*, 551–562. [CrossRef]

113. Rothhammer, V.; Kenison, J.E.; Tjon, E.; Takenaka, M.C.; de Lima, K.A.; Borucki, D.M.; Chao, C.-C.; Wilz, A.; Blain, M.; Healy, L.; et al. Sphingosine 1-phosphate receptor modulation suppresses pathogenic astrocyte activation and chronic progressive CNS inflammation. *Proc. Natl. Acad. Sci. USA* **2017**, *114*, 2012–2017. [CrossRef]

114. Karunakaran, I.; van Echten-Deckert, G. Sphingosine 1-phosphate—A double edged sword in the brain. *Biochim. Biophys. Acta Biomembr.* **2017**, *1859*, 1573–1582. [CrossRef] [PubMed]

115. Christoffersen, C.; Federspiel, C.K.; Borup, A.; Christensen, P.M.; Madsen, A.N.; Heine, M.; Nielsen, C.H.; Kjaer, A.; Holst, B.; Heeren, J.; et al. The Apolipoprotein M/S1P Axis Controls Triglyceride Metabolism and Brown Fat Activity. *Cell Rep.* **2018**, *22*, 175–188. [CrossRef] [PubMed]

Review

Sphingolipids in Type 1 Diabetes: Focus on Beta-Cells

Ewa Gurgul-Convey

Institute of Clinical Biochemistry, Hannover Medical School, Carl-Neuberg-Str.1, 30625 Hannover, Germany; Gurgul-Convey.Ewa@mh-hannover.de

Received: 30 June 2020; Accepted: 3 August 2020; Published: 4 August 2020

Abstract: Type 1 diabetes (T1DM) is a chronic autoimmune disease, with a strong genetic background, leading to a gradual loss of pancreatic beta-cells, which secrete insulin and control glucose homeostasis. Patients with T1DM require life-long substitution with insulin and are at high risk for development of severe secondary complications. The incidence of T1DM has been continuously growing in the last decades, indicating an important contribution of environmental factors. Accumulating data indicates that sphingolipids may be crucially involved in T1DM development. The serum lipidome of T1DM patients is characterized by significantly altered sphingolipid composition compared to nondiabetic, healthy probands. Recently, several polymorphisms in the genes encoding the enzymatic machinery for sphingolipid production have been identified in T1DM individuals. Evidence gained from studies in rodent islets and beta-cells exposed to cytokines indicates dysregulation of the sphingolipid biosynthetic pathway and impaired function of several sphingolipids. Moreover, a number of glycosphingolipids have been suggested to act as beta-cell autoantigens. Studies in animal models of autoimmune diabetes, such as the Non Obese Diabetic (NOD) mouse and the LEW.1AR1-iddm (IDDM) rat, indicate a crucial role of sphingolipids in immune cell trafficking, islet infiltration and diabetes development. In this review, the up-to-date status on the findings about sphingolipids in T1DM will be provided, the under-investigated research areas will be identified and perspectives for future studies will be given.

Keywords: type 1 diabetes; beta-cells; islets; insulin; cytokines; sphingolipids; S1P; animal models

---

1. Introduction

Sphingolipids (SLs) are a diverse family of lipid molecules playing a pivotal role in a number of autoimmune and inflammatory disorders [1–4]. The role of SLs in glucose homeostasis and insulin sensitivity is relatively well described in the context of metabolic-syndrome related type 2 diabetes (T2DM) [4–12]. In contrast, the importance of SLs in the beta-cell demise during autoimmune type 1 diabetes (T1DM) development has been so far less frequently addressed. Interestingly, a number of new investigations suggest that dietary fats and lipid metabolism may be considered as triggers that could induce or sensitize the autoimmunity onset in T1DM [13]. Polymorphisms in several genes encoding proteins involved in the SL pathway were recently linked to T1DM overt [14]. Moreover, profound changes in SL serum profiles upon autoimmunity development were detected in T1DM patients [14–20]. The last years have revealed that the enzymatic machinery and the system of receptors and transporters for bioactive SLs are significantly affected in pancreatic beta-cells by proinflammatory cytokines that are released from immune cells infiltrating islets [21]. SLs may be useful biomarkers for T1DM development [17]. In vitro studies of cytokine toxicity using genetically modified beta-cells, naturally occurring SLs and their analogues suggest that alterations of beta-cell SLs may affect insulin secretory capacity and beta-cell fate during T1DM development.

In this review various aspects of sphingolipid action and effects of the major proinflammatory cytokines on the SL pathway in pancreatic beta-cells will be discussed. Next, the engagement of SLs in the autoimmune reaction against beta-cells during T1DM development will be addressed. The present status of SL studies in animal models of autoimmune diabetes and an update on findings in T1DM patients will be summarized. Finally, perspectives, which should drive future research in the context of SLs and T1DM, will be presented.

2. Overview of Mechanisms of Beta-Cell Destruction in T1DM

Type 1 diabetes mellitus (T1DM) is an autoimmune disease with a strong genetic background, affecting millions of people worldwide, mostly in their childhood or early adolescence [22,23]. The incidence of T1DM has been significantly increasing in the last decades, similarly to other autoimmune diseases, indicating an important role of environmental factors [22,24]. During T1DM development pancreatic beta-cells are gradually destroyed due to an autoimmune reaction [22,25–29]. Beta-cells produce and supply our body with insulin, the most important anabolic hormone controlling blood glucose levels. The factors triggering the activation of immune cells, T-cells and macrophages, in T1DM remain unclear. It is speculated that certain food components (such as cow milk proteins or gluten), vitamin D3 deficiency, viral infections (e.g., enterovirus) and most recently saturated fats may trigger this response [13,22,26,30,31] (summarized in Figure 1). T1DM patients require a life-long substitution with insulin and are prone to severe secondary complications, such as cardiovascular dysfunction, nephropathy or retinopathy [22].

Immune cell activation is accompanied by dynamic changes in the gene expression of proteins related to inflammation and secretion of inflammatory mediators [26,27,30,32,33] (Figure 1). The infiltrate consists of a mixture of CD4+ and CD8+ T-cells, B-cells, macrophages and Natural Killer (NK) cells [31]. The activated immune cells infiltrating pancreatic islets convey their beta-cell directed cytotoxic effects via cell–cell interactions and through secretion of proinflammatory mediators. So far the research has focused predominantly on the three cytokines, namely IL-1β, TNFα and IFNγ, and their role in beta-cells is meanwhile relatively well characterized [25,27–31,33,34]. The intracellular effects of proinflammatory cytokines in beta-cells are pleiotropic and involve the activation of several transcription factors (NFκB, AP-1, Nova and others), disturbances of mRNA splicing, changes of the expression, activity and post-translational modifications of proteins [27,30,32,35]. Moreover, in beta-cells isolated from T1DM donors the upregulated mRNA and protein expression of Class I and Class II HLA were detected [36,37]. Several molecular processes have been associated with beta-cell death in T1DM including mitochondrial and endoplasmic reticulum (ER) stress, the generation of reactive oxygen species (ROS) and nitric oxide (NO$^\bullet$), the synthesis of inflammatory mediators, activation of immunoproteasome, no-go and nonsense-mediated RNA decay, dysregulation of calcium homeostasis and altered autophagy as well as the activation of the perforin-granzyme system [21,25,30,32,38–46] (summarized in Figure 1).

Beta-cells are particularly vulnerable to oxidative stress due to an imbalance in the enzymatic capacity responsible for generation and detoxification of ROS [40,42,44,45]. The superoxide dismutases, which dismutate superoxide radical anions to hydrogen peroxide (H_2O_2), are well expressed in the cytosolic (CuZnSOD) and in the mitochondrial compartment (MnSOD). This is in contrast to the enzymes catalyzing the detoxification of H_2O_2 to water, namely glutathione peroxidase (Gpx) and particularly catalase (Cat), which are very weakly expressed. Proinflammatory cytokines potentiate this imbalance specifically in the mitochondrial compartment through the induction of MnSOD expression and activity [40,44]. The increased generation rate of H_2O_2 together with the lack of an appropriate H_2O_2 detoxification system and a parallel increase of intracellular NO$^\bullet$ concentration are the perfect prerequisite for the formation of highly toxic hydroxyl radicals. This leads to intramitochondrial oxidative damage, caspase-9-dependent apoptosis induction, disturbed autophagy and beta-cell death [40,42]. Moreover, cytokines activate the ER stress response, leading to accumulation of unfolded/misfolded proteins, dysregulation of ER calcium homeostasis and induction of the proapoptotic transcription factor CHOP [47]. Additionally, cytokines induce alternative splicing and immunoproteasome activation in beta-cells, which may contribute to neoantigen formation, and aggravate the autoimmune reaction [47]. Several studies showed also that the vulnerability of beta-cells to proinflammatory cytokines is associated with an imbalance of the enzymatic capacity responsible for the generation of pro- and anti-inflammatory eicosanoids [39,48–50]. Our recent data indicate that MCPIP1 (monocyte chemotactic protein-induced protein 1), a strong regulator of the inflammatory response, which acts as a specific RNase, is strongly induced by proinflammatory cytokines in clonal beta-cells and upon diabetes development in the animal model of human T1DM [43]. The transcriptomic analysis of beta-cells from T1DM individuals suggests a significantly increased expression of the gene encoding MCPIP1 [37] (GEO Bioproject PRJNA497610 HLA Class II analysis of human pancreatic beta-cells). Our recent studies revealed that MCPIP1 regulates the beta-cell response to cytokine toxicity and the fine-tuning of its expression is essential for the beta-cell fate [43]. Interestingly, our preliminary observations indicate that a number of the sphingolipid pathway enzymes might be affected by this specific RNase, a mechanism that could contribute to cytokine effects on beta-cell sphingolipidome. Finally, proinflammatory cytokines dysregulate the beta-cell function by shutting down insulin biosynthesis and by disrupting glucose-stimulated insulin secretion (GSIS) [21,41,51–53].

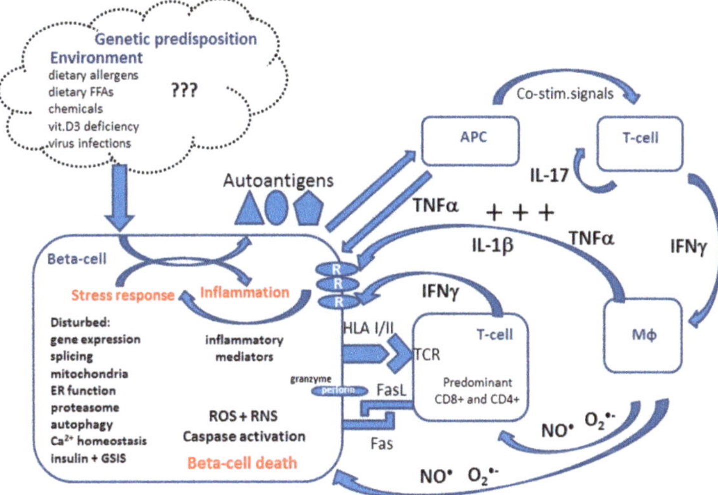

Figure 1. Model of cytokine-mediated beta-cell death in T1DM. In genetically predisposed individuals various environmental factors trigger the autoimmune response aimed at pancreatic beta-cells. Environmental triggers lead to beta-cell stress and release of autoantigens, which are processed and presented by antigen-presenting cells (APC). This leads to T-cell and macrophage (MΦ) activation. Consequently proinflammatory cytokines and radicals (NO•, nitric oxide and O₂•⁻, superoxide anion radicals) as well as perforin-granzyme mediators are released in the vicinity of beta-cells. Proinflammatory cytokines potentiate autoimmune reaction by stimulation of CD8+ and CD4+ T-cells. Activated immune cells interact with beta-cells via FasL-Fas and also via HLAI/II-TCR systems. The action of proinflammatory cytokines requires the binding and activation of cytokine receptors (R) on beta-cells. This accelerates the multifaceted stress response and induces inflammation in beta-cells. The aggravation of the autoimmunity is achieved by biosynthesis and release of inflammatory mediators from beta-cells. Beta-cells are particularly vulnerable to the stress response and inflammation due to their weak antioxidative and anti-inflammatory defense status. Cytokines induce reactive oxygen species (ROS) and reactive nitrogen species (RNS) formation in beta-cells. Both defects in the immune response and vulnerability of beta-cells participate in the execution of beta-cell demise during T1DM development (more details in text).

3. Biosynthesis of Sphingolipids

Sphingolipids are composed of a polar head group and two nonpolar tails. The core of SLs is the long-chain aliphatic amino alcohol sphingosine, which is O-linked to a polar head group (e.g., ethanolamine, serine or choline) and N-linked to an acyl group of various fatty acids. SLs are the most structurally diverse lipid family, considering the hundreds of possible head groups, dozens of long-chain bases, and many fatty acids, that can be used as building blocks. The maintenance of the so-called SL rheostat is crucial for the normal function of cells and cell survival, and is compromised by a network of enzymatic reactions depicted in Figure 2. Consequently, this promotes site-specific effects and downstream targets of various SLs.

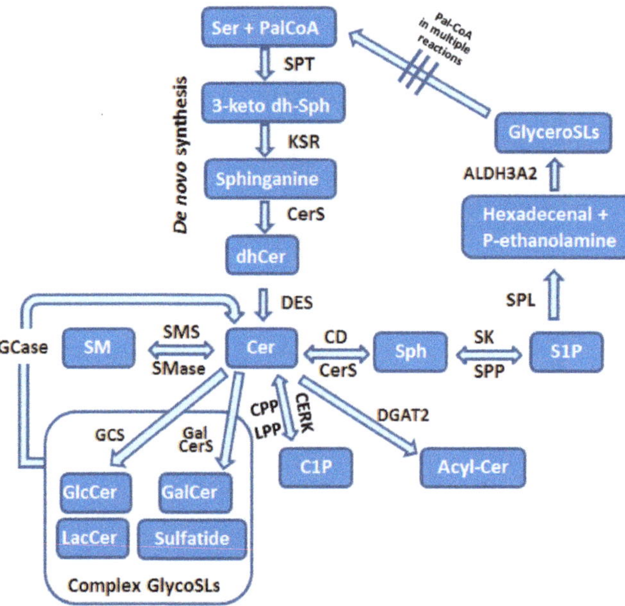

Figure 2. Transcriptomic data from islets and qRT-PCR results from beta-cell lines suggest that beta-cells express all genes regulating the SL pathway. The exact mRNA and protein expression level of various enzymatic components of the SL pathway in human beta-cells still needs to be characterized. Enzymes: ALDH2A3, fatty aldehyde desaturase A3, CD, ceramidase, CERK, ceramide kinase, CPP, ceramide 1-phosphate phosphatase, CerS, ceramide synthase, Des, sphingolipid-delta-4-desaturase, DGAT2, diacylglycerol acyltrasferase-2, GCS, GlcCer synthase, GalCerS, GalCer synthase, GCase, glycosidase, KSR, 3-keto sphingosine reductase, LPP, lysophospholipid phosphatase, SMS, sphingomyelin synthase; SMase, sphingomyelinase, SK, sphingosine kinase, SPP, sphingosine 1-phosphate phosphatase, SPT, serine palmitoyltransferase, SPL, sphingosine 1-phosphate lyase. Biomolecules: Cer, ceramide, C1P, ceramide 1-phosphate, dhCer, dihydroceramide, GlycoSLs, glycosphingolipids, GlyceroSLs, glycerosphingolipids, 3-keto dhSph, 3-keto dihydro sphingosine, Ser, serine, SM, sphingomyelin, Sph, sphingosine, S1P, sphingosine 1-phosphate, PalCoA, palmitoyl-coenzyme A.

The sphingolipid pathway starts with the de novo biosynthesis of 3-ketosphinganine in the ER from L-serine and palmitoyl-CoA in the rate-limiting reaction catalyzed by serine-palmitoyltransferase (SPT; Figure 2) [3,12,54,55]. SPT can use acyl-CoAs in the range of C12 to C18, but the most typically used is the C18-CoA. Besides serine, alanine and glycine may also be used under specific circumstances, resulting in the formation of a variety of atypical sphingoid bases, which represent approximately 15% of SLs in human plasma [56]. De novo synthesis of SLs is upregulated by substrate availability and can be downregulated by mammalian orosomucoid-like (ORMDL) proteins, which form a complex with SPT and inhibit its activity [57].

In the next step of the SL biosynthesis, 3-ketosphinganine undergoes a reduction to dihydrosphingosine by 3-keto-sphinganine reductase (KSR) that is thereafter N-acylated by the action of one of six ceramide synthases (CerS1-6) to form dihydroceramide in ER [3,54]. Each of CerS types uses specific acyl chains, with saturated or monounsaturated fatty acids (C14-C26). Dihydroceramides are then dehydrogenated to ceramides by dihydroceramide desaturase (DES) [54] (Figure 2). Ceramides are the central part of the SL pathway and can be used in numerous reactions (Figure 2). Cer can be utilized for formation of (i) sphingosine, (ii) sphingomyelin, (iii) complex glycosphingolipids, (iv) ceramide 1-phosphate (C1P) or (v) acylceramide [3,54] (Figure 2).

Ceramide is deacylated by ceramidases (CDs) to form sphingosine [3,54]. This reaction can take place in different subcellular organelles and is catalyzed by distinct subtypes of CDs [3]. Sphingosine is either reutilized for Cer generation by CerS or can be phosphorylated by one of two sphingosine kinases (SK1 and SK2) to form a potent bioactive SL called sphingosine 1-phosphate (S1P; Figure 2). SK1 localizes to plasma membrane and ER, while SK2 was found in the cytosol, ER, mitochondria and nucleus [58–60]. This differential spatial location of two SK isoforms defines the specialized and contradictory effects of each isoforms [58,59]. Cytosolic S1P can be dephosphorylated by ER-localized S1P phosphatases (SPP1 and SPP2), while at the cell surface S1P dephosphorylation is catalyzed by lipid phosphate phosphatases (LPP1-3). These reversible reactions are crucial for SL homeostasis within the cells, and enable refilling of sphingosine into the SL pathway [5].

The final step of SL metabolism is the irreversible degradation of S1P catalyzed by S1P lyase (SPL), an enzyme localized in ER with the catalytic center facing the cytosol (Figure 2) [3,61]. S1P is degraded to hexadecenal and phosphoethanolamine, which are intermediates in the phospholipid biosynthesis pathway. The potentially toxic, when accumulated in high amounts especially within the nucleus [62,63], hexadecenal is under normal conditions effectively metabolized by the fatty acid dehydrogenase ALDH3A2 to hexadecenoate [64]. Hexadecenoate can be further utilized for palmitoyl-CoA generation within the glycerolipid pathway [64]. Additionally to ER, SPL was also found by proteomics approaches to be present in the mitochondrial associated membranes (MAMs), the membrane crossroads of mitochondria and ER, which is involved in cell signaling and metabolite exchange [65]. These observations suggest a potential participation of SPL in unique functions of mitochondria and MAMs [66].

Ceramide can be transported to other subcellular locations by vesicular transport or with help of the ceramide transfer protein CERT [3,54]. In mammalian cells CERT function is regulated by SL levels through the PKD-dependent phosphorylation [67]. In the trans-Golgi Cer can be converted to sphingomyelin by sphingomyelin synthase (Figure 2) (SMS1) [3,54]. Another isoform of SMS, namely SMS2, is localized in plasma membrane [3]. Sphingomyelin can be metabolized back to Cer by sphingomyelinases (SMase), which are found in plasma membrane, the Golgi apparatus and in mitochondria-associated membranes (Figure 2) [3,54]. In the cis-Golgi Cer is converted by glucosylceramide synthase (GCS) to glucosylceramide (GluCer; Figure 2) [68]. Galactosylceramide (GalCer) is produced by GalCer synthase (Figure 2) [68]. The formation of complex glycosphingolipids (glycoSLs) requires the transport of GluCer to trans-Golgi network by four-phosphate adaptor protein 2 (FAPP2) that also regulates trafficking of vesicles from Golgi to plasma membrane [3]. Lactosylceramide (LacCer) is then produced by LacCer synthase transferring galactose from UDP-galactose to GluCer [68]. The more complex glycoSLs all use LacCer as a common backbone. The generation of multiple glycoSL species takes place by the action of specific enzymes catalyzing the stepwise addition of sugar monomers, which branch to form complex chains [68]. Complex glycoSLs are then transported from the Golgi to their target location, which is typically the plasma membrane [68]. Sphingomyelin and glycoSLs have been shown to be transported in a vesicle-dependent manner [3,68]. Both SM and glycoSLs undergo lysosomal degradation by glycosidases and acid SMase, respectively, that are responsible for the removal of head groups to form ceramides (Figure 2) [68].

Alternatively, ceramide can be also phosphorylated to ceramide 1-phosphate (C1P), a bioactive SL, by ceramide kinase (CERK) in the Golgi [3,69] (Figure 2). C1P may be converted back into ceramide by the action of one of LPPs or by C1P phosphatase (CPP) [3,69–71]. The irreversible reaction converting Cer to acylceramide (acylCer) is catalyzed by diacylglycerol acyltransferase-2 (DGAT2) [3] and is used to store Cer in lipid droplets [72] (Figure 2).

4. Overview of Major Functions of Bioactive Sphingolipids

Sphingolipids are constituents of cell membranes (about 50–60% of the membrane content) and important bioactive molecules acting as second messengers. They form lipid microdomains, important for cell signaling and participating in cell–cell interactions and absorption of extracellular lipids (e.g., fatty acids). SLs were shown to regulate calcium homeostasis, ROS formation, histone acetylation and activation of numerous transcription factors. Pathophysiological conditions result in alterations of the ratio between the structural and bioactive SL content, a phenomenon that has been shown to contribute to the regulation of the inflammatory response. Proinflammatory cytokines such as IL-1β and TNFα as well as several other stimuli (Fas ligand, phorbol esters, heat shock, oxidative stress and various chemotherapeutics) are well described disruptors of the SL homeostasis [3,54].

The best studied bioactive SLs are ceramide, sphingosine, S1P, C1P, as well as sulfatide. While ceramide and sphingosine are traditionally believed to exert proapoptotic signals, the action of S1P and C1P is largely context and tissue-dependent, with pro- and antiapoptotic activities described in various tissues. The intracellular concentrations of ceramides, sphingosine and S1P differ by an order of magnitude, with the lowest content of the least one. Therefore a small change in ceramide content might substantially impact S1P concentration. In the following sections the most important functions of major bioactive SLs in various cell types, which could be of relevance for beta-cell pathophysiology in T1DM, is briefly summarized. For a more detailed review on SLs in different tissues and disorders the reader is asked to refer to several excellent reviews addressing these topics [2,3,69].

4.1. Ceramide

Ceramide is implicated in a variety of cellular processes such as apoptosis, cell growth arrest, differentiation, cellular senescence, cell migration and adhesion [3,54,73]. Many proinflammatory mediators and oxidant agents have been shown to stimulate Cer generation, upon them T1DM-relevant cytokines (IL-1β, TNFα and IFNγ). The best described role of Cer is the regulation of apoptosis [3]. The main mechanism involved in Cer-induced apoptosis is the disruption of mitochondrial integrity and function [74,75]. The induction of mitochondrial apoptosis by ceramides was elegantly demonstrated by means of mitoCERT (CERT equipped with a MOM anchor), enabling Cer flow from ER to mitochondria. The translocation of ER ceramides to mitochondria resulted in the initiation of apoptosis [74]. Elevated levels of Cer were shown to control MOMP (mitochondrial outer membrane permeabilization) and thereby cytochrome c release, by fostering BAX/BAK oligomerization and formation of ceramide channels. Formation of Cer channels promotes the egress of proteins from the intermembrane space and ROS generation [74,75]. Cer contributes also to the suppression of mitochondrial electron transport chain and decreased ATP formation upon inflammation [76]. Recent studies identified the voltage-dependent anion channels VDAC1 and VDAC2 as mitochondrial ceramide binding proteins [77] and reported the expression of CerS within the mitochondrial compartment [77]. Moreover, the recent identification of a protein p17/PERMIT mediating ER-mitochondria tracking of newly translated CerS1 through MAMs to the mitochondrial outer membrane has been shown to be an event necessary for mitophagy induction [78]. CerS1/C18 ceramide generation was reported to foster LC3BII targeting of autophagolysosomes to mitochondria. Overexpression of CerS2, resulting in the formation of very long chain ceramides was also associated with mitophagy activation and mitochondrial dysfunction [78]. Furthermore, intramitochondrial accumulation of C16 generated by CerS6 was reported to induce mitochondrial fission and mitophagy (excellently reviewed by [73]). Additionally to these mitochondria-related effects, Cer was shown to regulate various signaling pathways (e.g., AKT/PKB or JNK) and to modulate different kinases (e.g., protein kinase C, PKC or protein phosphates PP1 and PP2A) [3]. Finally, elevated levels of Cer in the plasma membrane may increase membrane rigidity thereby stabilizing lipid rafts, which was shown to foster signal transduction [3,54].

4.2. Sphingosine

Similarly to Cer, sphingosine is thought to play an important role in cell cycle arrest and apoptosis [2,3]. Sphingosine was shown to inhibit the activity of PKC [2]. Moreover, sphingosine can bind to the anti-apoptotic protein 14-3-3 and inhibit its action by phosphorylation of the dimer interface [79]. Furthermore the ANP32a and ANP32b proteins have been shown to be targeted by sphingosine, and this is a mechanism by which sphingosine conveys its inhibitory effect on PP2A [80].

4.3. Sphingosine 1 Phosphate

Shingosine-1 phosphate is present in plasma (0.2–1 µM) and its concentration varies under different metabolic and pathological conditions. S1P is mostly bound to apolipoprotein M (ApoM, 50–70%) and albumin (ca. 30%), and recent reports point to an important role of HDL-bound S1P, especially in context of protection against cardiovascular disorders [81,82]. The main sources of blood S1P are erythrocytes and platelets [83]. The intracellular concentrations of S1P are much lower (nM, possibly due to a higher activity of enzymes involved in its turnover) and undergo severe changes under proinflammatory conditions. In most cell types S1P inhibits apoptosis, fosters cell proliferation and metabolism, and was shown to play a crucial role in immune cell trafficking and differentiation [1,2,84,85]. The deficiency of SPL promotes tumor growth and survival [86–88]. In neurons [89–95] and pancreatic beta-cells [21] intracellular S1P is involved in cytotoxic effects (see below). The effects of S1P are mediated by cell surface receptors and by activation of intracellular targets [1–3]. Five S1P receptors (S1PR1-S1PR5) were identified. They all belong to G protein-coupled receptors and are characterized by distinct mechanisms of action depending on the G subunit involved. Therefore the outcome of the receptor-dependent S1P action depends on the tissue-specific expression profile of S1PRs and its regulation upon various conditions [96]. Several proteins have been linked to cellular S1P export. Upon them the ABC transporter family and the Spns2 transporter are the best described [1,2,96]. Their cell type-specific expression determines S1P function in various tissues. The activation of S1PR2 was shown to stimulate cell-surface integrins and fibronectin matrix [96]. S1P has been shown to mediate the effect of cytokines on COX2 activation and PGE2 production [97]. Additionally to the classical receptors-dependent mediated S1P effects, the activation of intracellular target-dependent pathways participate in overall S1P action in cells. The zwitterionic head group of S1P, which is sufficiently water-soluble, enables S1P to flow between membranes and cellular organelles. Many molecules have been proposed as intracellular targets of S1P, among them the CIAP2 (cellular inhibitor of apoptosis 2), CerS2 and TRAF2 (TNF receptor associated factor 2) proteins [1–3,98]. Of note, the direct involvement of S1P in the regulation of TRAF2 activation was questioned by Etemadi and coworkers [99]. Moreover the SK2-mediated formation of S1P in the nucleus has been associated with changes in the epigenetic signature of cells due to the ability of S1P to directly bind and inhibit the histone deacetylases HDAC1 and HDAC2 [100,101]. Finally, many studies reported calcium as a possible second messenger in the intracellular action of S1P [21,89,93].

4.4. Ceramide-1 Phosphate

The function of ceramide 1-phosphate, similarly to S1P, is largely tissue and cell-context specific, with anti- and proinflammatory properties [70,102]. The enzymatic activity of CERK was shown to be upregulated by IL-1β, macrophage colony stimulating factors (M-CSF), calcium ionophore A23187, tyrosine-kinase pathway and agonists of nuclear peroxisome proliferator activated receptors (PPARs) [69]. C1P is believed to have strong mitogenic and antiapoptotic activities [69].

4.5. Sulfatide

Sulfatide (3-O-sulfogalactosylceramide) is a glycoSL that is synthesized from Cer by two transferases (ceramide galactosyltransferase and cerebroside sulfotransferase). Sulfatide is a multifunctional bioactive SL, which was demonstrated to play an important role in the nervous system, pancreas, immune system, as well as bacterial and virus infections (for more details please refer to [103]. Sulfatide is present predominantly in the nervous system and pancreatic islets [104]. The enzyme arylsulfatase A (ASA) specifically degrades it [103], though an ASA-independent degradation of sulfatide was also described. ASA activity requires the help of saposin B, which liberates sulfatide from membranes and makes it accessible for ASA [103]. Sulfatide is localized mainly in the Golgi apparatus, cellular membrane, and lysosomes; in pancreatic beta-cells it was found in insulin granules [103]. Moreover, sulfatide can activate binding of laminin to integrins and was described as a major L-selectin ligand, playing an essential role in monocyte infiltration in some organs [103].

5. Effects of Proinflammatory Cytokines on the Sphingolipid Pathway Enzymes in Pancreatic Beta-Cells

Most data available regarding the sphingolipid pathway in pancreatic beta-cells has been gained from murine and rodent models (RINm5F, INSE cells, MIN6 cell lines and mouse models). The extensive qRT-PCR analysis revealed the presence of all enzymes involved in the SL pathway in rodent beta-cells [21]. Pancreatic beta-cells are characterized by an imbalance of the enzymatic capacity for S1P formation (SK1 and SK2) and degradation (SPL) [21], which seems to play a particularly important role for cytokine toxicity (see Section 7).

Although a growing number of studies describing changes in the SL content upon T1DM development in humans are available, the SL rheostat of human beta-cells has not yet been characterized. Moreover detailed data about the expression pattern of the SL pathway enzymes in beta-cells in animal models of T1DM and in the human pancreas from T1DM individuals is still missing.

5.1. De Novo Synthesis

The enzymes catalyzing the first steps of the SL biosynthesis are particularly sensitive to cytokine action in rat insulin-secreting INS1E cells [21]. The subunits of SPT are affected differentially, namely the long chain 1 and long chain 2 subunits are strongly increased in response to proinflammatory cytokines (IL-1β, TNFα and IFNγ), while no changes in the expression level of the short chain subunit were detected [21]. Cytokines significantly stimulate the expression of DES in rat INS1E cells [21]. Interestingly, data from human islet transcriptome revealed that ORMDL3 is expressed in human islets and undergoes upregulation upon 24 h-exposure to a mixture of IL-1β and IFNγ [32]. Whether this expression can be accounted to beta-cells, what the contribution of TNFα on the expression of ORMDL3 is and if beta-cells express also other isoforms of ORMDL proteins requires further investigations. The ORMDL3 gene polymorphism was described in the recent studies involving T1DM individuals [14,105].

Recently a new mechanism regulating the expression of SPT has been described [106]. Using genome-wide CRISPR/Cas9 screening it was demonstrated that AHR (aryl hydrocarbon receptor) binds and activates the gene promoter of SPT [106]. Moreover, tissues from AHR KO mice were characterized by reduced expression of several other key genes in the SL biosynthetic pathway and decreased the SL content [106]. Interestingly, AHR is a transcription factor that has been recently shown to link the diet and gut microbiome alterations with islet autoimmunity [107], yielding the possibility that the observed activation of AHR in islets during T1DM development could stimulate the SPT expression and increase de novo SL biosynthesis in beta-cells. Whether this is indeed the case still needs to be experimentally proven.

5.2. Ceramide Metabolism Regulation

The best studied part of the SL biosynthesis pathway in beta-cells is the formation of Cer. IL-1β was the first cytokine shown to activate the sphingomyelin/ceramide pathway in rat insulin-producing RINm5F cells [108]. The studies revealed that a short exposure to IL-1β (2–5 min) could induce Cer and diacylglycerol (DAG) generation with a parallel rapid decrease of sphingomyelin content, indicating the activation of sphingomyelinase [108]. Another report evaluated the sphingomyelin hydrolysis in response to IL-1β exposure in rat and mouse islets as well as in RINm5F cells and failed to detect a significant effect of IL-1β within the time-frame of the experiment [109]. The elevation of Cer content in beta-cells in response to TNFα was demonstrated in mouse insulin-secreting MIN6 cells and it was suggested to be involved in TNFα-mediated apoptosis [110]. These observations were not confirmed in another insulin-secreting mouse cell line, the β-TC3 cells, in which cytokines (IL-1β, TNFα and IFNγ) failed to stimulate Cer formation [111]. The sphingomyelin content was around 50% decreased in β-TC3 cell line, similarly to RINm5F cells, however the authors showed that the kinetics of sphingomyelin hydrolysis within 4 h after cytokine addition did not differ between control and cytokine-treated cells [111]. The reasons for the contradictory findings could be related to methodological limitations of these early studies, time-specific effects of cytokine action and/or caused by a quick turnover of SLs in beta-cells.

More recent studies, involving modern mass spectrometry technology, confirmed an induction of neutral SMase and a parallel activation of iPLA2β leading to Cer accumulation in rat insulin-secreting INS1 cells exposed to proinflammatory cytokines (IL-1β + IFNγ) [112]. Our studies in INS1E cells revealed that also the mRNA expression of acid SMase is upregulated in response to a mixture of cytokines (IL-1β, TNFα and IFNγ) [21]. Beta-cell specific overexpression of iPLA2β in a mouse model (RIP-iPLA2β-Tg mice) results in upregulation of nSMase mRNA and protein expression, followed by decreased sphingomyelins with a parallel increase of Cer content [113]. This was accompanied by ER stress, mitochondrial damage and caspase-3 activation [112,113].

Ceramide content can also be upregulated by stimulation of Cer production by CerS. We showed that in beta-cells proinflammatory cytokines upregulate the mRNA expression of various types of CerS [21], strongly supporting a notion that under T1DM conditions an intra-beta-cell ceramide formation might be indeed induced. The most prevalent CerS isoforms in rodent beta-cells are CerS2 and CerS5/6, followed by CerS1 ([21], and Gurgul-Convey, unpublished). These isoforms are characterized by distinct subcellular localization (ER vs. mitochondria), as well as differences in the length of generated ceramides. Owning the role of ER stress and mitochondrial damage in cytokine toxicity to beta-cells, the cytokine-mediated induction of CerS expression may significantly contribute to toxic effects of cytokines to beta-cells. An increased Cer generation may particularly promote cytokine-induced mitochondria damage in beta-cells. The recent findings from the Brüning group strongly indicate such a scenario [114]. Using the CerS6 deficient mouse model the authors demonstrated that CerS6-derived C16 Cer, in contrast to CerS5-derived Cer, could promote mitochondrial fission and insulin resistance in obesity [114]. Using the CerS6 KO model exposed to STZ-mediated islet autoimmunity could enable in the future to assess the importance of mitochondrial Cer formation in the development of T1DM.

The involvement of mitochondrial ceramide in cytokine toxicity is indirectly suggested also by our observations in SPL-overexpressing INS1E cells [21]. Recently, we showed that the overexpression of SPL in beta-cells protects from cytokine toxicity by prevention of the ER-calcium leak, decreased mitochondrial calcium levels and promoting the expression of mitochondrial chaperones [21]. SPL deficiency was associated with increased Cer levels in other cell types [115]. Though the effects of SPL overexpression on Cer content in beta-cells still need to be analyzed, it seems plausible that the observed beta-cell protective effects of SPL overexpression may be at least partially related to decreased Cer levels, particularly in the mitochondrial compartment.

Another interesting aspect of mitochondrial Cer formation in the context of beta-cell fate in T1DM could involve sirtuins. Sirtuins are NAD+-dependent histone/protein deacetylases consisting of several subtypes that are present in various subcellular compartments and are involved in the regulation of oxidative stress through direct deacetylation of transcription factors controlling antioxidant genes [116]. Additionally, sirtuins are believed to be important regulators of glucose homeostasis, insulin secretion and mitochondrial biogenesis. Interestingly, the mitochondrial SIRT3 has been recently shown to regulate Cer generation in the brain in response to ischemia/reperfusion [76]. So far the role of altered expression of sirtuins, on mitochondrial formation of Cer in beta-cells has not been investigated.

Finally, cytokines were reported to induce the activity of aCD in INS1 cells and primary rat islets [117]. The pharmacological inhibition of aCD by NOE resulted in enhanced cytokine toxicity in INS1 cells [118]. In contrast to neutral CD [118], no changes on the mRNA and protein levels of aCD in INS1 cells and rat islets treated with cytokines were detected for up to 4 h [117]. We extended this study to a more mature beta-cell line, namely INS1E, and analyzed the mRNA expression pattern of acid and neutral CDs under longer time periods [21]. Our measurements revealed that a mixture of three major proinflammatory cytokines induce the upregulation of both nCD and aCD in INS1Ecells [21].

5.3. S1P Metabolism, Receptors and Transporters

We observed similar effects of high concentrations of IL-1β (as prevails in the first stage of islet autoimmunity) and of a mixture of the three major diabetogenic cytokines IL-1β, TNFα and IFNγ (as it occurs in the advance stage of insulitis) on the mRNA expression level of enzymes involved in S1P metabolism, indicating a crucial role of IL-1β [21]. This was in contrast to distinct effects of the acute (6 h) versus prolonged (24 h) cytokine exposure on the gene expression of enzymes involved in the SL metabolism, S1P receptors and S1P transporters [21].

Early reports identified four types of S1P receptors (S1P1, S1P2, S1P3 and S1P4) in mouse and rat islets, and in INS1 cells [119]. Recently we demonstrated the mRNA expression of S1PR2, 3 and 5, of which S1PR3 was the predominant subtype, in INS1E beta-cells [21]. S1PR2 and S1PR3 are coupled predominantly to Gq and activate phospholipase C (PLC) to induce Ca^{2+} mobilization through the production of inositol 1,4,5-trisphosphate [120,121], and induce activation of MAPK kinases [122]. S1PR5 was shown to interact with Gα subunits [123], which inhibit PLC activity. Beta-cells are equipped with various S1P transporters [21]. The Abca1 transporter is expressed at the highest level, followed by Abcc1 and a nearly 100-fold lower expression of Spns2 [21]. Acute exposure to proinflammatory cytokines results in a strong downregulation of S1PR3 (70% decrease), partially compensated by a mild upregulation of S1PR2. S1PR5 remains unchanged under such conditions [21]. Upon longer incubation with cytokines, the expression of all S1P receptors undergoes upregulation [21]. Similar observations were made in the case of S1P transporter mRNA expression [21]. The impact of these alterations of S1P receptor and transporter system in beta-cells under acute and chronic cytokine exposure on beta-cell fate has not yet been investigated, but it could provide interesting insights into the role of S1P in beta-cell vulnerability to cytokines.

Proinflammatory cytokines enhance the activity of SK in rat beta-cells [124,125]. Our gene expression data suggest that SK2 is the main isoform expressed in insulin-secreting cells and proinflammatory cytokines increase its expression [21]. Cytokines downregulate the expression of SPL in INS1E cells and rat islets, while they enhance the expression of SPP2 [21]. These observations suggest an increased rate of sphingosine generation in beta-cells upon cytokine exposure. This could foster increased Cer formation and/or accumulation in mitochondria or elevate the generation rate of S1P in mitochondria, nucleus und other specific locations. Since SPP has not yet been shown to localize in mitochondria and nucleus, the cytokine-induced S1P formation in these two subcellular compartments is likely unlimited in contrast to other compartments with a parallel cytokine-mediated overexpression of SPP. Thus cytokine action could affect subcellular SL gradients in beta-cells with high local Cer, sphingosine and/or S1P concentrations upon cytokine exposure.

Finally, the cytokine-mediated downregulation of SPL, which we observed in INS1E cells, does not coincidence with increased S1P concentration (Gurgul-Convey, unpublished). This observation further suggests a shift of the SL pathway to sphingosine/Cer formation in beta-cells exposed to cytokines. Such a phenomenon was observed in the Charcot–Marie–Tooth phenotype in humans that is characterized by SPL deficiency and was shown to be associated with elevated Cer levels [126,127].

How these changes on the expression level of enzymes of the SL pathway translate to the sphingolipid composition of beta-cells exposed to cytokines should be further investigated, since rearrangements of the beta-cell SL profile may have potentially remarkable consequences on the susceptibility of beta-cells towards proinflammatory cytokines. This could include alterations of the expression pattern of HLA Class I and II, acceleration of cytokine signaling or disturbed cell–cell interactions. Additionally an interesting question arises whether intracellular S1P may participate in the epigenetic regulation of genes relevant for beta-cell vulnerability to the autoimmune reaction in T1DM. Further studies are needed to characterize the changes of the SL enzymatic machinery in human beta-cells before and after T1DM onset.

6. Effects of Bioactive Sphingolipids on the Beta-Cell Function in T1DM

Multiple studies have demonstrated that various SLs may regulate the beta-cell secretory capacity [4,8,9,128–130]. These effects are conveyed by the activation of cell surface receptors, a regulation of ion channels or intersection with insulin production and folding. The abundance of sphingomyelin patches on beta-cell surface was additionally reported to modulate the insulin secretory capacity [131].

The inhibitory effects of Cer and its analogues on insulin production and secretion are well described [111,130,132,133]. So far no data is available on the effects of intracellular Cer production topology on GSIS disturbances in cytokine-treated beta-cells. In contrast to Cer, extracellular S1P is a potent stimulator of insulin secretion [21,129]. Upon the activation of its receptors, S1P potentiates GSIS, most likely by induction of cAMP generation [21,125]. Deletion of SK1 in INS1E cells results in defective insulin gene expression, lower insulin content and GSIS [134]. SK2 KO in MIN6 cells leads to higher GSIS, also in the presence of low glucose [129]. Our studies showed that depletion of intracellular S1P content by overexpression of SPL does not affect GSIS in the absence of proinflammatory cytokines [21]. SPL overexpression was however capable to partially protect against proinflammatory cytokine-mediated GSIS inhibition [21]. While the exact mechanism underlying this protective effect of SPL overexpression needs further investigation, we observed an increased protein expression of mitochondrial chaperones, which play an important role in ATP synthesis (mimitin and prohibitin 2) [21]. Earlier studies revealed that these mitochondrial proteins are essential for a proper GSIS [42,135]. siRNA-mediated inhibition of Phb2 in SPL-overexpressing beta-cells resulted in a partial loss of SPL-mediated protection [21]. Additionally, we observed an increased expression of BIP and Sec61a in ER of SPL-overexpressing INS1E cells that correlated with improved calcium homeostasis [21]. BIP and Sec61a cooperate to prevent a calcium leak from ER [136]. Therefore their enhanced expression in SPL-overexpressing INS1E cells could prevent cytokine-mediated disruption of calcium homeostasis. Moreover, improved BIP expression in SPL-overexpressing INS1E cells may help to secure unbiased insulin folding in cytokine-treated cells.

In the first hours of IL-1β action, which is not associated with cytotoxicity, a temporary potentiation of GSIS is observed [137]. Could cytokine-induced S1P generation be used by beta-cells for promoting this effect? In contrast, the acute phase of cytokine toxicity is accompanied by dysregulation of GSIS [21,51–53]. In this acute phase of cytokine toxicity the expression of SK2, S1P transporters and receptors expression is upregulated while SPL expression is downregulated. This is likely to result in decreased intracellular levels of S1P, due to a parallel high expression of SPP and CerS [21]. Since SK2 is not expressed on the plasma membrane, the SK2-derived S1P is not expected to be efficiently transported inside-out and activate its receptors on the beta-cell surface. Therefore it seems that the increased expression of S1P transport and the signaling system in beta-cells could serve as an adaptation strategy to make up for cytokine-mediated inhibition of GSIS by S1P-mediated cAMP generation and its potentiating effect on GSIS. Further studies are needed to describe the effects of cytokines on the SL content upon an acute phase of cytokine toxicity and the chronic incubation.

The results may help to understand the natural history of disturbances of beta-cell function in early and advance stages of cytokine toxicity and could enable designing preventive strategies.

Another twist to SL effects on beta-cell function was added by demonstration that synthaxin 4 (Stx4) is required for aSMase activity [138]. Stx4 is a plasma membrane–localized exocytosis protein that is crucial for GSIS [139]. Interestingly, Stx4 is a T1DM candidate protein [140]. Could therefore Stx-4-mediated changes in the beta-cell sphingomyelin content be involved in beta-cell dysfunction upon cytokine exposure?

Probably the best studied complex glycoSL in context of insulin production and secretion is sulfatide [103]. Interestingly, in T1DM patients the content of sulfatide was recently shown to be significantly reduced [14]. In pancreatic beta-cells sulfatide is present at the surface membrane of and in secretory granules [141], with a predominantly expressed C16:0 isoform [14]. Sulfatide is believed to promote proinsulin folding and to serve as a molecular chaperone for insulin [142]. The content of sulfatide in insulin granules decreases with rising metabolic activity of beta-cells [14]. Sulfatide is secreted together with insulin, facilitates rapid insulin monomerization and is crucial for insulin crystal preservation [142]. Moreover, sulfatide is required for normal insulin secretion through the activation of ATP-sensitive potassium ion channels and stimulation of calcium ion-dependent exocytosis [143]. In vivo administration of Zucker rats with C16:0 sulfatide resulted in significantly elevated GSIS without effects on glucose tolerance [144].

These results suggest that the place of origin and the type of SLs might be crucial for the final outcome of particular SLs on the beta-cell function. Further studies are needed to characterize which types of SLs are generated upon exposure to diabetogenic cytokines and how they influence insulin biosynthesis and beta-cell secretory capacity under T1DM conditions. The use of islets isolated from SK1, SK2 and SPL KO mouse models and exposed to cytokines will be useful in this context.

7. Effects of Bioactive Sphingolipids on Beta-Cell Fate in T1DM

Accumulative data indicates that SLs may be crucially involved in beta-cell death during T1DM development. Though the exact underpinning mechanisms remain unclear, evidence indicates that two elements may be of particular importance, namely the changes of the SL pattern of beta-cells and alterations of SL profiles in islet surroundings.

Toxic effects of Cer and sphingosine on beta-cells are well documented [4,5,110,111,118,130,133,145]. As discussed in the Section 5, proinflammatory cytokines were found to induce Cer formation, which was associated with apoptosis. Interestingly, overexpression of Stx4, which was shown to stimulate aSMase activity, reduces the chemokine ligand gene expression in beta-cells and protects beta-cells against cytokine toxicity [139]. Stx4 is a T1DM candidate gene [140]. Therefore a question arises whether the Stx-4-mediated changes in the beta-cell SL rheostat could be a link to islet autoimmunity by affecting cell membrane composition and thereby reducing the chemokine ligand presence on the beta-cell surface.

The role of S1P in cytokine toxicity to beta-cells has been extensively studied in the last decade. First it was demonstrated that extracellularly added S1P protects insulin-secreting INS1 cells and rat islets against cytokine toxicity [124,125]. The cytokine-mediated TUNEL staining, cytochrome c release and caspase-3 activation were reduced after treatment with nM concentrations of S1P. The S1P receptor antagonist BML-241 blocked this protective effect. Beneficial effects of S1P against cytokine toxicity were not associated with decreased cytokine-mediated iNOS expression or NO generation. This observation is particularly interesting in the context of human beta-cells, in which cytokines fail to induce the iNOS pathway and are nevertheless toxic [42,146]. How would extracellular S1P protect the beta-cell against cytokine toxicity? Laychock and colleagues showed that exposure of beta-cells to S1P leads to the stimulation of PLC activity, indicating the activation of Gq subunit of S1PRs in beta-cells [125]. Our own studies revealed that exposure of INS1E cells to S1P results in a rise of cAMP generation [21], extending the earlier observations that S1PR2 activation induces cAMP generation in other cell types [147–150]. Since cAMP conveys its cytoprotective effects in beta-cells via multiple mechanisms, including PKA activation and regulation of calcium homeostasis [40,151], it would be worth evaluating these pathways in detail by means of specific inhibitors and/or genetic modifications of S1P metabolism.

Could the addition of S1P to the culture medium for islets isolated for transplantation be implemented as a preservation method? Against this procedure points the fact that extracellular S1P was reported to increase insulin secretion, also in the absence of hyperglycemia [21]. Such a scenario could lead to depletion of beta-cell insulin capacity and lead to metabolic stress of isolated islets, making islets bathed in a S1P-containing medium less useful for transplantation. This is in contrast to another bioactive lipid compound, namely prostacyclin and its analogues, which also stimulate cAMP generation in beta-cells, however do not potentiate insulin secretion in the presence of low-glucose culture medium [151]. Furthermore, S1P has been shown to be implicated in islet allograft survival [152,153]. Fingolimod, a S1PR modulator, was demonstrated to enable long-term survival of islet allografts due to its effects on immune cell trafficking [152,153].

Interestingly, the action of intracellularly generated S1P in cytokine-treated beta-cells seems to be opposite to that of extracellular S1P [21]. Our data strongly indicates that intracellularly generated S1P participates in acute cytokine toxicity to beta-cells [21]. We observed an intermediate mRNA expression level of SPL in rodent beta-cells and islets as compared to other tissues [21]. This was downregulated in response to cytokines [21]. Overexpression of SPL protected insulin-secreting cells against cytokine-induced apoptosis [21]. SPL overexpression was accompanied by maintenance of calcium homeostasis, which is strongly impaired by the action of proinflammatory cytokines in beta-cells [21]. Additionally, SPL-overexpressing INS1E cells were protected against cytokine-mediated ER stress, as evident by a significant inhibition of CHOP expression after incubation with cytokines. Moreover, SPL overexpression reduced cytokine-mediated inhibition of cell proliferation and ATP content [21]. These protective effects were independent from the NFκB-iNOS pathway [21]. Furthermore, SPL overexpression provided protection against cytokine toxicity though it failed to downregulate cytokine-induced ROS generation. As mentioned above, we detected a higher expression of various ER and mitochondrial chaperones in SPL overexpressing INS1E cells, indicating that changes in intracellular S1P concentrations may indeed epigenetically regulate gene expression in beta-cells, like in other cell types [100,101]. Importantly, siRNA-mediated suppression of SPL expression resulted in opposite effects to those observed in SPL overexpressing INS1E cells [21]. Interestingly, though SPL overexpression has been reported to be implicated in toxic effects of hexadecenal accumulation in various cell types [62,63], in beta-cells SPL overexpression provided protective effects. The possible explanation to this phenomenon could be that beta-cells are rich in the enzyme responsible for hexadecenal detoxification, namely ALDH3A2 [21], enabling prevention of hexadecenal accumulation and toxic effects of SPL-overexpression in beta-cells. It will be important to evaluate the expression and activity of SPL in beta-cells chronically exposed to cytokines, and to investigate the effects of double-transfection approaches including SK1/SK2 and SPL to determine the role of intracellular S1P in more detail.

The involvement of intracellularly generated S1P in cytotoxic effects of cytokines in beta-cells seems to be very similar to the role of intracellular S1P in neurons as described in elegant studies by the Van Echten-Deckert group [89–95]. Interestingly, pancreatic beta-cells and neurons share multiple common features, though derived from distinct germ layers [154]. It is speculated that the endocrine and nervous systems developed from a common evolutionary ancestor [155,156]. Moreover, *Drosophila*-insulin-like peptides (Dilps), which are synthesized by neurons in flies, were shown to regulate energy metabolism similarly to mammalian insulin [157]. The way beta-cells biosynthesize and store insulin, and answer to external stimuli by insulin secretion mimics very closely the way neurons store and release neurotransmitters [154,156]. Many studies showed that beta-cells and neurons are characterized by similar gene expression patterns and spliceosome activity [158] (for more information please refer to [154]). It is therefore not surprising that beta-cells and neurons may also share similar sensitivity to intracellular S1P.

The mRNA expression of both SPP1 and SPP2 was found to be induced in beta-cells under acute exposure to cytokines [21]; what the impact of a chronic exposure to cytokines on SPP expression is or whether the expression/activation of other phosphatases would be of importance for beta-cell fate should be addressed in the future. Though exogenous Cer was shown to disrupt mitochondrial function [145] and induction of SMase was shown to be accompanied by increased Cer content, mitochondria damage and apoptosis [112], there is no direct evidence linking Cer accumulation in mitochondria to cytokine-mediated beta-cell apoptosis. Finally, several SL-enzyme knockout mouse models have been successfully characterized in context of T2DM [159–161] and many of them show a phenotype that could be interesting for T1DM research. For example mice lacking SPP2 display defective beta-cell proliferation, reduced islet mass and ER stress activation in beta-cells [159]. Exposure of such animal models to STZ, generation of beta-cell specific SPL knockout and knockin mouse models or development of SL-enzymes knockouts in Non Obese Diabetic (NOD) mice, will advance our understanding of T1DM development mechanisms.

Another interesting aspect of cytokine action on the SL pathway is the secretion of nCD via exosomes. Interestingly the low nontoxic cytokine concentrations [38,162] were shown to stimulate nCD, in a similar manner to high concentrations [163]. The lack of toxic effects of low-dose cytokine exposure correlated with a release of neutral ceramidase via exosomes [163] and the presence of nCD containing exosomes prevented apoptosis in INS1 cells incubated with high concentrations of cytokines [163]. The generation and secretion of S1P via nCD-rich exosomes was responsible for the activation of S1PR2 and the observed antiapoptotic effect. This study indicates that beta-cells may activate protective mechanisms at the beginning of the inflammatory response within islets, to prevent further damage caused by high concentrations of cytokines. Once this axis fails, the beta-cell apoptosis and destruction are accelerated.

Overall, this data indicates that proinflammatory cytokines may impact SL generation differentially upon acute and chronic exposure that could have potentially pronounced consequences for beta-cell viability and vulnerability to autoimmune insult. Future measurements of SL species and their distribution in cytokine-treated beta-cells should help to understand how these observed effects influence beta-cell vulnerability to cytokines.

8. Sphingolipids in Animal Models of Autoimmune Diabetes and Human T1DM

In this chapter the SL changes occurring in animal models of autoimmune diabetes and in human T1DM and their role for islet autoimmunity and for beta-cell survival will be addressed. As discussed above proinflammatory cytokines that are secreted by activated immune cells infiltrating islets during T1DM development were shown to affect the SL metabolic pathway of beta-cells. Changes in the SL content were associated with beta-cell dysfunction and death. Recent studies demonstrated an altered SL profiles in the blood of T1DM patients [14–16,18,130,164]. What are the consequences of these altered SL serum profiles on pancreatic beta-cell function and fate during T1DM development? Is the SL concentration around the islets altered? Is the SL composition of beta-cell distinct in T1DM individuals comparing with nondiabetic, healthy subjects?

8.1. Animal Models

While these questions remain at present open, the studies with the S1P receptor modulator Fingolimod/Gilenya (FTY720) may shed some light on this topic. Fingolimod was approved for the treatment of multiple sclerosis in over 40 countries [165]. Its effects were analyzed in multiple animal models of autoimmune diabetes, such as the NOD mouse, STZ-induced autoimmune diabetes in mice and in the rat model of human T1DM, the LEW.1AR1-iddm (IDDM) rat [166–168]. In all these animal models fingolimod treatment was reported to improve glycemia and prevent infiltration of islets. Long-term treatment with FTY720 proved to prevent the diabetes onset in IDDM rats and in NOD mice by reducing immune cell infiltration and cytokine-mediated beta-cell destruction [152,153,166–168].

The IDDM (LEW.1AR1-iddm) rat develops spontaneously autoimmune diabetes [169]. Pancreatic islets from fingolimod-treated IDDM rats are characterized by a well preserved architecture and dense insulin immunostaining [167]. This goes along with prevention of T-cell islet infiltration and reduced expression of proinflammatory cytokines in immune cells [167]. The remaining weak macrophage infiltration observed in the minority of islets in fingolimod-treated IDDM rats correlates with the increased beta-cell expression of MCP1, a well-known attractant for macrophages [167]. Interestingly, the infiltrating macrophages in fingolimod-treated IDDM rats do not express and release IL-1β and TNFα even after a prolonged time post fingolimod treatment, explaining the lack of beta-cell demise in these islets. The study by Jörns and colleagues was the first to show that fingolimod treatment may result in decreased expression of proinflammatory cytokines. Why did macrophages remain proinflammatory cytokine negative in response to the fingolimod treatment? Jörns and colleagues observed that the treatment with fingolimod leads to increased production of anti-inflammatory cytokines IL-4 and IL-10 [167], which leads to the activation of an anti-inflammatory M2 phenotype [170,171]. IL-4 was also shown to protect beta-cells against cytokine toxicity [38,172–174]. Thus the phenomenon of fingolimod-mediated macrophage phenotype rearrangements opens a new intriguing research area in the S1P biology.

Further insights into the role of SLs in T1DM development were gained in a well-established mouse model of autoimmune diabetes, the NOD mouse. The NOD mouse is characterized by a genetic susceptibility to autoimmune diabetes and spontaneously develops autoimmune reaction against pancreatic beta-cells. Environmental factors have also been shown to affect the diabetes incidence in this mouse model [175]. A continuous supplementation of L-serine, a precursor of the SL biosynthesis, was shown to reduce diabetes incidence and insulitis score in female NOD mice [176]. The authors observed significant changes in serum SLs, failed however to detect any significant effects on the pancreatic SL content. Such effects of L-serine should not, however, be excluded since the authors did not perform experiments in purified beta-cells. Administration of fingolimod prolonged the survival rate of islet allografts in diabetic mice [152,153]. Another study showed that the serum phospholipid and triglyceride composition might be associated with progression to T1DM both in NOD mice and humans [164]. The authors found that young female NOD mice who later progress to autoimmune diabetes exhibited the same lipidomic pattern as prediabetic children [164], confirming that the NOD mouse is a valuable model to study the role of SLs in T1DM development. Moreover, NOD thymocytes were found to be characterized by a lower level of S1PR1 and a decreased SPL mRNA and protein expression comparing with healthy mice [177]. These changes were suggested to participate in the T-cell migratory abnormalities observed in NOD mice during diabetes development [177]. S1P was shown to reduce CD4+ T-cell activation in NOD mice and to prevent vascular complications [178].

Administration of glycoSLs has been widely used to assess their effects on islet autoimmunity in NOD mice. For instance, ganglioside GM1 decreased the rate of islet infiltration, attenuated production of proinflammatory cytokines (IL-1β, TNFα and IFNγ) and increased the level of NGF in islets [177]. The protective effects of sulfatide against autoimmunity have been recognized already many years ago when Buschard and colleagues demonstrated that the treatment with sulfatide or its precursor GalCer prevents diabetes in NOD mice [179]. Sulfatide was reported to increase the population of CD3+CD25+ regulatory T-cells, while decreasing production of proinflammatory cytokines (for details refer to the excellent review [104]). These protective effects of GalCer and sulfatide against autoimmune reaction during T1DM development rely on their effects on NKT) cells. In T1DM individuals NKT cells are less frequent and display deficient IL-4 responses [180]. Similar observations were made in NOD mice [181]. Interestingly, α-GalCer was shown to activate NKT cells and prevent the onset and recurrence of T1DM in NOD mice [181]. Moreover, a sphingosine truncated derivative of α-GalCer, OCH, was reported to prevent insulitis and diabetes development in NOD mice more efficiently than its precursor, probably by enhancing the activity of NKT cells to produce IL-10 [182]. Rhost and coworkers observed that a fraction of NOD mice develop autoantibody reactivity to sulfatide, though they failed to demonstrate that sulfatide treatment reduces the diabetes incidence under the treatment scheme they undertook [183]. Using fenofibrate, which activates the sulfatide biosynthesis, Holm and colleagues were able to completely prevent diabetes development in NOD mice [14]. In the follow-up study they demonstrated that fenofibrate selectively elevates the pancreatic content of very-long-chain SLs in NOD mice and reduces the incidence of diabetes by around 50% [184]. Moreover they showed that fenofibrate treatment leads to remodeling of pancreatic lipidome with increased amount of lysoglycerophospholipids [184]. NOD mice treated with fenofibrate were characterized by more stable blood glucose and improved glucose tolerance [184]. Sulfatide was additionally shown to inhibit insulitis and to prevent diabetes in NOD mice by blockage of L-selectin [185]. Furthermore, the C16:0 isoform of sulfatide was reported to downregulate the production of proinflammatory cytokines [186]. In vitro experiments revealed that sulfatide has also the ability to reduce caspase-3/7-dependent apoptosis caused by exposure of insulin-secreting cells to IL-1β, IFNγ and TNFα [187].

Finally, in two additional animal models of autoimmune diabetes (STZ-induced diabetic rats and Akita diabetic mice) the serum concentration of S1P was shown to be significantly elevated in comparison to control animals [17], raising a question whether following the S1P content in blood could serve as a biomarker to track the disease progress.

8.2. Human T1DM

The incidence of T1DM is rising in the last decades in western countries; this suggests an important role of environmental factors in the pathogenesis of this disease. Accumulating evidence points to significant changes in serum metabolome proceeding T1DM development [14,15,18,130,164]. An increased risk of T1DM was described in response to prenatal exposure to perfluoroalkyl substances that modulate neonatal serum phospholipids [188]. Several phospholipids, particularly sphingomyelins and specific phosphocholines, which were shown to be significantly lower in the serum of children who later progress to T1DM [18] were downregulated by perfluoroalkyl substances [188]. Holm and colleagues identified polymorphisms in eight genes encoding proteins involved in the SL metabolism that contribute to the genetic predisposition to T1DM [14]. Single-nucleotide polymorphisms (SNPs) in the chromosome 17q12-q21 region with the gene coding for ORMDL3 were linked with T1DM [14,105]. Importantly, the level of these polymorphisms correlated with the degree of islet autoimmunity in patients with recent onset T1DM [14].

Interestingly alternations of S1P levels have been shown to control IL-17 production in human T-cells [189]. IL-17 arises as a crucial player in the autoimmunity development in T1DM [190,191]. Therefore a control of S1P levels and of the activation of S1PRs might represent an interesting intervention option that should be tested in the future.

With the respect to S1P action, it is important to mention that the cardioprotective effects of HDL have been shown to depend on its chaperoning function for S1P [81,82]. Dyslipidemia is typical in patients with long-term T1DM, and associates with an increased risk of cardiovascular events. How sphingolipids, or particularly S1P, contribute to the progression of diabetic complications in T1DM-patients requires further investigations.

Furthermore, an altered SL content in immune cells, peripheral blood mononuclear cells (PBMC) as well as in serum were reported in multiple studies with T1DM individuals [14,15,18,130,164]. The changes in the sphingomyelin content were defined as a new hallmark of progression to T1DM. Moreover sulfatide levels in newly onset of T1DM patients were found significantly lower than in healthy children [14]. Studies performed in the newborn infants, who later in life progress to T1DM, indicate that their lipidomic profiles are distinct from those of healthy infants [19,20]. The recently published clinical case report study by the Buschard group demonstrated that the treatment with fenofibrate (160 mg daily) initiated seven days after T1DM diagnosis resulted in a fast decline of insulin dose and long-term insulin-independency [192].

The ceramide pathways were found to be specifically associated with T1DM progression [18]. In T1DM progressors a lower content of L-serine was described, in line with the protective effects of L-serine administration against islet autoimmunity observed in the NOD mouse [176]. This was in contrast to glycoSLs such as GlcCer, LacCer and GalCer, which were significantly upregulated in T1DM patients. The authors concluded that these changes in the SL blood cell content could contribute to immune dysfunction in children, who later progress to T1DM [18].

Finally, an intriguing observation about a protective role of C1P on insulin signaling in diabetic kidney [193] might also apply to the phenomenon of insulin resistance in T1DM [194]. SMPDL3b (sphingomyelin phosphodiesterase acid-like 3b) is a lipid raft enzyme that regulates plasma membrane fluidity. Its enhanced expression was observed in diabetic kidney disease and was shown to affect the production of SLs resulting in decreased C1P content [193]. The SMPDL3b overexpression impaired insulin signaling by interfering with insulin receptor isoforms binding to caveolin-1 in the plasma membrane, which was rescued by supplementation with exogenous C1P [193]. Moreover S1P was shown to counteract insulin signaling in beta-cells through the activation of S1PR2 [128]. These studies are raising the question whether a lipid-based therapy might be an option for treatment of severe T1DM complications. Further studies involving human beta-cells, genetically modified beta-cell lines, and various T1DM animal models are needed to assess the role of C1P in T1DM development and beta-cell fate.

9. Sphingolipids as Autoantigens and Biomarkers in T1DM

The progression to T1DM is monitored mainly by evaluation of serological biomarkers (autoantibodies). The positivity of autoantibodies against beta-cells, the age of seroconversion and the positivity for multiple autoantibodies are predictors and major risk factors for the development of T1DM [195–197]. The primary islet autoantibodies are autoantibodies against insulin (IAA), insulinoma-associated antigen-2 (IA-2), glutamic acid decarboxylase (GAD), zinc-transporter 8 (ZnT8) and islet cell antibodies (ICA) [195–197]. These autoantibodies may appear at any age, but the peak of the first islet autoantibody is usually before the age of 3 years [195–198]. Though the vast majority of autoantibodies related to T1DM recognize peptide antigens, antibodies against lipid antigens have also been described [17]. It has been shown that around 60% of sera from children with T1DM react against antigens composed of lysophospholipids [199], with many epitopes directed against gangliosides and sulfatide.

Gangliosides are sialic acid containing glycolipids, which are formed of Cer and an oligosaccharide chain. They are associated with the plasma membrane and in some cell types, including pancreatic islet cells are also present in the cytosol membranes like those of secretory granules. They play an important role in cell–cell interactions. Gangliosides are targets of a variety of anti-islet autoantibodies (please refer to the excellent review [200]). Early studies in the STZ-mouse model of autoimmune diabetes revealed that administration of a mixture of gangliosides (Cronassial, 150 mg/kg body wt., 21% GM1, 40% GD1a, 16% GD1b, 19% GT1b and 5% others) can dampen islet inflammation, but is not able to stop beta-cell destruction [201,202]. Dotta and colleagues were the first to identify GM2-1 in islets that was a target for IgG islet cell autoantibodies [203]. The content of GM2-1 was significantly overexpressed in NOD mice and the antibodies against this ganglioside were found to strongly correlate with T1DM progression in relatives of T1DM individuals [203,204]. Moreover GM2-1 was found to co-localize with insulin granule, similar to other autoantigens. GT3, GD3 and GM2-1 have been shown to be associated with severity of autoimmune reaction in T1DM [204]. Finally, in a case-control analysis of the Croatian population an association of B4GALNT1 gene variations (an enzyme involved in the biosynthesis of GM2 and GD2) with T1DM was detected [205], a phenomenon confirmed by Holm and colleagues in a larger cohort study of T1DM individuals [14].

Interestingly, recent studies in brain cancer have shown that an ER ATP-dependent chaperone GRP94 regulates the ratio of GM2-GM3 gangliosides [206]. The GM2-activator protein, which is a cofactor of beta-hexosaminidase responsible for GM2 hydrolysis to GM3, was shown to be a client for GRP94. Recent studies from the Marzec group have described a reduced expression of GRP94 in beta-cells from T2DM individuals [207] and its role in the inducible proteasome activation-mediated proinsulin degradation [208]. The GRP94 deficiency could disable the proper activity of the GM2-activator protein and thus prevent GM2 hydrolysis to GM3, linking cytokine effects with increased GM2 expression and islet autoimmunity in T1DM.

Moreover, GM1 and GD1a gangliosides have been shown to modulate inflammatory effects of LPS [209] by prevention of TLR4 translocation into lipid rafts. Interestingly, the activation of TLR4 and TLR3 was shown to be activated in viral infections in islets of T1DM patients, as well as in animal models of autoimmune diabetes and beta-cell lines [210–213]. Thus, the ganglioside pattern in beta-cells may be of crucial importance for beta-cell susceptibility to viral infections that are considered as one of major triggers of T1DM development [31,36,210–218]. The generation of tools for influencing ganglioside pattern in islets may represent an interesting new possibility to protect beta-cells from cytokine- and inflammation-mediated toxicity.

Concerning the role of sulfatide in T1DM development, this glycoSL and its precursor GalCer were shown to be ligands for CD14, the macrophage scavenger receptor, in a subset of beta-cells [219]. Anti-sulfatide antibodies have been detected in prediabetic and newly diagnosed T1DM patients [220, 221] (excellently reviewed by Buschard [104]). Anti-sulfatide antibodies were shown to reduce insulin secretion and exocytosis from beta-cells [143].

Finally, sphingomyelin patches on plasma membrane act as epitopes for IC2, a monoclonal antibody that specifically recognizes the surface of beta-cells [222]. This raises the possibility that cell surface sphingomyelin pattern might be involved in the autoimmune reaction directed against beta-cells.

10. Conclusions and Perspectives

The unique beta-cell sphingolipid rheostat and limitations of self-regulation upon T1DM development might serve as one of the major underlying mechanisms involved in beta-cell dysfunction and death in T1DM, similarly to neurodegenerative disorders. Further studies, involving the modern techniques to track the SL flow and de novo SL biosynthesis (such as [223,224]) as well as SL analogues and inhibitors of the SL pathway enzymes, should enable identification of novel SL-related pathways and targets that are engaged in human beta-cell susceptibility to proinflammatory cytokines during T1DM development. The availability of the excellent model human beta-cell line, EndoC-βH1 beta-cells [225,226], for in vitro studies will drive the progress of studies on the role of SLs in T1DM in upcoming years. The impact of beta-cell S1P homeostasis and SPL needs further investigations in beta-cell specific SL enzyme KO models. Based on so-far obtained data/observations, it is expected that SLs will be likely shown to affect all major functions of beta-cells and will provide crucial insights into the activation of islet autoimmunity (illustrated in Figure 3).

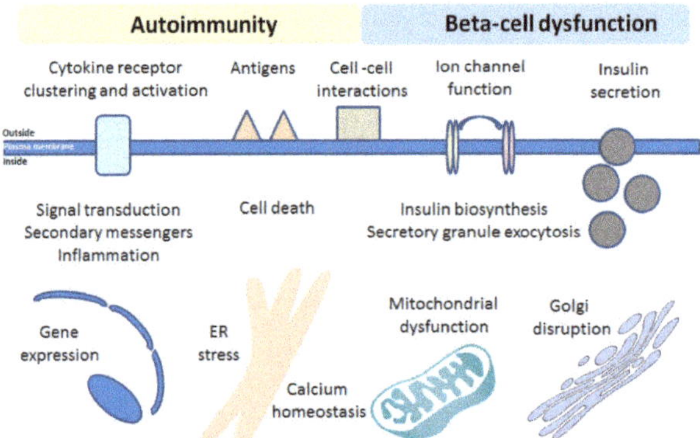

Figure 3. The possible involvement of sphingolipids in beta-cell biology during T1DM development. Rearrangements of SLs in response to the action of proinflammatory cytokines that are released by activated immune cells likely participate in islet autoimmunity, beta-cell dysfunction and death by multiple mechanisms.

Certainly, upcoming years will bring the full characterization of the effects of cytokines on SLs of beta-cells leading to exciting new insights into mechanisms underlying cytokine-mediated beta-cell death, which shall show us how to help the vulnerable beta-cell facing T1DM.

Funding: This research received no external funding.

Conflicts of Interest: The author declares no conflict of interest.

References

1. Maceyka, M.; Harikumar, K.B.; Milstien, S.; Spiegel, S. Sphingosine-1-phosphate signaling and its role in disease. *Trend Cell Biol.* **2012**, *22*, 50–60. [CrossRef] [PubMed]

2. Maceyka, M.; Spiegel, S. Sphingolipid metabolites in inflammatory disease. *Nature* **2014**, *510*, 58–67. [CrossRef] [PubMed]

3. Hannun, Y.A.; Obeid, L.M. Sphingolipids and their metabolism in physiology and disease. *Nat. Rev. Mol. Cell Biol.* **2018**, *19*, 175–191. [CrossRef]

4. Boslem, E.; Meikle, P.J.; Biden, T.J. Roles of ceramide and sphingolipids in pancreatic beta-cell function and dysfunction. *Islets* **2012**, *4*, 177–187. [CrossRef]

5. Hla, T.; Dannenberg, A.J. Sphingolipid signaling in metabolic disorders. *Cell Metab.* **2012**, *16*, 420–434. [CrossRef]

6. Boslem, E.; Weir, J.M.; MacIntosh, G.; Sue, N.; Cantley, J.; Meikle, P.J.; Biden, T.J. Alteration of endoplasmic reticulum lipid rafts contributes to lipotoxicity in pancreatic beta-cells. *J. Biol. Chem.* **2013**, *288*, 26569–26582. [CrossRef]

7. Boslem, E.; MacIntosh, G.; Preston, A.M.; Bartley, C.; Busch, A.K.; Fuller, M.; Laybutt, D.R.; Meikle, P.J.; Biden, T.J. A lipidomic screen of palmitate-treated MIN6 beta-cells links sphingolipid metabolites with endoplasmic reticulum (ER) stress and impaired protein trafficking. *Biochem. J.* **2011**, *435*, 267–276. [CrossRef]

8. Jessup, C.F.; Bonder, C.S.; Pitson, S.M.; Coates, P.T. The sphingolipid rheostat: A potential target for improving pancreatic islet survival and function. *Endocr. Metab. Immun. Disord. Drug Targ.* **2011**, *11*, 262–272. [CrossRef]

9. Veret, J.; Bellini, L.; Giussani, P.; Ng, C.; Magnan, C.; Le Stunff, H. Roles of sphingolipid metabolism in pancreatic beta-cell dysfunction induced by lipotoxicity. *J. Clin. Med.* **2014**, *3*, 646–662. [CrossRef]

10. Veret, J.; Coant, N.; Berdyshev, E.V.; Skobeleva, A.; Therville, N.; Bailbe, D.; Gorshkova, I.; Natarajan, V.; Portha, B.; Le Stunff, H. Ceramide synthase 4 and de novo production of ceramides with specific N-acyl chain lengths are involved in glucolipotoxicity-induced apoptosis of INS-1 beta-cells. *Biochem. J.* **2011**, *438*, 177–189. [CrossRef]

11. Veret, J.; Coant, N.; Gorshkova, I.A.; Giussani, P.; Fradet, M.; Riccitelli, E.; Skobeleva, A.; Goya, J.; Kassis, N.; Natarajan, V.; et al. Role of palmitate-induced sphingoid base-1-phosphate biosynthesis in INS-1 beta-cell survival. *BBA* **2013**, *1831*, 251–262. [CrossRef] [PubMed]

12. Gjoni, E.; Brioschi, L.; Cinque, A.; Coant, N.; Islam, M.N.; Ng, C.K.; Verderio, C.; Magnan, C.; Riboni, L.; Viani, P.; et al. Glucolipotoxicity impairs ceramide flow from the endoplasmic reticulum to the Golgi apparatus in INS-1 beta-cells. *PLoS ONE* **2014**, *9*, e110875. [CrossRef]

13. Dooley, J.; Tian, L.; Schonefeldt, S.; Delghingaro-Augusto, V.; Garcia-Perez, J.E.; Pasciuto, E.; Di Marino, D.; Carr, E.J.; Oskolkov, N.; Lyssenko, V.; et al. Genetic predisposition for beta-cell fragility underlies type 1 and type 2 diabetes. *Nat. Genet.* **2016**, *48*, 519–527. [CrossRef] [PubMed]

14. Holm, L.J.; Krogvold, L.; Hasselby, J.P.; Kaur, S.; Claessens, L.A.; Russell, M.A.; Mathews, C.E.; Hanssen, K.F.; Morgan, N.G.; Koeleman, B.P.C.; et al. Abnormal islet sphingolipid metabolism in type 1 diabetes. *Diabetologia* **2018**, *61*, 1650–1661. [CrossRef] [PubMed]

15. Oresic, M.; Simell, S.; Sysi-Aho, M.; Nanto-Salonen, K.; Seppanen-Laakso, T.; Parikka, V.; Katajamaa, M.; Hekkala, A.; Mattila, I.; Keskinen, P.; et al. Dysregulation of lipid and amino acid metabolism precedes islet autoimmunity in children who later progress to type 1 diabetes. *J. Exp. Med.* **2008**, *205*, 2975–2984. [CrossRef] [PubMed]

16. Wei, N.; Pan, J.; Pop-Busui, R.; Othman, A.; Alecu, I.; Hornemann, T.; Eichler, F.S. Altered sphingoid base profiles in type 1 compared to type 2 diabetes. *Lipid Health Dis.* **2014**, *13*, 161. [CrossRef] [PubMed]

17. Fox, T.E.; Bewley, M.C.; Unrath, K.A.; Pedersen, M.M.; Anderson, R.E.; Jung, D.Y.; Jefferson, L.S.; Kim, J.K.; Bronson, S.K.; Flanagan, J.M.; et al. Circulating sphingolipid biomarkers in models of type 1 diabetes. *J. Lipid Res.* **2011**, *52*, 509–517. [CrossRef]

18. Sen, P.; Dickens, A.M.; Lopez-Bascon, M.A.; Lindeman, T.; Kemppainen, E.; Lamichhane, S.; Ronkko, T.; Ilonen, J.; Toppari, J.; Veijola, R.; et al. Metabolic alterations in immune cells associate with progression to type 1 diabetes. *Diabetologia* **2020**, *63*, 1017–1031. [CrossRef]

19. Lamichhane, S.; Ahonen, L.; Dyrlund, T.S.; Dickens, A.M.; Siljander, H.; Hyoty, H.; Ilonen, J.; Toppari, J.; Veijola, R.; Hyotylainen, T.; et al. Circulating metabolites in progression to islet autoimmunity and type 1 diabetes. *Diabetologia* **2019**, *62*, 2287–2297. [CrossRef]

20. Lamichhane, S.; Kemppainen, E.; Trost, K.; Siljander, H.; Hyoty, H.; Ilonen, J.; Toppari, J.; Veijola, R.; Hyotylainen, T.; Knip, M.; et al. Cord-Blood Lipidome in Progression to Islet Autoimmunity and Type 1 Diabetes. *Biomolecules* **2019**, *9*, 33. [CrossRef]

21. Hahn, C.; Tyka, K.; Saba, J.D.; Lenzen, S.; Gurgul-Convey, E. Overexpression of sphingosine-1-phosphate lyase protects insulin-secreting cells against cytokine toxicity. *J. Biol. Chem.* **2017**, *292*, 20292–20304. [CrossRef] [PubMed]

22. Katsarou, A.; Gudbjornsdottir, S.; Rawshani, A.; Dabelea, D.; Bonifacio, E.; Anderson, B.J.; Jacobsen, L.M.; Schatz, D.A.; Lernmark, Å. Type 1 diabetes mellitus. *Nat. Rev. Dis. Primers* **2017**, *3*, 17016. [CrossRef] [PubMed]

23. Pociot, F.; Lernmark, A. Genetic risk factors for type 1 diabetes. *Lancet* **2016**, *387*, 2331–2339. [CrossRef]

24. Patterson, C.C.; Dahlquist, G.G.; Gyurus, E.; Green, A.; Soltesz, G.; Group, E.S. Incidence trends for childhood type 1 diabetes in Europe during 1989–2003 and predicted new cases 2005-20, a multicentre prospective registration study. *Lancet* **2009**, *373*, 2027–2033. [CrossRef]

25. Roep, B.O.; Peakman, M. Diabetogenic T lymphocytes in human type 1 diabetes. *Curr Opin. Immunol.* **2011**, *23*, 746–753. [CrossRef] [PubMed]

26. Coppieters, K.T.; von Herrath, M.G. The type 1 diabetes signature: Hardwired to trigger inflammation? *Diabetes* **2014**, *63*, 3581–3583. [CrossRef] [PubMed]

27. Eizirik, D.L.; Pasquali, L.; Cnop, M. Pancreatic β-cells in type 1 and type 2 diabetes mellitus: Different pathways to failure. *Nat. Rev. Endocrinol.* **2020**. [CrossRef]

28. Mandrup-Poulsen, T. Beta cell death and protection. *Ann. N. Y. Acad. Sci.* **2003**, *1005*, 32–42. [CrossRef]

29. Nerup, J.; Mandrup-Poulsen, T.; Helqvist, S.; Andersen, H.U.; Pociot, F.; Reimers, J.I.; Cuartero, B.G.; Karlsen, A.E.; Bjerre, U.; Lorenzen, T. On the pathogenesis of IDDM. *Diabetologia* **1994**, *37* (Suppl. 2), S82–S89. [CrossRef]

30. Eizirik, D.L.; Grieco, F.A. On the immense variety and complexity of circumstances conditioning pancreatic beta-cell apoptosis in type 1 diabetes. *Diabetes* **2012**, *61*, 1661–1663. [CrossRef]

31. van Belle, T.L.; Coppieters, K.T.; von Herrath, M.G. Type 1 diabetes: Etiology; immunology; and therapeutic strategies. *Physiol. Rev.* **2011**, *91*, 79–118. [CrossRef] [PubMed]

32. Eizirik, D.L.; Sammeth, M.; Bouckenooghe, T.; Bottu, G.; Sisino, G.; Igoillo-Esteve, M.; Ortis, F.; Santin, I.; Colli, M.L.; Barthson, J.; et al. The human pancreatic islet transcriptome: Expression of candidate genes for type 1 diabetes and the impact of pro-inflammatory cytokines. *PLoS Genet.* **2012**, *8*, e1002552. [CrossRef] [PubMed]

33. Horwitz, E.; Krogvold, L.; Zhitomirsky, S.; Swisa, A.; Fischman, M.; Lax, T.; Dahan, T.; Hurvitz, N.; Weinberg-Corem, N.; Klochendler, A.; et al. β-Cell DNA damage response promotes islet inflammation in type 1 diabetes. *Diabetes* **2018**, *67*, 2305–2318. [CrossRef] [PubMed]

34. Rabinovitch, A.; Suarez-Pinzon, W.L. Role of cytokines in the pathogenesis of autoimmune diabetes mellitus. *Rev. Endocr. Metab. Disord.* **2003**, *4*, 291–299. [CrossRef]

35. Villate, O.; Turatsinze, J.V.; Mascali, L.G.; Grieco, F.A.; Nogueira, T.C.; Cunha, D.A.; Nardelli, T.R.; Sammeth, M.; Salunkhe, V.A.; Esguerra, J.L.; et al. Nova1 is a master regulator of alternative splicing in pancreatic beta cells. *Nucl. Acid. Res.* **2014**, *42*, 11818–11830. [CrossRef]

36. Richardson, S.J.; Rodriguez-Calvo, T.; Gerling, I.C.; Mathews, C.E.; Kaddis, J.S.; Russell, M.A.; Zeissler, M.; Leete, P.; Krogvold, L.; Dahl-Jorgensen, K.; et al. Islet cell hyperexpression of HLA class I antigens: A defining feature in type 1 diabetes. *Diabetologia* **2016**, *59*, 2448–2458. [CrossRef]

37. Russell, M.A.; Redick, S.D.; Blodgett, D.M.; Richardson, S.J.; Leete, P.; Krogvold, L.; Dahl-Jorgensen, K.; Bottino, R.; Brissova, M.; Spaeth, J.M.; et al. HLA Class II antigen processing and presentation pathway components demonstrated by transcriptome and protein analyses of islet beta-cells from donors with type 1 diabetes. *Diabetes* **2019**, *68*, 988–1001. [CrossRef]

38. Souza, K.L.; Gurgul-Convey, E.; Elsner, M.; Lenzen, S. Interaction between pro-inflammatory and anti-inflammatory cytokines in insulin-producing cells. *J. Endocr.* **2008**, *197*, 139–150. [CrossRef]

39. Gurgul-Convey, E.; Lenzen, S. Protection against cytokine toxicity through endoplasmic reticulum and mitochondrial stress prevention by prostacyclin synthase overexpression in insulin-producing cells. *J. Biol. Chem.* **2010**, *285*, 11121–11128. [CrossRef]

40. Gurgul-Convey, E.; Mehmeti, I.; Lortz, S.; Lenzen, S. Cytokine toxicity in insulin-producing cells is mediated by nitro-oxidative stress-induced hydroxyl radical formation in mitochondria. *J. Mol. Med.* **2011**, *89*, 785–798. [CrossRef]

41. Hanzelka, K.; Skalniak, L.; Jura, J.; Lenzen, S.; Gurgul-Convey, E. Effects of the novel mitochondrial protein mimitin in insulin-secreting cells. *Biochem. J.* **2012**, *445*, 349–359. [CrossRef] [PubMed]

42. Gurgul-Convey, E.; Mehmeti, I.; Plötz, T.; Jörns, A.; Lenzen, S. Sensitivity profile of the human EndoC-betaH1 beta cell line to proinflammatory cytokines. *Diabetologia* **2016**, *59*, 2125–2133. [CrossRef] [PubMed]

43. Tyka, K.; Jörns, A.; Turatsinze, J.V.; Eizirik, D.L.; Lenzen, S.; Gurgul-Convey, E. MCPIP1 regulates the sensitivity of pancreatic beta-cells to cytokine toxicity. *Cell Death Dis.* **2019**, *10*, 29. [CrossRef] [PubMed]

44. Gurgul, E.; Lortz, S.; Tiedge, M.; Jörns, A.; Lenzen, S. Mitochondrial catalase overexpression protects insulin-producing cells against toxicity of reactive oxygen species and proinflammatory cytokines. *Diabetes* **2004**, *53*, 2271–2280. [CrossRef] [PubMed]

45. Lenzen, S. Oxidative stress: The vulnerable beta-cell. *Biochem. Soc. Trans.* **2008**, *36 Pt 3*, 343–347. [CrossRef]

46. Ghiasi, S.M.; Krogh, N.; Tyrberg, B.; Mandrup-Poulsen, T. The no-go and nonsense-mediated RNA decay pathways are regulated by inflammatory cytokines in insulin-producing cells and human islets and determine beta-cell insulin biosynthesis and survival. *Diabetes* **2018**, *67*, 2019–2037. [CrossRef]

47. Eizirik, D.L.; Cnop, M. ER stress in pancreatic beta cells: The thin red line between adaptation and failure. *Sci. Signal.* **2010**, *3*, pe7. [CrossRef]

48. Robertson, R.P. Eicosanoids as pluripotential modulators of pancreatic islet function. *Diabetes* **1988**, *37*, 367–370. [CrossRef]

49. Chambers, K.T.; Weber, S.M.; Corbett, J.A. PGJ2-stimulated beta-cell apoptosis is associated with prolonged UPR activation. *Am. J. Physiol. Endocr. Metab.* **2007**, *292*, E1052–E1061. [CrossRef]

50. Heitmeier, M.R.; Kelly, C.B.; Ensor, N.J.; Gibson, K.A.; Mullis, K.G.; Corbett, J.A.; Maziasz, T.J. Role of cyclooxygenase-2 in cytokine induced beta-cell dysfunction and damage by isolated rat and human islets. *J. Biol. Chem.* **2004**, *279*, 53145–53151. [CrossRef]

51. Ghiasi, S.M.; Dahllof, M.S.; Osmai, Y.; Osmai, M.; Jakobsen, K.K.; Aivazidis, A.; Tyrberg, B.; Perruzza, L.; Prause, M.C.B.; Christensen, D.P.; et al. Regulation of the beta-cell inflammasome and contribution to stress-induced cellular dysfunction and apoptosis. *Mol. Cell. Endocr.* **2018**, *478*, 106–114. [CrossRef] [PubMed]

52. Sandler, S.; Andersson, A.; Hellerström, C. Inhibitory effects of interleukin 1 on insulin secretion; insulin biosynthesis; and oxidative metabolism of isolated rat pancreatic islets. *Endocrinology* **1987**, *121*, 1424–1431. [CrossRef] [PubMed]

53. Southern, C.; Schulster, D.; Green, I.C. Inhibition of insulin secretion by interleukin-1 beta and tumour necrosis factor-alpha via an L-arginine-dependent nitric oxide generating mechanism. *FEBS Lett.* **1990**, *276*, 42–44. [CrossRef]

54. Hannun, Y.A.; Obeid, L.M. Principles of bioactive lipid signalling: Lessons from sphingolipids. *Nat. Rev. Mol. Cell Biol.* **2008**, *9*, 139–150. [CrossRef] [PubMed]

55. Hanada, K.; Kumagai, K.; Yasuda, S.; Miura, Y.; Kawano, M.; Fukasawa, M.; Nishijima, M. Molecular machinery for non-vesicular trafficking of ceramide. *Nature* **2003**, *426*, 803–809. [CrossRef]

56. Quehenberger, O.; Armando, A.M.; Brown, A.H.; Milne, S.B.; Myers, D.S.; Merrill, A.H.; Bandyopadhyay, S.; Jones, K.N.; Kelly, S.; Shaner, R.L.; et al. Lipidomics reveals a remarkable diversity of lipids in human plasma. *J. Lipid Res.* **2010**, *51*, 3299–3305. [CrossRef]

57. Kiefer, K.; Carreras-Sureda, A.; Garcia-Lopez, R.; Rubio-Moscardo, F.; Casas, J.; Fabrias, G.; Vicente, R. Coordinated regulation of the orosomucoid-like gene family expression controls de novo ceramide synthesis in mammalian cells. *J. Biol. Chem.* **2015**, *290*, 2822–2830. [CrossRef]

58. Maceyka, M.; Sankala, H.; Hait, N.C.; Le Stunff, H.; Liu, H.; Toman, R.; Collier, C.; Zhang, M.; Satin, L.S.; Merrill, A.H., Jr.; et al. SphK1 and SphK2; sphingosine kinase isoenzymes with opposing functions in sphingolipid metabolism. *J. Biol. Chem.* **2005**, *280*, 37118–37129. [CrossRef]

59. Strub, G.M.; Paillard, M.; Liang, J.; Gomez, L.; Allegood, J.C.; Hait, N.C.; Maceyka, M.; Price, M.M.; Chen, Q.; Simpson, D.C.; et al. Sphingosine-1-phosphate produced by sphingosine kinase 2 in mitochondria interacts with prohibitin 2 to regulate complex IV assembly and respiration. *FASEB J.* **2011**, *25*, 600–612. [CrossRef]

60. Ding, G.; Sonoda, H.; Yu, H.; Kajimoto, T.; Goparaju, S.K.; Jahangeer, S.; Okada, T.; Nakamura, S. Protein kinase D-mediated phosphorylation and nuclear export of sphingosine kinase 2. *J. Biol. Chem.* **2007**, *282*, 27493–27502. [CrossRef]

61. Saba, J.D.; Nara, F.; Bielawska, A.; Garrett, S.; Hannun, Y.A. The BST1 gene of Saccharomyces cerevisiae is the sphingosine-1-phosphate lyase. *J. Biol. Chem.* **1997**, *272*, 26087–26090. [CrossRef] [PubMed]

62. Kumar, A.; Byun, H.S.; Bittman, R.; Saba, J.D. The sphingolipid degradation product trans-2-hexadecenal induces cytoskeletal reorganization and apoptosis in a JNK-dependent manner. *Cell. Signal.* **2011**, *23*, 1144–1152. [CrossRef] [PubMed]

63. Upadhyaya, P.; Kumar, A.; Byun, H.S.; Bittman, R.; Saba, J.D.; Hecht, S.S. The sphingolipid degradation product trans-2-hexadecenal forms adducts with DNA. *Biochem. Biophys. Res. Commun.* **2012**, *424*, 18–21. [CrossRef] [PubMed]

64. Nakahara, K.; Ohkuni, A.; Kitamura, T.; Abe, K.; Naganuma, T.; Ohno, Y.; Zoeller, R.A.; Kihara, A. The Sjogren-Larsson syndrome gene encodes a hexadecenal dehydrogenase of the sphingosine 1-phosphate degradation pathway. *Mol. Cell* **2012**, *46*, 461–471. [CrossRef] [PubMed]

65. Hayashi, T.; Rizzuto, R.; Hajnoczky, G.; Su, T.P. MAM: More than just a housekeeper. *Trend Cell Biol.* **2009**, *19*, 81–88. [CrossRef] [PubMed]

66. Serra, M.; Saba, J.D. Sphingosine 1-phosphate lyase; a key regulator of sphingosine 1-phosphate signaling and function. *Adv. Enzyme Regul.* **2010**, *50*, 349–362. [CrossRef]

67. Hanada, K.; Kumagai, K.; Tomishige, N.; Kawano, M. CERT and intracellular trafficking of ceramide. *BBA* **2007**, *1771*, 644–653. [CrossRef]

68. Colley, K.J.; Varki, A.; Kinoshita, T. Cellular Organization of Glycosylation. In *Essentials of Glycobiology*, 3rd ed.; Varki, A., Cummings, R.D., Esko, J.D., Stanley, P., Hart, G.W., Eds.; Cold Spring Harbor Laboratory Press: Cold Spring Harbor, NY, USA, 2015; pp. 41–49.

69. Gomez-Munoz, A.; Presa, N.; Gomez-Larrauri, A.; Rivera, I.G.; Trueba, M.; Ordonez, M. Control of inflammatory responses by ceramide; sphingosine 1-phosphate and ceramide 1-phosphate. *Progr. Lipid Res.* **2016**, *61*, 51–62. [CrossRef]

70. Shinghal, R.; Scheller, R.H.; Bajjalieh, S.M. Ceramide 1-phosphate phosphatase activity in brain. *J. Neurochem.* **1993**, *61*, 2279–2285. [CrossRef]

71. Boudker, O.; Futerman, A.H. Detection and characterization of ceramide-1-phosphate phosphatase activity in rat liver plasma membrane. *J. Biol. Chem.* **1993**, *268*, 22150–22155.

72. Senkal, C.E.; Salama, M.F.; Snider, A.J.; Allopenna, J.J.; Rana, N.A.; Koller, A.; Hannun, Y.A.; Obeid, L.M. Ceramide is metabolized to acylceramide and stored in lipid droplets. *Cell Metab.* **2017**, *25*, 686–697. [CrossRef] [PubMed]

73. Fugio, L.B.; Coeli-Lacchini, F.B.; Leopoldino, A.M. Sphingolipids and mitochondrial dynamic. *Cells* **2020**, *9*, 581. [CrossRef] [PubMed]

74. Hernandez-Corbacho, M.J.; Salama, M.F.; Canals, D.; Senkal, C.E.; Obeid, L.M. Sphingolipids in mitochondria. *Biochim. Biophys. Acta Mol. Cell Biol. Lipid* **2017**, *1862*, 56–68. [CrossRef] [PubMed]

75. Colombini, M. Ceramide channels and their role in mitochondria-mediated apoptosis. *BBA* **2010**, *1797*, 1239–1244. [CrossRef]

76. Novgorodov, S.A.; Riley, C.L.; Keffler, J.A.; Yu, J.; Kindy, M.S.; Macklin, W.B.; Lombard, D.B.; Gudz, T.I. SIRT3 deacetylates ceramide synthases: Implications for mitochondrial dysfunction and brain injury. *J. Biol. Chem.* **2016**, *291*, 1957–1973. [CrossRef]

77. Dadsena, S.; Bockelmann, S.; Mina, J.G.M.; Hassan, D.G.; Korneev, S.; Razzera, G.; Jahn, H.; Niekamp, P.; Müller, D.; Schneider, M.; et al. Ceramides bind VDAC2 to trigger mitochondrial apoptosis. *Nat. Commun.* **2019**, *10*, 1832. [CrossRef]

78. Oleinik, N.; Kim, J.; Roth, B.M.; Selvam, S.P.; Gooz, M.; Johnson, R.H.; Lemasters, J.J.; Ogretmen, B. Mitochondrial protein import is regulated by p17/PERMIT to mediate lipid metabolism and cellular stress. *Sci. Adv.* **2019**, *5*, eaax1978. [CrossRef]

79. Woodcock, J.M.; Ma, Y.; Coolen, C.; Pham, D.; Jones, C.; Lopez, A.F.; Pitson, S.M. Sphingosine and FTY720 directly bind pro-survival 14-3-3 proteins to regulate their function. *Cell. Signal.* **2010**, *22*, 1291–1299. [CrossRef]

80. Habrukowich, C.; Han, D.K.; Le, A.; Rezaul, K.; Pan, W.; Ghosh, M.; Li, Z.; Dodge-Kafka, K.; Jiang, X.; Bittman, R.; et al. Sphingosine interaction with acidic leucine-rich nuclear phosphoprotein-32A (ANP32A) regulates PP2A activity and cyclooxygenase (COX)-2 expression in human endothelial cells. *J. Biol. Chem.* **2010**, *285*, 26825–26831. [CrossRef]

81. Keul, P.; Polzin, A.; Kaiser, K.; Gräler, M.; Dannenberg, L.; Daum, G.; Heusch, G.; Levkau, B. Potent anti-inflammatory properties of HDL in vascular smooth muscle cells mediated by HDL-S1P and their impairment in coronary artery disease due to lower HDL-S1P: A new aspect of HDL dysfunction and its therapy. *FASEB J.* **2019**, *33*, 1482–1495. [CrossRef]

82. Levkau, B. Cardiovascular effects of sphingosine-1-phosphate (S1P). *Handb. Exp. Pharmacol.* **2013**, 147–170. [CrossRef]

83. Fyrst, H.; Saba, J.D. An update on sphingosine-1-phosphate and other sphingolipid mediators. *Nat. Chem. Biol.* **2010**, *6*, 489–497. [CrossRef] [PubMed]

84. Maceyka, M.; Milstien, S.; Spiegel, S. Sphingosine-1-phosphate: The Swiss army knife of sphingolipid signaling. *J. Lipid Res.* **2009**, *50*, S272–S276. [CrossRef] [PubMed]

85. Strub, G.M.; Maceyka, M.; Hait, N.C.; Milstien, S.; Spiegel, S. Extracellular and intracellular actions of sphingosine-1-phosphate. *Adv. Exp. Med. Biol.* **2010**, *688*, 141–155. [CrossRef]

86. Degagne, E.; Pandurangan, A.; Bandhuvula, P.; Kumar, A.; Eltanawy, A.; Zhang, M.; Yoshinaga, Y.; Nefedov, M.; de Jong, P.J.; Fong, L.G.; et al. Sphingosine-1-phosphate lyase downregulation promotes colon carcinogenesis through STAT3-activated microRNAs. *J. Clin. Investig.* **2014**, *124*, 5368–5384. [CrossRef] [PubMed]

87. Degagne, E.; Saba, J.D. S1pping fire: Sphingosine-1-phosphate signaling as an emerging target in inflammatory bowel disease and colitis-associated cancer. *Clin. Exp. Gastroenterol.* **2014**, *7*, 205–214. [CrossRef] [PubMed]

88. Leong, W.I.; Saba, J.D. S1P metabolism in cancer and other pathological conditions. *Biochimie* **2010**, *92*, 716–723. [CrossRef]

89. Hagen, N.; Hans, M.; Hartmann, D.; Swandulla, D.; van Echten-Deckert, G. Sphingosine-1-phosphate links glycosphingolipid metabolism to neurodegeneration via a calpain-mediated mechanism. *Cell Death Diff.* **2011**, *18*, 1356–1365. [CrossRef]

90. Hagen, N.; Van Veldhoven, P.P.; Proia, R.L.; Park, H.; Merrill, A.H., Jr.; van Echten-Deckert, G. Subcellular origin of sphingosine 1-phosphate is essential for its toxic effect in lyase-deficient neurons. *J. Biol. Chem.* **2009**, *284*, 11346–11353. [CrossRef]

91. Karunakaran, I.; Alam, S.; Jayagopi, S.; Frohberger, S.J.; Hansen, J.N.; Kuehlwein, J.; Holbling, B.V.; Schumak, B.; Hubner, M.P.; Gräler, M.H.; et al. Neural sphingosine 1-phosphate accumulation activates microglia and links impaired autophagy and inflammation. *Glia* **2019**, *67*, 1859–1872. [CrossRef]

92. Karunakaran, I.; van Echten-Deckert, G. Sphingosine 1-phosphate–A double edged sword in the brain. *BBA Biomembr.* **2017**, *9 Pt B*, 1573–1582. [CrossRef]

93. Mitroi, D.N.; Deutschmann, A.U.; Raucamp, M.; Karunakaran, I.; Glebov, K.; Hans, M.; Walter, J.; Saba, J.; Gräler, M.; Ehninger, D.; et al. Sphingosine 1-phosphate lyase ablation disrupts presynaptic architecture and function via an ubiquitin- proteasome mediated mechanism. *Sci. Rep.* **2016**, *6*, 37064. [CrossRef] [PubMed]

94. Mitroi, D.N.; Karunakaran, I.; Gräler, M.; Saba, J.D.; Ehninger, D.; Ledesma, M.D.; van Echten-Deckert, G. SGPL1 (sphingosine phosphate lyase 1) modulates neuronal autophagy via phosphatidylethanolamine production. *Autophagy* **2017**, *13*, 885–899. [CrossRef] [PubMed]

95. van Echten-Deckert, G.; Alam, S. Sphingolipid metabolism—An ambiguous regulator of autophagy in the brain. *Biol. Chem.* **2018**, *399*, 837–850. [CrossRef] [PubMed]

96. Mendelson, K.; Evans, T.; Hla, T. Sphingosine 1-phosphate signalling. *Development* **2014**, *141*, 5–9. [CrossRef] [PubMed]

97. Pettus, B.J.; Kitatani, K.; Chalfant, C.E.; Taha, T.A.; Kawamori, T.; Bielawski, J.; Obeid, L.M.; Hannun, Y.A. The coordination of prostaglandin E2 production by sphingosine-1-phosphate and ceramide-1-phosphate. *Mol. Pharmacol.* **2005**, *68*, 330–335. [CrossRef]

98. Laviad, E.L.; Albee, L.; Pankova-Kholmyansky, I.; Epstein, S.; Park, H.; Merrill, A.H., Jr.; Futerman, A.H. Characterization of ceramide synthase 2, tissue distribution; substrate specificity; and inhibition by sphingosine 1-phosphate. *J. Biol. Chem.* **2008**, *283*, 5677–5684. [CrossRef]

99. Etemadi, N.; Chopin, M.; Anderton, H.; Tanzer, M.C.; Rickard, J.A.; Abeysekera, W.; Hall, C.; Spall, S.K.; Wang, B.; Xiong, Y.; et al. TRAF2 regulates TNF and NF-kappaB signalling to suppress apoptosis and skin inflammation independently of Sphingosine kinase 1. *eLife* **2015**, *4*, e10592. [CrossRef]

100. Hait, N.C.; Allegood, J.; Maceyka, M.; Strub, G.M.; Harikumar, K.B.; Singh, S.K.; Luo, C.; Marmorstein, R.; Kordula, T.; Milstien, S.; et al. Regulation of histone acetylation in the nucleus by sphingosine-1-phosphate. *Science* **2009**, *325*, 1254–1257. [CrossRef]

101. Spiegel, S.; Milstien, S.; Grant, S. Endogenous modulators and pharmacological inhibitors of histone deacetylases in cancer therapy. *Oncogene* **2012**, *31*, 537–551. [CrossRef]

102. Presa, N.; Gomez-Larrauri, A.; Rivera, I.G.; Ordonez, M.; Trueba, M.; Gomez-Munoz, A. Regulation of cell migration and inflammation by ceramide 1-phosphate. *BBA* **2016**, *1861*, 402–409. [CrossRef] [PubMed]

103. Takahashi, T.; Suzuki, T. Role of sulfatide in normal and pathological cells and tissues. *J. Lipid Res.* **2012**, *53*, 1437–1450. [CrossRef] [PubMed]

104. Buschard, K.; Blomqvist, M.; Osterbye, T.; Fredman, P. Involvement of sulfatide in beta cells and type 1 and type 2 diabetes. *Diabetologia* **2005**, *48*, 1957–1962. [CrossRef] [PubMed]

105. Verlaan, D.J.; Berlivet, S.; Hunninghake, G.M.; Madore, A.M.; Lariviere, M.; Moussette, S.; Grundberg, E.; Kwan, T.; Ouimet, M.; Ge, B.; et al. Allele-specific chromatin remodeling in the ZPBP2/GSDMB/ORMDL3 locus associated with the risk of asthma and autoimmune disease. *Am. J. Hum. Genet.* **2009**, *85*, 377–393. [CrossRef] [PubMed]

106. Majumder, S.; Kono, M.; Lee, Y.T.; Byrnes, C.; Li, C.; Tuymetova, G.; Proia, R.L. A genome-wide CRISPR/Cas9 screen reveals that the aryl hydrocarbon receptor stimulates sphingolipid levels. *J. Biol. Chem.* **2020**, *295*, 4341–4349. [CrossRef] [PubMed]

107. Miani, M.; Le Naour, J.; Waeckel-Enee, E.; Verma, S.C.; Straube, M.; Emond, P.; Ryffel, B.; van Endert, P.; Sokol, H.; Diana, J. Gut microbiota-stimulated innate lymphoid cells support beta-defensin 14 expression in pancreatic endocrine cells; preventing autoimmune diabetes. *Cell Metab.* **2018**, *28*, 557–572.e6. [CrossRef]

108. Welsh, N. Interleukin-1 beta-induced ceramide and diacylglycerol generation may lead to activation of the c-Jun NH2-terminal kinase and the transcription factor ATF2 in the insulin-producing cell line RINm5F. *J. Biol. Chem.* **1996**, *271*, 8307–8312. [CrossRef]

109. Kwon, G.; Bohrer, A.; Han, X.; Corbett, J.A.; Ma, Z.; Gross, R.W.; McDaniel, M.L.; Turk, J. Characterization of the sphingomyelin content of isolated pancreatic islets. Evaluation of the role of sphingomyelin hydrolysis in the action of interleukin-1 to induce islet overproduction of nitric oxide. *BBA* **1996**, *1300*, 63–72. [CrossRef]

110. Ishizuka, N.; Yagui, K.; Tokuyama, Y.; Yamada, K.; Suzuki, Y.; Miyazaki, J.; Hashimoto, N.; Makino, H.; Saito, Y.; Kanatsuka, A. Tumor necrosis factor alpha signaling pathway and apoptosis in pancreatic beta cells. *Metab. Clin. Exp.* **1999**, *48*, 1485–1492. [CrossRef]

111. Major, C.D.; Gao, Z.Y.; Wolf, B.A. Activation of the sphingomyelinase/ceramide signal transduction pathway in insulin-secreting beta-cells: Role in cytokine-induced beta-cell death. *Diabetes* **1999**, *48*, 1372–1380. [CrossRef]

112. Lei, X.; Bone, R.N.; Ali, T.; Zhang, S.; Bohrer, A.; Tse, H.M.; Bidasee, K.R.; Ramanadham, S. Evidence of contribution of iPLA2beta-mediated events during islet beta-cell apoptosis due to proinflammatory cytokines suggests a role for iPLA2beta in T1D development. *Endocrinology* **2014**, *155*, 3352–3364. [CrossRef] [PubMed]

113. Lei, X.; Bone, R.N.; Ali, T.; Wohltmann, M.; Gai, Y.; Goodwin, K.J.; Bohrer, A.E.; Turk, J.; Ramanadham, S. Genetic modulation of islet beta-cell iPLA(2)beta expression provides evidence for its impact on beta-cell apoptosis and autophagy. *Islets* **2013**, *5*, 29–44. [CrossRef] [PubMed]

114. Hammerschmidt, P.; Ostkotte, D.; Nolte, H.; Gerl, M.J.; Jais, A.; Brunner, H.L.; Sprenger, H.G.; Awazawa, M.; Nicholls, H.T.; Turpin-Nolan, S.M.; et al. CerS6-derived sphingolipids interact with Mff and promote mitochondrial fragmentation in obesity. *Cell* **2019**, *177*, 1536–1552.e23. [CrossRef] [PubMed]

115. Bektas, M.; Allende, M.L.; Lee, B.G.; Chen, W.; Amar, M.J.; Remaley, A.T.; Saba, J.D.; Proia, R.L. Sphingosine 1-phosphate lyase deficiency disrupts lipid homeostasis in liver. *J. Biol. Chem.* **2010**, *285*, 10880–10889. [CrossRef]

116. Imai, S.; Guarente, L. NAD+ and sirtuins in aging and disease. *Trend Cell Biol.* **2014**, *24*, 464–471. [CrossRef]

117. Zhu, Q.; Shan, X.; Miao, H.; Lu, Y.; Xu, J.; You, N.; Liu, C.; Liao, D.F.; Jin, J. Acute activation of acid ceramidase affects cytokine-induced cytotoxicity in rat islet beta-cells. *FEBS Lett.* **2009**, *583*, 2136–2141. [CrossRef]

118. Zhu, Q.; Jin, J.F.; Shan, X.H.; Liu, C.P.; Mao, X.D.; Xu, K.F.; Liu, C. Chronic activation of neutral ceramidase protects beta-cells against cytokine-induced apoptosis. *Acta Pharmacol. Sin.* **2008**, *29*, 593–599. [CrossRef]

119. Laychock, S.G.; Tian, Y.; Sessanna, S.M. Endothelial differentiation gene receptors in pancreatic islets and INS-1 cells. *Diabetes* **2003**, *52*, 1986–1993. [CrossRef]

120. Sato, K.; Kon, J.; Tomura, H.; Osada, M.; Murata, N.; Kuwabara, A.; Watanabe, T.; Ohta, H.; Ui, M.; Okajima, F. Activation of phospholipase C-Ca2+ system by sphingosine 1-phosphate in CHO cells transfected with Edg-3; a putative lipid receptor. *FEBS Lett.* **1999**, *443*, 25–30. [CrossRef]

121. Ancellin, N.; Hla, T. Differential pharmacological properties and signal transduction of the sphingosine 1-phosphate receptors EDG-1; EDG-3; and EDG-5. *J. Biol. Chem.* **1999**, *274*, 18997–19002. [CrossRef]

122. Gonda, K.; Okamoto, H.; Takuwa, N.; Yatomi, Y.; Okazaki, H.; Sakurai, T.; Kimura, S.; Sillard, R.; Harii, K.; Takuwa, Y. The novel sphingosine 1-phosphate receptor AGR16 is coupled via pertussis toxin-sensitive and -insensitive G-proteins to multiple signalling pathways. *Biochem. J.* **1999**, *337 Pt 1*, 67–75. [CrossRef]

123. Toman, R.E.; Spiegel, S. Lysophospholipid receptors in the nervous system. *Neurochem. Res.* **2002**, *27*, 619–627. [CrossRef]

124. Mastrandrea, L.D.; Sessanna, S.M.; Laychock, S.G. Sphingosine kinase activity and sphingosine-1 phosphate production in rat pancreatic islets and INS-1 cells: Response to cytokines. *Diabetes* **2005**, *54*, 1429–1436. [CrossRef] [PubMed]

125. Laychock, S.G.; Sessanna, S.M.; Lin, M.H.; Mastrandrea, L.D. Sphingosine 1-phosphate affects cytokine-induced apoptosis in rat pancreatic islet beta-cells. *Endocrinology* **2006**, *147*, 4705–4712. [CrossRef] [PubMed]

126. Atkinson, D.; Nikodinovic Glumac, J.; Asselbergh, B.; Ermanoska, B.; Blocquel, D.; Steiner, R.; Estrada-Cuzcano, A.; Peeters, K.; Ooms, T.; De Vriendt, E.; et al. Sphingosine 1-phosphate lyase deficiency causes Charcot-Marie-Tooth neuropathy. *Neurology* **2017**, *88*, 533–542. [CrossRef] [PubMed]

127. Prasad, R.; Hadjidemetriou, I.; Maharaj, A.; Meimaridou, E.; Buonocore, F.; Saleem, M.; Hurcombe, J.; Bierzynska, A.; Barbagelata, E.; Bergada, I.; et al. Sphingosine-1-phosphate lyase mutations cause primary adrenal insufficiency and steroid-resistant nephrotic syndrome. *J. Clin. Investig.* **2017**, *127*, 942–953. [CrossRef] [PubMed]

128. Japtok, L.; Schmitz, E.I.; Fayyaz, S.; Kramer, S.; Hsu, L.J.; Kleuser, B. Sphingosine 1-phosphate counteracts insulin signaling in pancreatic beta-cells via the sphingosine 1-phosphate receptor subtype 2. *FASEB J.* **2015**, *29*, 3357–3369. [CrossRef]

129. Cantrell Stanford, J.; Morris, A.J.; Sunkara, M.; Popa, G.J.; Larson, K.L.; Ozcan, S. Sphingosine 1-phosphate (S1P) regulates glucose-stimulated insulin secretion in pancreatic beta cells. *J. Biol. Chem.* **2012**, *287*, 13457–13464. [CrossRef]

130. Galadari, S.; Rahman, A.; Pallichankandy, S.; Galadari, A.; Thayyullathil, F. Role of ceramide in diabetes mellitus: Evidence and mechanisms. *Lipid Health Dis.* **2013**, *12*, 98. [CrossRef]

131. Kavishwar, A.; Moore, A. Sphingomyelin patches on pancreatic beta-cells are indicative of insulin secretory capacity. *J. Histochem. Cytochem.* **2013**, *61*, 910–919. [CrossRef]

132. Holland, W.L.; Summers, S.A. Sphingolipids, insulin resistance, and metabolic disease: New insights from in vivo manipulation of sphingolipid metabolism. *Endocr. Rev.* **2008**, *29*, 381–402. [CrossRef] [PubMed]

133. Sjöholm, A. Ceramide inhibits pancreatic beta-cell insulin production and mitogenesis and mimics the actions of interleukin-1 beta. *FEBS Lett.* **1995**, *367*, 283–286. [CrossRef]

134. Hasan, N.M.; Longacre, M.J.; Stoker, S.W.; Kendrick, M.A.; Druckenbrod, N.R.; Laychock, S.G.; Mastrandrea, L.D.; MacDonald, M.J. Sphingosine kinase 1 knockdown reduces insulin synthesis and secretion in a rat insulinoma cell line. *Arch. Biochem. Biophys.* **2012**, *518*, 23–30. [CrossRef]

135. Supale, S.; Thorel, F.; Merkwirth, C.; Gjinovci, A.; Herrera, P.L.; Scorrano, L.; Meda, P.; Langer, T.; Maechler, P. Loss of prohibitin induces mitochondrial damages altering beta-cell function and survival and is responsible for gradual diabetes development. *Diabetes* **2013**, *62*, 3488–3499. [CrossRef] [PubMed]

136. Schauble, N.; Lang, S.; Jung, M.; Cappel, S.; Schorr, S.; Ulucan, O.; Linxweiler, J.; Dudek, J.; Blum, R.; Helms, V.; et al. BiP-mediated closing of the Sec61 channel limits Ca^{2+} leakage from the ER. *EMBO J.* **2012**, *31*, 3282–3296. [CrossRef] [PubMed]

137. Eizirik, D.L.; Sandler, S.; Welsh, N.; Juntti-Berggren, L.; Berggren, P.O. Interleukin-1 beta-induced stimulation of insulin release in mouse pancreatic islets is related to diacylglycerol production and protein kinase C activation. *Mol. Cell. Endocrinol.* **1995**, *111*, 159–165. [CrossRef]

138. Perrotta, C.; Bizzozero, L.; Cazzato, D.; Morlacchi, S.; Assi, E.; Simbari, F.; Zhang, Y.; Gulbins, E.; Bassi, M.T.; Rosa, P.; et al. Syntaxin 4 is required for acid sphingomyelinase activity and apoptotic function. *J. Biol. Chem.* **2010**, *285*, 40240–40251. [CrossRef]

139. Oh, E.; Ahn, M.; Afelik, S.; Becker, T.C.; Roep, B.O.; Thurmond, D.C. Syntaxin 4 expression in pancreatic beta-cells promotes islet function and protects functional beta-cell mass. *Diabetes* **2018**, *67*, 2626–2639. [CrossRef]

140. Berchtold, L.A.; Störling, Z.M.; Ortis, F.; Lage, K.; Bang-Berthelsen, C.; Bergholdt, R.; Hald, J.; Brorsson, C.A.; Eizirik, D.L.; Pociot, F.; et al. Huntingtin-interacting protein 14 is a type 1 diabetes candidate protein regulating insulin secretion and beta-cell apoptosis. *Proc. Natl. Acad. Sci. USA* **2011**, *108*, E681–E688. [CrossRef]

141. Blomqvist, M.; Osterbye, T.; Mansson, J.E.; Horn, T.; Buschard, K.; Fredman, P. Sulfatide is associated with insulin granules and located to microdomains of a cultured beta cell line. *Glycoconj. J.* **2002**, *19*, 403–413. [CrossRef]

142. Osterbye, T.; Jorgensen, K.H.; Fredman, P.; Tranum-Jensen, J.; Kaas, A.; Brange, J.; Whittingham, J.L.; Buschard, K. Sulfatide promotes the folding of proinsulin; preserves insulin crystals; and mediates its monomerization. *Glycobiology* **2001**, *11*, 473–479. [CrossRef] [PubMed]

143. Buschard, K.; Hoy, M.; Bokvist, K.; Olsen, H.L.; Madsbad, S.; Fredman, P.; Gromada, J. Sulfatide controls insulin secretion by modulation of ATP-sensitive K(+)-channel activity and Ca(2+)-dependent exocytosis in rat pancreatic beta-cells. *Diabetes* **2002**, *51*, 2514–2521. [CrossRef] [PubMed]

144. Blomqvist, M.; Carrier, M.; Andrews, T.; Pettersson, K.; Mansson, J.E.; Rynmark, B.M.; Fredman, P.; Buschard, K. In vivo administration of the C16:0 fatty acid isoform of sulfatide increases pancreatic sulfatide and enhances glucose-stimulated insulin secretion in Zucker fatty (fa/fa) rats. *Diabetes Metab. Res. Rev.* **2005**, *21*, 158–166. [CrossRef] [PubMed]

145. Veluthakal, R.; Palanivel, R.; Zhao, Y.; McDonald, P.; Gruber, S.; Kowluru, A. Ceramide induces mitochondrial abnormalities in insulin-secreting INS-1 cells: Potential mechanisms underlying ceramide-mediated metabolic dysfunction of the beta cell. *Apoptosis* **2005**, *10*, 841–850. [CrossRef]

146. Brozzi, F.; Nardelli, T.R.; Lopes, M.; Millard, I.; Barthson, J.; Igoillo-Esteve, M.; Grieco, F.A.; Villate, O.; Oliveira, J.M.; Casimir, M.; et al. Cytokines induce endoplasmic reticulum stress in human; rat and mouse beta cells via different mechanisms. *Diabetologia* **2015**, *58*, 2307–2316. [CrossRef]

147. Michaud, J.; Im, D.S.; Hla, T. Inhibitory role of sphingosine 1-phosphate receptor 2 in macrophage recruitment during inflammation. *J. Immunol.* **2010**, *184*, 1475–1483. [CrossRef]

148. Sanchez, T.; Skoura, A.; Wu, M.T.; Casserly, B.; Harrington, E.O.; Hla, T. Induction of vascular permeability by the sphingosine-1-phosphate receptor-2 (S1P2R) and its downstream effectors ROCK and PTEN. *Arterioscl. Thrombos Vascul. Biol.* **2007**, *27*, 1312–1318. [CrossRef]

149. Sanchez, T.; Thangada, S.; Wu, M.T.; Kontos, C.D.; Wu, D.; Wu, H.; Hla, T. PTEN as an effector in the signaling of antimigratory G protein-coupled receptor. *Proc. Natl. Acad. Sci. USA* **2005**, *102*, 4312–4317. [CrossRef]

150. Windh, R.T.; Lee, M.J.; Hla, T.; An, S.; Barr, A.J.; Manning, D.R. Differential coupling of the sphingosine 1-phosphate receptors Edg-1; Edg-3; and H218/Edg-5 to the G(i); G(q); and G(12) families of heterotrimeric G proteins. *J. Biol. Chem.* **1999**, *274*, 27351–27358. [CrossRef]

151. Gurgul-Convey, E.; Hanzelka, K.; Lenzen, S. Mechanism of prostacyclin-induced potentiation of glucose-induced insulin secretion. *Endocrinology* **2012**, *153*, 2612–2622. [CrossRef]

152. Liu, L.; Wang, C.; He, X.; Shang, W.; Bi, Y.; Wang, D. Long-term effect of FTY720 on lymphocyte count and islet allograft survival in mice. *Microsurgery* **2007**, *27*, 300–304. [CrossRef] [PubMed]

153. Yin, N.; Zhang, N.; Xu, J.; Shi, Q.; Ding, Y.; Bromberg, J.S. Targeting lymphangiogenesis after islet transplantation prolongs islet allograft survival. *Transplantation* **2011**, *92*, 25–30. [CrossRef] [PubMed]

154. Arntfield, M.E.; van der Kooy, D. Beta-Cell evolution: How the pancreas borrowed from the brain: The shared toolbox of genes expressed by neural and pancreatic endocrine cells may reflect their evolutionary relationship. *Bioessays* **2011**, *33*, 582–587. [CrossRef] [PubMed]

155. LeRoith, D.; Delahunty, G.; Wilson, G.L.; Roberts, C.T., Jr.; Shemer, J.; Hart, C.; Lesniak, M.A.; Shiloach, J.; Roth, J. Evolutionary aspects of the endocrine and nervous systems. *Recent Prog. Horm. Res.* **1986**, *42*, 549–587. [CrossRef]

156. Le Roith, D.; Shiloach, J.; Roth, J. Is there an earlier phylogenetic precursor that is common to both the nervous and endocrine systems? *Peptides* **1982**, *3*, 211–215. [CrossRef]

157. Rulifson, E.J.; Kim, S.K.; Nusse, R. Ablation of insulin-producing neurons in flies: Growth and diabetic phenotypes. *Science* **2002**, *296*, 1118–1120. [CrossRef]

158. Juan-Mateu, J.; Rech, T.H.; Villate, O.; Lizarraga-Mollinedo, E.; Wendt, A.; Turatsinze, J.V.; Brondani, L.A.; Nardelli, T.R.; Nogueira, T.C.; Esguerra, J.L.; et al. Neuron-enriched RNA-binding proteins regulate pancreatic beta cell function and survival. *J. Biol. Chem.* **2017**, *292*, 3466–3480. [CrossRef]

159. Taguchi, Y.; Allende, M.L.; Mizukami, H.; Cook, E.K.; Gavrilova, O.; Tuymetova, G.; Clarke, B.A.; Chen, W.; Olivera, A.; Proia, R.L. Sphingosine-1-phosphate phosphatase 2 regulates pancreatic islet beta-cell endoplasmic reticulum stress and proliferation. *J. Biol. Chem.* **2016**, *291*, 12029–12038. [CrossRef]

160. Qi, Y.; Chen, J.; Lay, A.; Don, A.; Vadas, M.; Xia, P. Loss of sphingosine kinase 1 predisposes to the onset of diabetes via promoting pancreatic beta-cell death in diet-induced obese mice. *FASEB J.* **2013**, *27*, 4294–4304. [CrossRef]

161. Song, Z.; Wang, W.; Li, N.; Yan, S.; Rong, K.; Lan, T.; Xia, P. Sphingosine kinase 2 promotes lipotoxicity in pancreatic beta-cells and the progression of diabetes. *FASEB J.* **2019**, *33*, 3636–3646. [CrossRef]

162. Kacheva, S.; Lenzen, S.; Gurgul-Convey, E. Differential effects of proinflammatory cytokines on cell death and ER stress in insulin-secreting INS1E cells and the involvement of nitric oxide. *Cytokine* **2011**, *55*, 195–201. [CrossRef] [PubMed]

163. Zhu, Q.; Kang, J.; Miao, H.; Feng, Y.; Xiao, L.; Hu, Z.; Liao, D.F.; Huang, Y.; Jin, J.; He, S. Low-dose cytokine-induced neutral ceramidase secretion from INS-1 cells via exosomes and its anti-apoptotic effect. *FEBS J.* **2014**, *281*, 2861–2870. [CrossRef] [PubMed]

164. Sysi-Aho, M.; Ermolov, A.; Gopalacharyulu, P.V.; Tripathi, A.; Seppanen-Laakso, T.; Maukonen, J.; Mattila, I.; Ruohonen, S.T.; Vahatalo, L.; Yetukuri, L.; et al. Metabolic regulation in progression to autoimmune diabetes. *PLoS Comput. Biol.* **2011**, *7*, e1002257. [CrossRef] [PubMed]

165. Brinkmann, V.; Billich, A.; Baumruker, T.; Heining, P.; Schmouder, R.; Francis, G.; Aradhye, S.; Burtin, P. Fingolimod (FTY720): Discovery and development of an oral drug to treat multiple sclerosis. *Nat. Rev. Drug Disc.* **2010**, *9*, 883–897. [CrossRef] [PubMed]

166. Jörns, A.; Akin, M.; Arndt, T.; Terbish, T.; Zu Vilsendorf, A.M.; Wedekind, D.; Hedrich, H.J.; Lenzen, S. Anti-TCR therapy combined with fingolimod for reversal of diabetic hyperglycemia by beta cell regeneration in the LEW.1AR1-iddm rat model of type 1 diabetes. *J. Mol. Med.* **2014**, *92*, 743–755. [CrossRef]

167. Jörns, A.; Rath, K.J.; Terbish, T.; Arndt, T.; Meyer Zu Vilsendorf, A.; Wedekind, D.; Hedrich, H.J.; Lenzen, S. Diabetes prevention by immunomodulatory FTY720 treatment in the LEW.1AR1-iddm rat despite immune cell activation. *Endocrinology* **2010**, *151*, 3555–3565. [CrossRef]

168. Penaranda, C.; Tang, Q.; Ruddle, N.H.; Bluestone, J.A. Prevention of diabetes by FTY720-mediated stabilization of peri-islet tertiary lymphoid organs. *Diabetes* **2010**, *59*, 1461–1468. [CrossRef]

169. Lenzen, S.; Arndt, T.; Elsner, M.; Wedekind, D.; Jörns, A. Rat models of human type 1 diabetes. *Meth. Mol. Biol.* **2020**, *2128*, 69–85. [CrossRef]

170. MacKinnon, A.C.; Farnworth, S.L.; Hodkinson, P.S.; Henderson, N.C.; Atkinson, K.M.; Leffler, H.; Nilsson, U.J.; Haslett, C.; Forbes, S.J.; Sethi, T. Regulation of alternative macrophage activation by galectin-3. *J. Immunol.* **2008**, *180*, 2650–2658. [CrossRef]

171. O'Farrell, A.M.; Liu, Y.; Moore, K.W.; Mui, A.L. IL-10 inhibits macrophage activation and proliferation by distinct signaling mechanisms: Evidence for Stat3-dependent and -independent pathways. *EMBO J.* **1998**, *17*, 1006–1018. [CrossRef]

172. Kaminski, A.; Kaminski, E.R.; Morgan, N.G. Pre-incubation with interleukin-4 mediates a direct protective effect against the loss of pancreatic beta-cell viability induced by proinflammatory cytokines. *Clin. Exp. Immunol.* **2007**, *148*, 583–588. [CrossRef] [PubMed]

173. Kaminski, A.; Welters, H.J.; Kaminski, E.R.; Morgan, N.G. Human and rodent pancreatic beta-cells express IL-4 receptors and IL-4 protects against beta-cell apoptosis by activation of the PI3K and JAK/STAT pathways. *Biosci. Rep.* **2009**, *30*, 169–175. [CrossRef] [PubMed]

174. Russell, M.A.; Morgan, N.G. The impact of anti-inflammatory cytokines on the pancreatic beta-cell. *Islets* **2014**, *6*, e950547. [CrossRef] [PubMed]

175. Chen, D.; Thayer, T.C.; Wen, L.; Wong, F.S. Mouse models of autoimmune diabetes: The nonobese diabetic (NOD) mouse. *Meth. Mol. Biol.* **2020**, *2128*, 87–92. [CrossRef]

176. Holm, L.J.; Haupt-Jorgensen, M.; Larsen, J.; Giacobini, J.D.; Bilgin, M.; Buschard, K. L-serine supplementation lowers diabetes incidence and improves blood glucose homeostasis in NOD mice. *PLoS ONE* **2018**, *13*, e0194414. [CrossRef]

177. Lemos, J.P.; Smaniotto, S.; Messias, C.V.; Moreira, O.C.; Cotta-de-Almeida, V.; Dardenne, M.; Savino, W.; Mendes-da-Cruz, D.A. Sphingosine-1-phosphate receptor 1 is involved in non-obese diabetic mouse thymocyte migration disorders. *Int. J. Mol. Sci.* **2018**, *19*, 1446. [CrossRef]

178. Srinivasan, S.; Bolick, D.T.; Lukashev, D.; Lappas, C.; Sitkovsky, M.; Lynch, K.R.; Hedrick, C.C. Sphingosine-1-phosphate reduces CD4+ T-cell activation in type 1 diabetes through regulation of hypoxia-inducible factor short isoform I.1 and CD69. *Diabetes* **2008**, *57*, 484–493. [CrossRef]

179. Buschard, K.; Hanspers, K.; Fredman, P.; Reich, E.P. Treatment with sulfatide or its precursor; galactosylceramide; prevents diabetes in NOD mice. *Autoimmunity* **2001**, *34*, 9–17. [CrossRef]

180. Sharif, S.; Delovitch, T.L. Regulation of immune responses by natural killer T cells. *Arch. Immunol. Ther. Exp.* **2001**, *49* (Suppl. 1), S23–S31.

181. Sharif, S.; Arreaza, G.A.; Zucker, P.; Mi, Q.S.; Sondhi, J.; Naidenko, O.V.; Kronenberg, M.; Koezuka, Y.; Delovitch, T.L.; Gombert, J.M.; et al. Activation of natural killer T cells by alpha-galactosylceramide treatment prevents the onset and recurrence of autoimmune Type 1 diabetes. *Nat. Med.* **2001**, *7*, 1057–1062. [CrossRef]

182. Mizuno, M.; Masumura, M.; Tomi, C.; Chiba, A.; Oki, S.; Yamamura, T.; Miyake, S. Synthetic glycolipid OCH prevents insulitis and diabetes in NOD mice. *J. Autoimmun.* **2004**, *23*, 293–300. [CrossRef] [PubMed]

183. Rhost, S.; Lofbom, L.; Mansson, J.; Lehuen, A.; Blomqvist, M.; Cardell, S.L. Administration of sulfatide to ameliorate type I diabetes in non-obese diabetic mice. *Scand. J. Immunol.* **2014**, *79*, 260–266. [CrossRef] [PubMed]

184. Holm, L.J.; Haupt-Jorgensen, M.; Giacobini, J.D.; Hasselby, J.P.; Bilgin, M.; Buschard, K. Fenofibrate increases very-long-chain sphingolipids and improves blood glucose homeostasis in NOD mice. *Diabetologia* **2019**, *62*, 2262–2272. [CrossRef] [PubMed]

185. Suzuki, Y.; Toda, Y.; Tamatani, T.; Watanabe, T.; Suzuki, T.; Nakao, T.; Murase, K.; Kiso, M.; Hasegawa, A.; Tadano-Aritomi, K.; et al. Sulfated glycolipids are ligands for a lymphocyte homing receptor; L-selectin (LECAM-1); Binding epitope in sulfated sugar chain. *Biochem. Biophys. Res. Commun.* **1993**, *190*, 426–434. [CrossRef] [PubMed]

186. Buschard, K.; Diamant, M.; Bovin, L.E.; Mansson, J.E.; Fredman, P.; Bendtzen, K. Sulphatide and its precursor galactosylceramide influence the production of cytokines in human mononuclear cells. *APMIS* **1996**, *104*, 938–944. [CrossRef]

187. Roeske-Nielsen, A.; Dalgaard, L.T.; Mansson, J.E.; Buschard, K. The glycolipid sulfatide protects insulin-producing cells against cytokine-induced apoptosis; a possible role in diabetes. *Diabetes Metab. Res. Rev.* **2010**, *26*, 631–638. [CrossRef] [PubMed]

188. McGlinchey, A.; Sinioja, T.; Lamichhane, S.; Sen, P.; Bodin, J.; Siljander, H.; Dickens, A.M.; Geng, D.; Carlsson, C.; Duberg, D.; et al. Prenatal exposure to perfluoroalkyl substances modulates neonatal serum phospholipids; increasing risk of type 1 diabetes. *Environ. Int.* **2020**, *143*, 105935. [CrossRef]

189. Barra, G.; Lepore, A.; Gagliardi, M.; Somma, D.; Matarazzo, M.R.; Costabile, F.; Pasquale, G.; Mazzoni, A.; Gallo, C.; Nuzzo, G.; et al. Sphingosine Kinases promote IL-17 expression in human T lymphocytes. *Sci. Rep.* **2018**, *8*, 13233. [CrossRef]

190. Grieco, F.A.; Moore, F.; Vigneron, F.; Santin, I.; Villate, O.; Marselli, L.; Rondas, D.; Korf, H.; Overbergh, L.; Dotta, F.; et al. IL-17A increases the expression of proinflammatory chemokines in human pancreatic islets. *Diabetologia* **2014**, *57*, 502–511. [CrossRef]

191. Zheng, Z.; Zheng, F. A complex auxiliary: IL-17/Th17 signaling during type 1 diabetes progression. *Mol. Immunol.* **2019**, *105*, 16–31. [CrossRef]

192. Buschard, K.; Holm, L.J.; Feldt-Rasmussen, U. Insulin independence in newly diagnosed type 1 diabetes patient following fenofibrate treatment. *Case Rep. Med.* **2020**, *2020*, 6865190. [CrossRef] [PubMed]

193. Mitrofanova, A.; Mallela, S.K.; Ducasa, G.M.; Yoo, T.H.; Rosenfeld-Gur, E.; Zelnik, I.D.; Molina, J.; Varona Santos, J.; Ge, M.; Sloan, A.; et al. SMPDL3b modulates insulin receptor signaling in diabetic kidney disease. *Nat. Commun.* **2019**, *10*, 2692. [CrossRef] [PubMed]

194. Kaul, K.; Apostolopoulou, M.; Roden, M. Insulin resistance in type 1 diabetes mellitus. *Metab. Clin. Exp.* **2015**, *64*, 1629–1639. [CrossRef]

195. Mehdi, A.M.; Hamilton-Williams, E.E.; Cristino, A.; Ziegler, A.; Bonifacio, E.; Le Cao, K.A.; Harris, M.; Thomas, R. A peripheral blood transcriptomic signature predicts autoantibody development in infants at risk of type 1 diabetes. *JCI Insight* **2018**, *3*. [CrossRef]

196. Giannopoulou, E.Z.; Winkler, C.; Chmiel, R.; Matzke, C.; Scholz, M.; Beyerlein, A.; Achenbach, P.; Bonifacio, E.; Ziegler, A.G. Islet autoantibody phenotypes and incidence in children at increased risk for type 1 diabetes. *Diabetologia* **2015**, *58*, 2317–2323. [CrossRef] [PubMed]

197. Ziegler, A.G.; Rewers, M.; Simell, O.; Simell, T.; Lempainen, J.; Steck, A.; Winkler, C.; Ilonen, J.; Veijola, R.; Knip, M.; et al. Seroconversion to multiple islet autoantibodies and risk of progression to diabetes in children. *JAMA* **2013**, *309*, 2473–2479. [CrossRef]

198. Chiarelli, F.; Giannini, C.; Primavera, M. Prediction and prevention of type 1 diabetes in children. *Clin. Pediatric Endocrinol. Case Rep. Clin. Investig.* **2019**, *28*, 43–57. [CrossRef]

199. Bleich, D.; Polak, M.; Chen, S.; Swiderek, K.M.; Levy-Marchal, C. Sera from children with type 1 diabetes mellitus react against a new group of antigens composed of lysophospholipids. *Horm. Res.* **1999**, *52*, 86–94. [CrossRef]

200. Misasi, R.; Dionisi, S.; Farilla, L.; Carabba, B.; Lenti, L.; Di Mario, U.; Dotta, F. Gangliosides and autoimmune diabetes. *Diabet Metab. Rev.* **1997**, *13*, 163–179. [CrossRef]

201. Papaccio, G. Gangliosides prevent insulitis but not islet B cell destruction in low-dose streptozocin-treated mice. *Diabetes Res. Clin. Pract.* **1993**, *19*, 9–15. [CrossRef]

202. Papaccio, G.; Chieffi Baccari, G.; Mezzogiorno, V. In vivo effect of gangliosides on non-obese diabetic mice. *Acta Anat.* **1993**, *147*, 168–173. [CrossRef] [PubMed]

203. Dotta, F.; Previti, M.; Lenti, L.; Dionisi, S.; Casetta, B.; D'Erme, M.; Eisenbarth, G.S.; Di Mario, U. GM2-1 pancreatic islet ganglioside: Identification and characterization of a novel islet-specific molecule. *Diabetologia* **1995**, *38*, 1117–1121. [CrossRef] [PubMed]

204. Dionisi, S.; Dotta, F.; Diaz-Horta, O.; Carabba, B.; Viglietta, V.; Di Mario, U. Target antigens in autoimmune diabetes: Pancreatic gangliosides. *Ann. Ist. Super. Sanita* **1997**, *33*, 433–435. [PubMed]

205. Boraska, V.; Torlak, V.; Skrabic, V.; Kacic, Z.; Jaksic, J.; Stipancic, G.; Uroic, A.S.; Markotic, A.; Zemunik, T. Glycosyltransferase B4GALNT1 and type 1 diabetes in Croatian population: Clinical investigation. *Clin. Biochem.* **2009**, *42*, 819–822. [CrossRef] [PubMed]

206. Bedia, C.; Badia, M.; Muixi, L.; Levade, T.; Tauler, R.; Sierra, A. GM2-GM3 gangliosides ratio is dependent on GRP94 through down-regulation of GM2-AP cofactor in brain metastasis cells. *Sci. Rep.* **2019**, *9*, 14241. [CrossRef] [PubMed]

207. Ghiasi, S.M.; Dahlby, T.; Hede Andersen, C.; Haataja, L.; Petersen, S.; Omar-Hmeadi, M.; Yang, M.; Pihl, C.; Bresson, S.E.; Khilji, M.S.; et al. Endoplasmic reticulum chaperone glucose-regulated protein 94 is essential for proinsulin handling. *Diabetes* **2019**, *68*, 747–760. [CrossRef] [PubMed]

208. Khilji, M.S.; Bresson, S.E.; Verstappen, D.; Pihl, C.; Andersen, P.A.K.; Agergaard, J.B.; Dahlby, T.; Bryde, T.H.; Klindt, K.; Nielsen, C.K.; et al. The inducible beta5i proteasome subunit contributes to proinsulin degradation in GRP94-deficient beta-cells and is overexpressed in type 2 diabetes pancreatic islets. *Am. Physiol. Endocr. Metab.* **2020**, *318*, E892–E900. [CrossRef]

209. Nikolaeva, S.; Bayunova, L.; Sokolova, T.; Vlasova, Y.; Bachteeva, V.; Avrova, N.; Parnova, R. GM1 and GD1a gangliosides modulate toxic and inflammatory effects of *E. coli* lipopolysaccharide by preventing TLR4 translocation into lipid rafts. *BBA* **2015**, *1851*, 239–247. [CrossRef]

210. Colli, M.L.; Nogueira, T.C.; Allagnat, F.; Cunha, D.A.; Gurzov, E.N.; Cardozo, A.K.; Roivainen, M.; Op de Beeck, A.; Eizirik, D.L. Exposure to the viral by-product dsRNA or Coxsackievirus B5 triggers pancreatic beta cell apoptosis via a Bim/Mcl-1 imbalance. *PLoS Pathog.* **2011**, *7*, e1002267. [CrossRef]

211. Dogusan, Z.; Garcia, M.; Flamez, D.; Alexopoulou, L.; Goldman, M.; Gysemans, C.; Mathieu, C.; Libert, C.; Eizirik, D.L.; Rasschaert, J. Double-stranded RNA induces pancreatic beta-cell apoptosis by activation of the toll-like receptor 3 and interferon regulatory factor 3 pathways. *Diabetes* **2008**, *57*, 1236–1245. [CrossRef]

212. Liu, D.; Cardozo, A.K.; Darville, M.I.; Eizirik, D.L. Double-stranded RNA cooperates with interferon-gamma and IL-1 beta to induce both chemokine expression and nuclear factor-kappa B-dependent apoptosis in pancreatic beta-cells: Potential mechanisms for viral-induced insulitis and beta-cell death in type 1 diabetes mellitus. *Endocrinology* **2002**, *143*, 1225–1234. [CrossRef] [PubMed]

213. Roivainen, M.; Rasilainen, S.; Ylipaasto, P.; Nissinen, R.; Ustinov, J.; Bouwens, L.; Eizirik, D.L.; Hovi, T.; Otonkoski, T. Mechanisms of coxsackievirus-induced damage to human pancreatic beta-cells. *J. Clin. Endocr. Metab.* **2000**, *85*, 432–440. [CrossRef] [PubMed]

214. Kutlu, B.; Darville, M.I.; Cardozo, A.K.; Eizirik, D.L. Molecular regulation of monocyte chemoattractant protein-1 expression in pancreatic beta-cells. *Diabetes* **2003**, *52*, 348–355. [CrossRef] [PubMed]

215. Lind, K.; Richardson, S.J.; Leete, P.; Morgan, N.G.; Korsgren, O.; Flodström-Tullberg, M. Induction of an antiviral state and attenuated coxsackievirus replication in type III interferon-treated primary human pancreatic islets. *J. Virol.* **2013**, *87*, 7646–7654. [CrossRef]

216. Richardson, S.J.; Leet, E.P.; Bone, A.J.; Foulis, A.K.; Morgan, N.G. Expression of the enteroviral capsid protein VP1 in the islet cells of patients with type 1 diabetes is associated with induction of protein kinase R and downregulation of Mcl-1. *Diabetologia* **2013**, *56*, 185–193. [CrossRef] [PubMed]

217. Richardson, S.J.; Willcox, A.; Bone, A.J.; Morgan, N.G.; Foulis, A.K. Immunopathology of the human pancreas in type 1 diabetes. *Sem. Immunopathol.* **2011**, *33*, 9–21. [CrossRef] [PubMed]

218. Yeung, W.C.; Al-Shabeeb, A.; Pang, C.N.; Wilkins, M.R.; Catteau, J.; Howard, N.J.; Rawlinson, W.D.; Craig, M.E. Children with islet autoimmunity and enterovirus infection demonstrate a distinct cytokine profile. *Diabetes* **2012**, *61*, 1500–1508. [CrossRef]

219. Osterbye, T.; Funda, D.P.; Fundova, P.; Mansson, J.E.; Tlaskalova-Hogenova, H.; Buschard, K. A subset of human pancreatic beta cells express functional CD14 receptors: A signaling pathway for beta cell-related glycolipids; sulfatide and beta-galactosylceramide. *Diabetes Metab. Res. Rev.* **2010**, *26*, 656–667. [CrossRef]

220. Andersson, K.; Buschard, K.; Fredman, P.; Kaas, A.; Lidström, A.M.; Madsbad, S.; Mortensen, H.; Jan-Eric, M. Patients with insulin-dependent diabetes but not those with non-insulin-dependent diabetes have anti-sulfatide antibodies as determined with a new ELISA assay. *Autoimmunity* **2002**, *35*, 463–468. [CrossRef]

221. Blomqvist, M.; Kaas, A.; Mansson, J.E.; Formby, B.; Rynmark, B.M.; Buschard, K.; Fredman, P. Developmental expression of the type I diabetes related antigen sulfatide and sulfated lactosylceramide in mammalian pancreas. *J. Cell. Biochem.* **2003**, *89*, 301–310. [CrossRef]

222. Kavishwar, A.; Medarova, Z.; Moore, A. Unique sphingomyelin patches are targets of a beta-cell-specific antibody. *J. Lipid Res.* **2011**, *52*, 1660–1671. [CrossRef] [PubMed]

223. Wigger, D.; Gulbins, E.; Kleuser, B.; Schumacher, F. Monitoring the sphingolipid de novo synthesis by stable-isotope labeling and liquid chromatography-mass spectrometry. *Front. Cell Dev. Biol.* **2019**, *7*, 210. [CrossRef] [PubMed]

224. Parashuraman, S.; D'Angelo, G. Visualizing sphingolipid biosynthesis in cells. *Chem. Phys. Lipid* **2019**, *218*, 103–111. [CrossRef] [PubMed]

225. Ravassard, P.; Hazhouz, Y.; Pechberty, S.; Bricout-Neveu, E.; Armanet, M.; Czernichow, P.; Scharfmann, R. A genetically engineered human pancreatic beta cell line exhibiting glucose-inducible insulin secretion. *J. Clin. Investig.* **2011**, *121*, 3589–3597. [CrossRef]

226. Gurgul-Convey, E.; Kaminski, M.T.; Lenzen, S. Physiological characterization of the human EndoC-betaH1 beta-cell line. *Biochem. Biophys. Res. Commun.* **2015**, *464*, 13–19. [CrossRef]

MDPI

Review

New Insights into the Role of Sphingolipid Metabolism in Melanoma

Lorry Carrié [1,†], Mathieu Virazels [1,†], Carine Dufau [1,†], Anne Montfort [1], Thierry Levade [1,2], Bruno Ségui [1,‡] and Nathalie Andrieu-Abadie [1,*,‡]

[1] Centre de Recherches en Cancérologie de Toulouse, Equipe Labellisée Fondation ARC, Université Fédérale de Toulouse Midi-Pyrénées, Université Toulouse III Paul-Sabatier, Inserm 1037, 2 avenue Hubert Curien, CS 53717, 31037 Toulouse CEDEX 1, France; lorry.carrie@inserm.fr (L.C.); mathieu.virazels@inserm.fr (M.V.); carine.dufau@inserm.fr (C.D.); anne.montfort@inserm.fr (A.M.); thierry.levade@inserm.fr (T.L.); bruno.segui@inserm.fr (B.S.)

[2] Laboratoire de Biochimie Métabolique, CHU, 31059 Toulouse, France

* Correspondence: nathalie.andrieu@inserm.fr; Tel.: +33-582-7416-20

† These authors contributed equally to this work.

‡ These authors jointly supervised this work.

Received: 17 July 2020; Accepted: 24 August 2020; Published: 26 August 2020

Abstract: Cutaneous melanoma is a deadly skin cancer whose aggressiveness is directly linked to its metastatic potency. Despite remarkable breakthroughs in term of treatments with the emergence of targeted therapy and immunotherapy, the prognosis for metastatic patients remains uncertain mainly because of resistances. Better understanding the mechanisms responsible for melanoma progression is therefore essential to uncover new therapeutic targets. Interestingly, the sphingolipid metabolism is dysregulated in melanoma and is associated with melanoma progression and resistance to treatment. This review summarises the impact of the sphingolipid metabolism on melanoma from the initiation to metastatic dissemination with emphasis on melanoma plasticity, immune responses and resistance to treatments.

Keywords: cancer; ceramide; gangliosides; immunotherapy; metastasis; phenotype switching; sphingosine 1-phosphate

1. Introduction

Cutaneous melanoma is a skin cancer whose incidence is increasing significantly worldwide (Figure 1). Even though melanoma is not frequent, accounting for less than 5% of skin cancers, it can be very aggressive and causes more than 75% of all skin cancer deaths [1] (Figure 1). Despite significant improvement of treatment strategies in the last decade, owing both to the emergence of BRAF- or MEK-targeted therapies and checkpoint blockade immunotherapies (i.e., anti-cytotoxic T-lymphocyte-associated antigen-4 (CTLA4) and anti-programmed cell death-1 (PD-1)), the prognosis for patients with metastatic melanoma remains uncertain, predominantly due to treatment failures and recurrences [2]. Fortunately, melanoma is usually curable by excisional surgery if detected at an early stage, with a high five-year survival rate [3,4]. Thus, a better understanding of melanoma progression processes, before dissemination, is a major public health issue in order to discern new therapeutic targets.

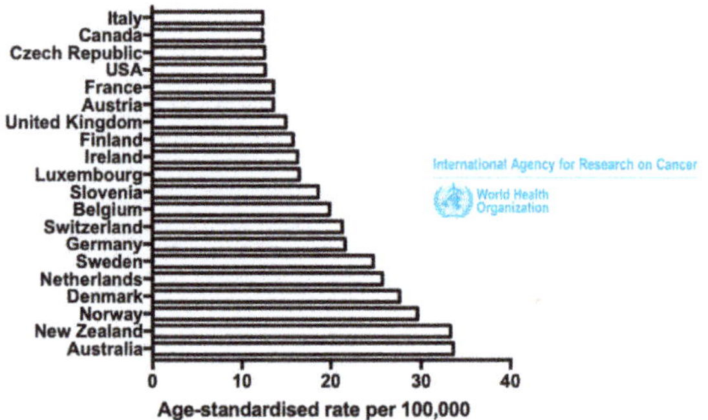

Figure 1. Cutaneous melanoma: the 19th most common cancer worldwide. Estimated age-standardised incidence rates of cutaneous melanoma in the most affected countries in 2018, for both sexes and all ages. Data from the International Agency for Research on Cancer (World Health Organisation).

Melanoma arises from melanocytes, i.e., melanin-producing neural crest-derived cells, which are located at the junction between the epidermis and the dermis [5]. The initial stage of melanomagenesis corresponds to a radial-growth phase (RGP), in which melanoma cells invade laterally but stay confined into the epidermis. This stage is followed by a vertical-growth phase (VGP), in which melanoma cells invade the dermis and are able to reach blood vessels. Then, an extravasation stage, corresponding to the release of melanoma cells from the blood circulation into new tissues, leads to the formation of metastatic niches [6–8]. Melanomagenesis requires at least two key events. The first is activating mutations in oncogenes such as *BRAF* or *NRAS*. BRAF mutations are found in 50% of melanoma patients (Figure 2) and the V600E mutation accounts for approximately 75% of all BRAF mutations detected in cutaneous melanoma [9]. Conversely, the most common NRAS mutations, i.e., Q61R and Q61K, affect about 20% of melanoma patients [10]. Since BRAF and NRAS mutations are mutually exclusives, these driver mutations lead to constitutive activation of the mitogen-activated protein kinase (MAPK) pathway and aberrant cancer cell proliferation in approximately 70% of patients (Figure 2) [11]. The second event is illustrated by the loss of expression of key tumour suppressor genes such as *PTEN* or cell cycle checkpoint regulators such as *CDKN2A*, which occur in 13% and 24% of patients, respectively (Figure 2). These genetic changes can bypass oncogene-induced senescence (OIS) processes and cause the immortalisation of tumour cells [4,7,12] (Figure 2).

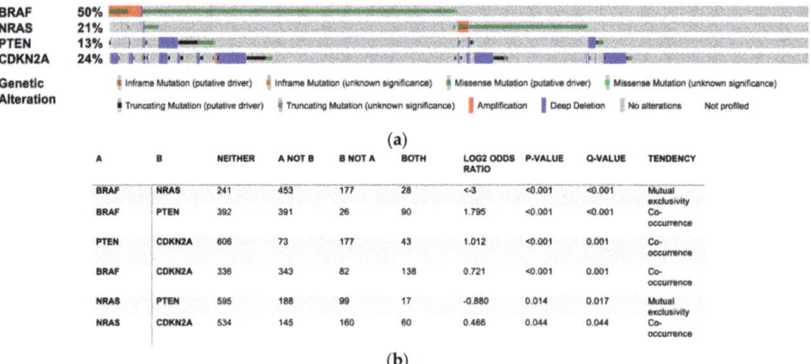

A	B	NEITHER	A NOT B	B NOT A	BOTH	LOG2 ODDS RATIO	P-VALUE	Q-VALUE	TENDENCY
BRAF	NRAS	241	453	177	28	<-3	<0.001	<0.001	Mutual exclusivity
BRAF	PTEN	392	391	26	90	1.795	<0.001	<0.001	Co-occurrence
PTEN	CDKN2A	606	73	177	43	1.012	<0.001	0.001	Co-occurrence
BRAF	CDKN2A	336	343	82	138	0.721	<0.001	0.001	Co-occurrence
NRAS	PTEN	595	188	99	17	-0.880	0.014	0.017	Mutual exclusivity
NRAS	CDKN2A	534	145	160	60	0.486	0.044	0.044	Co-occurrence

Figure 2. *BRAF*, *NRAS*, *PTEN* and *CDKN2A* are the most frequently mutated genes in cutaneous melanoma. Mutation rate, genetic alteration (**a**) and mutual exclusivity (**b**) for *BRAF*, *NRAS*, *PTEN* and *CDKN2A* mutations observed in 1635 samples from 1584 patients included in 12 studies analysed on cBioportal for cancer genomics (https://www.cbioportal.org).

Metabolic reprogramming is also crucial for melanomagenesis. Indeed, a shift from mitochondrial oxidative phosphorylation to cytoplasmic anaerobic glycolysis, known as Warburg effect, is required for metastatic dissemination [13]. The present review focuses on alterations in the metabolism of sphingolipids (SL). Interestingly, several key enzymes of the glycolytic pathway can be severely affected by changes in SL metabolism in melanoma. For instance, C16-ceramide, which is the major long-chain ceramide in melanocytes and melanoma cells, impairs pyruvate kinase, hexokinase and LDH activities, consequently altering cellular glycolysis and inhibiting melanoma progression [14].

How modulations of the SL metabolism affect dermatologic diseases have long been studied [15] and accumulating evidence demonstrates the presence of alterations in the ceramide metabolism in melanoma cells. This article aims at providing a comprehensive overview of the effects dysregulations of the SL metabolism have on melanomagenesis, melanoma progression, immunity and resistance to treatment, especially linked to the phenotype switching.

2. Alterations of Sphingolipid Metabolism in Melanoma

The main dysregulation affecting the SL metabolism in melanoma cells is a trend toward a reduction of ceramide, which promotes cell death (for review, see [16]). This is associated with changes in the expression and/or activity of a number of enzymes as well as the accumulation of tumour-promoting metabolites, which include sphingosine 1-phosphate (S1P) and gangliosides (Figure 3).

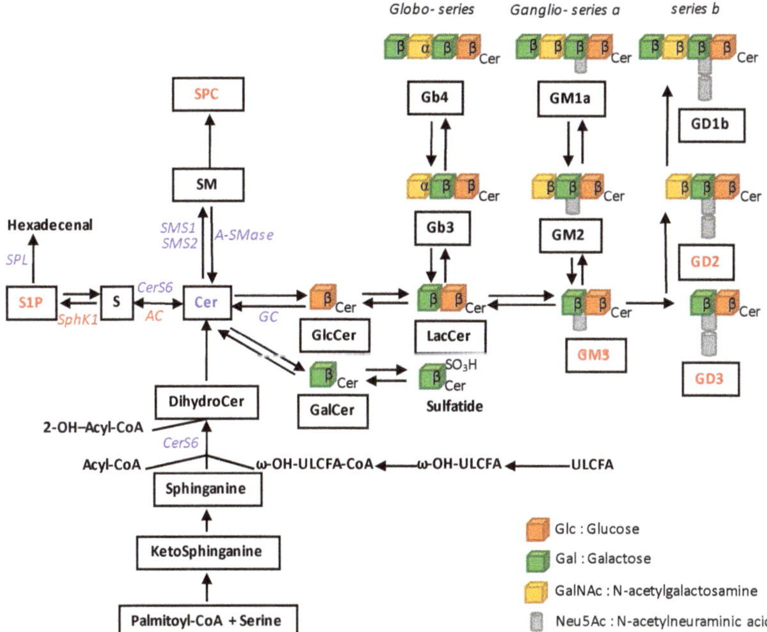

Figure 3. Multiple dysregulations of sphingolipid metabolism in melanoma. SL metabolites or SL-metabolising enzymes whose levels or expression are altered in melanoma, are mentioned. Decreases are indicated in blue and increases in red. AC, acid ceramidase; Cer, ceramide; CERS, ceramide synthase; CoA, coenzyme A; DihydroCer, dihydroceramide; GalCer, galactosylceramide; GC, glucosylceramidase; GlcCer, glucosylceramide; LacCer, lactosylceramide; S, sphingosine; S1P, sphingosine 1-phosphate; SM, sphingomyelin; SMases, sphingomyelinases; SMS, sphingomyelin synthase; SPC, sphingosylphosphorylcholine; SphK, sphingosine kinase; SPL, S1P lyase; ULCFA, ultralong chain fatty acids.

The impact of SL metabolism dysregulations in melanoma cell lines and/or patients is summarised in Table 1. For instance, a low expression of the ceramide synthase 6 (CerS6) in melanoma cells is related to malignant behaviours as demonstrated in WM35, WM451 and SKMEL28 human melanoma cell lines [17]. In addition, acid ceramidase (AC), encoded by the *ASAH1* gene, which hydrolyses ceramide into sphingosine, is expressed at high levels in melanocytes and proliferative melanoma cells, as observed in vitro as well as in biopsies from patients with stage II melanoma [18]. *ASAH1* expression was: (i) higher in human melanoma cell lines exhibiting a proliferative phenotype as compared to invasive ones; and (ii) reduced at the invasive front on tumour specimens from melanoma patients [19]. Sphingosine kinase 1 (SphK1), which phosphorylates sphingosine to produce sphingosine-1-phosphate (S1P), also shows increased expression and/or activity in melanoma cells compared to melanocytes, not only in human and murine cell lines [20,21] but also in human biopsies [22]. Collectively, these findings suggest a shift of the S1P-ceramide balance towards S1P production in melanoma cells. In accordance, the expression of *SGPL1* gene, encoding for S1P lyase (SPL), is downregulated in melanoma cell lines when compared to adult or juvenile melanocytes, suggesting that *SGPL1* might be downregulated during melanomagenesis [23].

Ceramide can also be converted into more complex SL, such as gangliosides. For instance, mono- and disialoganglioside levels are very high in human melanoma cells and tissues, especially GD3 [24]. Interestingly, levels of this latter ganglioside were correlated to the expression of the c-Yes tyrosine kinase whose activity is known to increase in melanoma cells as compared to melanocytes [25] and whose inhibition reduced the malignant potential of GD3⁺ melanoma cells only [26].

Alterations of the ceramide metabolism also include changes in the expression of sphingomyelin synthase 1 (SMS1), which is encoded by the *SGMS1* gene and catalyses the transformation of ceramide into sphingomyelin (SM) [27,28]. *SGMS1* is expressed at low levels in most of the human melanoma biopsies and low *SGMS1* expression is associated with a worse prognosis in metastatic melanoma patients. Of interest, a weak expression of *SMS1* was shown not to be associated with an intracellular accumulation of ceramide, most likely due to its conversion into glucosylceramide (GlcCer) through GlcCer synthase (GCS). Consequently, 6 out of 10 human melanoma cell lines exhibited higher levels of GlcCer than SM [29]. Moreover, SM can also be transformed into sphingosylphosphorylcholine (SPC) by a yet uncharacterised SM deacylase and SPC has been shown to stimulate regulators of melanomagenesis such as extracellular signal-regulated kinases (ERK), microphthalmia-associated transcription factor (MITF) and Akt/mTOR [30–33].

Table 1. Impact of SL and SL-metabolising enzymes dysregulations on melanoma cell lines and patients.

SL or SL-Metabolising Enzymes	Dysregulation	Cell Lines or Patients	Effects	Refs
CerS6	Decreased	WM35, WM451 and SKMEL28 human melanoma cells	Malignant behaviour	[17]
AC	Decreased	Proliferative and invasive human melanoma cells	Pro-invasive	[18,19]
SphK1	Increased	Murine and human melanoma cells and biopsies	Pro-tumoral Immunosuppressive signature	[20–22]
SPL	Decreased	Human melanoma cells	Resistance to chemotherapy Increased proliferation	[23]
GD3	Increased	GD3+ human melanoma cells with c-Yes inhibition	Reduced malignancy	[24–26]
SMS1	Decreased	Human biopsies	Worse prognosis	[29]

Table 1. *Cont.*

SL or SL-Metabolising Enzymes	Dysregulation	Cell Lines or Patients	Effects	Refs
SPC	Increased	Mel-Ab and human melanocytes	Stimulate melanomagenesis Hypopigmentation in melanocytes	[30–33]
A-SMase	Decreased	Primary melanomas and lymph node metastases Pigmented murine and human melanomas	Inverse correlation with melanin content	[34,35]

Finally, the expression of acid sphingomyelinase (A-SMase), which hydrolyses SM into ceramide, has been shown to be higher in benign nevi than in primary melanomas, and further reduced in the lymph-node metastases [34]. Moreover, a lower expression/activity of A-SMase was observed in hyper-pigmented murine and human melanomas as compared to the hypo-pigmented ones, suggesting an inverse correlation between A-SMase expression/activity and melanin content. In accordance, exogenous C2-ceramide decreased melanin content in melanocytes [35].

3. Role of the Sphingolipid Metabolism in Melanomagenesis

3.1. Do SL Metabolism Alterations Increase the Risk to Develop Melanoma?

A genome-wide association study has identified the 1q21.3 chromosomal region, containing LASS2 gene that encodes ceramide synthase 2 (CerS2), as a locus predisposing to cutaneous melanoma [36]. These observations suggest the involvement of some genetic determinants of the ceramide metabolism in melanoma predisposition. Interestingly, a defect in glucosylceramidase 1 (GBA1) resulting in Gaucher inborn disorder (GD), owing to a defect in GlcCer hydrolysis into ceramide, is associated with an increased risk of malignancies, including melanoma [37,38]. Indeed, accumulation of GlcCer as well as glucosylsphingosine, which arises from the cleavage of excess GlcCer by AC [39,40], occurs in macrophages and could severely alter the immune and inflammatory responses. This could create a favourable microenvironment to promote melanomagenesis (for review see [38]).

Another hypothetical link between glucosylceramidase (GC) and melanoma development is autophagy, which can be either cell protective, promoting their survival, or lethal, via induction of a programmed-cell death mechanism [41,42]. Defective autophagy has been reported in models of GC or saposin C deficiency [43]. Accumulation of GlcCer was associated to autophagy dysfunction in a drosophila model of GD that lacked the two fly GBA1 orthologues [44]. In accordance, GlcCer accumulation was also associated with autophagy impairment and defective autophagosome-lysosome fusion, resulting in autophagosome accumulation in induced pluripotent stem cells (iPSCs) derived from patients with GD [45]. Moreover, an hyperactivation of the autophagic inhibitor mTOR and a downregulation of the master regulator of lysosome function TFEB were reported in human neuroglioma cells treated with the GC inhibitor conduritol B epoxide [46]. Numerous studies have shown that impaired autophagy can favour melanoma development. Indeed, the activation of mTOR was associated with poor prognosis in melanoma patients [47]. In addition, ERK-induced TFEB phosphorylation impaired expression of autophagy-lysosome target genes in BRAF-mutated melanoma, which elicited the formation of TGF-β-dependent metastases [48].

All these findings indicate a possible link between GC deficiency and melanomagenesis, that may result from an altered immune response or disturbed autophagy. The underlying mechanisms remain, however, to be determined.

3.2. Sphingolipid Metabolism Modulates Melanoma Cell Proliferation and Survival

Transformation of normal melanocytes into melanoma cells is mediated by the activation of growth stimulatory pathways, typically leading to cellular proliferation as well as the inactivation of apoptotic and tumour suppressor pathways. The RAS-RAF-MEK-ERK pathway is one of the most important signalling pathways involved in melanoma cell growth and survival [49,50]. A constitutive activation of BRAF, mostly due to the substitution of valine by glutamic acid at position 600 (also known as V600E), affects about 50% of melanoma patients [9,51–53] (Figure 2). Therapies targeting the BRAF V600E mutation help advanced melanoma patients live longer [54,55]. Moreover, co-administration of BRAF (vemurafenib) and MEK (cobimetinib) inhibitors improves the progression-free survival [56] and extends the five-year overall survival by ~40% [57]. Unfortunately, most patients, including those who experience an initial tumour regression, exhibit disease progression within 6–8 months following the initiation of targeted therapy [58].

Multiple lines of evidence indicate that some SL-metabolising enzymes regulate melanoma cell proliferation and survival. First, SphK1 expression and activity are induced by ERK1/2 and AKT in numerous mammalian cells [59–61], including melanoma cells [20,21]. Moreover, SphK1 knockdown by siRNA decreased anchorage-dependent and -independent growth of human melanoma cells [20]. Similarly, targeting SphK1 using shRNA in B16F10 [62] or Yumm 1.7 [22] murine melanoma cells reduced tumour growth in syngeneic mice. Accordingly, the SphK1 inhibitor SKI-I, which increases the intracellular ceramide levels and decreases S1P levels in melanoma cells, resulted in a cell cycle arrest between G2/M and S phases as well as increased apoptotic cell death, caspase-3 activation and nuclear accumulation of cleaved PARP [20]. The intraperitoneal administration of SKI-I in mice harbouring melanoma also decreased tumour growth [20,22]. Consistently, the growth of B16F10 tumours is impaired in SphK1$^{-/-}$ mice as compared to wild-type animals [21].

S1P, which is mainly produced by SphK1 in melanoma cells, conveys oncogenic signals as an intracellular second messenger via a ligation of a family of G-protein coupled receptors (S1P1-5) expressed both on the malignant and their neighbouring cells [63]. We recently demonstrated that the melanoma cell-autonomous survival in response to the BRAF inhibitor vemurafenib is mediated by S1P1 and S1P3 [64].

Moreover, AC, which is expressed at high levels in proliferative melanoma cells, may also contribute to melanoma cell proliferation and survival. Indeed, AC inhibition by siRNA dramatically reduced the number of 501mel melanoma cells, as shown using short-term cell growth and colony formation assays [19]. Similarly, CRISPR/Cas9-mediated AC ablation in A375 melanoma cells blocked G1/S cell cycle progression, promoted senescence and apoptosis, resulting in reduced cell growth. These cells were unable to form spheroids and showed a lower replication rate as well as a decreased in their invasive capacity compared to controls. Mechanistically, AC ablation resulted in the accumulation of the saturated C14-, C16- and C18-ceramides and is accompanied by the down-regulation of MYC, CDK1, CHK1 and AKT [65]. In accordance, the inhibition of AC activity with a chemically stable AC inhibitor, named compound ARN14988, sensitised proliferative melanoma cells to the cytotoxic actions of various anti-tumour agents [18]. In line with these observations, we previously reported that the cytotoxic action of dacarbazine was accompanied with AC proteolysis in human melanoma cells [66]. Of interest, confocal immunofluorescence analyses revealed the nuclear localisation of AC in normal melanocytes, a phenomenon not observed in melanoma cells, suggesting that AC could activate proliferation pathways only in tumour cells [18].

By reducing ceramide levels, AC, in concert with SphK1, favours melanoma cell proliferation. This was confirmed using short-chain C2-ceramide, which was reported to inhibit AKT and ERK activation as well as proliferation in Malme-3M melanoma cells [67]. In addition, the GCS inhibitor PDMP, which increases intracellular C16-ceramide levels, inhibited cell proliferation, migration and invasion of WM35 and WM451 human melanoma cells. The effect of PDMP was associated with the inhibition of key enzymes from the glycolysis pathway including the pyruvate kinase, hexokinase and lactic acid dehydrogenase. Strikingly, the treatment of melanoma cells with exogenous C16-ceramide neither altered melanoma cell growth nor migration and invasion. In contrast, exogenous C16-ceramide was shown to promote glycolysis. This opposite effect could be explained by the reduction of endogenous C16-ceramide levels, which was induced by exogenous C16-ceramide treatment [14].

Finally, GCS, which catalyses the first committed step in the synthesis of most glycosphingolipids, i.e., the transfer of glucose to ceramide to form GlcCer, is also able to control tumorigenic capability of melanoma cells. Indeed, antisense oligonucleotide targeting the Ugcg gene, encoding GCS, reduced tumorigenicity of MEB4 murine melanoma cells [68]. Similarly, the inhibitor of GCS, OGT2378, inhibited MEB4 melanoma tumour growth in a syngeneic, orthotopic murine model [69]. In agreement with these findings, we previously showed that overexpression of GBA2 in melanoma cells, an enzyme able to degrade GlcCer into ceramide, reduced tumour cell growth both in vitro and in vivo by triggering ER stress-induced apoptosis [70]. Of note, GBA2 gene is downregulated in melanoma cells as compared to melanocytes [70]. Altogether, these observations demonstrate that the transformation of ceramide into GlcCer facilitates melanoma cell proliferation.

Interestingly, sialic acid-containing glycosphingolipids, i.e., gangliosides, can also regulate melanoma cell proliferation. First, treatment of SKMEL-28 melanoma cells with the anti-GD3 antibody R24 reduced their growth in vitro and decreased their tumorigenicity when injected in immunodeficient mice [71]. Second, Furukawa et al. demonstrated that GD3 increased the proliferation of GD3 synthase-overexpressing melanoma cells [72]. In these settings, GD3 mediated the convergence of several pro-tumoral signals, including those induced by hepatocyte growth factor (HGF) and the receptor tyrosine kinase c-MET, notably promoting cell proliferation [73].

Altogether, these data illustrate that SL metabolism alterations, which redirect ceramide metabolism towards S1P or GD3 production, can promote melanoma cell proliferation and survival in response to drugs.

4. Role of the Sphingolipid Metabolism in Melanoma Progression

4.1. SL Metabolism Regulates Melanoma Cell Adhesion

Cell junctions, which are crucial for the communication between neighbouring cells and with the extracellular environment, can be divided into three major classes: anchoring junctions (including adherens junctions, desmosomes, hemidesmosomes and focal adhesions), tight junctions and gap junctions. In the epidermis, cadherins are the major adhesion molecules, especially involved in the composition of desmosomes and adherens junctions [74], whereas integrins are the major component of hemidesmosomes [75]. Among cadherins, E-cadherin mediates the adhesion between melanocytes and keratinocytes allowing keratinocytes to control cell growth and dendricity of melanocytes [76]. E-cadherin expression is lost in melanoma cells during the first steps of tumour progression [77]. Interestingly, when E-cadherin expression is restored, keratinocytes recover control of melanoma cells thus preventing tumour progression [78].

Here, we review studies documenting the role of the SL metabolism in the control of the expression of adhesion molecules as well as melanoma cell adhesion capacity. Previous reports have shown that E-cadherin loss was observed in SphK1-overexpressing cancer cells [79,80]. S1P-induced E-cadherin downregulation could be mediated by S1P2 and S1P3, as shown in alveolar epithelial cells [81] and lung fibroblasts [82]. This phenomenon could be indirect in melanoma cells as the SphK1/S1P pathway is able to stimulate TGF-β1 production [62], which may trans-activate S1P2 and S1P3 [82]. Interestingly, overexpression of S1P2, but not S1P1, in B16F10 melanoma cells resulted in the inhibition of the small GTPase Rac activity as well as tumour progression in mice [83]. Importantly, Rac is crucial to create E-cadherin-dependent cell-cell contacts [84].

Moreover, downregulation of AC in melanoma cells induced E-cadherin loss and, inversely, increased expression of the epithelial-mesenchymal transition (EMT)-associated protein TWIST1, which is in accordance with a more aggressive phenotype [19].

Finally, GD3 was shown to favour the recruitment of integrins through glycolipid-enriched microdomains in GD3 synthase-overexpressing melanoma cells. Under these conditions, melanoma cell adhesion to the extracellular matrix (ECM) was increased [85]. Similarly, Ohmi et al. demonstrated that cell adhesion increased in GD2 synthase-overexpressing melanoma cells as compared control cells [86].

Thus, SL alterations appear to impact on melanoma cell adhesion, particularly through E-cadherin loss, which promotes melanoma progression.

4.2. SL Metabolism as a Determinant of Melanoma Plasticity

To colonise distant organs, tumour cells need, besides losing their cell junctions, to acquire invasive capacities. In skin cancers, EMT plays a key role in this process. This fundamental mechanism allows epithelial cells to gain mesenchymal features, increasing their migration and invasion abilities (for review, see [87]). However, unlike other skin cancers, melanoma does not arise from epithelial cells but from neural crest-derived melanocytes. For this reason, EMT stricto sensu cannot be considered in melanoma progression. Nevertheless, an EMT-like phenomenon has been described, in which melanoma cells can dynamically and reversibly switch between a proliferative and an invasive state; this is known as "phenotype switching". Indeed, the microarray analysis of DNA from different human melanoma cell lines allowed Hoek et al. to determine a transcriptional signature representative of metastatic cell behaviour. The authors indeed demonstrated that MITF is one crucial actor in this switch, particularly in maintaining the proliferative state [88].

MITF represents a melanocytic lineage-specific transcription factor that regulates melanocyte differentiation, function and survival as well as melanoma progression [89]. MITF regulates pigment cell-specific transcription of genes encoding melanogenic enzymes such as TYR, DCT and TYRP1, as well as proteins involved in melanosome formation and maturation such as Melan-A, Premelanosome Protein and G Protein-Coupled Receptor 143. As a matter of fact, MITF expression and activity are modulated by a range of activators and suppressors operating at transcriptional, post-transcriptional and post-translational levels (for review, see [90]). MITF function has been tightly connected to melanoma cell plasticity. It is now well accepted that melanoma cells expressing moderate to high levels of MITF proliferate rapidly and are poorly invasive, whereas melanoma cells characterised by low MITF levels grow more slowly and are more invasive. Thereby, low levels of MITF correlate with a worse prognosis for melanoma patients [19,89,91].

Interestingly, numerous studies showed that the SL metabolism regulates MITF expression. Firstly, A-SMase expression has been shown to induce ERK-mediated MITF degradation by the proteasome. Therefore, the loss of A-SMase observed during melanoma progression accounts for the upregulation of MITF as well as for some of its downstream targets CDK2, Bcl-2 and c-MET [34]. Secondly, AC ablation by the CRISPR-Cas9 technology, which is associated with the accumulation of long-chain saturated ceramides, led to a strong downregulation of MITF expression in human A375 melanoma cells, reducing their ability to form cancer-initiating cells and to undergo self-renewal [65]. Furthermore, exogenous addition of C2-ceramide was shown to reduce MITF expression in human melanocytes [35]. Reciprocally, we demonstrated that MITF expression increased in AC-overexpressing melanoma cells. However, at variance with Lai et al., we observed that melanoma cells expressing AC at high levels displayed a proliferative phenotype as compared to cells with low expression of AC that exhibited high mobility and gain of mesenchymal features [19]. Moreover, using a ChIP-Seq database, we identified AC as a new target of MITF, demonstrating that MITF and AC are part of a positive feedback loop.

SL metabolism could also modulate MITF levels by acting on signalling pathways known to regulate its expression in melanoma cells. For instance, canonical Wnt signalling through the Wnt/β-catenin pathway is a critical activator of MITF expression in melanoma cells [92], and deactivation correlates with a higher metastatic potential [88]. Interestingly, exogenous sphingosine has been shown to reduce nuclear and cytosolic β-catenin expression in SW480 and T84 colon cancer cells [93]. In accordance, pharmacological inhibition of SphK1 with SKI-II was associated to a decreased β-catenin expression in human hepatoma carcinoma [94]. As expected, FTY720, a sphingosine analogue known to inhibit SphK1 [95], led to the reduction of β-catenin as well as MITF expression in melanocytes [96].

Moreover, a switch in EMT-associated transcription factors (EMT-TFs) occurs in melanoma and drives tumour progression. This dynamic network includes Snail, Zeb and Twist families, which are major repressors of epithelial genes, and, conversely, major activators of mesenchymal genes (for review, see [87]). In particular, a reduced expression of ZEB2 and SNAIL2 in favour of an increased expression in ZEB1 and TWIST1 was linked to E-cadherin loss, increased invasion properties and poor clinical outcomes in human melanoma [97]. The EMT-TFs switch was associated with a reduction of MITF expression. Indeed, ZEB1 and TWIST1 have been shown to downregulate MITF whereas SNAIL2 or ZEB2 induce MITF expression, demonstrating that these EMT-TFs act as key players in melanoma phenotype switching.

Recent studies identified a strong connection between SL metabolism and EMT-TFs. Indeed, reduced expression of CerS6, which decreased the levels of intracellular C16-ceramide, was associated with an increased expression of SNAIL2 in SW480 colon cancer cells [98]. Hence, by controlling the expression of EMT-TFs or by altering plasma membrane fluidity, C16-ceramide could affect cancer cell motility [99]. Another SL-metabolising enzyme could also account for the effects of ceramide on EMT-TFs. Indeed, SMS2, which produces SM from ceramide, seems to stimulate the expression of mesenchymal markers and enhance migration and invasion of MCF-7 and MDA-MB-231 breast cancer cell lines. Interestingly, SMS2 expression was higher in metastatic breast cancer than in non-metastatic tumours. Mechanistically, SMS2 was shown to activate the canonical TGFβ/SMAD signalling pathway leading to the expression of its downstream target Snail [100]. In addition, previous studies have demonstrated that some ganglioside-metabolising enzymes are connected with EMT-FTs and gangliosides play a critical role in EMT [101]. For instance, ZEB1 was reported to be a direct regulator of the GM3 synthase gene (St3gal5) in mammary epithelial NM18 cells. ZEB1 also impaired the expression of miR-200a, a microRNA targeting the 3′UTR GM3 synthase mRNA. Knockdown of GM3 synthase partly mimicked the effects of ZEB1 inhibition, leading to increased expression of cell junction components such as E-cadherin as well as intercellular adhesion [102]. Moreover, overexpressing TWIST or SNAIL1 in transformed human mammary epithelial cells enhanced the expression of GD3 synthase [103]. GD3 synthase knockdown reduced breast cancer stem cell-associated properties and completely abrogated tumour formation in vivo. In accordance, other studies have shown that ceramide glycosylation by GCS was enhanced in breast [104] and colon [105] cancer stem cells and GCS inhibition significantly decreased the expression of ZEB1 and β-catenin [105]. Whether GCS, a ganglioside-metabolising enzyme, CerS6 or SMS2 modulates EMT-TFs in melanoma cells remains to be evaluated.

Furthermore, numerous studies have shown that TGF-β is a strong promoter of EMT in many tumours [106,107], including melanoma [88]. TGF-β signalling inhibits MITF expression through PAX3 repression and GLI2 activation. Many studies reported that the SphK1/S1P and the TGF-β signalling are interconnected. Indeed, through its binding to S1P receptors, S1P was shown to promote TGF-β receptor trans-activation leading to Smad phosphorylation and cell migration [108–110]. Additionally, in different cancers including melanoma, S1P was reported to increase TGF-β expression and secretion [62,111,112]. Inversely, TGF-β was able to increase SphK expression and activity, which were essential to control the effects of TGF-β on extracellular matrix remodelling, cell migration and invasion [113,114].

Finally, TEAD transcription factors (TEADs) were identified as key regulators of the invasive state in melanoma [115]. TEADs need coactivators such as YAP and TAZ, which are known effectors of the Hippo pathway. This pathway has been shown to modulate Wnt and TGF-β signalling and confer pro-invasive properties in melanoma [116]. Strikingly, multiple studies have identified S1P as an activator of YAP through S1P2 signalling [117–119]. Similarly, a recent study demonstrated that inhibition of SphK1 using the PF-543 inhibitor could inhibit TGFβ-induced activation of YAP [120]. As anticipated from its close structural similarity with S1P, SPC also regulated the Hippo pathway via S1P2, in a rather unclear manner as it could both inhibit and activate YAP [121].

To summarise, tight connections have been reported between SL metabolism and key players of the phenotype switching in melanoma including transcription factors such as MITF, EMT-TFs, TEADs and fundamental signalling pathways such as Wnt, TGF-β and Hippo. These observations further highlight the importance of the SL metabolism in melanoma progression.

4.3. SL Metabolism as a Major Regulator of Melanoma Aggressiveness

As discussed above, melanoma aggressiveness depends on the balance between its proliferative potential and migratory/invasive properties. S1P was reported either to activate or to inhibit melanoma cell migration depending on S1P receptor subtypes. Indeed, whereas S1P inhibited cell migration, with the concomitant inhibition of Rac and stimulation of RhoA, in S1P2-expressing B16F10 cells, it stimulated cell migration of S1P1-overexpressing cells, demonstrating a receptor subtype-specific action of S1P on melanoma cells [83]. The inhibitory effects of S1P were reversed by the S1P2-selective antagonist JTE013, which stimulated Rac and migration of B16F10 cells overexpressing either S1P1 or S1P3 [122]. Similar results were obtained in B16F10 cells treated with SPC instead of S1P [122].

In addition, AC overexpression in melanoma cells decreased tumour cell motility, whereas AC silencing had the opposite effect, as was observed for MITF [19]. We recently demonstrated that low AC expression was associated to increased FAK phosphorylation and relocation at focal adhesions instead of cytoplasm. This phenomenon led to increased expression of integrin β5 (ITGβ5) and integrin αV (ITGαV), which play a critical roles in the migratory and invasive capacity of cancer cells. As a result, the melanoma invasive behaviour induced by AC inhibition was reduced using an ITGαVβ5 blocking antibody [19].

Another study reported that lung metastases were reduced in A-SMase-deficient mice injected with B16-F10 melanoma cells. Treating B16F10 cells with exogenous A-SMase or C16-ceramide before inoculation restored lung metastatic lesions in A-SMase-deficient mice. Mechanistically, melanoma cells were shown to activate A-SMase in platelets, leading to ceramide production, which favoured the clustering and activation of α5β1 integrins at the surface of melanoma cells and therefore tumour cell adhesion in the lungs [123]. As the expression of melanoma A-SMase negatively correlates with tumour aggressiveness [34], one could speculate that A-SMase expression acts as a key factor that controls melanoma cell invasion and adhesion into the metastatic niches.

Gangliosides also likely contribute to melanoma cell dissemination. As a matter of fact, the level of GM3, which has been described as one of the major gangliosides in melanoma [124], increased in murine metastatic melanoma [125], suggesting a role for GM3 in tumour aggressiveness. Indeed, the addition of GM3 to B16LuF1 melanoma cells, i.e., B16-melanoma cells of lower metastatic potential to lungs, increased their dissemination capacity once injected in mice [126]. Liu et al. also described that de-N-acetyl GM3 (d-GM3), a derivative of ganglioside GM3, was mainly found in metastatic melanomas but not in benign nevi or most primary melanomas. d-GM3 expressing melanoma cells possess increased migratory and invasive capacities as compared to melanoma cells lacking d-GM3. Mechanistically, d-GM3 stimulated MMP-2 expression via the urokinase-like plasminogen activator (uPA) receptor [127].

Some studies indicated that GD3 also stimulates melanoma cell invasion. Indeed, human melanoma GD3-positive cells showed a markedly increased cell invasion potential as compared to GD3-negative cells. The invasive activity induced by GD3 was shown to be mediated by p130Cas or paxillin, two components of the focal adhesion cytoskeleton [72]. More recently, Ohmi et al. compared the effect of GD3 and GD2 on melanoma progression. Using GD3-^{high} or GD2-^{high} melanoma cells, obtained by overexpressing the respective glycosyltransferases involved in their production, they demonstrated that GD2 enhanced the adhesion properties of melanoma cells, while GD3 stimulated their invasive capacities. These findings led the authors to propose that GD2 would rather act at the primary and metastatic sites in order to promote cell proliferation and dissemination, while GD3 would favour melanoma cell invasion in order to reach a metastatic niche [86].

Finally, emerging literature indicates that tumour exosomes actively participate in tumour invasiveness and favour the formation of pre-metastatic niches in various types of cancer, including melanoma [128–130]. Exosomes are small extracellular vesicles (EVs) that originate from the fusion of multivesicular bodies with the plasma membrane and convey their cargo towards target cells. They carry transmembrane and cytosolic proteins, DNA and small RNAs [131]. In vitro, melanoma-derived exosomes were shown to promote the EMT-like processes in primary melanocytes.

This effect occurred in an autocrine/paracrine fashion and was mediated by the microRNA Let-7i [132]. In mice, B16-F10-derived exosomes demonstrated preferential homing to lymph nodes and facilitated the seeding of intravenously injected parental cells [133]. In particular, Peinado et al. showed that, through the receptor c-MET, B16-F10-derived exosomes can educate bone marrow-derived cells, promoting angiogenesis, vascular leakiness, the growth of primary tumours and metastasis [128].

It is well known that ceramide, generated by the neutral SMase2 (nSMase2) on the cytosolic leaflet of endosomal membranes, is involved in the budding of exosomes [134]. Mechanistically, the cone-shaped structure of ceramide could induce spontaneous negative curvature by creating an area difference between the membrane leaflets [135]. Moreover, a decrease in the activity of nSMase2 induced by the GW4869 compound, resulted in the reduced release of exosomal miRNAs [136]. By controlling exosomal miRNA secretion, nSMase2 is able to promote angiogenesis as well as metastasis [137]. Furthermore, Kajimoto et al. also showed that S1P, produced by SphK2 but not Sphk1, can regulate the cargo content in exosomes [138,139] probably through the G$\beta\gamma$ subunit of Gi proteins coupled with S1P1 [140]. Whether nSMase2- and/or SphK2-dependent exosome formation modulates melanoma progression remains to be investigated.

All these findings clearly establish a close relationship between SL metabolism and melanoma invasion and suggest that SL metabolism could be therapeutically targeted in order to improve the outcome of melanoma patients.

5. Role of SL Metabolism in the Immune Response to Melanoma

Melanoma cells harbour an aberrant antigenic profile, which allows for an anti-tumour immune response [141]. Despite their high immunogenicity, melanoma cells eventually evade the immune system, grow and metastasise [142,143]. A growing body of evidence in the literature indicates that SLs regulate various immune processes. Thus, deciphering the role of SL metabolism in melanoma immune escape is of great clinical interest.

5.1. S1P in Lymphocyte Traffic and Differentiation

Lymphocytes sense S1P concentration via S1P1 [144,145] allowing their egress from the thymus and lymph nodes to peripheral tissues [146]. S1P1 expression is modulated cyclically during lymphocyte traffic, depending on the local S1P concentration: it is downregulated in the blood, upregulated in secondary lymphoid organs (SLO) and downregulated again in the lymph [147]. CD69, an early activation marker on lymphocyte surface, induces S1P1 internalisation and degradation [148], sequestering lymphocytes in SLO [149] and peripheral tissues [150]. S1P1 downregulation is necessary to establish a long-term memory in the skin [151,152] and CD69 is one of the markers (with CD103) for tissue-resident memory cells (TRM) [153], which play a critical role in melanoma immunosurveillance [154–156].

Drouillard et al. proposed that the S1P1/S1P2 ratio dictates the migration of T cells, as S1P2 inhibited the chemo-attraction of peripheral T cells [157]. Sic et al. reported that human B cells also migrate towards S1P in an S1P1-dependent manner that is inhibited by CD69 expression [158]. Interestingly, egress of natural killer (NK) cells from the bone marrow and SLO is mediated by the expression of S1P5 [159,160], which is regulated by T-box transcription factor TBX21 [161]. Increased S1P5 expression as well as downregulation of CXCR4 during NK differentiation is necessary for their egress from the bone marrow [162].

SphK activity and S1P1 expression were shown to mediate differentiation of CD4$^+$ T cells to Th1 cells and inhibit induced Treg (iTreg) generation [163]. In accordance, in T cell-specific S1p1-transgenic mice, S1P1 oriented the differentiation of CD4$^+$ towards the Th1 lineage when antigen-activated. Moreover, S1P1 overexpression impaired the maintenance of Foxp3 expression in naïve TGF-β-treated CD4$^+$ T cells. The differentiation of naïve CD4$^+$ T cells towards Th1 or iTreg appeared to be reciprocal, driven by the S1P1-mTOR axis, and dependent on the SphK activity as demonstrated with the SphK inhibitors N,N,-dimethylsphingosine (DMS) and SKI. Similarly, CD4$^+$ T cells deficient for Sphk1 showed a lesser Foxp3 expression when cultured with IL-2 and TGF-β [164].

5.2. S1P Impairs the Immune Response in Melanoma

We reported that melanoma SphK1 plays a key role in the recruitment and phenotypic switch of TAM notably promoting their commitment to a pro-tumoral M2-like phenotype [62]. Moreover, we recently showed that high SphK1 expression in melanoma cells was associated with shorter progression-free and overall survivals in melanoma patients treated with anti-PD-1-based immunotherapy. In mice, SphK1 knockdown in melanoma tumours potently reduced the production of a number of immunosuppressive cytokines including TGF-β [22,62], limiting Treg tumour infiltration. Under these conditions, the response of melanoma cells to anti-PD-1 or anti-CTLA-4-based immunotherapy highly increased [22]. Interestingly, Chakraborty et al. also reported that tumour-infiltrating lymphocytes display higher Sphk1 expression as compared to splenocytes in B16-F10-bearing mice [164]. Melanoma antigen-specific T cells deficient for Sphk1 (pMel-SphK1$^{-/-}$ T cells) were shown to maintain a central memory phenotype and have a reduced propensity to differentiate into Treg as compared to wild-type T cells (pMel T cells). Tumour growth was significantly slower upon adoptive transfer of pMel-SphK1$^{-/-}$ T cells, as compared to mice injected with wild-type pMel T cells [164].

Recent findings also show that S1P secretion, via the S1P transporter Spinster Homologue 2 (Spns2), reduced CD8$^+$ T cell function and therefore promoted lung metastasis. Indeed, the deletion of Spns2, either globally or in a lymphatic endothelial cell-specific manner, was associated with an increased ratio of effector T-cells to immunosuppressive Tregs, in the lungs of Spns2-deficient mice intravenously injected with B16F10 or HCmel12 murine melanoma cells. This resulted in a reduced pulmonary metastatic burden as compared to what was observed in wild-type animals [165].

5.3. Ceramide and Its Derivatives in the Immune Response

Several studies have shown that ceramide metabolism could regulate the immune response in different melanoma models. Firstly, A-SMase-deficient B16-F1 melanoma cells engrafted in mice display an inflammatory TME and are infiltrated by high levels MDSCs and Tregs and low levels of DCs. A-SMase overexpression in these cells restores CD8$^+$ and CD4$^+$ T cells and DCs infiltration while reducing levels of infiltrating MDSCs and Tregs, thereby reducing tumour growth [166].

Secondly, KRN7000, a synthetic alpha-galactosylceramide [167], showed promising results in enhancing NK, NKT, CD8$^+$ T cells and M1 infiltration in the syngeneic murine B16 metastatic melanoma model [168] but further investigation needs to be conducted.

Thirdly, it was recently reported that liposomes enriched in C2-ceramide were shown to reprogram the immune TME in a PKCζ-dependent manner in B16-F10-bearing mice. Under these conditions, TAMs shifted towards an M1 phenotype and CD8$^+$ and Th1 cells infiltration was enhanced while intra-tumour MDSCs and Tregs levels were reduced [169].

Finally, gangliosides also represent attractive targets for immunotherapies as they are abundant in melanoma cells [24] and recognised by NKT cells [170]. GM2 [171] and N-glycolyl GM3 (NGcGM3), in particular, have been the main gangliosides used as targets for the development of anti-melanoma antibodies and vaccine [172,173]. In addition, gangliosides are also known to be shed by melanoma cells [174,175] and exert a pro-apoptotic effect on DCs [176,177]. 3F8, a monoclonal anti-GD2 mAb, demonstrated anti-proliferative and pro-apoptotic activity in human melanoma cell lines [178] but clinical studies focused on neuroblastoma [179] and medulloblastoma [180] patients. More recently, GD2 has been considered as a promising target for the treatment of melanoma patients using either CAR-T cell therapy [181] or the immunocytokine hu14.18-IL2, an anti-GD2 humanised mAb linked to two molecules of IL-2 and administered to patients with recurrent resectable stage III or IV melanoma [182].

5.4. Melanoma-Derived Exosomes Are Vectors of Immunosuppression

As mentioned above, Trajkovic et al. showed that the production of ceramide by nSMase 2 was part of the mechanisms involved in exosome budding [134]. In addition to favouring progression and metastasis, melanoma-derived small extracellular vesicles, often defined as exosomes, also carry immunomodulatory molecules that impair anti-tumour immune responses. In vitro, exosomes released from B16F0 murine melanoma cells inhibited the proliferation of T cells by delivering PTPN11(SHP-2) mRNA and protein [183]. B16F10-derived exosomes can also activate the mitochondrial apoptotic pathway of CD4$^+$ T cells in vitro and in vivo, thereby increasing tumour growth and reducing T cell infiltration [184]. The authors proposed that the miRNA cargo of exosomes (e.g., miR-690) inhibited the expression of anti-apoptotic proteins in CD4$^+$ T cells. In addition, small EVs produced by A375 human melanoma cells were shown to be able to reduce MHC class I molecules to the cell surface of primary human monocytes and THP-1 cells and downregulate the expression of endogenous MHC class I and II molecules in DCs [185]. Exosomes from metastatic melanoma-derived cell lines inhibited TCR signalling and cytokines secretion in CD8$^+$ T cells by transferring an array of miRNA cargo [186]. Moreover, tumour-derived exosomes harvested from melanoma patients' plasma were shown to induce the apoptosis, inhibit proliferation and decrease the activation of CD8$^+$ T cells. They were also able to downregulate NKG2D expression on NK cells [187]. In metastatic melanoma patients, circulating exosomal PD-L1 suppressed CD8$^+$ T cell activity. The authors reported that the pre-treatment level of circulating exosomal PD-L1 was a better predictor of clinical response to anti-PD-1 therapy than total circulating PD-L1 [188]. Importantly, Poggio et al. showed that Pdl1 knockout or exosome depletion by knocking out Smpd3, the gene encoding nSMase2, was sufficient to restore the anti-tumour immune response and to induce an efficient anti-tumour immune-memory response in the murine TRAMP-C2 prostate cancer model [189].

The relationship between SL metabolism and exosome-mediated immunosuppression in melanoma is not well understood, yet it could be a major mechanism of resistance to immunotherapy and thus deserves further investigation.

6. Potential Therapeutic Strategies for Melanoma Patients

Historically, when tumour resection was not possible or failed, chemotherapy was used to treat melanoma. dacarbazine (DTIC) has been approved as first-line treatment for advanced-stage melanoma and has remained for more than 30 years the standard chemotherapy despite no clear overall survival benefits [190–192].

The identification of melanoma driven mutations such as BRAF V600E allowed for a real breakthrough in the treatment of patients with metastatic melanoma. The emergence of BRAF targeted agents such as vemurafenib [54] and dabrafenib [55] allowed tremendous progresses in the field of personalised medicine and demonstrated survival benefits in metastatic melanoma patients as compared to dacarbazine-treated patients. Subsequently, the MEK inhibitor was also approved as treatment for this pathology as it showed survival benefits for patients displaying the BRAF V600E mutation [193]. Combination therapy using BRAF and MEK inhibitors such as cobimetinib [56] is nowadays one of the first line treatment for patients with BRAF V600E metastatic melanoma. This treatment results in higher rates as well as extended duration of response and decreases the cutaneous toxicities observed with the BRAF inhibitor monotherapy. Unfortunately, such therapeutic approaches remain constrained by the inevitable emergence of resistance to single-pathway blockade [194].

Immune checkpoint blockade (ICB) was the first therapeutic strategy to provide sustained responses and survival for advanced melanoma patients, even after treatment discontinuation [195–197]. Administering monoclonal antibodies targeting the immune checkpoint PD-1, alone or in combination with anti-CTLA-4 blocking antibodies is, to date, the standard of care for advanced melanoma patients. Independently of BRAF mutation status, patients treated with the anti-PD-1 and anti-CTLA-4 combo achieve at five years a progression-free survival and an overall survival of 36% and 52%, respectively [198]. Unfortunately, half of the patients do not respond or develop early resistance to ICB and exhibit severe immune-related adverse events (IRAE) [199]. Although treatment discontinuation due to adverse events seems not to affect the outcome for patients treated with ICB combination therapy [198], IRAEs tend to be associated with a better outcome for patients treated with anti-PD1 monotherapy [200].

Targeted therapies and ICB have deeply changed therapeutic management of patients with metastatic melanoma but all these therapeutic approaches still need improvement. Understanding the mechanisms that underlie resistance to these treatments is of utmost importance to improve the outcome of melanoma patients.

Here, we review some studies, which identified alterations in the SL metabolism as a cause of melanoma resistance to treatment (Table 2) and other studies using SL-related molecules as monotherapy or combined therapy to fight melanoma (Table 3).

Table 2. SL-metabolising enzymes regulate the response of melanoma to therapy.

Targeted SL-Metabolising Enzyme	Melanoma Cells	Experimental Strategy	Treatment	Effects on Drug Sensitivity	Refs
A-SMase	B16-W6_pSIL10	shRNA	Cisplatin (chemotherapy)	Low A-SMase is associated with reduced mTOR-related autophagy and resistance to cisplatin	[201]
AC	A375	AC overexpression	Dacarbazine (chemotherapy)	AC overexpression confers resistance to dacarbazine	[66]
AC	G361 A375	ARN14988 ARN398 (AC inhibitor)	5-FU (chemotherapy)	AC inhibition sensitises G361 cells (proliferative phenotype) but not A375 cells (invasive phenotype) to chemotherapeutic drugs	[18]
SphK1	SK-Mel-28 A375	FTY720	Cisplatin (chemotherapy)	SK1 inhibition increases cisplatin-induced apoptosis through a downregulation of the PI3K/AKT/mTOR pathway and decreases EGFR expression	[202]
SphK1	UACC 903	siRNA	Staurosporine (Apoptosis inducing agent)	Downregulation of Sphk1 sensitises cells to staurosporine-induced apoptosis through AKT inhibition, and G0/G1 phase cell cycle arrest	[20]
SphK1	A375 (overexpression) Mel-2a (downregulation)	SphK1 overexpression or downregulation	Doxorubicin (chemotherapy)	Sphk1 overexpression induces resistance to doxorubicin-induced apoptosis whereas its downregulation by siRNA increases melanoma cell sensitivity to the treatment	[203]
SphK1	WM115 SK-Mel-28	FTY720	Vemurafenib (BRAF inhibitor)	SK1 inhibition increases vemurafenib-induced apoptosis	[204]
SphK1	WM9	SKI-I	Vemurafenib (BRAF inhibitor)	Sphk1 inhibition blocks BRAFi-resistant melanoma cell growth by reducing MITF and Bcl-2 expression	[64]
SphK1	B16-F10	PF-543	ICB Adoptive transfer of melanoma antigen-specific T cells	Sphk1 inhibition in T cells maintains Tcm phenotype, reduces Treg induction and synergises with anti-PD1 treatment	[164]

Table 2. *Cont.*

Targeted SL-Metabolising Enzyme	Melanoma Cells	Experimental Strategy	Treatment	Effects on Drug Sensitivity	Refs
SphK1	Yumm 1.7	shRNA	ICB	SphK1 downregulation enhances ICB therapy efficacy by reducing Treg infiltration	[22]
GCS	B16	PDMP	Genistein (Apoptosis inducing agent)	Ceramide accumulation enhances genistein-induced apoptosis and growth inhibition through JNK activation and AKT inhibition	[205]

Table 3. SL-related molecules used therapy in melanoma.

SL-Related Treatment	Models	Associated Drug	Effects	Refs
Nanoliposomal ceramide	UACC 903 cells 1205 Lu cells Xenografts in nude mice	Sorafenib	Inhibition of melanoma cell growth by targeting both PI3K and MAPK signalling	[206]
Nanoliposomal ceramide	1205 Lu cells In vitro experiments	None	Reduction of integrin affinity and inhibition of melanoma cell migration through PI3K and PKCζ, tumour-suppressive activities	[207]
KRN7000	B16 melanoma cell graft intravenously injected in mice	None	Increase of lifespan of mice	[208]
OGT2378 (GCS inhibitor)	B16 derived MEB4 melanoma cell graft in female C57BL/6 mice	None	Inhibition of tumour growth and reduction of established tumours	[69]
Intra-muscular GM3/VSSP vaccine	Phase I clinical trial: 26 patients with advanced (stage III and IV) melanoma	Adjuvant Montanide Isa 51	GM3/VSSP vaccine induces anti-GM3 IgM response in 44% of patients. Serum reactivity against melanoma cells and tumour biopsies is reported	[209]
L612-HuMAb (Human monoclonal antibody that binds to GM3)	Phase I clinical trial: 9 patients with advanced (stage IV) melanoma	None	L612 HuMAb induces significant antitumour activity in melanoma patients	[210]

7. Conclusions

As developed in this review, changes in SL metabolism that contribute to melanomagenesis, tumour progression and therapeutic resistance are multiple (Figure 4).

The action of sphingolipids (via the enzymes that control their metabolism, and transporters) is likely mediated by modifications in key regulatory processes including the phenotypic switch and EV-mediated cell-cell communication. Interestingly, these metabolic alterations could be envisioned as potential biomarkers and be exploited to better characterise tumour progression in melanoma patients. As a matter of fact, we identified that a reduced expression of SMS1 was significantly associated with a worse prognosis in metastatic melanoma [29]. AC was also identified as a potential biomarker for the prognosis of melanoma [211]. Moreover, we recently demonstrated that human invasive melanoma cells had lower AC levels and activity than proliferative melanoma cells [19]. In accordance, high AC expression was observed in node-negative stage II melanomas [18].

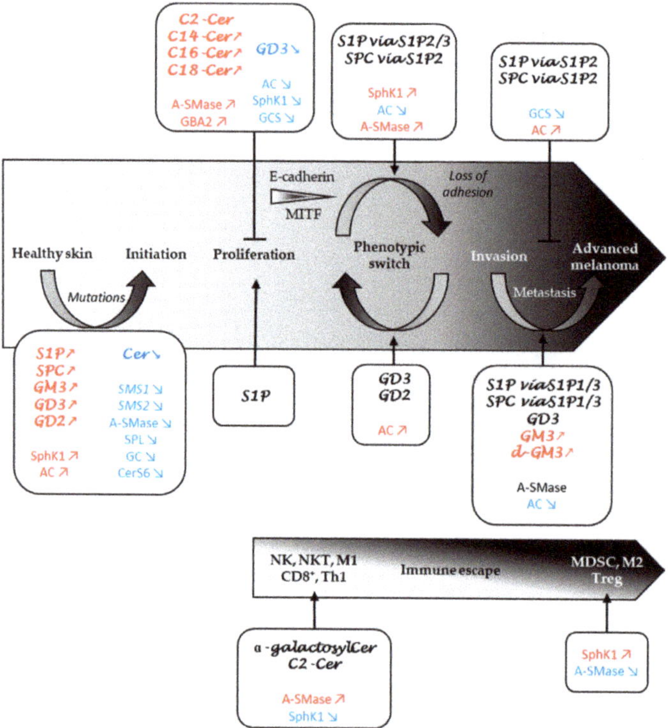

Figure 4. Role of sphingolipid metabolism in melanoma progression and immune response. SL metabolites and SL-metabolising enzymes whose levels and expression are increased, decreased or implicated are marked in red, blue or black, respectively. AC, acid ceramidase; CD8+, CD8+ T cells; Cer, ceramide; CerS, ceramide synthase; GC, glucosylceramidase; GCS, glucosylceramide synthase; M1, M1 macrophages; M2, M2 macrophages; MDSC, myeloid-derived suppressor cells; NK, natural killer cells; NKT, natural killer T cells; S1P, sphingosine 1-phosphate; S1P1/2/3, S1P receptor type 1/2/3; SMases, sphingomyelinases; SMS, sphingomyelin synthase; SPC, sphingosylphosphorylcholine; SphK, sphingosine kinase; SPL, S1P lyase; Th1, Th1 CD4+ T cells; Treg, tegulatory T cells.

It is also interesting to note that a strong association between increased serum levels of gangliosides and high Breslow index or high Clark level as well as the presence of ulceration has been reported in melanoma patients, suggesting that circulating gangliosides may serve as potential markers for melanoma staging [212].

Monitoring the expression of SL-metabolising enzymes as well as SL levels could also be used to track the response to therapy in melanoma. Indeed, we previously showed that AC expression was associated to the response of melanoma cells to dacarbazine. Whereas overexpression of AC conferred resistance to dacarbazine, AC downregulation sensitised tumour cells to the drug [66]. DTIC triggered AC degradation and this effect was accompanied with an increased ceramide/S1P ratio. Our recent results also reveal that a distinct SL profile, i.e., a tendency for increased very long-chain ceramide species, was observed in the plasma of patients with melanoma who achieve a response to a BRAF-targeted therapy as compared with patients with progressive disease [64]. Finally, we recently discovered that melanoma patients with low SphK1 expression had significantly longer progression-free survival and overall survival than those with high SphK1 expression and patients with high SphK1 expression mostly failed to respond to anti-PD-1 therapy. These findings support the hypothesis that SphK1 expression represents a potential biomarker to predict tumour progression and resistance to anti-PD-1 in metastatic melanoma patients [22].

It would now be of great interest to evaluate the possible association between these SL metabolic alterations and the mutation status of oncogenes such as BRAF or NRAS as well as immune responses in metastatic melanoma patients. This will be performed in patients treated with anti-PD-1 in combination or not with anti-CTLA-4 in a prospective clinical trial (IMMUSPHINX: NCT03627026) we are currently conducting in our institute.

Funding: This research was funded by INSERM, Paul Sabatier University, Fondation Association pour la Recherche sur le Cancer [B. Ségui (R19179BB) and N. Andrieu-Abadie (R18167BB)], Société Française de Dermatologie [N. Andrieu-Abadie (R18126BB)], Fondation Toulouse Cancer Santé [B. Ségui (R19225BB)] and Institut National du Cancer [N. Andrieu-Abadie (R19243BP)]. The work also received funding from the Transcan-2 Research Program, which is a transnational R&D program jointly funded by national funding organisations within the framework of the ERA-NET Transcan-2 [N. Andrieu-Abadie (TRANS201601250)]. The APC was funded by Fondation Association pour la Recherche sur le Cancer [N. Andrieu-Abadie (R18167BB)]. L.C. is a recipient of a fellowship from Fondation pour la Recherche Médicale.

Conflicts of Interest: The authors declare no conflict of interest.

References

1. Gershenwald, J.E.; Guy, G.P. Stemming the Rising Incidence of Melanoma: Calling Prevention to Action. *JNCI J. Natl. Cancer Inst.* **2016**, *108*. [CrossRef]

2. Zhu, Z.; Liu, W.; Gotlieb, V. The rapidly evolving therapies for advanced melanoma—Towards immunotherapy, molecular targeted therapy, and beyond. *Crit. Rev. Oncol. Hematol.* **2016**, *99*, 91–99. [CrossRef] [PubMed]

3. Crowson, A.N.; Magro, C.M.; Mihm, M.C. Prognosticators of melanoma, the melanoma report, and the sentinel lymph node. *Mod. Pathol.* **2006**, *19*, S71–S87. [CrossRef] [PubMed]

4. Shain, A.H.; Bastian, B.C. From melanocytes to melanomas. *Nat. Rev. Cancer* **2016**, *16*, 345–358. [CrossRef]

5. Chang, A.E.; Karnell, L.H.; Menck, H.R. The National Cancer Data Base report on cutaneous and noncutaneous melanoma: A summary of 84,836 cases from the past decade. The American College of Surgeons Commission on Cancer and the American Cancer Society. *Cancer* **1998**, *83*, 1664–1678. [CrossRef]

6. Herlyn, M.; Thurin, J.; Balaban, G.; Bennicelli, J.L.; Bondi, E.; Guerry, D.; Nowell, P.; Clark, W.H.; Koprowski, H. Characteristics of Cultured Human Melanocytes Isolated from Different Stages of Tumor Progression. *Cancer Res.* **1985**, *45*, 8.

7. Liu, J.; Fukunaga-Kalabis, M.; Li, L.; Herlyn, M. Developmental pathways activated in melanocytes and melanoma. *Arch. Biochem. Biophys.* **2014**, *563*, 13–21. [CrossRef]

8. Hsu, M.-Y.; Meier, F.; Herlyn, M. Melanoma development and progression: A conspiracy between tumor and host. *Differentiation* **2002**, *70*, 522–536. [CrossRef]

9. Davies, H.; Bignell, G.R.; Cox, C.; Stephens, P.; Edkins, S.; Clegg, S.; Teague, J.; Woffendin, H.; Garnett, M.J.; Bottomley, W.; et al. Mutations of the BRAF gene in human cancer. *Nature* **2002**, *417*, 949–954. [CrossRef]

10. Akbani, R.; Akdemir, K.C.; Aksoy, B.A.; Albert, M.; Ally, A.; Amin, S.B.; Arachchi, H.; Arora, A.; Auman, J.T.; Ayala, B.; et al. Genomic Classification of Cutaneous Melanoma. *Cell* **2015**, *161*, 1681–1696. [CrossRef]

11. Hodis, E.; Watson, I.R.; Kryukov, G.V.; Arold, S.T.; Imielinski, M.; Theurillat, J.-P.; Nickerson, E.; Auclair, D.; Li, L.; Place, C.; et al. A Landscape of Driver Mutations in Melanoma. *Cell* **2012**, *150*, 251–263. [CrossRef] [PubMed]

12. Lian, C.G.; Murphy, G. The Genetic Evolution of Melanoma. *N. Engl. J. Med.* **2016**, *374*, 993–996. [CrossRef]

13. Abildgaard, C.; Guldberg, P. Molecular drivers of cellular metabolic reprogramming in melanoma. *Trends Mol. Med.* **2015**, *21*, 164–171. [CrossRef] [PubMed]

14. Liu, R.; Cao, K.; Tang, Y.; Liu, J.; Li, J.; Chen, J.; Wang, S.; Chen, Z.; Zhou, J. C16:0 ceramide effect on melanoma malignant behavior and glycolysis depends on its intracellular or exogenous location. *Am. J. Transl. Res.* **2020**, *12*, 1123–1135. [PubMed]

15. Borodzicz, S.; Rudnicka, L.; Mirowska-Guzel, D.; Cudnoch-Jedrzejewska, A. The role of epidermal sphingolipids in dermatologic diseases. *Lipids Health Dis.* **2016**, *15*, 13. [CrossRef]

16. Garandeau, D.; Mrad, M.; Levade, T.; Perrotta, C.; Andrieu-Abadie, N.; Diab-Assaf, M. Dysregulation of Sphingolipid Metabolism in Melanoma: Roles in Pigmentation, Cell Survival and Tumor Progression. In *Bioactive Sphingolipids in Cancer Biology and Therapy*; Hannun, Y.A., Luberto, C., Mao, C., Obeid, L.M., Eds.; Springer International Publishing: Cham, Switzerland, 2015; pp. 123–139. ISBN 978-3-319-20749-0.

17. Tang, Y.; Cao, K.; Wang, Q.; Chen, J.; Liu, R.; Wang, S.; Zhou, J.; Xie, H. Silencing of CerS6 increases the invasion and glycolysis of melanoma WM35, WM451 and SK28 cell lines via increased GLUT1-induced downregulation of WNT5A. *Oncol. Rep.* **2016**, *35*, 2907–2915. [CrossRef]

18. Realini, N.; Palese, F.; Pizzirani, D.; Pontis, S.; Basit, A.; Bach, A.; Ganesan, A.; Piomelli, D. Acid Ceramidase in Melanoma: EXPRESSION, LOCALIZATION, AND EFFECTS OF PHARMACOLOGICAL INHIBITION. *J. Biol. Chem.* **2016**, *291*, 2422–2434. [CrossRef]

19. Leclerc, J.; Garandeau, D.; Pandiani, C.; Gaudel, C.; Bille, K.; Nottet, N.; Garcia, V.; Colosetti, P.; Pagnotta, S.; Bahadoran, P.; et al. Lysosomal acid ceramidase ASAH1 controls the transition between invasive and proliferative phenotype in melanoma cells. *Oncogene* **2019**, *38*, 1282–1295. [CrossRef]

20. Madhunapantula, S.V.; Hengst, J.; Gowda, R.; Fox, T.E.; Yun, J.K.; Robertson, G.P. Targeting sphingosine kinase-1 to inhibit melanoma: Targeting SPHK1 in melanomas. *Pigment Cell Melanoma Res.* **2012**, *25*, 259–274. [CrossRef]

21. Albinet, V.; Bats, M.-L.; Huwiler, A.; Rochaix, P.; Chevreau, C.; Ségui, B.; Levade, T.; Andrieu-Abadie, N. Dual role of sphingosine kinase-1 in promoting the differentiation of dermal fibroblasts and the dissemination of melanoma cells. *Oncogene* **2014**, *33*, 3364–3373. [CrossRef]

22. Imbert, C.; Montfort, A.; Fraisse, M.; Marcheteau, E.; Gilhodes, J.; Martin, E.; Bertrand, F.; Marcellin, M.; Burlet-Schiltz, O.; Peredo, A.G.d.; et al. Resistance of melanoma to immune checkpoint inhibitors is overcome by targeting the sphingosine kinase-1. *Nat. Commun.* **2020**, *11*, 437. [CrossRef] [PubMed]

23. Colie, S.; Van Veldhoven, P.P.; Kedjouar, B.; Bedia, C.; Albinet, V.; Sorli, S.-C.; Garcia, V.; Djavaheri-Mergny, M.; Bauvy, C.; Codogno, P.; et al. Disruption of Sphingosine 1-Phosphate Lyase Confers Resistance to Chemotherapy and Promotes Oncogenesis through Bcl-2/Bcl-xL Upregulation. *Cancer Res.* **2009**, *69*, 9346–9353. [CrossRef] [PubMed]

24. Portoukalian, J.; Zwingelstein, G.; Doré, J.F. Lipid composition of human malignant melanoma tumors at various levels of malignant growth. *Eur. J. Biochem.* **1979**, *94*, 19–23. [CrossRef] [PubMed]

25. Loganzo, F.; Dosik, J.S.; Zhao, Y.; Vidal, M.J.; Nanus, D.M.; Sudol, M.; Albino, A.P. Elevated expression of protein tyrosine kinase c-Yes, but not c-Src, in human malignant melanoma. *Oncogene* **1993**, *8*, 2637–2644. [PubMed]

26. Hamamura, K.; Tsuji, M.; Hotta, H.; Ohkawa, Y.; Takahashi, M.; Shibuya, H.; Nakashima, H.; Yamauchi, Y.; Hashimoto, N.; Hattori, H.; et al. Functional Activation of Src Family Kinase Yes Protein Is Essential for the Enhanced Malignant Properties of Human Melanoma Cells Expressing Ganglioside GD3. *J. Biol. Chem.* **2011**, *286*, 18526–18537. [CrossRef] [PubMed]

27. Huitema, K.; van den Dikkenberg, J.; Brouwers, J.F.H.M.; Holthuis, J.C.M. Identification of a family of animal sphingomyelin synthases. *EMBO J.* **2004**, *23*, 33–44. [CrossRef]

28. Yamaoka, S.; Miyaji, M.; Kitano, T.; Umehara, H.; Okazaki, T. Expression Cloning of a Human cDNA Restoring Sphingomyelin Synthesis and Cell Growth in Sphingomyelin Synthase-defective Lymphoid Cells. *J. Biol. Chem.* **2004**, *279*, 18688–18693. [CrossRef]

29. Bilal, F.; Montfort, A.; Gilhodes, J.; Garcia, V.; Riond, J.; Carpentier, S.; Filleron, T.; Colacios, C.; Levade, T.; Daher, A.; et al. Sphingomyelin Synthase 1 (SMS1) Downregulation Is Associated With Sphingolipid Reprogramming and a Worse Prognosis in Melanoma. *Front. Pharmacol.* **2019**, *10*, 443. [CrossRef]

30. Higuchi, K.; Kawashima, M.; Ichikawa, Y.; Imokawa, G. Sphingosylphosphorylcholine is a Melanogenic Stimulator for Human Melanocytes. *Pigment Cell Res.* **2003**, *16*, 670–678. [CrossRef]

31. Kim, D.-S.; Park, S.-H.; Kwon, S.-B.; Park, E.-S.; Huh, C.-H.; Youn, S.-W.; Park, K.-C. Sphingosylphosphorylcholine-induced ERK activation inhibits melanin synthesis in human melanocytes. *Pigment Cell Res.* **2006**, *19*, 146–153. [CrossRef]

32. Jeong, H.-S.; Lee, S.H.; Yun, H.-Y.; Baek, K.J.; Kwon, N.S.; Park, K.-C.; Kim, D.-S. Involvement of mTOR signaling in sphingosylphosphorylcholine-induced hypopigmentation effects. *J. Biomed. Sci.* **2011**, *18*, 55. [CrossRef] [PubMed]

33. Jeong, H.-S.; Park, K.-C.; Kim, D.-S. PP2A and DUSP6 are involved in sphingosylphosphorylcholine-induced hypopigmentation. *Mol. Cell. Biochem.* **2012**, *367*, 43–49. [CrossRef]

34. Bizzozero, L.; Cazzato, D.; Cervia, D.; Assi, E.; Simbari, F.; Pagni, F.; De Palma, C.; Monno, A.; Verdelli, C.; Querini, P.R.; et al. Acid sphingomyelinase determines melanoma progression and metastatic behaviour via the microphtalmia-associated transcription factor signalling pathway. *Cell Death Differ.* **2014**, *21*, 507–520. [CrossRef] [PubMed]

35. Kim, D.-S.; Kim, S.-Y.; Chung, J.-H.; Kim, K.-H.; Eun, H.-C.; Park, K.-C. Delayed ERK activation by ceramide reduces melanin synthesis in human melanocytes. *Cell. Signal.* **2002**, *14*, 779–785. [CrossRef]

36. Amos, C.I.; Wang, L.-E.; Lee, J.E.; Gershenwald, J.E.; Chen, W.V.; Fang, S.; Kosoy, R.; Zhang, M.; Qureshi, A.A.; Vattathil, S.; et al. Genome-wide association study identifies novel loci predisposing to cutaneous melanoma†. *Hum. Mol. Genet.* **2011**, *20*, 5012–5023. [CrossRef]

37. Astudillo, L.; Therville, N.; Colacios, C.; Ségui, B.; Andrieu-Abadie, N.; Levade, T. Glucosylceramidases and malignancies in mammals. *Biochimie* **2016**, *125*, 267–280. [CrossRef]

38. Dubot, P.; Astudillo, L.; Therville, N.; Sabourdy, F.; Stirnemann, J.; Levade, T.; Andrieu-Abadie, N. Are Glucosylceramide-Related Sphingolipids Involved in the Increased Risk for Cancer in Gaucher Disease Patients? Review and Hypotheses. *Cancers* **2020**, *12*, 475. [CrossRef]

39. Ferraz, M.J.; Marques, A.R.A.; Appelman, M.D.; Verhoek, M.; Strijland, A.; Mirzaian, M.; Scheij, S.; Ouairy, C.M.; Lahav, D.; Wisse, P.; et al. Lysosomal glycosphingolipid catabolism by acid ceramidase: Formation of glycosphingoid bases during deficiency of glycosidases. *FEBS Lett.* **2016**, *590*, 716–725. [CrossRef]

40. Flanagan, J.; Ranes, B.; Brignol, N.; Hamler, R.; Clark, S. The origins of glucosylsphingosine in Gaucher disease. *Mol. Genet. Metab.* **2013**, *108*, S40–S41. [CrossRef]

41. Lai, M.; La Rocca, V.; Amato, R.; Freer, G.; Pistello, M. Sphingolipid/Ceramide Pathways and Autophagy in the Onset and Progression of Melanoma: Novel Therapeutic Targets and Opportunities. *Int. J. Mol. Sci.* **2019**, *20*, 3436. [CrossRef]

42. Jiang, W.; Ogretmen, B. Autophagy paradox and ceramide. *Biochim. Biophys. Acta BBA Mol. Cell Biol. Lipids* **2014**, *1841*, 783–792. [CrossRef] [PubMed]

43. Sun, Y.; Liou, B.; Ran, H.; Skelton, M.R.; Williams, M.T.; Vorhees, C.V.; Kitatani, K.; Hannun, Y.A.; Witte, D.P.; Xu, Y.-H.; et al. Neuronopathic Gaucher disease in the mouse: Viable combined selective saposin C deficiency and mutant glucocerebrosidase (V394L) mice with glucosylsphingosine and glucosylceramide accumulation and progressive neurological deficits. *Hum. Mol. Genet.* **2010**, *19*, 1088–1097. [CrossRef] [PubMed]

44. Kinghorn, K.J.; Grönke, S.; Castillo-Quan, J.I.; Woodling, N.S.; Li, L.; Sirka, E.; Gegg, M.; Mills, K.; Hardy, J.; Bjedov, I.; et al. A Drosophila Model of Neuronopathic Gaucher Disease Demonstrates Lysosomal-Autophagic Defects and Altered mTOR Signalling and Is Functionally Rescued by Rapamycin. *J. Neurosci.* **2016**, *36*, 11654–11670. [CrossRef] [PubMed]

45. Panicker, L.M.; Miller, D.; Park, T.S.; Patel, B.; Azevedo, J.L.; Awad, O.; Masood, M.A.; Veenstra, T.D.; Goldin, E.; Stubblefield, B.K.; et al. Induced pluripotent stem cell model recapitulates pathologic hallmarks of Gaucher disease. *Proc. Natl. Acad. Sci. USA* **2012**, *109*, 18054–18059. [CrossRef]

46. Brown, R.A.; Voit, A.; Srikanth, M.P.; Thayer, J.A.; Kingsbury, T.J.; Jacobson, M.A.; Lipinski, M.M.; Feldman, R.A.; Awad, O. mTOR hyperactivity mediates lysosomal dysfunction in Gaucher's disease iPSC-neuronal cells. *Dis. Model. Mech.* **2019**, *12*, dmm038596. [CrossRef]

47. Pópulo, H.; Soares, P.; Faustino, A.; Rocha, A.S.; Silva, P.; Azevedo, F.; Lopes, J.M. mTOR pathway activation in cutaneous melanoma is associated with poorer prognosis characteristics: Letter to the Editor. *Pigment Cell Melanoma Res.* **2011**, *24*, 254–257. [CrossRef]

48. Li, S.; Song, Y.; Quach, C.; Guo, H.; Jang, G.-B.; Maazi, H.; Zhao, S.; Sands, N.A.; Liu, Q.; In, G.K.; et al. Transcriptional regulation of autophagy-lysosomal function in BRAF-driven melanoma progression and chemoresistance. *Nat. Commun.* **2019**, *10*, 1693. [CrossRef]

49. Dhillon, A.S.; Hagan, S.; Rath, O.; Kolch, W. MAP kinase signalling pathways in cancer. *Oncogene* **2007**, *26*, 3279–3290. [CrossRef]

50. Smalley, K.S.M. A pivotal role for ERK in the oncogenic behaviour of malignant melanoma? *Int. J. Cancer* **2003**, *104*, 527–532. [CrossRef]

51. Gorden, A.; Osman, I.; Gai, W.; He, D.; Huang, W.; Davidson, A.; Houghton, A.N.; Busam, K.; Polsky, D. Analysis of BRAF and N-RAS mutations in metastatic melanoma tissues. *Cancer Res.* **2003**, *63*, 3955–3957.

52. Wellbrock, C.; Karasarides, M.; Marais, R. The RAF proteins take centre stage. *Nat. Rev. Mol. Cell Biol.* **2004**, *5*, 875–885. [CrossRef] [PubMed]

53. Curtin, J.A.; Patel, H.N.; Cho, K.-H.; LeBoit, P.E. Distinct Sets of Genetic Alterations in Melanoma. *N. Engl. J. Med.* **2005**, *13*. [CrossRef] [PubMed]

54. Chapman, P.B.; Hauschild, A.; Robert, C.; Haanen, J.B.; Ascierto, P.; Larkin, J.; Dummer, R.; Garbe, C.; Testori, A.; Maio, M.; et al. Improved Survival with Vemurafenib in Melanoma with BRAF V600E Mutation. *N. Engl. J. Med.* **2011**, *364*, 2507–2516. [CrossRef] [PubMed]

55. Hauschild, A.; Grob, J.-J.; Demidov, L.V.; Jouary, T.; Gutzmer, R.; Millward, M.; Rutkowski, P.; Blank, C.U.; Miller, W.H.; Kaempgen, E.; et al. Dabrafenib in BRAF-mutated metastatic melanoma: A multicentre, open-label, phase 3 randomised controlled trial. *Lancet* **2012**, *380*, 358–365. [CrossRef]

56. Larkin, J.; Ascierto, P.A.; Dréno, B.; Atkinson, V.; Liszkay, G.; Maio, M.; Mandalà, M.; Demidov, L.; Stroyakovskiy, D.; Thomas, L.; et al. Combined vemurafenib and cobimetinib in BRAF-mutated melanoma. *N. Engl. J. Med.* **2014**, *371*, 1867–1876. [CrossRef] [PubMed]

57. Ribas, A.; Daud, A.; Pavlick, A.C.; Gonzalez, R.; Lewis, K.D.; Hamid, O.; Gajewski, T.F.; Puzanov, I.; Wongchenko, M.; Rooney, I.; et al. Extended 5-Year Follow-up Results of a Phase Ib Study (BRIM7) of Vemurafenib and Cobimetinib in BRAF -Mutant Melanoma. *Clin. Cancer Res.* **2020**, *26*, 46–53. [CrossRef]

58. Patel, H.; Yacoub, N.; Mishra, R.; White, A.; Yuan, L.; Alanazi, S.; Garrett, J.T. Current Advances in the Treatment of BRAF-Mutant Melanoma. *Cancers* **2020**, *12*, 482. [CrossRef]

59. Pitson, S.M.; Xia, P.; Leclercq, T.M.; Moretti, P.A.B.; Zebol, J.R.; Lynn, H.E.; Wattenberg, B.W.; Vadas, M.A. Phosphorylation-dependent translocation of sphingosine kinase to the plasma membrane drives its oncogenic signalling. *J. Exp. Med.* **2005**, *201*, 49–54. [CrossRef]

60. Leclercq, T.; Pitson, S. Cellular signalling by sphingosine kinase and sphingosine 1-phosphate. *IUBMB Life Int. Union Biochem. Mol. Biol. Life* **2006**, *58*, 467–472. [CrossRef]

61. Francy, J.M.; Nag, A.; Conroy, E.J.; Hengst, J.A.; Yun, J.K. Sphingosine kinase 1 expression is regulated by signaling through PI3K, AKT2, and mTOR in human coronary artery smooth muscle cells. *Biochim. Biophys. Acta BBA Gene Struct. Expr.* **2007**, *1769*, 253–265. [CrossRef]

62. Mrad, M.; Imbert, C.; Garcia, V.; Rambow, F.; Therville, N.; Carpentier, S.; Ségui, B.; Levade, T.; Azar, R.; Marine, J.-C.; et al. Downregulation of sphingosine kinase-1 induces protective tumor immunity by promoting M1 macrophage response in melanoma. *Oncotarget* **2016**, *7*, 71873–71886. [CrossRef] [PubMed]

63. Pyne, N.J.; Pyne, S. Sphingosine 1-phosphate and cancer. *Nat. Rev. Cancer* **2010**, *10*, 489–503. [CrossRef]

64. Garandeau, D.; Noujarède, J.; Leclerc, J.; Imbert, C.; Garcia, V.; Bats, M.-L.; Rambow, F.; Gilhodes, J.; Filleron, T.; Meyer, N.; et al. Targeting the Sphingosine 1-Phosphate Axis Exerts Potent Antitumor Activity in BRAFi-Resistant Melanomas. *Mol. Cancer Ther.* **2019**, *18*, 289–300. [CrossRef]

65. Lai, M.; Realini, N.; La Ferla, M.; Passalacqua, I.; Matteoli, G.; Ganesan, A.; Pistello, M.; Mazzanti, C.M.; Piomelli, D. Complete Acid Ceramidase ablation prevents cancer-initiating cell formation in melanoma cells. *Sci. Rep.* **2017**, *7*, 7411. [CrossRef]

66. Bedia, C.; Casas, J.; Andrieu-Abadie, N.; Fabriàs, G.; Levade, T. Acid Ceramidase Expression Modulates the Sensitivity of A375 Melanoma Cells to Dacarbazine. *J. Biol. Chem.* **2011**, *286*, 28200–28209. [CrossRef] [PubMed]

67. Han, W.S.; Yoo, J.Y.; Youn, S.W.; Kim, D.S.; Park, C.; Kim, S.Y.; Kim, K.H. Effects of C2-ceramide on the Malme-3M melanoma cell line. *J. Dermatol. Sci.* **2002**, *10*. [CrossRef]

68. Deng, W.; Li, R.; Guerrera, M.; Liu, Y.; Ladisch, S. Transfection of glucosylceramide synthase antisense inhibits mouse melanoma formation. *Glycobiology* **2002**, *12*, 145–152. [CrossRef]

69. Weiss, M.; Hettmer, S.; Smith, P.; Ladisch, S. Inhibition of melanoma tumor growth by a novel inhibitor of glucosylceramide synthase. *Cancer Res.* **2003**, *63*, 3654–3658. [PubMed]

70. Sorli, S.; Colié, S.; Albinet, V.; Dubrac, A.; Touriol, C.; Guilbaud, N.; Bedia, C.; Fabriàs, G.; Casas, J.; Ségui, B.; et al. The nonlysosomal β-glucosidase GBA2 promotes endoplasmic reticulum stress and impairs tumorigenicity of human melanoma cells. *FASEB J.* **2013**, *27*, 489–498. [CrossRef]

71. Nakano, J.; Raj, B.K.; Asagami, C.; Lloyd, K.O. Human melanoma cell lines deficient in GD3 ganglioside expression exhibit altered growth and tumorigenic characteristics. *J. Invest. Dermatol.* **1996**, *107*, 543–548. [CrossRef]

72. Hamamura, K.; Furukawa, K.; Hayashi, T.; Hattori, T.; Nakano, J.; Nakashima, H.; Okuda, T.; Mizutani, H.; Hattori, H.; Ueda, M.; et al. Ganglioside GD3 promotes cell growth and invasion through p130Cas and paxillin in malignant melanoma cells. *Proc. Natl. Acad. Sci. USA* **2005**, *102*, 11041–11046. [CrossRef] [PubMed]

73. Furukawa, K.; Kambe, M.; Miyata, M.; Ohkawa, Y.; Tajima, O.; Furukawa, K. Ganglioside GD3 induces convergence and synergism of adhesion and hepatocyte growth factor/Met signals in melanomas. *Cancer Sci.* **2014**, *105*, 52–63. [CrossRef]

74. Li, G.; Satyamoorthy, K.; Herlyn, M. Dynamics of Cell Interactions and Communications during Melanoma Development. *Crit. Rev. Oral Biol. Med.* **2002**, *13*, 62–70. [CrossRef] [PubMed]

75. Walko, G.; Castañón, M.J.; Wiche, G. Molecular architecture and function of the hemidesmosome. *Cell Tissue Res.* **2015**, *360*, 529–544. [CrossRef] [PubMed]

76. Haass, N.K.; Smalley, K.S.M.; Herlyn, M. The Role of Altered Cell–Cell Communication in Melanoma Progression. *J. Mol. Histol.* **2003**, *35*, 309–318. [CrossRef] [PubMed]

77. Haass, N.K.; Smalley, K.S.M.; Li, L.; Herlyn, M. Adhesion, migration and communication in melanocytes and melanoma. *Pigment Cell Res.* **2005**, *18*, 150–159. [CrossRef]

78. Hsu, M.-Y.; Meier, F.E.; Nesbit, M.; Hsu, J.-Y.; Van Belle, P.; Elder, D.E.; Herlyn, M. E-Cadherin Expression in Melanoma Cells Restores Keratinocyte-Mediated Growth Control and Down-Regulates Expression of Invasion-Related Adhesion Receptors. *Am. J. Pathol.* **2000**, *156*, 1515–1525. [CrossRef]

79. Tamashiro, P.M.; Furuya, H.; Shimizu, Y.; Kawamori, T. Sphingosine kinase 1 mediates head & neck squamous cell carcinoma invasion through sphingosine 1-phosphate receptor 1. *Cancer Cell Int.* **2014**, *14*, 76. [CrossRef] [PubMed]

80. Liu, H.; Ma, Y.; He, H.-W.; Zhao, W.-L.; Shao, R.-G. SPHK1 (sphingosine kinase 1) induces epithelial-mesenchymal transition by promoting the autophagy-linked lysosomal degradation of CDH1/E-cadherin in hepatoma cells. *Autophagy* **2017**, *13*, 900–913. [CrossRef] [PubMed]

81. Milara, J.; Navarro, R.; Juan, G.; Peiró, T.; Serrano, A.; Ramón, M.; Morcillo, E.; Cortijo, J. Sphingosine-1-phosphate is increased in patients with idiopathic pulmonary fibrosis and mediates epithelial to mesenchymal transition. *Thorax* **2012**, *67*, 147–156. [CrossRef]

82. Kono, Y.; Nishiuma, T.; Nishimura, Y.; Kotani, Y.; Okada, T.; Nakamura, S.; Yokoyama, M. Sphingosine Kinase 1 Regulates Differentiation of Human and Mouse Lung Fibroblasts Mediated by TGF-β1. *Am. J. Respir. Cell Mol. Biol.* **2007**, *37*, 395–404. [CrossRef]

83. Yamaguchi, H.; Kitayama, J.; Takuwa, N.; Arikawa, K.; Inoki, I.; Takehara, K.; Nagawa, H.; Takuwa, Y. Sphingosine-1-phosphate receptor subtype-specific positive and negative regulation of Rac and haematogenous metastasis of melanoma cells. *Biochem. J.* **2003**, *374*, 715–722. [CrossRef] [PubMed]

84. Braga, V.M.M.; Machesky, L.M.; Hall, A.; Hotchin, N.A. The Small GTPases Rho and Rac Are Required for the Establishment of Cadherin-dependent Cell–Cell Contacts. *J. Cell Biol.* **1997**, *137*, 1421–1431. [CrossRef] [PubMed]

85. Ohkawa, Y.; Miyazaki, S.; Hamamura, K.; Kambe, M.; Miyata, M.; Tajima, O.; Ohmi, Y.; Yamauchi, Y.; Furukawa, K.; Furukawa, K. Ganglioside GD3 Enhances Adhesion Signals and Augments Malignant Properties of Melanoma Cells by Recruiting Integrins to Glycolipid-enriched Microdomains. *J. Biol. Chem.* **2010**, *285*, 27213–27223. [CrossRef] [PubMed]

86. Ohmi, Y.; Kambe, M.; Ohkawa, Y.; Hamamura, K.; Tajima, O.; Takeuchi, R.; Furukawa, K.; Furukawa, K. Differential roles of gangliosides in malignant properties of melanomas. *PLoS ONE* **2018**, *13*, e0206881. [CrossRef] [PubMed]

87. Hodorogea, A.; Calinescu, A.; Antohe, M.; Balaban, M.; Nedelcu, R.I.; Turcu, G.; Ion, D.A.; Badarau, I.A.; Popescu, C.M.; Popescu, R.; et al. Epithelial-Mesenchymal Transition in Skin Cancers: A Review. *Anal. Cell. Pathol.* **2019**, *2019*, 1–11. [CrossRef]

88. Hoek, K.S.; Schlegel, N.C.; Brafford, P.; Sucker, A.; Ugurel, S.; Kumar, R.; Weber, B.L.; Nathanson, K.L.; Phillips, D.J.; Herlyn, M.; et al. Metastatic potential of melanomas defined by specific gene expression profiles with no BRAF signature. *Pigment Cell Res.* **2006**, *19*, 290–302. [CrossRef]

89. Levy, C.; Khaled, M.; Fisher, D.E. MITF: Master regulator of melanocyte development and melanoma oncogene. *Trends Mol. Med.* **2006**, *12*, 406–414. [CrossRef]

90. Goding, C.R.; Arnheiter, H. MITF—The first 25 years. *Genes Dev.* **2019**, *33*, 983–1007. [CrossRef]

91. Levy, C.; Lee, Y.-N.; Nechushtan, H.; Schueler-Furman, O.; Sonnenblick, A.; Hacohen, S.; Razin, E. Identifying a common molecular mechanism for inhibition of MITF and STAT3 by PIAS3. *Blood* **2006**, *107*, 2839–2845. [CrossRef]

92. Yasumoto, K.; Takeda, K.; Saito, H.; Watanabe, K.; Takahashi, K.; Shibahara, S. Microphthalmia-associated transcription factor interacts with LEF-1, a mediator of Wnt signaling. *EMBO J.* **2002**, *21*, 2703–2714. [CrossRef] [PubMed]

93. Schmelz, E.M.; Roberts, P.C.; Kustin, E.M.; Lemonnier, L.A.; Sullards, M.C.; Dillehay, D.L.; Merrill, A.H. Modulation of intracellular beta-catenin localization and intestinal tumorigenesis in vivo and in vitro by sphingolipids. *Cancer Res.* **2001**, *61*, 6723–6729. [PubMed]

94. Liu, H.; Zhang, C.-X.; Ma, Y.; He, H.-W.; Wang, J.-P.; Shao, R.-G. SphK1 inhibitor SKI II inhibits the proliferation of human hepatoma HepG2 cells via the Wnt5A/β-catenin signaling pathway. *Life Sci.* **2016**, *151*, 23–29. [CrossRef] [PubMed]

95. White, C.; Alshaker, H.; Cooper, C.; Winkler, M.; Pchejetski, D. The emerging role of FTY720 (Fingolimod) in cancer treatment. *Oncotarget* **2016**, *7*, 23106–23127. [CrossRef]

96. Lee, J.E.; Kim, S.Y.; Jeong, Y.-M.; Yun, H.-Y.; Baek, K.J.; Kwon, N.S.; Park, K.-C.; Kim, D.-S. The regulatory mechanism of melanogenesis by FTY720, a sphingolipid analogue: The regulation of melanogenesis by FTY720. *Exp. Dermatol.* **2011**, *20*, 237–241. [CrossRef]

97. Caramel, J.; Papadogeorgakis, E.; Hill, L.; Browne, G.J.; Richard, G.; Wierinckx, A.; Saldanha, G.; Osborne, J.; Hutchinson, P.; Tse, G.; et al. A Switch in the Expression of Embryonic EMT-Inducers Drives the Development of Malignant Melanoma. *Cancer Cell* **2013**, *24*, 466–480. [CrossRef]

98. Lu, P.; White-Gilbertson, S.; Nganga, R.; Kester, M.; Voelkel-Johnson, C. Expression of the SNAI2 transcriptional repressor is regulated by C 16 -ceramide. *Cancer Biol. Ther.* **2019**, *20*, 922–930. [CrossRef]

99. Edmond, V.; Dufour, F.; Poiroux, G.; Shoji, K.; Malleter, M.; Fouqué, A.; Tauzin, S.; Rimokh, R.; Sergent, O.; Penna, A.; et al. Downregulation of ceramide synthase-6 during epithelial-to-mesenchymal transition reduces plasma membrane fluidity and cancer cell motility. *Oncogene* **2015**, *34*, 996–1005. [CrossRef]

100. Zheng, K.; Chen, Z.; Feng, H.; Chen, Y.; Zhang, C.; Yu, J.; Luo, Y.; Zhao, L.; Jiang, X.; Shi, F. Sphingomyelin synthase 2 promotes an aggressive breast cancer phenotype by disrupting the homoeostasis of ceramide and sphingomyelin. *Cell Death Dis.* **2019**, *10*, 157. [CrossRef]

101. Levade, T.; Andrieu-Abadie, N.; Micheau, O.; Legembre, P.; Ségui, B. Sphingolipids modulate the epithelial-mesenchymal transition in cancer. *Cell Death Discov.* **2015**, *1*, 15001. [CrossRef]

102. Mathow, D.; Chessa, F.; Rabionet, M.; Kaden, S.; Jennemann, R.; Sandhoff, R.; Gröne, H.-J.; Feuerborn, A. Zeb1 affects epithelial cell adhesion by diverting glycosphingolipid metabolism. *EMBO Rep.* **2015**, *16*, 321–331. [CrossRef] [PubMed]

103. Battula, V.L.; Shi, Y.; Evans, K.W.; Wang, R.-Y.; Spaeth, E.L.; Jacamo, R.O.; Guerra, R.; Sahin, A.A.; Marini, F.C.; Hortobagyi, G.; et al. Ganglioside GD2 identifies breast cancer stem cells and promotes tumorigenesis. *J. Clin. Investig.* **2012**, *122*, 2066–2078. [CrossRef]

104. Gupta, V.; Bhinge, K.N.; Hosain, S.B.; Xiong, K.; Gu, X.; Shi, R.; Ho, M.-Y.; Khoo, K.-H.; Li, S.-C.; Li, Y.-T.; et al. Ceramide glycosylation by glucosylceramide synthase selectively maintains the properties of breast cancer stem cells. *J. Biol. Chem.* **2012**, *287*, 37195–37205. [CrossRef] [PubMed]

105. Hosain, S.B.; Khiste, S.K.; Uddin, M.B.; Vorubindi, V.; Ingram, C.; Zhang, S.; Hill, R.A.; Gu, X.; Liu, Y.-Y. Inhibition of glucosylceramide synthase eliminates the oncogenic function of p53 R273H mutant in the epithelial-mesenchymal transition and induced pluripotency of colon cancer cells. *Oncotarget* **2016**, *7*, 60575–60592. [CrossRef]

106. Hao, Y.; Baker, D.; ten Dijke, P. TGF-β-Mediated Epithelial-Mesenchymal Transition and Cancer Metastasis. *Int. J. Mol. Sci.* **2019**, *20*, 2767. [CrossRef] [PubMed]

107. Tsubakihara, Y.; Moustakas, A. Epithelial-Mesenchymal Transition and Metastasis under the Control of Transforming Growth Factor β. *Int. J. Mol. Sci.* **2018**, *19*, 3672. [CrossRef] [PubMed]

108. Xin, C.; Ren, S.; Kleuser, B.; Shabahang, S.; Eberhardt, W.; Radeke, H.; Schäfer-Korting, M.; Pfeilschifter, J.; Huwiler, A. Sphingosine 1-Phosphate Cross-activates the Smad Signaling Cascade and Mimics Transforming Growth Factor-β-induced Cell Responses. *J. Biol. Chem.* **2004**, *279*, 35255–35262. [CrossRef] [PubMed]

109. Sauer, B.; Vogler, R.; von Wenckstern, H.; Fujii, M.; Anzano, M.B.; Glick, A.B.; Schäfer-Korting, M.; Roberts, A.B.; Kleuser, B. Involvement of Smad Signaling in Sphingosine 1-Phosphate-mediated Biological Responses of Keratinocytes. *J. Biol. Chem.* **2004**, *279*, 38471–38479. [CrossRef]

110. Radeke, H.H.; von Wenckstern, H.; Stoidtner, K.; Sauer, B.; Hammer, S.; Kleuser, B. Overlapping Signaling Pathways of Sphingosine 1-Phosphate and TGF-β in the Murine Langerhans Cell Line XS52. *J. Immunol.* **2005**, *174*, 2778–2786. [CrossRef]

111. Zeng, Y.; Yao, X.; Chen, L.; Yan, Z.; Liu, J.; Zhang, Y.; Feng, T.; Wu, J.; Liu, X. Sphingosine-1-phosphate induced epithelial-mesenchymal transition of hepatocellular carcinoma via an MMP-7/syndecan-1/TGF-β autocrine loop. *Oncotarget* **2016**, *7*, 63324–63337. [CrossRef]

112. Liu, Y.-N.; Zhang, H.; Zhang, L.; Cai, T.-T.; Huang, D.-J.; He, J.; Ni, H.-H.; Zhou, F.-J.; Zhang, X.-S.; Li, J. Sphingosine 1 phosphate receptor-1 (S1P1) promotes tumor-associated regulatory T cell expansion: Leading to poor survival in bladder cancer. *Cell Death Dis.* **2019**, *10*, 50. [CrossRef]

113. Miller, A.V.; Alvarez, S.E.; Spiegel, S.; Lebman, D.A. Sphingosine Kinases and Sphingosine-1-Phosphate Are Critical for Transforming Growth Factor β-Induced Extracellular Signal-Regulated Kinase 1 and 2 Activation and Promotion of Migration and Invasion of Esophageal Cancer Cells. *Mol. Cell. Biol.* **2008**, *28*, 4142–4151. [CrossRef] [PubMed]

114. Yamanaka, M.; Shegogue, D.; Pei, H.; Bu, S.; Bielawska, A.; Bielawski, J.; Pettus, B.; Hannun, Y.A.; Obeid, L.; Trojanowska, M. Sphingosine Kinase 1 (SPHK1) Is Induced by Transforming Growth Factor-β and Mediates TIMP-1 Up-regulation. *J. Biol. Chem.* **2004**, *279*, 53994–54001. [CrossRef] [PubMed]

115. Verfaillie, A.; Imrichova, H.; Atak, Z.K.; Dewaele, M.; Rambow, F.; Hulselmans, G.; Christiaens, V.; Svetlichnyy, D.; Luciani, F.; Van den Mooter, L.; et al. Decoding the regulatory landscape of melanoma reveals TEADS as regulators of the invasive cell state. *Nat. Commun.* **2015**, *6*, 6683. [CrossRef] [PubMed]

116. Nallet-Staub, F.; Marsaud, V.; Li, L.; Gilbert, C.; Dodier, S.; Bataille, V.; Sudol, M.; Herlyn, M.; Mauviel, A. Pro-invasive activity of the Hippo pathway effectors YAP and TAZ in cutaneous melanoma. *J. Invest. Dermatol.* **2014**, *134*, 123–132. [CrossRef]

117. Miller, E.; Yang, J.; DeRan, M.; Wu, C.; Su, A.I.; Bonamy, G.M.C.; Liu, J.; Peters, E.C.; Wu, X. Identification of Serum-Derived Sphingosine-1-Phosphate as a Small Molecule Regulator of YAP. *Chem. Biol.* **2012**, *19*, 955–962. [CrossRef]

118. Yu, F.-X.; Zhao, B.; Panupinthu, N.; Jewell, J.L.; Lian, I.; Wang, L.H.; Zhao, J.; Yuan, H.; Tumaneng, K.; Li, H.; et al. Regulation of the Hippo-YAP Pathway by G-Protein-Coupled Receptor Signaling. *Cell* **2012**, *150*, 780–791. [CrossRef]

119. Pors, S.E.; Harðardóttir, L.; Olesen, H.Ø.; Riis, M.L.; Jensen, L.B.; Andersen, A.S.; Cadenas, J.; Grønning, A.P.; Colmorn, L.B.; Dueholm, M.; et al. Effect of sphingosine-1-phosphate on activation of dormant follicles in murine and human ovarian tissue. *Mol. Hum. Reprod.* **2020**, gaaa022. [CrossRef]

120. Huang, L.S.; Sudhadevi, T.; Fu, P.; Punathil-Kannan, P.-K.; Ebenezer, D.L.; Ramchandran, R.; Putherickal, V.; Cheresh, P.; Zhou, G.; Ha, A.W.; et al. Sphingosine Kinase 1/S1P Signaling Contributes to Pulmonary Fibrosis by Activating Hippo/YAP Pathway and Mitochondrial Reactive Oxygen Species in Lung Fibroblasts. *Int. J. Mol. Sci.* **2020**, *21*, 2064. [CrossRef]

121. Kemppainen, K.; Wentus, N.; Lassila, T.; Laiho, A.; Törnquist, K. Sphingosylphosphorylcholine regulates the Hippo signaling pathway in a dual manner. *Cell. Signal.* **2016**, *28*, 1894–1903. [CrossRef]

122. Arikawa, K.; Takuwa, N.; Yamaguchi, H.; Sugimoto, N.; Kitayama, J.; Nagawa, H.; Takehara, K.; Takuwa, Y. Ligand-dependent Inhibition of B16 Melanoma Cell Migration and Invasion via Endogenous S1P 2 G Protein-coupled Receptor: REQUIREMENT OF INHIBITION OF CELLULAR RAC ACTIVITY. *J. Biol. Chem.* **2003**, *278*, 32841–32851. [CrossRef]

123. Carpinteiro, A.; Becker, K.A.; Japtok, L.; Hessler, G.; Keitsch, S.; Požgajovà, M.; Schmid, K.W.; Adams, C.; Müller, S.; Kleuser, B.; et al. Regulation of hematogenous tumor metastasis by acid sphingomyelinase. *EMBO Mol. Med.* **2015**, *7*, 714–734. [CrossRef]

124. Tsuchida, T.; Saxton, R.E.; Morton, D.L.; Irie, R.F. Gangliosides of human melanoma. *J. Natl. Cancer Inst.* **1987**, *78*, 45–54. [CrossRef] [PubMed]

125. Sawada, M.; Moriya, S.; Shineha, R.; Satomi, S.; Miyagi, T. Comparative study of sialidase activity and G(M3) content in B16 melanoma variants with different metastatic potential. *Acta Biochim. Pol.* **1998**, *45*, 343–349. [CrossRef] [PubMed]

126. Saha, S.; Mohanty, K.C. Enhancement of metastatic potential of mouse B16-melanoma cells to lung after treatment with gangliosides of B-16-melanoma cells of higher metastatic potential to lung. *Indian J. Exp. Biol.* **2003**, *41*, 1253–1258. [PubMed]

127. Liu, J.-W.; Sun, P.; Yan, Q.; Paller, A.S.; Gerami, P.; Ho, N.; Vashi, N.; Le Poole, I.C.; Wang, X.-Q. De-N-acetyl GM3 Promotes Melanoma Cell Migration and Invasion through Urokinase Plasminogen Activator Receptor Signaling-Dependent MMP-2 Activation. *Cancer Res.* **2009**, *69*, 8662–8669. [CrossRef]

128. Peinado, H.; Alečković, M.; Lavotshkin, S.; Matei, I.; Costa-Silva, B.; Moreno-Bueno, G.; Hergueta-Redondo, M.; Williams, C.; García-Santos, G.; Ghajar, C.M.; et al. Melanoma exosomes educate bone marrow progenitor cells toward a pro-metastatic phenotype through MET. *Nat. Med.* **2012**, *18*, 883–891. [CrossRef]

129. Costa-Silva, B.; Aiello, N.M.; Ocean, A.J.; Singh, S.; Zhang, H.; Thakur, B.K.; Becker, A.; Hoshino, A.; Mark, M.T.; Molina, H.; et al. Pancreatic cancer exosomes initiate pre-metastatic niche formation in the liver. *Nat. Cell Biol.* **2015**, *17*, 816–826. [CrossRef]

130. Hoshino, A.; Costa-Silva, B.; Shen, T.-L.; Rodrigues, G.; Hashimoto, A.; Tesic Mark, M.; Molina, H.; Kohsaka, S.; Di Giannatale, A.; Ceder, S.; et al. Tumour exosome integrins determine organotropic metastasis. *Nature* **2015**, *527*, 329–335. [CrossRef]

131. Raposo, G.; Stoorvogel, W. Extracellular vesicles: Exosomes, microvesicles, and friends. *J. Cell Biol.* **2013**, *200*, 373–383. [CrossRef]

132. Xiao, D.; Barry, S.; Kmetz, D.; Egger, M.; Pan, J.; Rai, S.N.; Qu, J.; McMasters, K.M.; Hao, H. Melanoma cell–derived exosomes promote epithelial–mesenchymal transition in primary melanocytes through paracrine/autocrine signaling in the tumor microenvironment. *Cancer Lett.* **2016**, *376*, 318–327. [CrossRef] [PubMed]

133. Hood, J.L.; San, R.S.; Wickline, S.A. Exosomes Released by Melanoma Cells Prepare Sentinel Lymph Nodes for Tumor Metastasis. *Cancer Res.* **2011**, *71*, 3792–3801. [CrossRef] [PubMed]

134. Trajkovic, K.; Hsu, C.; Chiantia, S.; Rajendran, L.; Wenzel, D.; Wieland, F.; Schwille, P.; Brugger, B.; Simons, M. Ceramide Triggers Budding of Exosome Vesicles into Multivesicular Endosomes. *Science* **2008**, *319*, 1244–1247. [CrossRef] [PubMed]

135. López-Montero, I.; Vélez, M.; Devaux, P.F. Surface tension induced by sphingomyelin to ceramide conversion in lipid membranes. *Biochim. Biophys. Acta BBA Biomembr.* **2007**, *1768*, 553–561. [CrossRef]

136. Kosaka, N.; Iguchi, H.; Yoshioka, Y.; Takeshita, F.; Matsuki, Y.; Ochiya, T. Secretory Mechanisms and Intercellular Transfer of MicroRNAs in Living Cells. *J. Biol. Chem.* **2010**, *285*, 17442–17452. [CrossRef] [PubMed]

137. Kosaka, N.; Iguchi, H.; Hagiwara, K.; Yoshioka, Y.; Takeshita, F.; Ochiya, T. Neutral Sphingomyelinase 2 (nSMase2)-dependent Exosomal Transfer of Angiogenic MicroRNAs Regulate Cancer Cell Metastasis. *J. Biol. Chem.* **2013**, *288*, 10849–10859. [CrossRef]

138. Kajimoto, T.; Okada, T.; Miya, S.; Zhang, L.; Nakamura, S. Ongoing activation of sphingosine 1-phosphate receptors mediates maturation of exosomal multivesicular endosomes. *Nat. Commun.* **2013**, *4*, 2712. [CrossRef]

139. Mohamed, N.N.I.; Okada, T.; Kajimoto, T.; Nakamura, S.-I. Essential Role of Sphingosine Kinase 2 in the Regulation of Cargo Contents in the Exosomes from K562 Cells. *Kobe J. Med. Sci.* **2018**, *63*, E123–E129.

140. Kajimoto, T.; Mohamed, N.N.I.; Badawy, S.M.M.; Matovelo, S.A.; Hirase, M.; Nakamura, S.; Yoshida, D.; Okada, T.; Ijuin, T.; Nakamura, S. Involvement of Gβγ subunits of G i protein coupled with S1P receptor on multivesicular endosomes in F-actin formation and cargo sorting into exosomes. *J. Biol. Chem.* **2018**, *293*, 245–253. [CrossRef]

141. Schumacher, T.N.; Schreiber, R.D. Neoantigens in cancer immunotherapy. *Science* **2015**, *348*, 69–74. [CrossRef]

142. Dunn, G.P.; Bruce, A.T.; Ikeda, H.; Old, L.J.; Schreiber, R.D. Cancer immunoediting: From immunosurveillance to tumor escape. *Nat. Immunol.* **2002**, *3*, 991–998. [CrossRef]

143. Kim, R.; Emi, M.; Tanabe, K. Cancer immunoediting from immune surveillance to immune escape. *Immunology* **2007**, *121*, 1–14. [CrossRef] [PubMed]

144. Mandala, S.; Hadju, R.; Bergstrom, J.; Quackenbush, E.; Xie, J.; Milligan, J.; Thornton, R.; Shei, G.-J.; Card, D. Alteration of lymphocyte trafficking by sphingosine-1-phosphate receptor agonists. *Science* **2002**, *296*, 346–349. [CrossRef] [PubMed]

145. Matloubian, M.; Lo, C.G.; Cinamon, G.; Lesneski, M.J.; Xu, Y.; Brinkmann, V.; Allende, M.L.; Proia, R.L.; Cyster, J.G. Lymphocyte egress from thymus and peripheral lymphoid organs is dependent on S1P receptor 1. *Nature* **2004**, *427*, 355–360. [CrossRef] [PubMed]

146. Resop, R.S.; Douaisi, M.; Craft, J.; Jachimowski, L.C.M.; Blom, B.; Uittenbogaart, C.H. Sphingosine-1-phosphate/sphingosine-1-phosphate receptor 1 signaling is required for migration of naive human T cells from the thymus to the periphery. *J. Allergy Clin. Immunol.* **2016**, *138*, 551–557.e8. [CrossRef] [PubMed]

147. Lo, C.G.; Xu, Y.; Proia, R.L.; Cyster, J.G. Cyclical modulation of sphingosine-1-phosphate receptor 1 surface expression during lymphocyte recirculation and relationship to lymphoid organ transit. *J. Exp. Med.* **2005**, *201*, 291–301. [CrossRef]

148. Bankovich, A.J.; Shiow, L.R.; Cyster, J.G. CD69 Suppresses Sphingosine 1-Phosophate Receptor-1 (S1P 1) Function through Interaction with Membrane Helix 4. *J. Biol. Chem.* **2010**, *285*, 22328–22337. [CrossRef]

149. Shiow, L.R.; Rosen, D.B.; Brdičková, N.; Xu, Y.; An, J.; Lanier, L.L.; Cyster, J.G.; Matloubian, M. CD69 acts downstream of interferon-α/β to inhibit S1P1 and lymphocyte egress from lymphoid organs. *Nature* **2006**, *440*, 540–544. [CrossRef]

150. Mackay, L.K.; Braun, A.; Macleod, B.L.; Collins, N.; Tebartz, C.; Bedoui, S.; Carbone, F.R.; Gebhardt, T. Cutting Edge: CD69 Interference with Sphingosine-1-Phosphate Receptor Function Regulates Peripheral T Cell Retention. *J. Immunol.* **2015**, *194*, 2059–2063. [CrossRef]

151. Skon, C.N.; Lee, J.-Y.; Anderson, K.G.; Masopust, D.; Hogquist, K.A.; Jameson, S.C. Transcriptional downregulation of S1pr1 is required for the establishment of resident memory CD8+ T cells. *Nat. Immunol.* **2013**, *14*, 1285–1293. [CrossRef]

152. Amsen, D.; van Gisbergen, K.P.J.M.; Hombrink, P.; van Lier, R.A.W. Tissue-resident memory T cells at the center of immunity to solid tumors. *Nat. Immunol.* **2018**, *19*, 538–546. [CrossRef] [PubMed]

153. Gebhardt, T.; Wakim, L.M.; Eidsmo, L.; Reading, P.C.; Heath, W.R.; Carbone, F.R. Memory T cells in nonlymphoid tissue that provide enhanced local immunity during infection with herpes simplex virus. *Nat. Immunol.* **2009**, *10*, 524–530. [CrossRef]

154. Park, S.L.; Buzzai, A.; Rautela, J.; Hor, J.L.; Hochheiser, K.; Effern, M.; McBain, N.; Wagner, T.; Edwards, J.; McConville, R.; et al. Tissue-resident memory CD8+ T cells promote melanoma–immune equilibrium in skin. *Nature* **2019**, *565*, 366–371. [CrossRef] [PubMed]

155. Menares, E.; Gálvez-Cancino, F.; Cáceres-Morgado, P.; Ghorani, E.; López, E.; Díaz, X.; Saavedra-Almarza, J.; Figueroa, D.A.; Roa, E.; Quezada, S.A.; et al. Tissue-resident memory CD8+ T cells amplify anti-tumor immunity by triggering antigen spreading through dendritic cells. *Nat. Commun.* **2019**, *10*, 4401. [CrossRef] [PubMed]

156. Hochheiser, K.; Aw Yeang, H.X.; Wagner, T.; Tutuka, C.; Behren, A.; Waithman, J.; Angel, C.; Neeson, P.J.; Gebhardt, T.; Gyorki, D.E. Accumulation of CD103 + CD8 + T cells in a cutaneous melanoma micrometastasis. *Clin. Transl. Immunol.* **2019**, *8*. [CrossRef]

157. Drouillard, A.; Neyra, A.; Mathieu, A.-L.; Marçais, A.; Wencker, M.; Marvel, J.; Belot, A.; Walzer, T. Human Naive and Memory T Cells Display Opposite Migratory Responses to Sphingosine-1 Phosphate. *J. Immunol.* **2018**, *200*, 551–557. [CrossRef]

158. Sic, H.; Kraus, H.; Madl, J.; Flittner, K.-A.; von Münchow, A.L.; Pieper, K.; Rizzi, M.; Kienzler, A.-K.; Ayata, K.; Rauer, S.; et al. Sphingosine-1-phosphate receptors control B-cell migration through signaling components associated with primary immunodeficiencies, chronic lymphocytic leukemia, and multiple sclerosis. *J. Allergy Clin. Immunol.* **2014**, *134*, 420–428.e15. [CrossRef]

159. Walzer, T.; Chiossone, L.; Chaix, J.; Calver, A.; Carozzo, C.; Garrigue-Antar, L.; Jacques, Y.; Baratin, M.; Tomasello, E.; Vivier, E. Natural killer cell trafficking in vivo requires a dedicated sphingosine 1-phosphate receptor. *Nat. Immunol.* **2007**, *8*, 1337–1344. [CrossRef]

160. Drouillard, A.; Mathieu, A.-L.; Marçais, A.; Belot, A.; Viel, S.; Mingueneau, M.; Guckian, K.; Walzer, T. S1PR5 is essential for human natural killer cell migration toward sphingosine-1 phosphate. *J. Allergy Clin. Immunol.* **2018**, *141*, 2265–2268.e1. [CrossRef]

161. Jenne, C.N.; Enders, A.; Rivera, R.; Watson, S.R.; Bankovich, A.J.; Pereira, J.P.; Xu, Y.; Roots, C.M.; Beilke, J.N.; Banerjee, A.; et al. T-bet–dependent S1P5 expression in NK cells promotes egress from lymph nodes and bone marrow. *J. Exp. Med.* **2009**, *206*, 2469–2481. [CrossRef]

162. Mayol, K.; Biajoux, V.; Marvel, J.; Balabanian, K.; Walzer, T. Sequential desensitization of CXCR4 and S1P5 controls natural killer cell trafficking. *Blood* **2011**, *118*, 4863–4871. [CrossRef] [PubMed]

163. Liu, G.; Yang, K.; Burns, S.; Shrestha, S.; Chi, H. The S1P1-mTOR axis directs the reciprocal differentiation of TH1 and Treg cells. *Nat. Immunol.* **2010**, *11*, 1047–1056. [CrossRef] [PubMed]

164. Chakraborty, P.; Vaena, S.G.; Thyagarajan, K.; Chatterjee, S.; Al-Khami, A.; Selvam, S.P.; Nguyen, H.; Kang, I.; Wyatt, M.W.; Baliga, U.; et al. Pro-Survival Lipid Sphingosine-1-Phosphate Metabolically Programs T Cells to Limit Anti-tumor Activity. *Cell Rep.* **2019**, *28*, 1879–1893.e7. [CrossRef] [PubMed]

165. Sanger Mouse Genetics Project; van der Weyden, L.; Arends, M.J.; Campbell, A.D.; Bald, T.; Wardle-Jones, H.; Griggs, N.; Velasco-Herrera, M.D.C.; Tüting, T.; Sansom, O.J.; et al. Genome-wide in vivo screen identifies novel host regulators of metastatic colonization. *Nature* **2017**, *541*, 233–236. [CrossRef] [PubMed]

166. Assi, E.; Cervia, D.; Bizzozero, L.; Capobianco, A.; Pambianco, S.; Morisi, F.; De Palma, C.; Moscheni, C.; Pellegrino, P.; Clementi, E.; et al. Modulation of Acid Sphingomyelinase in Melanoma Reprogrammes the Tumour Immune Microenvironment. *Mediators Inflamm.* **2015**, *2015*, 1–13. [CrossRef]

167. Coerdt, I.; Full-Scharrer, G.; Hösl, E. Long-term results in surgical treatment of cleft lips and palates. *Prog. Pediatr. Surg.* **1977**, *10*, 1–3.

168. Nakagawa, R.; Serizawa, I.; Motoki, K.; Sato, M.; Ueno, H.; Iijima, R.; Nakamura, H.; Shimosaka, A.; Koezuka, Y. Antitumor activity of alpha-galactosylceramide, KRN7000, in mice with the melanoma B16 hepatic metastasis and immunohistological study of tumor infiltrating cells. *Oncol. Res.* **2000**, *12*, 51–58. [CrossRef]

169. Ghosh, S.; Juin, S.K.; Nandi, P.; Majumdar, S.B.; Bose, A.; Baral, R.; Sil, P.C.; Majumdar, S. PKCζ mediated anti-proliferative effect of C2 ceramide on neutralization of the tumor microenvironment and melanoma regression. *Cancer Immunol. Immunother. CII* **2020**, *69*, 611–627. [CrossRef]

170. Tiwary, S.; Berzofsky, J.A.; Terabe, M. Altered Lipid Tumor Environment and Its Potential Effects on NKT Cell Function in Tumor Immunity. *Front. Immunol.* **2019**, *10*, 2187. [CrossRef]

171. Bay, S.; Fort, S.; Birikaki, L.; Ganneau, C.; Samain, E.; Coïc, Y.-M.; Bonhomme, F.; Dériaud, E.; Leclerc, C.; Lo-Man, R. Induction of a melanoma-specific antibody response by a monovalent, but not a divalent, synthetic GM2 neoglycopeptide. *ChemMedChem* **2009**, *4*, 582–587. [CrossRef]

172. Fernandez, L.E.; Gabri, M.R.; Guthmann, M.D.; Gomez, R.E.; Gold, S.; Fainboim, L.; Gomez, D.E.; Alonso, D.F. NGcGM3 ganglioside: A privileged target for cancer vaccines. *Clin. Dev. Immunol.* **2010**, *2010*, 814397. [CrossRef]

173. Pérez, K.; Osorio, M.; Hernández, J.; Carr, A.; Fernández, L.E. NGcGM3/VSSP vaccine as treatment for melanoma patients. *Hum. Vaccines Immunother.* **2013**, *9*, 1237–1240. [CrossRef] [PubMed]

174. Bernhard, H.; Meyer zum Büschenfelde, K.H.; Dippold, W.G. Ganglioside GD3 shedding by human malignant melanoma cells. *Int. J. Cancer* **1989**, *44*, 155–160. [CrossRef] [PubMed]

175. Portoukalian, J.; Zwingelstein, G.; Abdul-Malak, N.; Doré, J.F. Alteration of gangliosides in plasma and red cells of humans bearing melanoma tumors. *Biochem. Biophys. Res. Commun.* **1978**, *85*, 916–920. [CrossRef]

176. Péguet-Navarro, J.; Sportouch, M.; Popa, I.; Berthier, O.; Schmitt, D.; Portoukalian, J. Gangliosides from human melanoma tumors impair dendritic cell differentiation from monocytes and induce their apoptosis. *J. Immunol. Baltim. Md 1950* **2003**, *170*, 3488–3494. [CrossRef]

177. Bennaceur, K.; Popa, I.; Chapman, J.A.; Migdal, C.; Péguet-Navarro, J.; Touraine, J.-L.; Portoukalian, J. Different mechanisms are involved in apoptosis induced by melanoma gangliosides on human monocyte-derived dendritic cells. *Glycobiology* **2009**, *19*, 576–582. [CrossRef]

178. Tsao, C.-Y.; Sabbatino, F.; Cheung, N.-K.V.; Hsu, J.C.-F.; Villani, V.; Wang, X.; Ferrone, S. Anti-proliferative and pro-apoptotic activity of GD2 ganglioside-specific monoclonal antibody 3F8 in human melanoma cells. *Oncoimmunology* **2015**, *4*, e1023975. [CrossRef]

179. Kushner, B.H.; Cheung, I.Y.; Modak, S.; Basu, E.M.; Roberts, S.S.; Cheung, N.-K. Humanized 3F8 Anti-GD2 Monoclonal Antibody Dosing With Granulocyte-Macrophage Colony-Stimulating Factor in Patients With Resistant Neuroblastoma: A Phase 1 Clinical Trial. *JAMA Oncol.* **2018**, *4*, 1729–1735. [CrossRef]

180. Kramer, K.; Pandit-Taskar, N.; Humm, J.L.; Zanzonico, P.B.; Haque, S.; Dunkel, I.J.; Wolden, S.L.; Donzelli, M.; Goldman, D.A.; Lewis, J.S.; et al. A phase II study of radioimmunotherapy with intraventricular 131 I-3F8 for medulloblastoma. *Pediatr. Blood Cancer* **2018**, *65*. [CrossRef]

181. Yu, J.; Wu, X.; Yan, J.; Yu, H.; Xu, L.; Chi, Z.; Sheng, X.; Si, L.; Cui, C.; Dai, J.; et al. Anti-GD2/4-1BB chimeric antigen receptor T cell therapy for the treatment of Chinese melanoma patients. *J. Hematol. Oncol. J. Hematol. Oncol.* **2018**, *11*, 1. [CrossRef]

182. Albertini, M.R.; Yang, R.K.; Ranheim, E.A.; Hank, J.A.; Zuleger, C.L.; Weber, S.; Neuman, H.; Hartig, G.; Weigel, T.; Mahvi, D.; et al. Pilot trial of the hu14.18-IL2 immunocytokine in patients with completely resectable recurrent stage III or stage IV melanoma. *Cancer Immunol. Immunother. CII* **2018**, *67*, 1647–1658. [CrossRef] [PubMed]

183. Wu, Y.; Deng, W.; McGinley, E.C.; Klinke, D.J. Melanoma exosomes deliver a complex biological payload that upregulates PTPN11 to suppress T lymphocyte function. *Pigment Cell Melanoma Res.* **2017**, *30*, 203–218. [CrossRef] [PubMed]

184. Zhou, J.; Yang, Y.; Wang, W.; Zhang, Y.; Chen, Z.; Hao, C.; Zhang, J. Melanoma-released exosomes directly activate the mitochondrial apoptotic pathway of CD4+ T cells through their microRNA cargo. *Exp. Cell Res.* **2018**, *371*, 364–371. [CrossRef] [PubMed]

185. Düchler, M.; Czernek, L.; Peczek, L.; Cypryk, W.; Sztiller-Sikorska, M.; Czyz, M. Melanoma-Derived Extracellular Vesicles Bear the Potential for the Induction of Antigen-Specific Tolerance. *Cells* **2019**, *8*, 665. [CrossRef] [PubMed]

186. Vignard, V.; Labbé, M.; Marec, N.; André-Grégoire, G.; Jouand, N.; Fonteneau, J.-F.; Labarrière, N.; Fradin, D. MicroRNAs in Tumor Exosomes Drive Immune Escape in Melanoma. *Cancer Immunol. Res.* **2020**, *8*, 255–267. [CrossRef] [PubMed]

187. Sharma, P.; Diergaarde, B.; Ferrone, S.; Kirkwood, J.M.; Whiteside, T.L. Melanoma cell-derived exosomes in plasma of melanoma patients suppress functions of immune effector cells. *Sci. Rep.* **2020**, *10*, 92. [CrossRef]

188. Chen, G.; Huang, A.C.; Zhang, W.; Zhang, G.; Wu, M.; Xu, W.; Yu, Z.; Yang, J.; Wang, B.; Sun, H.; et al. Exosomal PD-L1 contributes to immunosuppression and is associated with anti-PD-1 response. *Nature* **2018**, *560*, 382–386. [CrossRef]

189. Poggio, M.; Hu, T.; Pai, C.-C.; Chu, B.; Belair, C.D.; Chang, A.; Montabana, E.; Lang, U.E.; Fu, Q.; Fong, L.; et al. Suppression of Exosomal PD-L1 Induces Systemic Anti-tumor Immunity and Memory. *Cell* **2019**, *177*, 414–427.e13. [CrossRef]

190. Falkson, C.I.; Ibrahim, J.; Kirkwood, J.M.; Coates, A.S.; Atkins, M.B.; Blum, R.H. Phase III trial of dacarbazine versus dacarbazine with interferon alpha-2b versus dacarbazine with tamoxifen versus dacarbazine with interferon alpha-2b and tamoxifen in patients with metastatic malignant melanoma: An Eastern Cooperative Oncology Group study. *J. Clin. Oncol. Off. J. Am. Soc. Clin. Oncol.* **1998**, *16*, 1743–1751. [CrossRef]

191. Middleton, M.R.; Grob, J.J.; Aaronson, N.; Fierlbeck, G.; Tilgen, W.; Seiter, S.; Gore, M.; Aamdal, S.; Cebon, J.; Coates, A.; et al. Randomized phase III study of temozolomide versus dacarbazine in the treatment of patients with advanced metastatic malignant melanoma. *J. Clin. Oncol. Off. J. Am. Soc. Clin. Oncol.* **2000**, *18*, 158–166. [CrossRef]

192. Avril, M.F.; Aamdal, S.; Grob, J.J.; Hauschild, A.; Mohr, P.; Bonerandi, J.J.; Weichenthal, M.; Neuber, K.; Bieber, T.; Gilde, K.; et al. Fotemustine compared with dacarbazine in patients with disseminated malignant melanoma: A phase III study. *J. Clin. Oncol. Off. J. Am. Soc. Clin. Oncol.* **2004**, *22*, 1118–1125. [CrossRef] [PubMed]

193. Flaherty, K.T.; Robert, C.; Hersey, P.; Nathan, P.; Garbe, C.; Milhem, M.; Demidov, L.V.; Hassel, J.C.; Rutkowski, P.; Mohr, P.; et al. Improved survival with MEK inhibition in BRAF-mutated melanoma. *N. Engl. J. Med.* **2012**, *367*, 107–114. [CrossRef] [PubMed]

194. Tian, Y.; Guo, W. A Review of the Molecular Pathways Involved in Resistance to BRAF Inhibitors in Patients with Advanced-Stage Melanoma. *Med. Sci. Monit. Int. Med. J. Exp. Clin. Res.* **2020**, *26*, e920957. [CrossRef] [PubMed]

195. Schadendorf, D.; Hodi, F.S.; Robert, C.; Weber, J.S.; Margolin, K.; Hamid, O.; Patt, D.; Chen, T.-T.; Berman, D.M.; Wolchok, J.D. Pooled Analysis of Long-Term Survival Data From Phase II and Phase III Trials of Ipilimumab in Unresectable or Metastatic Melanoma. *J. Clin. Oncol. Off. J. Am. Soc. Clin. Oncol.* **2015**, *33*, 1889–1894. [CrossRef] [PubMed]

196. Robert, C.; Schachter, J.; Long, G.V.; Arance, A.; Grob, J.J.; Mortier, L.; Daud, A.; Carlino, M.S.; McNeil, C.; Lotem, M.; et al. Pembrolizumab versus Ipilimumab in Advanced Melanoma. *N. Engl. J. Med.* **2015**, *372*, 2521–2532. [CrossRef] [PubMed]

197. Robert, C.; Long, G.V.; Brady, B.; Dutriaux, C.; Maio, M.; Mortier, L.; Hassel, J.C.; Rutkowski, P.; McNeil, C.; Kalinka-Warzocha, E.; et al. Nivolumab in previously untreated melanoma without BRAF mutation. *N. Engl. J. Med.* **2015**, *372*, 320–330. [CrossRef]

198. Larkin, J.; Chiarion-Sileni, V.; Gonzalez, R.; Grob, J.-J.; Rutkowski, P.; Lao, C.D.; Cowey, C.L.; Schadendorf, D.; Wagstaff, J.; Dummer, R.; et al. Five-Year Survival with Combined Nivolumab and Ipilimumab in Advanced Melanoma. *N. Engl. J. Med.* **2019**, *381*, 1535–1546. [CrossRef]

199. Postow, M.A. Managing immune checkpoint-blocking antibody side effects. *Am. Soc. Clin. Oncol. Educ. Book* **2015**, 76–83. [CrossRef]

200. Suo, A.; Chan, Y.; Beaulieu, C.; Kong, S.; Cheung, W.Y.; Monzon, J.G.; Smylie, M.; Walker, J.; Morris, D.; Cheng, T. Anti-PD1-Induced Immune-Related Adverse Events and Survival Outcomes in Advanced Melanoma. *Oncologist* **2020**, *25*, 438–446. [CrossRef]

201. Cervia, D.; Assi, E.; De Palma, C.; Giovarelli, M.; Bizzozero, L.; Pambianco, S.; Di Renzo, I.; Zecchini, S.; Moscheni, C.; Vantaggiato, C.; et al. Essential role for acid sphingomyelinase-inhibited autophagy in melanoma response to cisplatin. *Oncotarget* **2016**, *7*. [CrossRef]

202. Ishitsuka, A.; Fujine, E.; Mizutani, Y.; Tawada, C.; Kanoh, H.; Banno, Y.; Seishima, M. FTY720 and cisplatin synergistically induce the death of cisplatin-resistant melanoma cells through the downregulation of the PI3K pathway and the decrease in epidermal growth factor receptor expression. *Int. J. Mol. Med.* **2014**, *34*, 1169–1174. [CrossRef] [PubMed]

203. Bektas, M.; Jolly, P.S.; Müller, C.; Eberle, J.; Spiegel, S.; Geilen, C.C. Sphingosine kinase activity counteracts ceramide-mediated cell death in human melanoma cells: Role of Bcl-2 expression. *Oncogene* **2005**, *24*, 178–187. [CrossRef] [PubMed]

204. Takahashi, T.; Abe, N.; Kanoh, H.; Banno, Y.; Seishima, M. Synergistic effects of vemurafenib and fingolimod (FTY720) in vemurafenib-resistant melanoma cell lines. *Mol. Med. Rep.* **2018**, *18*, 5151–5158. [CrossRef] [PubMed]

205. Ji, C.; Yang, Y.-L.; He, L.; Gu, B.; Xia, J.-P.; Sun, W.-L.; Su, Z.-L.; Chen, B.; Bi, Z.-G. Increasing ceramides sensitizes genistein-induced melanoma cell apoptosis and growth inhibition. *Biochem. Biophys. Res. Commun.* **2012**, *421*, 462–467. [CrossRef] [PubMed]

206. Tran, M.A.; Smith, C.D.; Kester, M.; Robertson, G.P. Combining nanoliposomal ceramide with sorafenib synergistically inhibits melanoma and breast cancer cell survival to decrease tumor development. *Clin. Cancer Res. Off. J. Am. Assoc. Cancer Res.* **2008**, *14*, 3571–3581. [CrossRef] [PubMed]

207. Zhang, P.; Fu, C.; Hu, Y.; Dong, C.; Song, Y.; Song, E. C6-ceramide nanoliposome suppresses tumor metastasis by eliciting PI3K and PKCζ tumor-suppressive activities and regulating integrin affinity modulation. *Sci. Rep.* **2015**, *5*, 9275. [CrossRef]

208. Kobayashi, E.; Motoki, K.; Uchida, T.; Fukushima, H.; Koezuka, Y. KRN7000, a novel immunomodulator, and its antitumor activities. *Oncol. Res.* **1995**, *7*, 529–534.

209. Guthmann, M.D.; Bitton, R.J.; Carnero, A.J.L.; Gabri, M.R.; Cinat, G.; Koliren, L.; Lewi, D.; Fernandez, L.E.; Alonso, D.F.; Gómez, D.E.; et al. Active specific immunotherapy of melanoma with a GM3 ganglioside-based vaccine: A report on safety and immunogenicity. *J. Immunother.* **2004**, *27*, 442–451. [CrossRef]

210. Irie, R.F.; Ollila, D.W.; O'Day, S.; Morton, D.L. Phase I pilot clinical trial of human IgM monoclonal antibody to ganglioside GM3 in patients with metastatic melanoma. *Cancer Immunol. Immunother. CII* **2004**, *53*, 110–117. [CrossRef]

211. Musumarra, G.; Barresi, V.; Condorelli, D.F.; Scirè, S. A bioinformatic approach to the identification of candidate genes for the development of new cancer diagnostics. *Biol. Chem.* **2003**, *384*, 321–327. [CrossRef]

212. Ene (Nicolae), C.-D.; Nicolae, I. Gangliosides and Antigangliosides in Malignant Melanoma. In *Melanoma—Current Clinical Management and Future Therapeutics*; Murph, M., Ed.; Intech: London, UK, 2015; ISBN 978-953-51-2036-0.

Review

Transcriptional Regulation of Sphingosine Kinase 1

Joseph Bonica [1], Cungui Mao [2,3], Lina M. Obeid [1,2,3,†] and Yusuf A. Hannun [1,2,3,*]

[1]   Department of Pharmacology, Stony Brook University, Stony Brook, NY 11794, USA;
      joseph.bonica@stonybrook.edu (J.B.); lina.obeid@stonybrookmedicine.edu (L.M.O.)
[2]   Department of Medicine, Stony Brook University, Stony Brook, NY 11794, USA;
      cungui.mao@stonybrookmedicine.edu
[3]   Cancer Center, Stony Brook University, Stony Brook, NY 11794, USA
*   Correspondence: yusuf.hannun@stonybrookmedicine.edu; Tel.: +1-631-444-8067
†   Deceased.

Received: 5 October 2020; Accepted: 5 November 2020; Published: 8 November 2020

Abstract: Once thought to be primarily structural in nature, sphingolipids have become increasingly appreciated as second messengers in a wide array of signaling pathways. Sphingosine kinase 1, or SK1, is one of two sphingosine kinases that phosphorylate sphingosine into sphingosine-1-phosphate (S1P). S1P is generally pro-inflammatory, pro-angiogenic, immunomodulatory, and pro-survival; therefore, high SK1 expression and activity have been associated with certain inflammatory diseases and cancer. It is thus important to develop an understanding of the regulation of SK1 expression and activity. In this review, we explore the current literature on SK1 transcriptional regulation, illustrating a complex system of transcription factors, cytokines, and even micro-RNAs (miRNAs) on the post transcriptional level.

Keywords: sphingosine kinase 1; SK1; microRNA; transcription factor; hypoxia; long non-coding RNA

1. Introduction

Sphingolipids are a class of cellular lipids that are involved in both maintaining cell structure and mediating cellular signaling processes. Sphingomyelins are important in cell membrane structure, while ceramides, sphingosine, and sphingosine-1-phosphate (S1P) are involved in cell signaling and are thus referred to as bioactive sphingolipids [1,2]. Ceramides can be produced by one of three pathways. The de novo pathway begins with the activity of serine-palmitoyl transferase to eventually generate ceramide. The hydrolytic pathway generates ceramide (and sphingosine) from the hydrolysis of complex sphingolipids. The salvage pathway involves re-generating ceramide and sphingolipids via salvage of sphingosine generated in the lysosome and then re-incorporated into ceramide. All three bioactive sphingolipids are known to have important signaling consequences in both healthy and diseased cell states. Ceramide and sphingosine are known to induce cell death, senescence, and/or differentiation and are generally antiproliferative [1,2]. S1P, on the other hand, is best known to increase cell survival, proliferation, angiogenesis, migration and invasion, and immune cell egress [2–4]. S1P typically signals through one of five G-protein coupled receptors, named S1PR1-5, although there is evidence it also signals intracellularly [5]. Due to the delicate and sometimes opposing nature of the signaling processes utilizing sphingolipids, the absolute and relative concentration of these lipids are tightly controlled [6]. As such, the enzymes involved in the metabolism of sphingolipids are also closely regulated, with their expression and activities modulated by the concentrations of lipids in the cell.

Sphingosine kinases catalyze the phosphorylation of sphingosine into S1P. There exist two known major isoforms of sphingosine kinase, products of the two distinct genes, sphingosine kinase 1 (SK1) and sphingosine kinase 2 (SK2). SK1 is typically localized to and more active at the plasma membrane [7,8], while SK2 is localized to the nucleus upon activation [9,10]. SK1 is the more closely studied of the two isozymes, and understanding its regulation and activity is important due to the pro-proliferative and pro-survival activities of its product S1P; this is especially true in cancer. SK1 is known to be highly upregulated in many cancers, such as breast cancer, colon cancer, head and neck cancer, and glioblastoma [11–14], and its upregulation is associated strongly with poor prognosis and increased cancer metastasis [15–19]. SK1 has also been shown to contribute to chemoresistance in cancer. In CML, for instance, higher expression of SK1 led to disruption in the ratio between C18 ceramide and S1P, contributing to imatinib resistance [20]. In melanoma, high SK1 expression is shown to contribute to resistance to immune checkpoint inhibitors [21]. Metabolically, SK1 drives the formation of S1P while also serving to clear sphingosine, and thus providing an exit from the sphingolipid metabolic network. As such, SK1 may also regulate the levels of ceramide and possibly other upstream sphingolipids. SK1 is thus considered a particularly key regulator of the levels of bioactive sphingolipids. Therefore, it is important to understand the mechanisms of SK1 regulation, to generate further study in cancer, and to identify potential drug targets.

SK1 regulation is well studied and has been discussed in several reviews [7,22,23]. However, it should be noted that work looking specifically at transcriptional regulation of the enzyme is somewhat limited. The gene for sphingosine kinase 1 is located on chromosome 17q25.2 [24]. Work on the rat sphk1 gene showed 6 exons, although 6 alternative first exons were also detected [25]. The human SPHK1 gene is 7 exons in length and 11,276 bp long.

The current body of work reveals that SK1 expression is regulated by several different transcription factors, indicating the enzyme's importance in different signaling pathways. These pathways encompass such important conditions as neuronal growth, hypoxia, ischemia, and cancer. It is known that three primary isoforms of SK1 exist, namely, SK1a, SK1b, and SK1c [25–27]. The expression and distribution of these isoforms is partly governed by the methylation of CpG islands in the SK1 promoter [25]. At 384 amino acids, SK1a is the smallest isoform and the most highly expressed; SK1b is 398 amino acids long, and SK1c is 470 [27]. While not much is known about differences in activity or biological effects of these isoforms, there is evidence that SK1a and SK1b interact with different proteins in breast cancer [28]. Additionally, SK1b appears to be resistant to the SK1 inhibitor Ski in prostate cancer cells [29]. Otherwise, the three isoforms differ only in the size of their N-terminal regions [24]. There is also increasing evidence that SK1 regulation is partially governed by microRNAs, or miRNAs, and that dysregulation of miRNAs in cancer is responsible for higher expression of SK1. In this review, we discuss what is known about SK1 transcriptional and post-transcriptional regulation, what signaling pathways effect and are affected by SK1 regulation, and what further work needs to be done to fully understand regulation of this important enzyme.

2. Transcriptional Regulation-Sp1

Much work on SK1 transcriptional regulation implicates specificity protein 1, or Sp1, as an important transcription factor. In rat PC-12 cells, it was shown that neuronal growth factor (NGF) led to increased SK1 expression by increasing the expression of Sp1 [30]. In these cells, Sp1 was shown to interact specifically with exon 1d of the SK1 gene. Sp1 has also been shown to regulate SK1 in humans; for instance, Sp1 has been demonstrated to be packaged and transported in exosomes to upregulate SK1 in nearby cells, which protects from ischemia/perfusion injury [31]. Here, Sp1 is shown to interact with the region 0.5–0.6 kb upstream of the SK1 promoter (Figure 1). In a hepatocellular cancer model, blocking of Sp1 activity and expression with the known Sp1 inhibitor peretinoin decreased SK1 levels both in vitro and in vivo [32]. Interestingly, Sp1 is overexpressed in several cancers, such as breast, pancreatic, lung, glioma, and thyroid [33]. Many of these cancers also show upregulated SK1. Much like SK1, higher levels of Sp1 in these cancers are correlated with increased severity, stage, angiogenesis, and metastasis. This indicates an important relationship between SK1 and Sp1 and implies that increased Sk1 activity and expression can be used as a readout for conditions that increase Sp1 activity and/or expression.

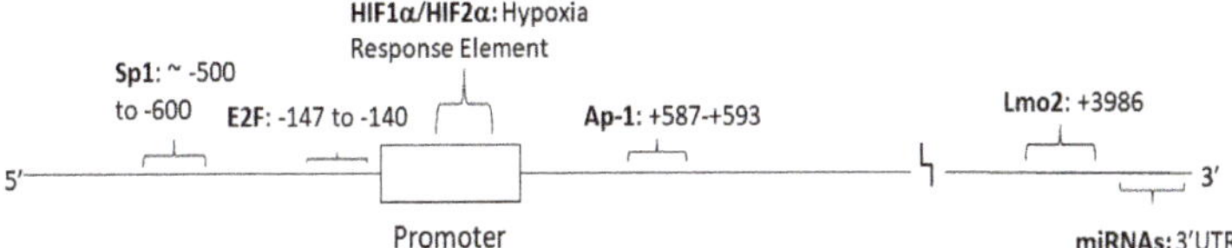

Figure 1. Location of transcription factor binding on the SK1 Gene. Approximate locations of *SPHK1* transcription factors in relation to the *SPHK1* promoter region. Transcription factors have been shown to bind both upstream and downstream of the promoter. E2F is known to be important in regulating SK1 in cancer, while Sp1 is associated with upregulating SK1 in cancer and in response to neuronal growth factors. In hypoxic conditions, HIF2a and Lmo2 were shown to upregulate SK1, especially in cancer models. AP-1 upregulates SK1 in response to the cytokine IL-1b. MicroRNAs (miRNAs) have been shown to bind to certain sequences in the 3'UTR of the *SPHK1* transcript.

3. Hypoxia and Ischemia

Due to S1P's importance in angiogenesis, SK1 has long been studied as an important element in response to hypoxia. As oxygen levels decline, the body needs more blood vessels to move blood and oxygen quickly. This need for further angiogenesis serves to stimulate the upregulation of SK1 in hypoxic conditions. In endothelial cells, hypoxia is curiously shown to regulate SK1 but not SK2 [34], highlighting the former's importance in angiogenesis. SK1 has a hypoxia response element (HRE) in its promoter (Figure 1), and as such, it responds to hypoxia sensitive transcription factors [34]. The regulation of SK1 in response to hypoxia has been demonstrated several times in living systems. For instance, SK1 has been shown to be upregulated in arteries after short periods of hypoxia [35]. Acute and chronic hypoxia have also been shown to upregulate SK1 in human pulmonary smooth muscle cells [36]. SK1 is also implicated in the regulation of two main hypoxia-induced transcription factors (HIFs), HIF1a and HIF2a [37].

Ischemia, which often leads to hypoxia, is known to regulate SK1, leading to both protective and deleterious effects on the involved system. For instance, SK1 was demonstrated to be substantially upregulated in a mouse stroke model, and its upregulation increased inflammatory response and poorer outcome [38]. A large increase in SK1 expression was similarly shown in the area of stroke lesion [39]. SK1 has also been shown to be upregulated in activated microglia, and to play a key role in the inflammatory response to cerebral ischemia-reperfusion (IR) [39–42]. Cerebral IR was shown in brain tissues to upregulate SK1, which in turn increased IL-17A expression in primary microglia [40]. The neuronal injury following cerebral IR in this model was reduced via administration of the SK1 inhibitor PF-543, cementing SK1's role in driving inflammatory injury in this system. Further work demonstrated that SK1 affects IL-17 expression via upstream effects on TRAF2 expression and NFκB activation [41]. Crosstalk between SK1 and TLR2 has also been implicated in the inflammatory response to cerebral IR, as both were shown to be upregulated in microglia after cerebral IR [42]. In addition to PF-543, treatment of mice with fingolimod (a pro-drug functional antagonist of S1PR1) reduced hemorrhagic transformation and stroke injury [39]. These results suggest that targeting SK1 represents a potential treatment option for stroke and post-stroke injury, which have limited treatment options. Interestingly, SK1 upregulation induced by the anesthetic isoflurane is actually protective against intestinal injury in a renal ischemia model known to result in intestinal injury [43]. A separate renal IR injury model in mice showed that the increase in SK1 expression mediated the severity of the injury [44]. Upregulation of SK1 via exposure to conditioned media from mesenchymal stem cells in endothelial colony-forming cells potentiated the revascularization of endothelial colony-forming cells, considered an exciting means of treating infarct damage [45]. However, SK1 is also shown to be upregulated after myocardial infarction, and it contributes to dysfunctional cardiac remodeling and heart failure [46]. The specific transcription factors and regulatory pathways upstream of SK1 upregulation in ischemia do not appear to be widely studied; however, and future research in this area can improve our understanding of SK1's various apparent roles in the ischemic response.

There is evidence that SK2 expression, on the other hand, offers protection against ischemic injury in a potentially compensatory manner. For instance, SK2 has been shown to be important in hypoxic preconditioning, and its activity and expression were required for the protective effects of such conditioning [47]. In cardiomyocytes, hypoxic preconditioning was responsible for increased expression of SK2, which was required to prevent apoptosis in ischemic conditions [48].

Hypoxia is an important response in cancer, due to often low oxygen conditions in the tumor; as such, hypoxic pathways are considered potential therapeutic targets in cancer [49]. SK1 is known to have several roles in cancer-induced hypoxia. Interestingly, most evidence points to HIF2alpha and not HIF1alpha as the primary regulator of SK1 in hypoxia [50,51]. In the U87MG glioma cell line, knockdown of HIF2a but not HIF1a led to downregulation of SK1 [50]. Transfection of cells with the SK1 promoter, however, led to upregulation of SK1 in response to CoCl2 treatment to simulate hypoxia, and several HRE elements were identified in the SK1 promoter [50]. HIF2a has also been associated with SK1 upregulation in clear cell renal carcinoma (ccRC) cells, where again knockdown of HIF2a led to reduced SK1 expression [51].There also exists evidence of a feedback loop between expression of HIF2a and SK1 in certain systems; a different model of ccRC from the one above seems to implicate SK1 in HIF2a regulation [52]. Another transcription factor important in angiogenesis, Lim domain only 2 transcription factor (Lmo2), has been shown to directly upregulate SK1 by binding to a sequence +3986 bases from the promoter [53]. This demonstrates the importance of SK1 in the hypoxic response, and especially the hypoxic cancer response.

4. Cytokines

SK1 and S1P are known to be involved in several inflammatory signaling pathways [23], and SK1 expression and activity are themselves regulated by several cytokines. In glioblastoma, it was found that SK1 is upregulated via the cytokine IL-1 through a JNK/c-Jun dependent pathway [26]. Since additional data show that high expression of SK1 correlates negatively with glioblastoma prognosis [17], understanding the mechanism of SK1 upregulation in response to cytokines is vital to understanding the disease severity. Interestingly, IL-1B only seems to upregulate the SK1a and SK1c isoforms in this system, with little effect on SK1b. Here, SK1 was shown to be upregulated by the binding of transcription factor AP-1 to the first intron at +587 to +593 [26].

In leukemia macrophage THP-1 cells, LPS initiated toll-like receptor 4 signaling stimulated SK1 expression, which led to the accumulation of IL-6 [54]. While evidence exists for SK1's involvement in inflammatory pathways, evidence for cytokines themselves, leading to SK1 upregulation continues to be somewhat limited. It should be noted though, that cytokines have been also shown to activate SK1 through non-transcriptional mechanisms TNFα, for instance, is known to increase SK1 activity by inducing the phosphorylation of Ser 225, necessary for the trafficking of SK1 to the plasma membrane [55]. SK1 has also been shown to be activated by tumor necrosis factor receptor-associated factor 2 (TRAF2) via direct binding to the protein [56].

Transforming growth factor-B (TGF-B) signaling is known to upregulate and activate SK1, and SK1 activity has been shown to be important in many TGF-B dependent pathways. Treatment of fibroblasts with TGF-B led to increased SK1 expression, which was required for TIMP-1 upregulation [57,58]. TGF-B induced upregulation of SK1 has been shown to be important in coronary artery disease and liver fibrosis [58,59]. In breast cancer, TGF-B driven SK1 upregulation has been implicated in bone metastasis [60]. Despite the apparent link between the well-studied TGF-B signaling pathway and SK1 upregulation, the exact transcription factors and regulatory elements governing SK1's response to TGF-B are currently not well established.

5. E2F and Long Non-Coding RNAs

In certain cancers, there is evidence of SK1 expression being under the control of transcription factors known as E2F transcription factors. This family of eight transcription factors is known to regulate proliferation, apoptosis, and stress responses [61]. Increased expression of E2Fs result in oncogenic activity in several cancers [62]. In head and neck squamous cell carcinoma, SK1 expression was shown to be downstream of E2F7, which is a direct transcription factor of SK1 [63]. In this cancer, E2F7-driven upregulation of SK1 is associated with anthracycline resistance; inhibition of E2F7 activity reduced SK1 levels and sensitized tumor cells to anthracycline chemotherapies. However, not all E2F family transcription factors regulate SK1 expression through direct actions on its promoter. For instance, the transcription factor E2F1 has been shown in liver carcinoma to regulate SK1 via the long non-coding RNA called HULC, or highly upregulated in liver cancer [64]. The connection between E2F transcription factors and SK1 regulation, especially in cancer, is one that merits further investigation based on these intriguing data.

Long non-coding RNAs, or lncRNAs, are RNA molecules of at least 200 nucleotides in length that do not code for any proteins [65]. While this type of RNA is incompletely understood, it is believed that they are largely regulatory in nature and known to be involved in epigenetic regulation [65,66]. While lncRNA's role in disease is not completely understood, there exists compelling evidence that they are dysregulated in certain cancers [67]. In the case discussed above, the lncRNA HULC sequesters micro-RNA 107, a miRNA targeting E2F1; this preserves E2F1 and allows it to bind to the promoter [68]. As shown in Figure 1, E2F1 binds to the SK1 promoter at −147 to −140, stimulating expression. HULC's activity in maintaining high SK1 levels is not limited to the liver, however, it has also been shown to upregulate SK1 in non-small cell lung carcinoma [69]. In that case, higher levels of SK1 prevent apoptosis and stimulate proliferation in the cancer cells.

Occasionally, lncRNAs are not transcribed from intergenic regions but from the antisense strand of a gene itself [66–68]. The SK1 gene, *SPHK1*, has its own antisense lncRNA called Khps1, of which several different subtypes exist [69]. Khps1 has been shown in rats to govern CpG island methylation of the *Sphk1* gene, and thus governing tissue differential expression of SK1 [69]. Khps1 has also been found to regulate SK1 by directly associating with histones and altering their structure [70]. However, work on the disease-specific effects of Khps1 on SK1 expression remains limited.

6. MicroRNAs

A growing body of work implicates microRNAs, or miRNAs, in SK1 regulation in both healthy and diseased states. MiRNAs are short oligonucleotides (typically ~22 nt in length), which degrade mRNAs by binding to their 3'untranslated region (3'UTR), targeting them for processing by the RISC [71]. Much of this regulation is relevant to cancer, where miRNAs are found to be frequently dysregulated. MiRNA dysregulation has, in fact, been linked to several diverse types of cancer [72]. There is also evidence of separate dysregulation of miRNA in the tumor microenvironment [73].

Several different miRNAs have been implicated in SK1 regulation, especially in cancer (Table 1). MiR-124, for instance, has been shown to regulate SK1 expression in a variety of cancers, with implications in invasion and metastasis, proliferation, and tumor formation. The 3'-UTR binding site of miR-124 in SK1 has been defined. Typically, miR-124 is downregulated in cancer, which leads to increased expression of SK1 and associated biology. In ovarian cancer, for instance, miR-124 overexpression in cell lines was shown to lead to SK1 degradation, which leads to reduced tumor invasion and migration [74]. This loss in invasive potential was restored with overexpression of SK1 in the cells. It has also been demonstrated that ovarian cancer-associated fibroblasts revert to normal fibroblasts when exposed to exosomes containing miR-124, and this is mediated via the degradation of SK1 in the CAFs [75]. MiR-124 also regulates SK1 in gastric cancer, leading to reduced proliferation and tumorigenicity [76]. MiR-124 has been shown to directly target SK1 in an osteosarcoma model, affecting proliferation, invasion, and matrix metalloprotease expression [77]. Outside of cancer, miR-124 promotes cell death in myocardial infarction by downregulating SK1 [78].

Table 1. microRNAs Known to Downregulate SK1.

MicroRNA	Disease Association	Effects Due to SK1 Downregulation
miR-124	Ovarian Cancer [74]; gastric cancer [76], osteosarcoma [75,77], myocardial infarction [78]	Reduced migration and invasion [74,75]; reduced fibroblast cancer association [76], cell death [78]
miR-506	Hepatocellular carcinoma [79]; pancreatic cancer [80]; osteosarcoma [81].	Reduced angiogenesis [79]; enhanced chemosensitivity [80], mesenchymal-to-epithelial transition [81]
miR-125b	Bladder cancer [82], Alzheimer's disease [68].	Reduced proliferation [82]; increased cell death and disease severity [68]
miR-613	Bladder Cancer [83]	Reduced proliferation, migration, EMT [83]
miR-659-3p	Colon cancer [84], leukemia [68].	Chemosensitivity [84], reduced proliferation [68]
miR-1-3p	Hypoxia in pulmonary smooth muscle cells [85]	Upregulation of Sk1 [85]

MiR-506 has also been established as an SK1 regulator in several different systems. In hepatocellular carcinoma, for instance, MiR-506 downregulated SK1 and inhibited angiogenesis [79]. This work also established a negative correlation between MiR-506 expression and SK1 expression in hepatocellular carcinoma tissues. MiR-506 is also downregulated in pancreatic cancer, and restoring its expression disrupts the SK1/AKT/NFkB axis [80]. This change both enhanced cancer cell death and increased sensitivity to gemcitabine, a chemotherapeutic commonly used in the treatment of pancreatic cancer. In osteosarcoma, a close relative of miR-506 called miR-506-3p inhibits SK1, leading to reduced invasion. Interestingly, the expression of this miRNA even causes the cells to lose metastatic potential via mesenchymal-to-epithelial transition [81].

While much work has focused on those two miRNAs in regulating SK1 expression, evidence has linked other miRNAs to SK1 regulation as well. MiR125b, for instance, has been shown to regulate SK1 expression in bladder cancer [82] and Alzheimer's disease [68]. In the former case, overexpression of miR-125b reduced SK1 levels and proliferation [82]; in the latter case, expression of miR-125b was correlated to more severe disease, as SK1 was downregulated and cell death increased. In bladder cancer, the miRNA-613 is downregulated, leading to upregulation of SK1 and increased proliferation, migration, and EMT [83]. In colon cancer, miR-659-3p reduced SK1 and sensitized them to cisplatin [84], while in K562 leukemia cells, miR-659-3p reduced proliferation by targeting SK1 [68]. Hypoxic conditions downregulated expression of miR-1-3p in pulmonary smooth muscle cells, which led to upregulation of SK1 [85].

7. Discussion

Sphingosine kinase 1, or SK1, is a key enzyme in the sphingolipid pathway, as it converts the pro-death and pro-senescence lipid sphingosine into the pro-survival S1P. The dysregulation of SK1 has been associated with severity in several diseases, especially cancer. Higher expression of SK1 in several types of cancer is associated with poor survival and increased disease severity. High expression of SK1 increases cancer severity by driving proliferation, angiogenesis, metastasis, and chemoresistance through increased production of S1P.

While regulation of SK1 via post-translational modifications such as phosphorylation and proteolysis is well known, the means for transcriptional regulation of SK1 are somewhat less established. It is known that the transcription factor Sp1 upregulates SK1 under certain conditions, which is important in neuronal growth and possibly cancer. Hypoxia is also known to be vital to SK1 upregulation, with HIF2a being the primary transcriptional factor of SK1 in this case. In ischemia, upregulation of SK1 seems to have complex effects on recovery after injury, with the enzyme correlating with better or worse prognosis depending on the system. Despite these varying effects in ischemia, the transcription factors regulating SK1 expression in response to ischemia are not well defined. Further study of the transcriptional regulation of SK1 in ischemia can help broaden understanding of the development of ischemic injury and help establish SK1 as a drug target or possible upstream drug targets in this system. Cytokine signaling has also been shown to regulate SK1 expression, although again, the transcription factors governing this are not well established.

In cancer, the E2F family of transcription factors has recently been demonstrated to affect SK1 expression, which in turn was shown to improve chemoresistance and angiogenesis in tumors. Further work elucidating the connection between E2F transcription factors and SK1 in cancer would go a long way towards understanding the mechanism of these transcription factors in regulating disease. E2Fs have been shown to sometimes regulate SK1 expression via microRNAs, or miRNAs. Indeed, several different miRNAs have been found to regulate SK1 expression, and many of these miRNAs are downregulated in cancer.

Transcriptional upregulation of SK1 in several different diseases makes it an attractive therapeutic target. Some work has been done exploring the effect of small molecule inhibitors of known SK1 transcription factors on certain diseases. For instance, the pan-E2F inhibitor HLM006474 was shown to induce cell death in models of metastatic melanoma and lung cancer [86,87]. However, no work has been done to assess how disruption of SK1 expression may be related to these effects. The acyclic retinoid peretinoin has been shown to prevent carcinogenesis in liver fibroblasts by downregulating SK1 via Sp1 inhibition [32]. However, since transcription factors typically regulate the expression of several genes and since the effects of SK1 in several systems are well established, targeting of SK1 and the S1P pathway is more likely to be an effective treatment option. Experimentally, treatment with both the SK1 inhibitor PF-543 and the S1PR1antagonist fingolimod seemed to alleviate neuronal injury. Several in vivo studies have been conducted using various SK1 inhibitors, probing their effects on diseases such as asthma, sickle-cell anemia, multiple sclerosis, myocardial infarction, arthritis, and several cancers [88].

However, few clinical trials of SK1 inhibitors have been conducted. One such trial, looking at the putative SK1 inhibitor safingol in conjunction with cisplatin in solid tumors, showed that the inhibitor was well tolerated in patients [89]. Despite these safety data, little subsequent work has been done to measure the efficacy of SK1 inhibitors in human patients. In fact, a search of clinical trials occurring currently in the United States on clinicaltrials.gov reveals only 8 clinical trials, and they are mostly probing inhibitors of SK2 rather than SK1. Such work would be especially welcome in cancer, where SK1's role is well established and where there remains a substantial need for targeted therapies. Further investigation should also be done on the role of SK1 in ischemia injury, as the initial results appear promising, and there is an enormous lack of pharmacological options for treatment.

While considerable efforts have been applied to understanding SK1 regulation post-translationally, many elements of SK1 transcriptional regulation remain poorly understood. Some direct regulation via transcription factors has already been established, and we should examine SK1 levels in conditions where it is known that one of these transcription factors is more active. Post-transcriptional regulation also continues to be studied, and research into SK1 regulation via miRNAs is growing particularly rapidly. As we further elucidate just how miRNAs are regulated and effect certain disease states, so should we further study how miRNA downregulation of SK1 contributes to the disease state.

Author Contributions: Writing—original draft preparation: J.B.; writing—reading and editing: J.B., C.M., and Y.A.H.; relevant research and literature study of SK1: J.B. and L.M.O. All authors have read and agreed to the published version of the manuscript.

Funding: This work was partly supported by NIH grant R01 GM130878.

Conflicts of Interest: The authors declare no conflict of interest.

References

1. Ogretmen, B.; Hannun, Y.A. Biologically active sphingolipids in cancer pathogenesis and treatment. *Nat. Rev. Cancer* **2004**, *4*, 604–616. [CrossRef]
2. Hannun, Y.A.; Obeid, L.M. Principles of bioactive lipid signalling: Lessons from sphingolipids. *Nat. Rev. Mol. Cell Biol.* **2008**, *9*, 139–150. [CrossRef]
3. Lee, O.-H.; Lee, Y.-M.; Lee, Y.M.; Moon, E.-J.; Lee, D.-J.; Kim, J.-H.; Kim, K.-W.; Kwon, Y.-G. Sphingosine 1-Phosphate Induces Angiogenesis: Its Angiogenic Action and Signaling Mechanism in Human Umbilical Vein Endothelial Cells. *Biochem. Biophys. Res. Commun.* **1999**, *264*, 743–750. [CrossRef]
4. Pappu, R.; Schwab, S.R.; Cornelissen, I.; João, P.P.; Regard, J.B.; Xu, Y.; Camerer, E.; Zheng, Y.-W.; Huang, Y.; Cyster, J.G.; et al. Promotion of Lymphocyte Egress into Blood and Lymph by Distinct Sources of Sphingosine-1-Phosphate. *Science* **2007**, *316*, 295–298. [CrossRef] [PubMed]
5. Payne, S.G.; Milstien, S.; Spiegel, S. Sphingosine-1-phosphate: Dual messenger functions. *FEBS Lett.* **2002**, *531*, 54–57. [CrossRef]
6. Maceyka, M.; Payne, S.G.; Milstien, S.; Spiegel, S. Sphingosine kinase, sphingosine-1-phosphate, and apoptosis. *Biochim. Biophys. Acta BBA Mol. Cell Biol. Lipids* **2002**, *1585*, 193–201. [CrossRef]
7. Pulkoski-Gross, M.J.; Obeid, L.M. Molecular mechanisms of regulation of sphingosine kinase 1. *Biochim. Biophys. Acta Mol. Cell Biol. Lipids* **2018**, *1863*, 1413–1422. [CrossRef] [PubMed]
8. Johnson, K.R.; Becker, K.P.; Facchinetti, M.M.; Hannun, Y.A.; Obeid, L.M. Pkc-dependent activation of sphingosine kinase 1 and translocation to the plasma membrane: Extracellular release of sphingosine-1-phosphate induced by phorbol 12-myristate 13-acetate (pma). *J. Biol. Chem.* **2002**, *277*, 35257–35262. [CrossRef]
9. Ding, G.; Sonoda, H.; Yu, H.; Kajimoto, T.; Goparaju, S.K.; Jahangeer, S.; Okada, T.; Nakamura, S. Protein Kinase D-mediated Phosphorylation and Nuclear Export of Sphingosine Kinase 2. *J. Biol. Chem.* **2007**, *282*, 27493–27502. [CrossRef]
10. Dominguez, G.; Maddelein, M.-L.; Pucelle, M.; Nicaise, Y.; Maurage, C.-A.; Duyckaerts, C.; Cuvillier, O.; Delisle, M.-B. Neuronal sphingosine kinase 2 subcellular localization is altered in Alzheimer's disease brain. *Acta Neuropathol. Commun.* **2018**, *6*, 25. [CrossRef]
11. Kawamori, T.; Osta, W.; Johnson, K.R.; Pettus, B.J.; Bielawski, J.; Tanaka, T.; Wargovich, M.J.; Reddy, B.S.; Hannun, Y.A.; Obeid, L.M.; et al. Sphingosine kinase 1 is up-regulated in colon carcinogenesis. *Faseb J.* **2006**, *20*, 386–388. [CrossRef] [PubMed]
12. Shirai, K.; Kaneshiro, T.; Wada, M.; Furuya, H.; Bielawski, J.; Hannun, Y.A.; Obeid, L.M.; Ogretmen, B.; Kawamori, T. A Role of Sphingosine Kinase 1 in Head and Neck Carcinogenesis. *Cancer Prev. Res.* **2011**, *4*, 454–462. [CrossRef]
13. Nava, V.E.; Hobson, J.P.; Murthy, S.; Milstien, S.; Spiegel, S. Sphingosine kinase type 1 promotes estrogen-dependent tumorigenesis of breast cancer MCF-7 cells. *Exp. Cell Res.* **2002**, *281*, 115–127. [CrossRef]
14. Estrada-Bernal, A.; Estrada-Bernal, A.; Lawler, S.E.; Nowicki, M.O.; Ray Chaudhury, A.; Van Brocklyn, J.R. The role of sphingosine kinase-1 in EGFRvIII-regulated growth and survival of glioblastoma cells. *J. Neurooncol.* **2011**, *102*, 353–366. [CrossRef]
15. Ruckhäberle, E.; Rody, A.; Engels, K.; Gaetje, R.; von Minckwitz, G.; Schiffmann, S.; Grösch, S.; Geisslinger, G.; Holtrich, U.; Karn, T.; et al. Microarray analysis of altered sphingolipid metabolism reveals prognostic significance of sphingosine kinase 1 in breast cancer. *Breast Cancer Res. Treat.* **2008**, *112*, 41–52. [CrossRef] [PubMed]
16. Li, J.; Guan, H.Y.; Gong, L.Y.; Song, L.B.; Zhang, N.; Wu, J.; Yuan, J.; Zheng, Y.J.; Huang, Z.S.; Li, M. Clinical significance of sphingosine kinase-1 expression in human astrocytomas progression and overall patient survival. *Clin. Cancer Res.* **2008**, *14*, 6996–7003. [CrossRef]
17. Van Brocklyn, J.R.; Jackson, C.A.; Pearl, D.K.; Kotur, M.S.; Snyder, P.J.; Prior, T.W. Sphingosine kinase-1 expression correlates with poor survival of patients with glioblastoma multiforme: Roles of sphingosine kinase isoforms in growth of glioblastoma cell lines. *J. Neuropathol. Exp. Neurol.* **2005**, *64*, 695–705. [CrossRef]
18. Acharya, S.; Yao, J.; Li, P.; Zhang, C.; Lowery, F.J.; Zhang, Q.; Guo, H.; Qu, J.; Yang, F.; Wistuba, I.I.; et al. Sphingosine Kinase 1 Signaling Promotes Metastasis of Triple-Negative Breast Cancer. *Cancer Res.* **2019**, *79*, 4211–4226. [CrossRef]

19. Bae, G.E.; Do, S.I.; Kim, K.; Park, J.H.; Cho, S.; Kim, H.S. Increased Sphingosine Kinase 1 Expression Predicts Distant Metastasis and Poor Outcome in Patients with Colorectal Cancer. *Anticancer Res.* **2019**, *39*, 663–670. [CrossRef] [PubMed]
20. Baran, Y.; Salas, A.; Senkal, C.E.; Gunduz, U.; Bielawski, J.; Obeid, L.M.; Ogretmen, B. Alterations of Ceramide/Sphingosine 1-Phosphate Rheostat Involved in the Regulation of Resistance to Imatinib-induced Apoptosis in K562 Human Chronic Myeloid Leukemia Cells. *J. Biol. Chem.* **2007**, *282*, 10922–10934. [CrossRef]
21. Imbert, C.; Montfort, A.; Fraisse, M.; Marcheteau, E.; Gilhodes, J.; Martin, E.; Bertrand, F.; Marcellin, M.; Burlet-Schiltz, O.; de Peredo, A.G.; et al. Resistance of melanoma to immune checkpoint inhibitors is overcome by targeting the sphingosine kinase-1. *Nat. Commun.* **2020**, *11*, 437. [CrossRef] [PubMed]
22. Heffernan-Stroud, L.A.; Obeid, L.M. Sphingosine kinase 1 in cancer. *Adv. Cancer Res.* **2013**, *117*, 201–235.
23. Snider, A.J.; Gandy, K.A.O.; Obeid, L.M. Sphingosine kinase: Role in regulation of bioactive sphingolipid mediators in inflammation. *Biochimie* **2010**, *92*, 707–715. [CrossRef]
24. Haddadi, N.; Lin, Y.; Simpson, A.M.; Nassif, N.T.; McGowan, E.M. "Dicing and Splicing" Sphingosine Kinase and Relevance to Cancer. *Int. J. Mol. Sci.* **2017**, *18*, 1891. [CrossRef]
25. Imamura, T.; Ohgane, J.; Ito, S.; Ogawa, T.; Hattori, N.; Tanaka, S.; Shiota, K. CpG island of rat sphingosine kinase-1 gene: Tissue-dependent DNA methylation status and multiple alternative first exons. *Genomics* **2001**, *76*, 117–125. [CrossRef]
26. Paugh, B.S.; Bryan, L.; Paugh, S.W.; Wilczynska, K.M.; Alvarez, S.M.; Singh, S.K.; Kapitonov, D.; Rokita, H.; Wright, S.; Griswold-Prenner, I.; et al. Interleukin-1 Regulates the Expression of Sphingosine Kinase 1 in Glioblastoma Cells. *J. Biol. Chem.* **2009**, *284*, 3408–3417. [CrossRef]
27. Venkataraman, K.; Thangada, S.; Michaud, J.; Oo, M.L.; Ai, Y.; Lee, Y.-M.; Wu, M.; Parikh, N.S.; Khan, F.; Proia, R.L.; et al. Extracellular export of sphingosine kinase-1a contributes to the vascular S1P gradient. *Biochem. J.* **2006**, *397*, 461–471. [CrossRef]
28. Yagoub, D.; Wilkins, M.R.; Lay, A.J.; Kaczorowski, D.C.; Hatoum, D.; Bajan, S.; Hutvagner, G.; Lai, J.H.; Wu, W.; Martiniello-Wilks, R.; et al. Sphingosine kinase 1 isoform-specific interactions in breast cancer. *Mol. Endocrinol.* **2014**, *28*, 1899–1915. [CrossRef]
29. Lim, K.G.; Tonelli, F.; Berdyshev, E.; Gorshkova, I.; Leclercq, T.; Pitson, S.M.; Bittman, R.; Pyne, S.; Pyne, N.J. Inhibition kinetics and regulation of sphingosine kinase 1 expression in prostate cancer cells: Functional differences between sphingosine kinase 1a and 1b. *Int. J. Biochem. Cell Biol.* **2012**, *44*, 1457–1464. [CrossRef]
30. Sobue, S.; Hagiwara, K.; Banno, Y.; Tamiya-Koizumi, K.; Suzuki, M.; Takagi, A.; Kojima, T.; Asano, H.; Nozawa, Y.; Murate, T. Transcription factor specificity protein 1 (Sp1) is the main regulator of nerve growth factor-induced sphingosine kinase 1 gene expression of the rat pheochromocytoma cell line, PC12. *J. Neurochem.* **2005**, *95*, 940–949. [CrossRef]
31. Yuan, X.; Li, D.; Chen, X.; Han, C.; Xu, L.; Huang, T.; Dong, Z.; Zhang, M. Extracellular vesicles from human-induced pluripotent stem cell-derived mesenchymal stromal cells (hiPSC-MSCs) protect against renal ischemia/reperfusion injury via delivering specificity protein (SP1) and transcriptional activating of sphingosine kinase 1 and inhibiting necroptosis. *Cell Death Dis.* **2017**, *8*, 3200.
32. Funaki, M.; Kitabayashi, J.; Shimakami, T.; Nagata, N.; Sakai, Y.; Takegoshi, K.; Okada, H.; Murai, K.; Shirasaki, T.; Oyama, T.; et al. Peretinoin, an acyclic retinoid, inhibits hepatocarcinogenesis by suppressing sphingosine kinase 1 expression in vitro and in vivo. *Sci. Rep.* **2017**, *7*, 16978. [CrossRef] [PubMed]
33. Beishline, K.; Azizkhan-Clifford, J. Sp1 and the 'hallmarks of cancer'. *FEBS J.* **2015**, *282*, 224–258. [CrossRef]
34. Schwalm, S.; Döll, F.; Römer, I.; Bubnova, S.; Pfeilschifter, J.; Huwiler, A. Sphingosine kinase-1 is a hypoxia-regulated gene that stimulates migration of human endothelial cells. *Biochem. Biophys. Res. Commun.* **2008**, *368*, 1020–1025. [CrossRef]
35. Alganga, H.; Almabrouk, T.A.M.; Katwan, O.J.; Daly, C.J.; Pyne, S.; Pyne, N.J.; Kennedy, S. Short Periods of Hypoxia Upregulate Sphingosine Kinase 1 and Increase Vasodilation of Arteries to Sphingosine 1-Phosphate (S1P) via S1P$_3$. *J. Pharmacol. Exp. Ther.* **2019**, *371*, 63–74. [CrossRef]
36. Ahmad, M.; Long, J.S.; Pyne, N.J.; Pyne, S. The effect of hypoxia on lipid phosphate receptor and sphingosine kinase expression and mitogen-activated protein kinase signaling in human pulmonary smooth muscle cells. *Prostaglandins Other Lipid Mediat.* **2006**, *79*, 278–286. [CrossRef]
37. Cuvillier, O.; Ader, I. Hypoxia-Inducible Factors and Sphingosine 1-Phosphate Signaling. *Anti-Cancer Agents Med. Chem.* **2011**, *11*, 854–862. [CrossRef]

38. Zheng, S.; Wei, S.; Wang, X.; Xu, Y.; Xiao, Y.; Liu, H.; Jia, J.; Cheng, J. Sphingosine kinase 1 mediates neuroinflammation following cerebral ischemia. *Exp. Neurol.* **2015**, *272*, 160–169. [CrossRef] [PubMed]

39. Matsumoto, N.; Yamashita, T.; Shang, J.; Feng, T.; Osakada, Y.; Sasaki, R.; Tadokoro, K.; Nomura, E.; Tsunoda, K.; Omote, Y. Up-regulation of sphingosine-1-phosphate receptors and sphingosine kinase 1 in the peri-ischemic area after transient middle cerebral artery occlusion in mice. *Brain Res.* **2020**, *1739*, 146831. [CrossRef]

40. Lv, M.; Zhang, D.; Dai, D.; Zhang, W.; Zhang, L. Sphingosine kinase 1/sphingosine-1-phosphate regulates the expression of interleukin-17A in activated microglia in cerebral ischemia/reperfusion. *Inflamm. Res.* **2016**, *65*, 551–562. [CrossRef]

41. Su, D.; Cheng, Y.; Li, S.; Dai, D.; Zhang, W.; Lv, M. Sphk1 mediates neuroinflammation and neuronal injury via TRAF2/NF-κB pathways in activated microglia in cerebral ischemia reperfusion. *J. Neuroimmunol.* **2017**, *305*, 35–41. [CrossRef]

42. Sun, W.; Ding, Z.; Xu, S.; Su, Z.; Li, H. Crosstalk between TLR2 and Sphk1 in microglia in the cerebral ischemia/reperfusion-induced inflammatory response. *Int. J. Mol. Med.* **2017**, *40*, 1750–1758.

43. Kim, M.; Park, S.W.; Kim, M.; D'Agati, V.D.; Lee, H.T. Isoflurane activates intestinal sphingosine kinase to protect against renal ischemia-reperfusion-induced liver and intestine injury. *Anesthesiology* **2011**, *114*, 363–373. [CrossRef]

44. Park, S.W.; Kim, M.; Kim, M.; D'Agati, V.D.; Thomas Lee, H. Sphingosine kinase 1 protects against renal ischemia–reperfusion injury in mice by sphingosine-1-phosphate$_1$ receptor activation. *Kidney Int.* **2011**, *80*, 1315–1327. [CrossRef]

45. Poitevin, S.; Cussac, D.; Leroyer, A.S.; Albinet, V.; Sarlon-Bartoli, G.; Guillet, B.; Hubert, L.; Andrieu-Abadie, N.; Couderc, B.; Parini, A. Sphingosine kinase 1 expressed by endothelial colony-forming cells has a critical role in their revascularization activity. *Cardiovasc. Res.* **2014**, *103*, 121–130. [CrossRef]

46. Zhang, F.; Xia, Y.; Yan, W.; Zhang, H.; Zhou, F.; Zhao, S.; Wang, W.; Zhu, D.; Xin, C.; Yan, L.; et al. Sphingosine 1-phosphate signaling contributes to cardiac inflammation, dysfunction, and remodeling following myocardial infarction. *Am. J. Physiol. Heart Circ. Physiol.* **2016**, *310*, H250–H261. [CrossRef]

47. Wacker, B.K.; Perfater, J.L.; Gidday, J.M. Hypoxic preconditioning induces stroke tolerance in mice via a cascading HIF, sphingosine kinase, and CCL2 signaling pathway. *J. Neurochem.* **2012**, *123*, 954–962. [CrossRef]

48. Zhang, R.; Li, L.; Yuan, L.; Zhao, M. Hypoxic preconditioning protects cardiomyocytes against hypoxia/reoxygenation-induced cell apoptosis via sphingosine kinase 2 and FAK/AKT pathway. *Exp. Mol. Pathol.* **2016**, *100*, 51–58. [CrossRef]

49. Harris, A.L. Hypoxia—A key regulatory factor in tumour growth. *Nat. Rev. Cancer* **2002**, *2*, 38–47. [CrossRef]

50. Anelli, V.; Gault, C.R.; Cheng, A.B.; Obeid, L.M. Sphingosine kinase 1 is up-regulated during hypoxia in u87mg glioma cells: Role of hypoxia-inducible factors 1 and 2. *J. Biol. Chem.* **2008**, *283*, 3365–3375. [CrossRef]

51. Salama, M.F.; Carroll, B.; Adada, M.; Pulkoski-Gross, M.; Hannun, Y.A.; Obeid, L.M. A novel role of sphingosine kinase-1 in the invasion and angiogenesis of VHL mutant clear cell renal cell carcinoma. *FASEB J.* **2015**, *29*, 2803–2813. [CrossRef] [PubMed]

52. Bouquerel, P.; Malavaud, B.; Pyronnet, S.; Martineau, Y.; Ader, I.; Cuvillier, O. Essential role for SphK1/S1P signaling to regulate hypoxia-inducible factor 2α expression and activity in cancer. *Oncogenesis* **2016**, *5*, e209. [CrossRef]

53. Matrone, G.; Meng, S.; Gu, Q.; Lv, J.; Fang, L.; Chen, F.; Cook, J. Lmo2 (LIM-Domain-Only 2) Modulates Sphk1 (Sphingosine Kinase) and Promotes Endothelial Cell Migration. *Arterioscler. Thromb. Vasc. Biol.* **2017**, *37*, 1860–1868. [CrossRef]

54. Pchejetski, D.; Nunes, J.; Coughlan, K.; Lall, H.; Pitson, S.M.; Waxman, J.; Sumbayev, V.V. The involvement of sphingosine kinase 1 in LPS-induced Toll-like receptor 4-mediated accumulation of HIF-1α protein, activation of ASK1 and production of the pro-inflammatory cytokine IL-6. *Immunol. Cell Biol.* **2011**, *89*, 268–274. [CrossRef]

55. Pitson, S.M.; Moretti, P.A.; Zebol, J.R.; Lynn, H.E.; Xia, P.; Vadas, M.A.; Wattenberg, B.W. Activation of sphingosine kinase 1 by ERK1/2-mediated phosphorylation. *Embo J.* **2003**, *22*, 5491–5500. [CrossRef]

56. Xia, P.; Wang, L.; Moretti, P.A.B.; Albanese, N.; Chai, F.; Pitson, S.M.; D'Andrea, R.J.; Gamble, J.R.; Vadas, M.A. Sphingosine Kinase Interacts with TRAF2 and Dissects Tumor Necrosis Factor-α Signaling. *J. Biol. Chem.* **2002**, *277*, 7996–8003. [CrossRef]

57. Yamanaka, M.; Shegogue, D.; Pei, H.; Bu, S.; Bielawska, A.; Bielawski, J.; Pettus, B.; Hannun, Y.A.; Obeid, L.; Trojanowska, M. Sphingosine Kinase 1 (SPHK1) Is Induced by Transforming Growth Factor-β and Mediates TIMP-1 Up-regulation. *J. Biol. Chem.* **2004**, *279*, 53994–54001. [CrossRef]

58. Wang, S.; Wang, S.; Zhang, Q.; Wang, Y.; You, B.; Meng, Q.; Zhang, S.; Li, X.; Ge, Z. Transforming Growth Factor β1 (TGF-β1) Appears to Promote Coronary Artery Disease by Upregulating Sphingosine Kinase 1 (SPHK1) and Further Upregulating Its Downstream TIMP-1. *Med. Sci. Monit. Int. Med. J. Exp. Clin. Res.* **2018**, *24*, 7322–7328. [CrossRef] [PubMed]

59. Yang, L.; Chang, N.; Liu, X.; Han, Z.; Zhu, T.; Li, C.; Yang, L.; Li, L. Bone marrow-derived mesenchymal stem cells differentiate to hepatic myofibroblasts by transforming growth factor-β1 via sphingosine kinase/sphingosine 1-phosphate (S1P)/S1P receptor axis. *Am. J. Pathol.* **2012**, *181*, 85–97. [CrossRef]

60. Stayrook, K.R.; Mack, J.K.; Cerabona, D.; Edwards, D.F.; Bui, H.H.; Niewolna, M.; Fournier, P.G.; Mohammad, K.S.; Waning, D.L.; Guise, T.A. TGFβ-Mediated induction of SphK1 as a potential determinant in human MDA-MB-231 breast cancer cell bone metastasis. *Bonekey Rep.* **2015**, *4*, 719. [CrossRef]

61. DeGregori, J.; David, G.J. Distinct and Overlapping Roles for E2F Family Members in Transcription, Proliferation and Apoptosis. *Curr. Mol. Med.* **2006**, *6*, 739–748.

62. Kent, L.N.; Leone, G. The broken cycle: E2F dysfunction in cancer. *Nat. Rev. Cancer* **2019**, *19*, 326–338. [CrossRef]

63. Hazar-Rethinam, M.; de Long, L.M.; Gannon, O.M.; Topkas, E.; Boros, S.; Vargas, A.C.; Dzienis, M.; Mukhopadhyay, P.; Simpson, F.; Endo-Munoz, L.; et al. A Novel E2F/Sphingosine Kinase 1 Axis Regulates Anthracycline Response in Squamous Cell Carcinoma. *Clin. Cancer Res.* **2015**, *21*, 417–427. [CrossRef]

64. Lu, Z.; Lu, Z.; Xiao, Z.; Liu, F.; Cui, M.; Li, W.; Yang, Z.; Li, J.; Ye, L.; Zhang, X. Long non-coding RNA HULC promotes tumor angiogenesis in liver cancer by up-regulating sphingosine kinase 1 (SPHK1). *Oncotarget* **2016**, *7*, 241–254. [CrossRef]

65. Ponting, C.P.; Oliver, P.L.; Reik, W. Evolution and Functions of Long Noncoding RNAs. *Cell* **2009**, *136*, 629–641. [CrossRef]

66. Boon, R.A.; Jaé, N.; Holdt, L.; Dimmeler, S. Long Noncoding RNAs: From Clinical Genetics to Therapeutic Targets? *J. Am. Coll. Cardiol.* **2016**, *67*, 1214–1226. [CrossRef]

67. Moran, V.A.; Perera, R.J.; Khalil, A.M. Emerging functional and mechanistic paradigms of mammalian long non-coding RNAs. *Nucleic Acids Res.* **2012**, *40*, 6391–6400. [CrossRef]

68. Liu, Z.; He, C.; Qu, Y.; Chen, X.; Zhu, H.; Xiang, B. MiR-659-3p regulates the progression of chronic myeloid leukemia by targeting SPHK1. *Int. J. Clin. Exp. Pathol.* **2018**, *11*, 2470–2478.

69. Imamura, T.; Yamamoto, S.; Ohgane, J.; Hattori, N.; Tanaka, S.; Shiota, K. Non-coding RNA directed DNA demethylation of Sphk1 CpG island. *Biochem. Biophys. Res. Commun.* **2004**, *322*, 593–600. [CrossRef]

70. Postepska-Igielska, A.; Giwojna, A.; Gasri-Plotnitsky, L.; Schmitt, N.; Dold, A.; Ginsberg, D.; Grummt, I. LncRNA Khps1 Regulates Expression of the Proto-oncogene SPHK1 via Triplex-Mediated Changes in Chromatin Structure. *Mol. Cell* **2015**, *60*, 626–636. [CrossRef] [PubMed]

71. Kim, V.N.; Han, J.; Siomi, M.C. Biogenesis of small RNAs in animals. *Nat. Rev. Mol. Cell Biol.* **2009**, *10*, 126–139. [CrossRef]

72. Iorio, M.V.; Croce, C.M. MicroRNAs in cancer: Small molecules with a huge impact. *J. Clin. Oncol. Off. J. Am. Soc. Clin. Oncol.* **2009**, *27*, 5848–5856. [CrossRef]

73. Schoepp, M.; Ströse, A.J.; Haier, J. Dysregulation of miRNA Expression in Cancer Associated Fibroblasts (CAFs) and Its Consequences on the Tumor Microenvironment. *Cancers* **2017**, *9*, 54. [CrossRef] [PubMed]

74. Zhang, H.; Wang, Q.; Zhao, Q.; Di, W. MiR-124 inhibits the migration and invasion of ovarian cancer cells by targeting SphK1. *J. Ovarian Res.* **2013**, *6*, 84. [CrossRef]

75. Zhang, Y.; Cai, H.; Chen, S.; Sun, D.; Zhang, D.; He, Y. Exosomal transfer of miR-124 inhibits normal fibroblasts to cancer-associated fibroblasts transition by targeting sphingosine kinase 1 in ovarian cancer. *J. Cell. Biochem.* **2019**, *120*, 13187–13201. [CrossRef]

76. Xia, J.; Wu, Z.; Yu, C.; He, W.; Zheng, H.; He, Y.; Jian, W.; Chen, L.; Zhang, L.; Li, W. miR-124 inhibits cell proliferation in gastric cancer through down-regulation of SPHK1. *J. Pathol.* **2012**, *227*, 470–480. [CrossRef]

77. Zhou, Y.; Han, Y.; Zhang, Z.; Shi, Z.; Zhou, L.; Liu, X.; Jia, X. MicroRNA-124 upregulation inhibits proliferation and invasion of osteosarcoma cells by targeting sphingosine kinase 1. *Hum. Cell* **2017**, *30*, 30–40. [CrossRef]

78. Liu, B.F.; Chen, Q.; Zhang, M.; Zhu, Y.K. MiR-124 promotes ischemia-reperfusion induced cardiomyocyte apoptosis by targeting sphingosine kinase 1. *Eur. Rev. Med. Pharmacol. Sci.* **2019**, *23*, 7049–7058.

79. Lu, Z.; Zhang, W.; Gao, S.; Jiang, Q.; Xiao, Z.; Ye, L.; Zhang, X. MiR-506 suppresses liver cancer angiogenesis through targeting sphingosine kinase 1 (SPHK1) mRNA. *Biochem. Biophys. Res. Commun.* **2015**, *468*, 8–13. [CrossRef]

80. Li, J.; Wu, H.; Li, W.; Yin, L.; Guo, S.; Xu, X.; Ouyang, Y.; Zhao, Z.; Liu, S.; Tian, Y. Downregulated miR-506 expression facilitates pancreatic cancer progression and chemoresistance via SPHK1/Akt/NF-κB signaling. *Oncogene* **2016**, *35*, 5501–5514. [CrossRef]

81. Wang, D.; Bao, F.; Teng, Y.; Li, Q.; Li, J. MicroRNA-506-3p initiates mesenchymal-to-epithelial transition and suppresses autophagy in osteosarcoma cells by directly targeting SPHK1. *Biosci. Biotechnol. Biochem.* **2019**, *83*, 836–844. [CrossRef]

82. Zhao, X.; He, W.; Li, J.; Huang, S.; Wan, X.; Luo, H.; Wu, D. MiRNA-125b inhibits proliferation and migration by targeting SphK1 in bladder cancer. *Am. J. Transl. Res.* **2015**, *7*, 2346–2354.

83. Yu, H.; Duan, P.; Zhu, H.; Rao, D. miR-613 inhibits bladder cancer proliferation and migration through targeting SphK1. *Am. J. Transl. Res.* **2017**, *9*, 1213–1221. [PubMed]

84. Li, S.; Fang, Y.; Qin, H.; Fu, W.; Zhang, X. miR-659-3p is involved in the regulation of the chemotherapy response of colorectal cancer via modulating the expression of SPHK1. *Am. J. Cancer Res.* **2016**, *6*, 1976–1985.

85. Sysol, J.R.; Chen, J.; Singla, S.; Zhao, S.; Comhair, S.; Natarajan, V.; Machado, R. Micro-RNA-1 is decreased by hypoxia and contributes to the development of pulmonary vascular remodeling via regulation of sphingosine kinase 1. *Am. J. Physiol. Lung Cell Mol. Physiol.* **2018**, *314*, L461–L472. [CrossRef]

86. Kurtyka, C.A.; Chen, L.; Cress, W.D. E2F Inhibition Synergizes with Paclitaxel in Lung Cancer Cell Lines. *PLoS ONE* **2014**, *9*, e96357. [CrossRef]

87. Rouaud, F.; Hamouda-Takaya, N.; Cerezo, M.; Abbe, P.; Zangari, J.; Hofman, V.; Ohanna, M.; Mograbi, B.; El-Hachem, N.; Benfodda, Z.; et al. E2F1 inhibition mediates cell death of metastatic melanoma. *Cell Death Dis.* **2018**, *9*, 527. [CrossRef]

88. Pitman, M.R.; Costabile, M.; Pitson, S.M. Recent advances in the development of sphingosine kinase inhibitors. *Cell. Signal.* **2016**, *28*, 1349–1363. [CrossRef] [PubMed]

89. Dickson, M.A.; Carvajal, R.; Merrill, A.; Gonen, M.; Cane, L.; Schwartz, G. A phase I clinical trial of safingol in combination with cisplatin in advanced solid tumors. *Clin. Cancer Res.* **2011**, *17*, 2484–2492. [CrossRef]

Publisher's Note: MDPI stays neutral with regard to jurisdictional claims in published maps and institutional affiliations.

Article

Dissecting $G_{q/11}$-Mediated Plasma Membrane Translocation of Sphingosine Kinase-1

Kira Vanessa Blankenbach [1,†], Ralf Frederik Claas [1,†], Natalie Judith Aster [1,†], Anna Katharina Spohner [1], Sandra Trautmann [2], Nerea Ferreirós [2], Justin L. Black [3], John J. G. Tesmer [4], Stefan Offermanns [5], Thomas Wieland [6] and Dagmar Meyer zu Heringdorf [1,*]

[1] Institut für Allgemeine Pharmakologie und Toxikologie, Universitätsklinikum, Goethe-Universität, Theodor-Stern-Kai 7, 60590 Frankfurt am Main, Germany; kira.blankenbach@web.de (K.V.B.); frederik.claas@gmx.de (R.F.C.); s5599083@stud.uni-frankfurt.de (N.J.A.); spohner@em.uni-frankfurt.de (A.K.S.)

[2] Institut für Klinische Pharmakologie, Universitätsklinikum, Goethe-Universität, Theodor-Stern-Kai 7, 60590 Frankfurt am Main, Germany; trautmann@med.uni-frankfurt.de (S.T.); ferreirosbouzas@em.uni-frankfurt.de (N.F.)

[3] Department of Biochemistry and Biophysics, University of North Carolina, Chapel Hill, NC 27599, USA; justinblack8@gmail.com

[4] Departments of Biological Sciences and of Medicinal Chemistry and Molecular Pharmacology, Purdue University West Lafayette, West Lafayette, IN 47907-2054, USA; jtesmer@purdue.edu

[5] Abteilung für Pharmakologie, Max-Planck-Institut für Herz- und Lungenforschung, 61231 Bad Nauheim, Germany; offermanns.pharmacology@mpi-bn.mpg.de

[6] Experimentelle Pharmakologie Mannheim, European Center for Angioscience, Universität Heidelberg, 68167 Mannheim, Germany; thomas.wieland@medma.uni-heidelberg.de

* Correspondence: heringdorf@med.uni-frankfurt.de; Tel.: +49-69-6301-3906

† These authors contributed equally to this work.

Received: 25 August 2020; Accepted: 27 September 2020; Published: 29 September 2020

Abstract: Diverse extracellular signals induce plasma membrane translocation of sphingosine kinase-1 (SphK1), thereby enabling inside-out signaling of sphingosine-1-phosphate. We have shown before that G_q-coupled receptors and constitutively active $G\alpha_{q/11}$ specifically induced a rapid and long-lasting SphK1 translocation, independently of canonical G_q/phospholipase C (PLC) signaling. Here, we further characterized $G_{q/11}$ regulation of SphK1. SphK1 translocation by the M_3 receptor in HEK-293 cells was delayed by expression of catalytically inactive G-protein-coupled receptor kinase-2, p63Rho guanine nucleotide exchange factor (p63RhoGEF), and catalytically inactive $PLC\beta_3$, but accelerated by wild-type $PLC\beta_3$ and the $PLC\delta$ PH domain. Both wild-type SphK1 and catalytically inactive SphK1-G82D reduced M_3 receptor-stimulated inositol phosphate production, suggesting competition at $G\alpha_q$. Embryonic fibroblasts from $G\alpha_{q/11}$ double-deficient mice were used to show that amino acids W263 and T257 of $G\alpha_q$, which interact directly with $PLC\beta_3$ and p63RhoGEF, were important for bradykinin B_2 receptor-induced SphK1 translocation. Finally, an AIXXPL motif was identified in vertebrate SphK1 (positions 100–105 in human SphK1a), which resembles the $G\alpha_q$ binding motif, ALXXPI, in $PLC\beta$ and p63RhoGEF. After M_3 receptor stimulation, SphK1-A100E-I101E and SphK1-P104A-L105A translocated in only 25% and 56% of cells, respectively, and translocation efficiency was significantly reduced. The data suggest that both the AIXXPL motif and currently unknown consequences of $PLC\beta$/$PLC\delta$(PH) expression are important for regulation of SphK1 by $G_{q/11}$.

Keywords: sphingosine kinase; sphingosine-1-phosphate; G-protein-coupled receptors; $G\alpha_{q/11}$

1. Introduction

Sphingosine-1-phosphate (S1P) is a multifunctional lipid mediator involved in organismal development and homeostasis of the immune, cardiovascular, nervous, and metabolic systems [1]. The metabolism of S1P is evolutionarily highly conserved, comprising sphingosine kinases (SphK), lipid phosphate phosphatases, S1P phosphatases, and S1P lyase [2]. In vertebrates, S1P activates five G-protein-coupled receptors (S1P-GPCRs), $S1P_{1-5}$ [3]. These receptors differentially couple to G_i, $G_{q/11}$, and $G_{12/13}$ proteins, and thereby regulate cell proliferation, survival, migration, adhesion, and Ca^{2+}-dependent functions [1]. S1P-GPCRs are ubiquitously expressed and implicated for example in angiogenesis, maintenance of vascular tone and permeability, and immune cell trafficking. Accordingly, S1P-GPCRs play a role in autoimmunity, inflammation, fibrosis, and cancer [1]. Beyond these well-established effects of extracellular S1P, several roles and targets have been described for intracellular S1P. Examples include endocytic membrane trafficking [4,5], Ca^{2+} mobilization, regulation of histone deacetylases, and mitochondrial respiration (reviewed in [6]). Of note, all of these studies show highly localized signaling by SphK, supporting the early conclusion that SphK localization is a key to function [7]. There are two SphK isoforms, which are derived from different genes and differ in tissue expression, structure, subcellular localization, regulation, and function. SphK2 has been observed in the cytosol, ER, mitochondria, and nucleus, whereas SphK1 is mainly found in the cytosol and may translocate to the plasma membrane upon stimulation [7–10]. Thus, SphK1 seems to be poised to generate S1P for cellular export, and thereby trigger S1P-GPCR cross-activation, which is known as inside-out signaling. A prominent example for inside-out signaling by S1P is observed in fibroblasts, in which SphK1 activation by platelet-derived growth factor (PDGF) leads to cross-activation of the $S1P_1$ receptor, which then mediates PDGF-induced cell migration [11]. Numerous other studies have shown the importance of the SphK/S1P axis for auto- and paracrine S1P signaling, and diverse transporters mediating S1P export have been identified (reviewed in [1,12]).

SphK1 is regulated transcriptionally, translationally, and post-translationally by many different pathways [8–10]. Acute activation of SphK1, with or without membrane translocation, can be induced by growth factors (for example PDGF, epidermal growth factor, nerve growth factor), cytokines (for example tumor necrosis factor-α, interleukin-1β, transforming growth factor-β), immunoglobulin receptors, or GPCRs (for review, see [8–10]). Several mechanisms have been described for plasma membrane translocation of SphK1. Phorbol-12-myristate-13-acetate (PMA)-induced translocation [13] involved phosphorylation of SphK1 at S225 by extracellular signal-regulated kinases (ERK)-1/2 [14]. In contrast, SphK1 translocation by oncogenic K-Ras was dependent on ERK but independent of S225 phosphorylation [15]. SphK1 translocation by both PMA and oncogenic Ras required calcium-and-integrin-binding-protein-1 (CIB1) [16]. Another pathway for acute activation and translocation of SphK1 is phosphatidic acid production [17]. Membrane binding of SphK1 has been attributed to a hydrophobic patch involving L194, F197, and L198 [4]. Furthermore, a highly positively charged site composed of K27, K29, and R186 was shown to form a single contiguous interface with the hydrophobic patch, mediating electrostatic interactions of SphK1 with membranes [18]. Finally, based on SphK1 crystal structures [19,20], Adams et al. have suggested that a dimeric quaternary structure may play a role in curvature-dependent targeting of SphK1 to the plasma membrane, and suggested how phosphorylation at S225 and protein binding to the C-terminus may potentially unmask membrane association determinants in SphK1 [21].

Our own studies have focused on regulation of SphK by GPCRs. Whereas overall SphK activity can be stimulated via G_i as well as via G_q pathways (reviewed in [22]), we have shown that specifically G_q-coupled receptors induce a rapid and long-lasting translocation of SphK1 to the plasma membrane [23,24]. SphK1 translocation was further induced by overexpression of constitutively active $G\alpha_q$ and $G\alpha_{11}$, but not $G\alpha_i$, $G\alpha_{12}$, or $G\alpha_{13}$ [23]. Importantly, G_q-mediated SphK1 translocation was independent of phosphorylation at S225, because SphK1-S225A translocated after stimulation of the M_3 receptor in HEK-293 cells or the B_2 receptor in C2C12 myoblasts, similarly to the wild-type enzyme [23,25]. Classical $G_{q/11}$/phospholipase C (PLC) signaling pathways were not involved in SphK1 targeting. Thus, neither cell-permeable diacylglycerol analogues or PMA, which induce activation of protein kinase C, nor thapsigargin or ionomycin, which induce increases in $[Ca^{2+}]_i$, were able to induce SphK1 translocation to the extent that it was induced by M_3 receptor stimulation. Even a combined pretreatment with PMA plus ionomycin for about 8 min, which caused a minor SphK1 translocation by itself, did not prevent a subsequent marked SphK1 translocation stimulated by the M_3 receptor. Furthermore, the involvement of Ca^{2+}/calmodulin, phospholipase D, tyrosine kinases, Rho kinase, and mitogen-activated protein kinase kinase was ruled out by specific inhibitors [23].

We, therefore, studied the regulation of SphK1 by $G_{q/11}$ signaling pathways in more detail. We identified and characterized a motif conserved in vertebrate SphK1, with similarities to $G\alpha_{q/11}$ binding motifs in direct G_q effectors, which is required for G_q-mediated SphK1 translocation. We also showed that PLCβ, beyond its canonical downstream effectors, is important in inducing the most rapid membrane SphK1 translocation upon GPCR stimulation, and that this is mimicked by the pleckstrin homology (PH) domain of PLCδ_1.

2. Materials and Methods

2.1. Materials

Carbachol, bradykinin, and fatty-acid-free bovine serum albumin (BSA) were purchased from Sigma-Aldrich (Sigma-Aldrich Chemie GmbH, Taufkirchen, Germany). S1P was from Biomol GmbH (Hamburg, Germany). All other materials were from previously described sources [24,26].

2.2. Plasmids

The 3xHA-S1P$_1$ in pcDNA3.1 was obtained from the Missouri S&T cDNA Resource Center (Rolla, MO, USA). $G\alpha_{qi5}$-G66D was kindly provided by Dr. Evi Kostenis (University of Bonn, Bonn, Germany) [27]. The plasmid for expression of $G\alpha_q$-YFP was a kind gift from Dr. Catherine Berlot (Weis Center for Research, Danville, PA, USA) [28]. Plasmids for expression of $G\alpha_{i2}$, $G\alpha_q$ wild-type, $G\alpha_q$-Q209L-EE, $G\alpha_q$-T257E, $G\alpha_q$-Y261N, $G\alpha_q$-W263D, $G\alpha_q$-D321A, $G\alpha_q$-Y356K, $G\alpha_{15}$-Q212L-EE, G-protein-coupled-receptor-kinase-2 (GRK2)-K220R, and the bradykinin B_2 receptor have been described previously [23,29–33].

Plasmids for expression of murine YFP-SphK1 (YFP-mSphK1), human GFP-SphK1 (GFP-hSphK1), human SphK1-G82D (hSphK1-G82D), and human mCherry-SphK1 (mCherry-hSphK1) have been described before [23,24]. Human SphK1-cerulean (hSphK1-cerulean) is a synthetic sequence with optimized codons (Mr. Gene, Regensburg, Germany) deduced from human SphK1 (GenBank accession number AF200328.1) and cerulean fluorescent protein (GenBank accession number ACO48272.1), which was cloned into the pcDNA3.1 vector using HindIII and XhoI. SphK1-F197A-L198Q-GFP and SphK1-L194Q-GFP were kindly provided by Dr. Pietro De Camilli (Yale University School of Medicine, New Haven, CT, USA) [4]. For experiments with the SphK1 mutants, mCherry-hSphK1-A100E-I101E and mCherry-hSphK1-P104A-L105A, these mutants and a second construct of mCherry-hSphK1 wild-type were designed according to the human SphK1 sequence described in GenBank accession number NM_001142601.2. All three constructs were in pmCherry-C1 vector (Clontech/Takara Bio Europe, Saint-Germain-en-Laye, France) and obtained from Proteogenix (Schiltigheim, France). Full-length p63Rho guanine nucleotide exchange factor (p63RhoGEF) in pmCherry-C1 vector was a kind gift from Dr. Dorus Gadella (University of Amsterdam, Amsterdam, The Netherlands; Addgene plasmid #67896; http://n2t.net/addgene:67896; RRID:Addgene_67896) [34]. PLCβ_3 (GenBank accession number NM_000932), C-terminally tagged with TurboGFP in pCMV6-AC-GFP vector, was obtained from OriGene Technologies (product #RG224268; Rockville, MD 20850, USA). PLCβ_3-H332A-GFP in pCMV6-AC-GFP vector was obtained from Proteogenix (Schiltigheim, France). PLCδ_1(PH)-CFP was a kind gift from Dr. Michael Schäfer (University of Leipzig, Leipzig, Germany) [35].

2.3. Cell Culture and Transfection

HEK-293 cells stably expressing the M_3 muscarinic acetylcholine receptor were cultured in Dulbecco's modified Eagle's medium (DMEM/F12) supplemented with 10% fetal calf serum, 100 U/mL penicillin G, and 0.1 mg/mL streptomycin as described [23]. Stock cultures of HEK-293 cells were grown in the presence of 0.5 mg/mL G418. Mouse embryonic fibroblasts (MEFs) from CIB1-deficient mice were made by J.L. Black in the laboratory of Dr. Leslie V. Parise (University of North Carolina at Chapel Hill, NC, USA) [36,37]. These MEFs, along with MEFs from $G\alpha_{q/11}$ double-deficient mice [38], were cultured in DMEM/F12 medium with 10% fetal calf serum, 100 U/mL penicillin G, and 0.1 mg/mL streptomycin. Transfection of HEK-293 cells was performed with Lipofectamine 2000 (Invitrogen GmbH, Karlsruhe, Germany), while MEFs were transfected with Turbofect (Fermentas, St. Leon-Rot, Germany) according to the manufacturer's instructions. For microscopy, the cells were seeded onto poly-L-lysine-coated 8-well slides (μ-slide; ibidi GmbH, Martinsried, Germany). Before experiments, the cells were kept in serum-free medium overnight.

2.4. Measurement of SphK1 Translocation

SphK1 translocation was analyzed using fluorescently labelled SphK1 constructs and confocal laser scanning microscopy as described recently [24]. Cells grown on poly-L-lysine-treated 8-well slides (μ-slide; ibidi GmbH, Martinsried, Germany) were incubated in Hank´s balanced salt solution (HBSS) containing 118 mM NaCl, 5 mM KCl, 1 mM $CaCl_2$, 1 mM $MgCl_2$, 5 mM glucose, and 15 mM HEPES, pH 7.4. Fluorescence microscopy was performed with a Zeiss LSM510 Meta inverted confocal laser scanning microscope equipped with a Plan-Apochromat 63×/1.4 oil immersion objective (Carl Zeiss MicroImaging GmbH, Göttingen, Germany). The following excitation (ex) laser lines and emission (em) filter sets were used: CFP and cerulean: ex 458 nm, em band-pass 465–510 nm; GFP: ex 488 nm, em long-pass 505 nm; GFP in combination with mCherry: ex 488 nm, em band-pass 505–530; YFP: ex 514, em band pass 525–600; mCherry: ex 561 nm, em band-pass 575–630 nm. Translocation half-times were determined by measuring the fluorescence intensity within defined cytosolic regions and fitting exponential functions to the translocation-induced decay in cytosolic fluorescence intensity. For estimation of translocation efficiency, the translocated fraction was calculated from these exponential curves as % decay in cytosolic fluorescence. While the average translocation half-time was ~5 s throughout all experiments, the translocated fraction values were comparatively variable between experiments because their calculation was influenced to a certain degree by cell shape changes. For quantification of SphK1 plasma membrane localization in cells co-transfected with $G\alpha_q$-Q209L-EE (Figure 6G,H), we measured the fluorescence profiles of individual cells using the ZEN software (Carl Zeiss MicroImaging GmbH, Göttingen, Germany), and calculated the ratios of plasma membrane and cytosolic fluorescence intensities.

2.5. Inositol Phosphate Production

Inositol phosphate production was measured as described recently [24]. Briefly, HEK-293 cells labelled with 1 μCi/mL myo-2-[^{3}H]-inositol (23.75 Ci/mmol; Perkin Elmer Life and Analytical Sciences, Rodgau-Jügesheim, Germany) were stimulated with 100 μM carbachol in HBSS containing LiCl for 20 min at 37 °C. The reaction was stopped by addition of 2 mL ice-cold methanol. The cells were scraped from the dishes, 1 mL H_2O and 2 mL chloroform were added, and the aqueous phase was transferred to Poly-Prep AG 1-X8 columns (Bio-Rad, Hercules, CA, USA). After washing with H_2O and 50 mM ammonium formate, inositol phosphates were eluted with 5 mL of 1 M ammonium formate and 0.1 M formic acid. The radioactivity was measured by liquid scintillation counting.

2.6. Western Blotting

Cells grown to near confluence on 6 cm-dishes were lysed, the proteins were separated by SDS gel electrophoresis and blotted onto polyvinylidene difluoride membranes. The SphK1 antibody, directed against the C-terminus of human SphK1, was a kind gift from Drs. Andrea Huwiler (University of Bern, Bern, Switzerland) and Josef Pfeilschifter (Goethe-University Frankfurt, Frankfurt, Germany) [39]. Anti-mCherry antibody (ab125096) was from Abcam (Cambridge, UK). Anti-β-actin (A5441) was from Sigma-Aldrich Chemie GmbH (Taufkirchen, Germany), HRP-conjugated secondary antibodies were from GE Healthcare (Freiburg, Germany), and the enhanced chemiluminescence system was from Millipore Corporation (Billerica, MA, USA).

2.7. High-Performance Liquid Chromatography Tandem Mass Spectrometry

S1P (d18:1) concentrations were determined by high-performance liquid chromatography tandem mass spectrometry as described recently [24].

2.8. Data Analysis and Presentation

Fluorescence images were edited with the ZEN software (Carl Zeiss MicroImaging GmbH, Göttingen, Germany). Statistical tests, curve fitting, and calculations of translocation half-times were done with Prism-5 (GraphPad Software, San Diego, California, USA). Averaged data are expressed as means ± SD or means ± SEM from the indicated number (n) of cells, samples, or experiments, respectively.

3. Results and Discussion

Similar to our previous publications [23,25], we observed a rapid and long-lasting translocation of both murine and human SphK1 to the plasma membrane upon stimulation of the M_3 muscarinic acetylcholine receptor in HEK-293 cells (Figures 1, 2, 5, and 6). To confirm the involvement of $G\alpha_{q/11}$ in this system, we analyzed the influence of a catalytically inactive mutant of GRK2, GRK2-K220R. GRK2-K220R is unable to phosphorylate G protein-coupled receptors but directly binds $G\alpha_{q/11}$ and $G\beta\gamma$, and acts as a $G\alpha_{q/11}/G\beta\gamma$ scavenger [29] (see also [40]). When co-expressed with YFP-mSphK1, GRK2-K220R strongly delayed, reduced, and in several cells even fully prevented translocation of the enzyme by the M_3 receptor (Figure 1A–C). Together with our previous observation that constitutively active $G\alpha_q$ induced SphK1 translocation, this result indicated that M_3 receptor-induced translocation was mediated by $G\alpha_q$.

Another binding partner and direct effector of $G\alpha_{q/11}$ is p63RhoGEF [41,42]. It is known that p63RhoGEF activates RhoA but also competes with PLCβ for activated $G\alpha_{q/11}$, and vice versa [41]. Therefore, we tested the influence of p63RhoGEF on SphK1 translocation. As shown in Figure 1D–F, overexpressed mCherry-p63RhoGEF strongly delayed and reduced M_3 receptor-induced translocation of GFP-hSphK1. This result indicated that SphK1 is not activated by p63RhoGEF downstream signaling and matches our previous observation that a Rho kinase inhibitor did not prevent $G_{q//11}$–mediated SphK1 translocation [23]. The data obtained with GRK2-K220R and p63RhoGEF can be explained by competition of the different $G\alpha_{q/11}$ effectors for binding at the active $G\alpha_q$ or $G\alpha_{11}$ subunit. Thus, the data suggest two possibilities: (1) that SphK1 is activated downstream of PLCβ, and (2) that SphK1 directly interacts with and is translocated by active $G\alpha_{q/11}$. To analyze these possibilities further, we next tested the influence of PLCβ on SphK1 translocation.

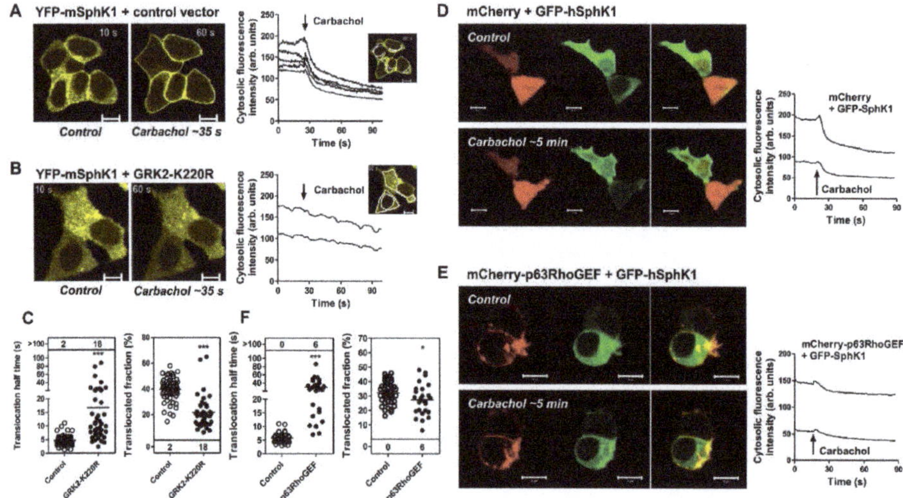

Figure 1. Influence of catalytically inactive GRK2 and p63RhoGEF on M_3 receptor-induced SphK1 translocation. HEK-293 cells stably expressing the M_3 muscarinic acetylcholine receptor were transfected as indicated and translocation of SphK1 was monitored by live cell imaging with a confocal laser scanning microscope. (**A–C**) Cells were transfected with YFP-mSphK1 and either catalytically inactive GRK2 (GRK2-K220R) or control vector. Time series were acquired with ~3 images/s and 100 µM carbachol was added after ~25 s. Images were taken from representative time series at 10 and 60 s, thus showing localization of YFP-mSphK1 before and ~35 s after stimulation with carbachol, respectively. The line graphs show the corresponding time courses of cytosolic fluorescence intensity, measured in the indicated cytosolic regions. (**C**) The SphK1 translocation half-times were measured by fitting exponential curves to the translocation-induced decay in cytosolic fluorescence intensity. Each dot represents a single cell. Translocation half-times >100 depict the number of cells which did not respond. Note: *** $p < 0.0001$ in *t*-tests with Welch's correction for unequal variances. (**D–F**) Cells were transfected with GFP-hSphK1 (green) and either mCherry or mCherry-p63RhoGEF (red). Representative images were taken at high spatial resolution immediately before and after the acquisition of time series, during which only GFP fluorescence was monitored with 1 image/s. The line graphs showing time courses of cytosolic fluorescence intensity correspond to the cells shown in the images. Carbachol was added at the indicated time points. (**F**) SphK1 translocation half-time was measured as described in (C). Note: * $p < 0.05$, *** $p < 0.0001$ in *t*-tests with Welch's correction for unequal variances. Micrometer bars, 10 µm.

As shown in Figure 2A–D, overexpression of catalytically inactive PLCβ$_3$-H332A-GFP caused a significant delay in M_3 receptor-induced translocation of mCherry-hSphK1. In contrast, wild-type PLCβ$_3$-GFP slightly but significantly accelerated translocation of mCherry-hSphK1. While the inhibitory effect of PLCβ$_3$-H332A could be caused both by competition at Gα$_q$ and by its activity as a GTPase-activating protein (GAP), the acceleration by wild-type PLCβ$_3$ shows that its GAP activity was surmounted by its stimulatory effect. Interestingly, acceleration of M_3 receptor-induced SphK1 translocation was also observed upon expression of the PH domain of PLCδ$_1$, which serves as a sensor for phosphatidylinositol-4,5-bisphosphate (PI(4,5)P$_2$) [43]. As described [35,43], PLCδ$_1$(PH)-CFP was localized at the plasma membrane in unstimulated cells and rapidly translocated to the cytosol after stimulation of the M_3 receptor (Figure 2E). The velocity of PLCδ$_1$(PH)-CFP translocation was not altered by co-expression of YFP-mSphK1; however, translocation of YFP-mSphK1 was significantly accelerated by co-expression of PLCδ$_1$(PH)-CFP (Figure 2F). Since the PH domain of PLCδ$_1$ does not induce DAG and IP$_3$ production, its effect on SphK1 is rather due to PI(4,5)P$_2$ binding or other cellular effects, for example competition with other PI(4,5)P$_2$ binding proteins. In fact, PLCδ$_1$(PH) has been shown to reduce the amount of phosphatidylinositol-4-phosphate-5-kinase (PIP5K) at the plasma membrane, and thereby the cellular level of PI(4,5)P$_2$ [44]. According to this report, expression of PLCδ_1(PH) will rather decrease than increase PLCβ catalytic activity, but nevertheless accelerated SphK1 translocation. Importantly, RhoA, the downstream effector of p63RhoGEF, activates type I PIP5K [45]. Thus, it is possible that (over)expression of PLCβ_3 or PLCδ_1(PH) accelerated SphK1 translocation while p63RhoGEF delayed SphK1 translocation by decreasing and enhancing PI(4,5)P$_2$ levels, respectively. However, other tools manipulating PI(4,5)P$_2$ levels such as the multiple pathway inhibitor genistein [46] did not alter SphK1 translocation velocity in our cells ([23] and data not shown). Consequently, the mechanisms by which (over)expression of PLCβ_3 and PLCδ_1(PH) accelerate SphK1 translocation remain unclear.

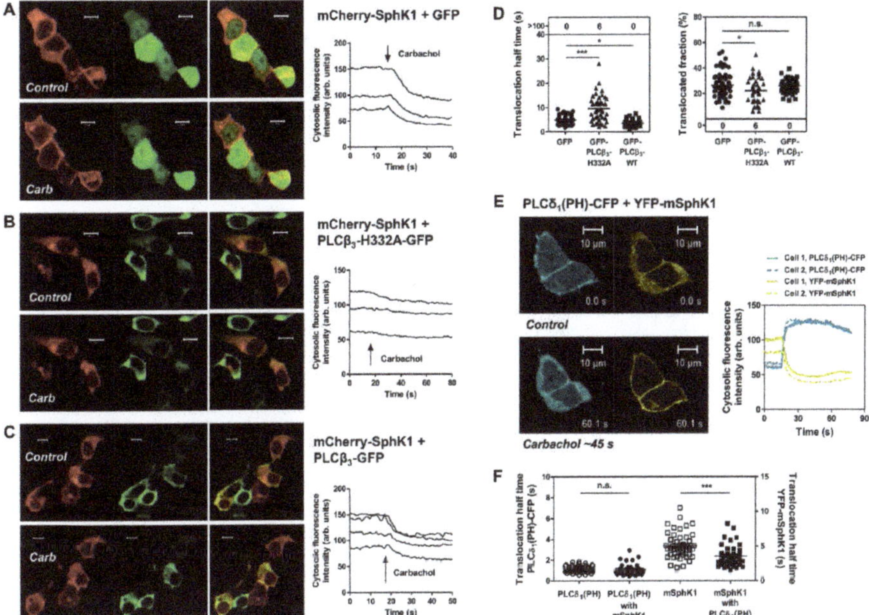

Figure 2. Influence of PLCβ_3 and PLCδ(PH) on M$_3$ receptor-induced SphK1 translocation. HEK-293 cells stably expressing the M$_3$ receptor were transfected as indicated. (**A–D**) Cells were transfected with mCherry-hSphK1 (red) and GFP (**A**), catalytically inactive PLCβ_3 (PLCβ_3-H332A-GFP) (**B**) or PLCβ_3 wild-type (PLCβ_3-WT) (**C**). Representative images were taken at high spatial resolution immediately before and after the acquisition of time series during which only mCherry fluorescence was monitored at 1 image/s. The line graphs showing time courses of cytosolic fluorescence intensity correspond to the cells shown in the images. Carbachol was added at the indicated time points. (**D**) SphK1 translocation half-times were measured by fitting exponential curves to the decay in cytosolic fluorescence. The translocated fraction represents the decay in cytosolic fluorescence intensity in % of initial fluorescence. Each dot represents a single cell. Translocation half-times >100 depict the number of cells which did not respond. Note: n.s., not significant; * p <0.05, *** p < 0.0001 in t-tests with Welch's correction for unequal variances. (**E,F**) Cells were transfected with PLCδ(PH)-CFP (cyan), YFP-mSphK1 (yellow), or both. Carbachol-induced translocation of the two proteins was studied in time series with ~3–4 images/s. (**E**) Images were taken from a representative time series with double-transfected cells before and ~45 s after addition of carbachol. The line graph shows the time course of cytosolic fluorescence intensity for both CFP and YFP. (**F**) Translocation half-times for PLCδ(PH)-CFP and YFP-mSphK1, both alone and in combination. Note: n.s., not significant; *** p < 0.0001 in t-tests with Welch's correction for unequal variances. Micrometer bars, 10 μm.

Interestingly, not all members of the $G\alpha_q$ subfamily ($G\alpha_q$, $G\alpha_{11}$, $G\alpha_{14}$, and $G\alpha_{15/16}$, with $G\alpha_{15}$ being the murine homologue of human $G\alpha_{16}$) interact with the established targets in the same manner. For example, $G\alpha_{15/16}$ does not bind to GRK2, and binds to but does not activate p63RhoGEF (reviewed in [40]). With the aim of possibly separating PLCβ activation and SphK1 translocation, we expressed constitutively active $G\alpha_{15}$-Q212L-EE and studied its influence on SphK1 localization. As shown in Figure 3A, mCherry-SphK1 was strongly localized at the plasma membrane in $G\alpha_{15}$-Q212L-EE-transfected cells. Thus, PLCβ activation and SphK1 translocation could not be separately targeted by using $G\alpha_{15/16}$.

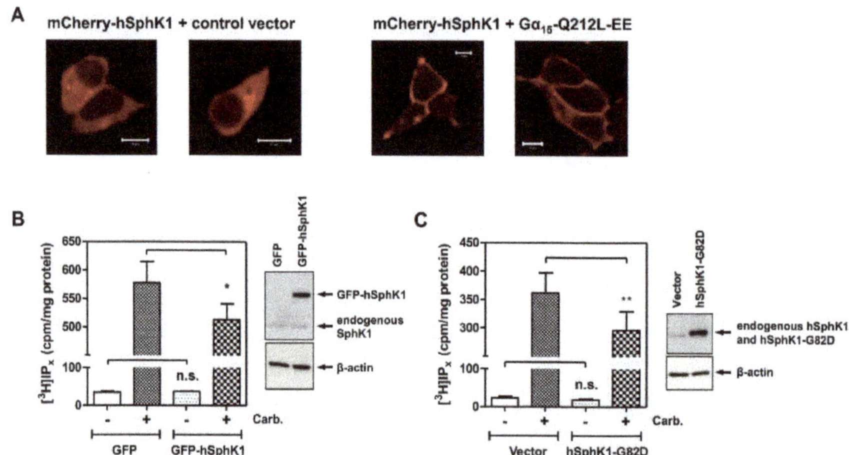

Figure 3. (**A**) SphK1 translocation by constitutively active $G\alpha_{15}$. HEK-293 cells stably expressing the M_3 receptor were transfected with mCherry-hSphK1 and either control vector or $G\alpha_{15}$-Q212L-EE. Shown are two representative images each. Bars, 10 μm. (**B,C**) Influence of SphK1 overexpression on M_3 receptor-induced inositol phosphate production. HEK-293 cells were transfected with GFP or GFP-hSphK1 (**B**), or with the pcDNA3.1 vector or hSphK1-G82D (**C**). The formation of [³H]inositol phosphates ([³H]IP$_x$) was measured in [³H]inositol-labelled cells stimulated with 100 μM carbachol (Carb.) for 20 min in the presence of LiCl. The expression of GFP-hSphK1 and hSphK1-G82D was confirmed with an anti-SphK1 antibody. The data are means ± SEM from $n = 9$ (B) or $n = 8$ (C) independent experiments, each performed in triplicate. Note: n.s., not significant; * $p < 0.05$, ** $p = 0.01$ in paired t-test.

To further analyze the mutual interactions of SphK1 and PLCβ, we next studied the influence of overexpressed SphK1 on M_3 receptor-induced accumulation of [³H]inositol phosphates in [³H]inositol-labelled cells. As shown in Figure 3B, GFP-SphK1 had no influence on basal inositol phosphate production, but significantly reduced M_3 receptor-stimulated inositol phosphate production. Moreover, SphK1-G82D, which is a catalytically inactive mutant [47], had the same effect (Figure 3C), indicating that it was due to protein–protein interactions and independent of S1P signaling. This result indeed suggests that SphK1 competed with PLCβ for $G\alpha_{q/11}$ in the context of a living cell, although it remains possible that SphK1 binds to PLCβ, thereby reducing its activity.

Next, we used embryonic fibroblasts from mice deficient in both $G\alpha_q$ and the related $G\alpha_{11}$ [38] to study structural requirements of $G\alpha_{q/11}$ for inducing SphK1 translocation. In $G\alpha_{q/11}$ double-deficient MEFs, both hSphK1-cerulean and YFP-mSphK1 were localized in the cytosol, and their localization did not change upon stimulation of the B_2 bradykinin receptor unless $G\alpha_q$-YFP or $G\alpha_q$ wild-type were co-transfected (Figure 4). Of note, in cells expressing $G\alpha_q$ wild-type or $G\alpha_q$-YFP, SphK1 was localized to a small part at the plasma membrane, even under control conditions (Figure 4B,E). This was not the case in cells lacking both $G\alpha_q$ and $G\alpha_{11}$ (Figure 4A,D). Using these cells, we confirmed that stimulation of the G_i-coupled $S1P_1$ receptor did not induce SphK1 translocation, even when $S1P_1$, GFP-hSphK1, and $G\alpha_{i2}$ were co-transfected (Figure 4C). However, expression of the chimeric G-protein, $G\alpha_{qi5}$-G66D, which links G_i-coupled receptors to G_q signaling pathways [27], enabled $S1P_1$ to induce SphK1 translocation (Figure 4C). In cells co-expressing the B_2 receptor, YFP-mSphK1, and $G\alpha_q$ wild-type, 10 μM bradykinin induced translocation of SphK1 with an average half-time of 5.8 ± 0.6 s (mean ± SEM, n = 31 cells; Figure 4E,G). Several $G\alpha_q$ mutants were able to fully restore B_2 receptor-induced SphK1 translocation in $G\alpha_{q/11}$ double-deficient MEFs, with translocation half-times that did not significantly differ from that of $G\alpha_q$ wild-type. These were $G\alpha_q$-Y261N ($t_{1/2}$ = 7.9 ± 1.4 s, n = 17), $G\alpha_q$-D321A ($t_{1/2}$ = 4.1 ± 0.5 s, n = 14), and $G\alpha_q$-Y356K ($t_{1/2}$ = 4.5 ± 0.5 s, n = 17) (all means ± SEM; Figure 4G). In cells expressing $G\alpha_q$-W263D, B_2 receptor-stimulated SphK1 translocation was significantly delayed, with a half-time of 10.8 ± 1.1 s (mean ± SEM, n = 18 cells; Figure 4G). Importantly, SphK1 translocation was very slow in cells expressing $G\alpha_q$-T257E and typically visible only after 2–3 min, for which reason the average translocation half-time was not determined (Figure 4F). Taken together, the amino acids T257 and W263 of $G\alpha_q$ are important for targeting of SphK1. Interestingly, $G\alpha_q$-T257, $G\alpha_q$-Y261, and $G\alpha_q$-W263 are implicated in $G\alpha_q$/GRK2 interaction, as mutation of these residues abolished $G\alpha_q$ binding to GRK2 [31,48]. Furthermore, PLCβ activation was completely inhibited by mutation of $G\alpha_q$-R256/T257 to alanines [49]. Finally, mutants $G\alpha_q$-Y261N and $G\alpha_q$-W263D had reduced binding to p63RhoGEF, while $G\alpha_q$-T257E neither bound nor activated p63RhoGEF [33]. Thus, amino acid T257 of $G\alpha_q$ plays a major role in binding or activation of PLCβ, p63RhoGEF, and GRK2, and likewise is important for SphK1 translocation. This result is in agreement with both hypotheses: (1) that SphK1 is activated downstream of PLCβ, and (2) that SphK1 competes with PLCβ, p63RhoGEF, and GRK2 for the same $G\alpha_q$ binding site.

Next, we wondered which structural elements in SphK1 were required for $G\alpha_q$-mediated translocation of the enzyme. As described above, SphK1 membrane translocation by PMA and oncogenic Ras involved CIB1, which binds to the calmodulin binding site of SphK1 [16,50]. Amino acids L194, F197, and L198 of hSphK1 were important for CIB1 binding, and the double mutant hSphK1-F197A-L198Q had reduced CIB1 binding and did not translocate to the plasma membrane in response to PMA, while its catalytic activity remained nearly intact [16,50]. Another study localized L194, F197, and L198 within a hydrophobic patch on the surface of SphK1 and demonstrated that hSphK1-L194Q and hSphK1-F197A-L198Q did not bind to acidic liposomes in vitro and were not recruited to tubular membrane invaginations induced by cholesterol extraction in living cells [4]. Hence, this hydrophobic patch is regarded as essential for curvature-sensitive membrane binding of SphK1 [9,21]. We show here that the two hSphK1 mutants, hSphK1-L194Q-GFP and hSphK1-F197A-L198Q-GFP, did not visibly translocate to the plasma membrane in response to M_3 receptor activation in HEK-293 cells (Figure 5A–C). Interestingly, while usually there were only low levels of wild-type SphK1 in the nuclei of transfected HEK-293 cells, there was significant fluorescence in the nuclei of cells transfected with hSphK1-L194Q-GFP (Figure 5B). Furthermore, the mutant, hSphK1-F197A-L198Q-GFP, was strongly localized to the nuclei, with some cells expressing even more fluorescence in the nucleus than in the cytoplasm (Figure 5C). This observation suggests that mutations in this region might possibly disrupt a nuclear export sequence, although the two known nuclear export sequences in hSphK1 comprise amino acids 147–155 and 161–169 [51]. Because of the mentioned involvement of hSphK1-L194, -F197, and -L198 in CIB1 binding [16], we furthermore studied $G_{q/11}$-dependent SphK1 translocation in CIB1-deficient MEFs. As shown in Figure 5D, the B_2 receptor was clearly able to induce GFP-hSphK1 translocation in cells lacking CIB1. In addition, the translocation half-time was not altered (data not shown). Taken together, we demonstrate that this region comprising L194, F197, and L198 in hSphK1 is important for $G_{q/11}$-dependent SphK1 translocation, very likely because this hydrophobic patch is required for membrane binding. CIB1, however, does not play a role in G_q-mediated SphK1 translocation.

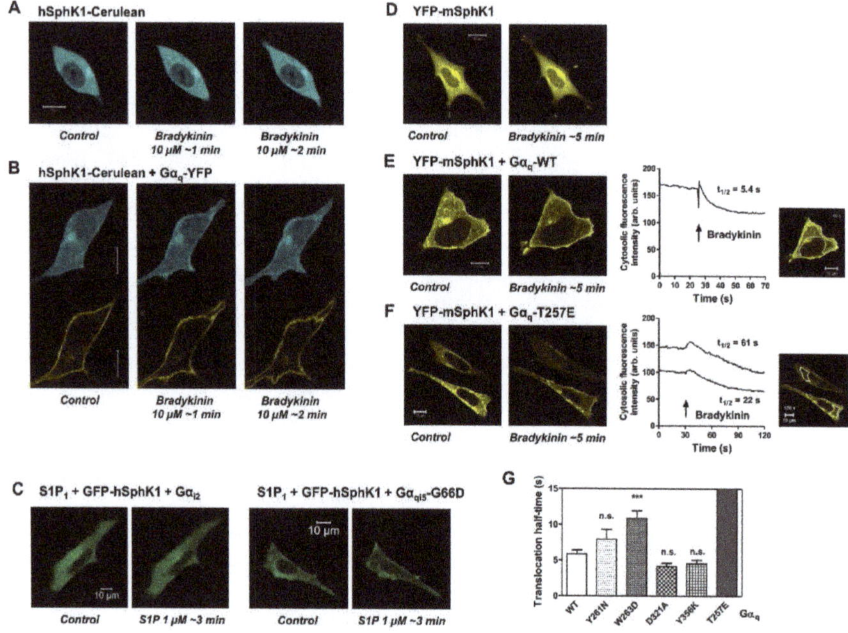

Figure 4. Identification of Gα$_q$ residues required for G$_{q/11}$-mediated SphK1 translocation. (**A–G**) SphK1 translocation was analyzed in Gα$_{q/11}$-double-deficient MEFs. (**A,B**) The cells were co-transfected with the bradykinin B$_2$ receptor and hSphK1-cerulean without (**A**) and with (**B**) Gα$_q$-YFP. Images were taken before and after addition of 10 μM bradykinin as indicated. (**C**) The cells were co-transfected with the S1P$_1$ receptor, GFP-hSphK1, and either Gα$_{i2}$ or Gα$_{qi5}$-G66D as indicated. Images were taken before and ~3 min after addition of 1 μM S1P. (**D–G**) The cells were co-transfected with the B$_2$ receptor, YFP-mSphK1, and Gα$_q$ wild-type (Gα$_q$-WT) or various Gα$_q$ mutants as indicated. Images were taken at a high resolution before and ~5 min after addition of 10 μM bradykinin. The time course of SphK1 translocation was measured by taking series of images at lower spatial resolution at ~1 image/400 ms. Cytosolic fluorescence intensity was measured in selected regions as indicated in the inserts in (**E,F**), and translocation half-times were calculated by fitting exponential curves to the decay in the cytosolic fluorescence intensity. (**G**) Translocation half-times obtained with the various Gα$_q$ mutants. Data are means ± SEM; $n = 31$ (Gα$_q$-WT); $n = 14$–18 (Gα$_q$ mutants). Note: n.s., not significant; *** $p < 0.0001$ in one-way ANOVA followed by Bonferroni´s post-test. The micrometer bars represent 10 μm.

PLCβ and p63RhoGEF bind to the effector binding site of Gα$_q$ primarily via their conserved ALXXPI motifs (X represents any amino acid) [40]. Although both enzymes have additional domains that contribute to the interaction with active Gα$_q$, mutation of the conserved leucine in this motif (L859 in human PLCβ$_3$, L475 in human p63RhoGEF) is sufficient to eliminate Gα$_q$ binding [40]. Similarly, although GRK2 lacks the ALXXPI motif, it contains a structurally equivalent leucine (L118) that is essential for Gα$_q$ binding [40]. Interestingly, there is a similar motif, AIXXPL, in vertebrate SphK1, with isoleucine instead of leucine in position 2 and leucine instead of isoleucine in position 6 of this motif (Figure 6A). Comparison of different SphK1 homologues shows the conservation of this motif among vertebrates, with small variations concerning the leucine/isoleucine substitutions, such as ALXXPL in *Gallus gallus* and AIXXPI in *Xenopus laevis* (Figure 6A). We did not find such a motif in non-vertebrate SphK, such as *Drosophila melanogaster* or *Caenorhabditis elegans* SphK, in agreement with the current view that S1P-GPCRs, which are ultimately targeted by SphK1 plasma membrane translocation, have evolved with the vertebrates (see [1]). To study the functional importance of the AIXXPL motif, we generated the mutants hSphK1-A100E-I101E and hSphK1-P104A-L105A as fusion proteins with *N*-terminal mCherry. When expressed in HEK-293 cells, both mCherry-hSphK1-A100E-I101E and mCherry-hSphK1-P104A-L105A were detected by an anti-mCherry antibody and by an antibody directed against the C-terminus of human SphK1, and had the same molecular weight as mCherry-hSphK1 wild-type (Figure 6B). The double bands seen in Figure 6 were also seen with mCherry alone, and thus caused by the fluorescent tag (Figure 6B). Furthermore, expression of all the SphK1 mutants elevated intracellular S1P concentrations, as measured by high-performance liquid chromatography tandem mass spectrometry. In cells transfected with mCherry, the concentration of S1P was 1050 ± 150 pg/mg protein ($n = 8$), while expression of mCherry-hSphK1 wild-type increased S1P to 1500 ± 200 pg/mg protein ($n = 6$; $p < 0.001$). Cells expressing mCherry-hSphK1-A100E-I101E had S1P concentrations of 1300 ± 160 pg/mg protein ($n = 9$; $p < 0.05$), and cells expressing mCherry-hSphK1-P104A-L105A had 1500 ± 190 pg/mg protein ($n = 9$; $p < 0.001$) (all values represent means $\pm$ SD, with significance tested in one-way ANOVA). These results indicated that all of the mutants were catalytically active.

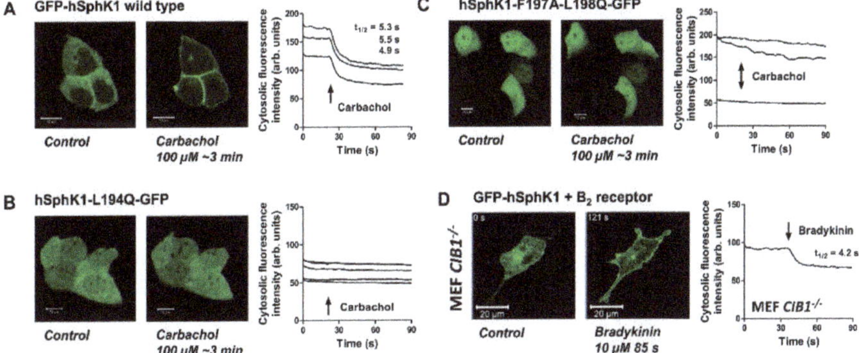

Figure 5. Role of the hydrophobic patch in SphK1 for $G_{q/11}$-mediated SphK1 translocation. (**A–C**) HEK-293 cells stably expressing the M_3 receptor were transfected with GFP-hSphK1, hSphK1-L194Q-GFP, or hSphK1-F197A-L198Q-GFP. Translocation of SphK1 mutants was studied upon stimulation of the cells with 100 µM carbachol. Shown are images before and after stimulation, and time courses of cytosolic fluorescence from representative experiments. In (**C**), the cytosolic fluorescence of the cell in the upper left could not be evaluated because of the strong change in cell shape. The micrometer bars represent 10 µm. (**D**) Role of CIB1 for G_q-mediated SphK1 translocation. MEFs from CIB1-deficient mice were transfected with GFP-hSphK1 and the B_2 receptor. Shown are images and time courses of cytosolic fluorescence from a representative time series during which 10 µM bradykinin was added after 35 s. The two images, thus, show localization of GFP-hSphK1 before and 85 s after addition of bradykinin. Micrometer bars, 20 µm.

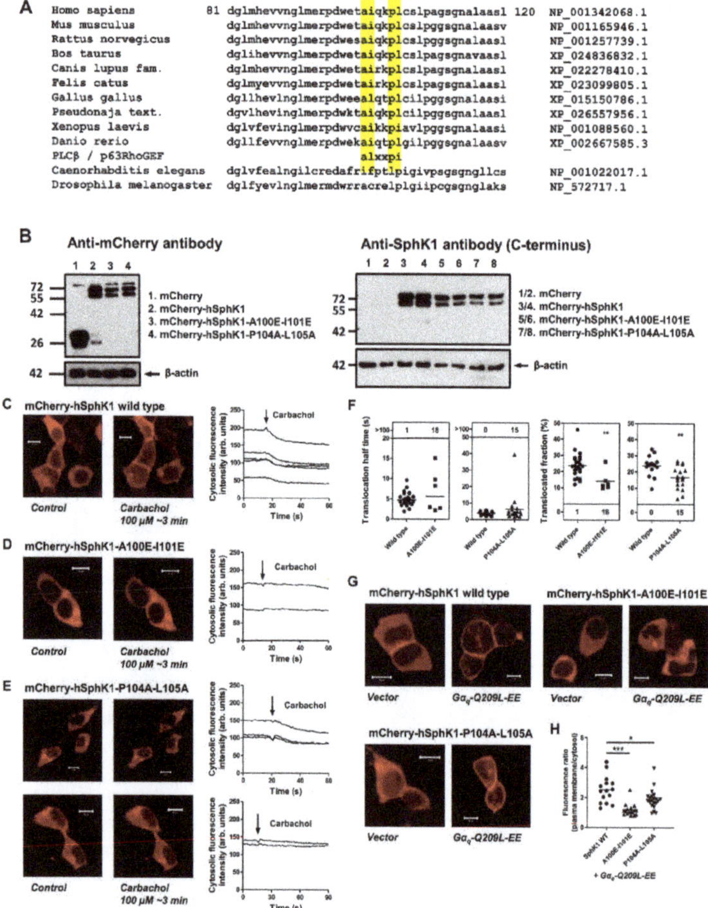

Figure 6. Identification of structural elements in SphK1 required for $G_{q/11}$-mediated translocation. (**A**) Sequence alignment of diverse SphK1 homologues between amino acids 81–120 of human SphK1. (**B**) Western blot analysis of mCherry-hSphK1-A100E-I101E and mCherry-hSphK1-P104A-L105A expressed in HEK-293 cells. (**C–E**) Analysis of localization and translocation of mCherry-hSphK1-A100E-I101E and mCherry-hSphK1-P104A-L105A in HEK-293 cells stably expressing the M_3 receptor. The cells were stimulated with 100 µM carbachol as indicated. Shown are images before and after stimulation, and time courses of cytosolic fluorescence from representative experiments. Since translocation of mCherry-hSphK1-P104A-L105A was variable, two examples are shown here for this mutant. (**F**) Summary of translocation half-times and translocated fractions from (**C–E**). Each dot represents a single cell. Translocation half-times >100 depict the number of cells which did not respond. Note: ** $p < 0.01$ in unpaired t-test. (**G,H**) Influence of constitutively active $G\alpha_q$ on subcellular localization of hSphK1 mutants. HEK-293 cells were transfected with mCherry-hSphK1 wild-type, mCherry-hSphK1-A100E-I101E, or mCherry-hSphK1-P104A-L105A, plus either control vector or $G\alpha_q$-Q209L-EE as indicated. (**G**) Representative images. All micrometer bars, 10 µm. (**H**) Quantification of SphK1 plasma membrane localization was performed by measuring the fluorescence profiles of individual cells and calculating the plasma membrane/cytosol fluorescence ratios. Each dot represents a single cell. Note: * $p < 0.05$, *** $p < 0.0001$ in one-way ANOVA followed by Dunnett's multiple comparisons test.

Next, we studied plasma membrane translocation of the two mutants in response to M_3 receptor activation in HEK-293 cells. In unstimulated cells, both mCherry-hSphK1-A100E-I101E and mCherry-hSphK1-P104A-L105A were localized in the cytosol of the cells and only a small part was sometimes seen in the nucleus, similar to mCherry-SphK1 wild-type (Figure 6C–E). Interestingly, carbachol-induced translocation of mCherry-hSphK1-A100E-I101E occurred in only 6 of 24 cells (25%), while mCherry-hSphK1 wild-type translocated in 25 of 26 cells (96%) in this set of experiments (Figure 6F). The translocation half-times of mCherry-hSphK1-A100E-I101E in the 6 cells with translocation were not significantly different from those of mCherry-hSphK1 wild-type (Figure 6F). However, translocation efficiency, measured as the % decrease in cytosolic fluorescence, was significantly lower with mCherry-hSphK1-A100E-I101E than with the wild-type enzyme (14.0 ± 2.5%, n = 25 versus 23.6 ± 1.4%, n = 6; mean ± SEM; Figure 6D,F). The other mutant, mCherry-hSphK1-P104A-L105A, translocated in 19 of 34 cells stimulated with carbachol (56%), while mCherry-hSphK1 wild-type translocated in 100% of cells in this set of experiments (Figure 6E,F). Again, in the cells which had a response, translocation efficiency of this mutant was significantly reduced (16.5 ± 1.6%, n = 19 versus 23.4 ± 1.5%, n = 16; mean ± SEM), while translocation half-times were not significantly different when compared to the wild-type enzyme (Figure 6F). In cells co-transfected with constitutively active $G\alpha_q$-Q209L-EE, mCherry-hSphK1-A100E-I101E remained cytosolic in the majority of cells, while mCherry-hSphK1-P104A-L105A was localized at the plasma membrane to a variable extent (Figure 6G). Quantification of the plasma membrane/cytosol fluorescence ratios revealed that $G\alpha_q$-Q209L-EE-induced membrane attachment of both mutants was significantly reduced compared to the wild-type enzyme, and that the A100E-I101E mutant was again more affected (Figure 6H). Thus, the results for the two mutants support a role for the AIXXPL motif in G_q targeting of SphK1. G_q-mediated translocation was much more strongly affected by mutation of A100 and I101 to glutamate than by mutation of P104 and L105 to alanine. This might be explained by the stronger disruption of the domain by the negative charges of the two glutamates, or by a higher relevance of AI compared to PL. Intriguingly, within the AIXXPL motif, I101 corresponds to L859 in human PLCβ3 and L475 in human p63RhoGEF, which are most important for G_q binding [40]. We think that the mutations did not disrupt the general structure of hSphK1, as (1) the molecular weight and subcellular localization in unstimulated cells were normal, (2) both mutants were able to translocate at least in some cells, and (3) S1P concentrations were elevated in cells expressing SphK1-A100E-I101E and SphK1-P104A-L104A indicating catalytic activity. The remaining responses of the mutants might be due to incomplete disruption of the domain, contribution of other parts of the enzyme (see below), or to a second parallel pathway that might involve phosphorylation.

Taken together, we present functional data showing that SphK1 translocation by $G_{q/11}$-coupled receptors is prevented by (over)expression of diverse $G\alpha_{q/11}$ effectors or binding partners, suggesting that SphK1 is targeted either via PLCβ or directly by activated $G\alpha_{q/11}$. The fact that expression of catalytically inactive hSphK1-G82D reduced receptor-stimulated inositol phosphate production argues in favor of the latter possibility, although the role of PLCβ in this scenario remains unclear. Furthermore, we show that SphK1's conserved AIXXPL motif is involved in translocation of the enzyme by $G\alpha_q$. Further studies are required to examine whether this motif indeed mediates direct interaction of SphK1 with $G\alpha_{q/11}$. We do not exclude that there are other structural elements in SphK1 that directly or indirectly interact with $G\alpha_q$. In fact, we had shown before that both the N-terminus and C-terminus of hSphK1 (except for the TRAF2 binding site) were required for M_3 receptor-induced translocation [23]. The fragment with N-terminal deletion of the first 110 amino acids (hSphK1$^{111-384}$), thus without AIXXPL motif, did not translocate. The fragment with C-terminal deletion of 27 amino acids in addition to the TRAF2 binding site (hSphK1$^{1-350}$) also did not translocate, suggesting that there are other unknown elements in this region required for interaction with $G\alpha_{q/11}$, the membrane, or other regulatory proteins (see also discussion in [21]).

There are numerous examples of the importance of SphK1 in $G_{q/11}$ signaling and functional responses. $G_{q/11}$-coupled receptors engaging SphK1 include not only muscarinic receptors [52] and bradykinin receptors [25], but also protease-activated receptors [53,54], angiotensin receptors [55], or histamine receptors [56,57], just to name a few. These examples show that SphK1 was involved in $G_{q/11}$-dependent regulation of the vascular endothelium and smooth muscle, $G_{q/11}$-mediated myogenic differentiation of skeletal muscle, or $G_{q/11}$-regulated inflammatory responses. From its eminent role in vascular regulation, it was hypothesized recently that SphK1 might be a therapeutic target in pulmonary hypertension [58], in addition to its roles in inflammation, fibrosis, and cancer [10]. Constitutively active $G_{q/11}$ proteins act as oncogenes [59], and thus given the role of SphK1 in cancer, it will be interesting to unravel a possible interconnection. Another important theme is the emerging role of SphK1 in epithelial–mesenchymal transition [60,61]; however, it should be kept in mind that besides $G_{q/11}$ proteins, there are many other pathways regulating SphK1 expression and activity, and which may be important in this context. Nevertheless, we strongly believe that unravelling the mechanism(s) by which $G_{q/11}$ regulates SphK1 will further help to understand the functional roles of this enzyme and facilitate its targeting by potential therapeutics.

Author Contributions: Conceptualization, D.M.z.H., T.W.; acquisition of the data, K.V.B., R.F.C., N.J.A., A.K.S., D.M.z.H., S.T., N.F.; resources, J.L.B., J.J.G.T., S.O., T.W., D.M.z.H.; evaluation and discussion of the data, K.V.B., R.F.C., N.J.A., A.K.S., J.J.G.T., S.O., T.W., D.M.z.H.; writing, review and editing, K.V.B., R.F.C., J.J.G.T., T.W., D.M.z.H. All authors have read and agreed to the published version of the manuscript.

Funding: This work was funded by the Deutsche Forschungsgemeinschaft (FOG-784; SFB1039 TP A04, B07, and Z01), the LOEWE Lipid Signaling Forschungszentrum Frankfurt (LiFF), and National Institutes of Health grants HL071818 and CA221289 to J.J.G.T.

Acknowledgments: We thank Andrea Huwiler (University of Bern, Switzerland) and Josef Pfeilschifter (Goethe-University Frankfurt, Germany) for the SphK1 antibody, and Catherine Berlot (Weis Center for Research, Danville, PA, USA) for the $G\alpha_q$-YFP expression plasmid. The expert technical assistance of Luise Reinsberg, Agnes Rudowski, and Nicole Kämpfer-Kolb is gratefully acknowledged.

Conflicts of Interest: The authors declare no conflict of interest.

References

1. Cartier, A.; Hla, T. Sphingosine 1-phosphate: Lipid signaling in pathology and therapy. *Science* **2019**, *366*. [CrossRef] [PubMed]

2. Hannun, Y.A.; Obeid, L.M. Sphingolipids and their metabolism in physiology and disease. *Nat. Rev. Mol. Cell Biol.* **2018**, *19*, 175–191. [CrossRef] [PubMed]

3. Blaho, V.; Chun, J.; Jonnalagadda, D.; Kihara, Y.; Mizuno, H.; Mpamhanga, C.; Spiegel, S.; Tan, V. Lysophospholipid (S1P) receptors (version 2019.4) in the IUPHAR/BPS Guide to Pharmacology Database. *GtoPdb CITE* **2019**, *2019*. [CrossRef]

4. Shen, H.; Giordano, F.; Wu, Y.; Chan, J.; Zhu, C.; Milosevic, I.; Wu, X.; Yao, K.; Chen, B.; Baumgart, T.; et al. Coupling between endocytosis and sphingosine kinase 1 recruitment. *Nat. Cell Biol.* **2014**, *16*, 652–662. [CrossRef] [PubMed]

5. Lima, S.; Milstien, S.; Spiegel, S. Sphingosine and Sphingosine Kinase 1 Involvement in Endocytic Membrane Trafficking. *J. Biol. Chem.* **2017**, *292*, 3074–3088. [CrossRef]

6. Maceyka, M.; Harikumar, K.B.; Milstien, S.; Spiegel, S. Sphingosine-1-phosphate signaling and its role in disease. *Trends Cell Biol.* **2012**, *22*, 50–60. [CrossRef]

7. Siow, D.; Wattenberg, B. The compartmentalization and translocation of the sphingosine kinases: Mechanisms and functions in cell signaling and sphingolipid metabolism. *Crit. Rev. Biochem. Mol. Biol.* **2011**, *46*, 365–375. [CrossRef]

8. Chan, H.; Pitson, S.M. Post-translational regulation of sphingosine kinases. *Biochim. Biophys. Acta* **2013**, *1831*, 147–156. [CrossRef]

9. Pulkoski-Gross, M.J.; Obeid, L.M. Molecular mechanisms of regulation of sphingosine kinase 1. *Biochim. Biophys. Acta Mol. Cell Biol. Lipids* **2018**, *1863*, 1413–1422. [CrossRef]

10. Pyne, S.; Adams, D.R.; Pyne, N.J. Sphingosine Kinases as Druggable Targets. *Handb. Exp. Pharmacol.* **2018**. [CrossRef]

11. Hobson, J.P.; Rosenfeldt, H.M.; Barak, L.S.; Olivera, A.; Poulton, S.; Caron, M.G.; Milstien, S.; Spiegel, S. Role of the sphingosine-1-phosphate receptor EDG-1 in PDGF-induced cell motility. *Science* **2001**, *291*, 1800–1803. [CrossRef] [PubMed]

12. Spiegel, S.; Maczis, M.A.; Maceyka, M.; Milstien, S. New insights into functions of the sphingosine-1-phosphate transporter SPNS2. *J. Lipid Res.* **2019**, *60*, 484–489. [CrossRef] [PubMed]

13. Johnson, K.R.; Becker, K.P.; Facchinetti, M.M.; Hannun, Y.A.; Obeid, L.M. PKC-dependent activation of sphingosine kinase 1 and translocation to the plasma membrane. Extracellular release of sphingosine-1-phosphate induced by phorbol 12-myristate 13-acetate (PMA). *J. Biol. Chem.* **2002**, *277*, 35257–35262. [CrossRef] [PubMed]

14. Pitson, S.M.; Moretti, P.A.B.; Zebol, J.R.; Lynn, H.E.; Xia, P.; Vadas, M.A.; Wattenberg, B.W. Activation of sphingosine kinase 1 by ERK1/2-mediated phosphorylation. *EMBO J.* **2003**, *22*, 5491–5500. [CrossRef] [PubMed]

15. Gault, C.R.; Eblen, S.T.; Neumann, C.A.; Hannun, Y.A.; Obeid, L.M. Oncogenic K-Ras regulates bioactive sphingolipids in a sphingosine kinase 1-dependent manner. *J. Biol. Chem.* **2012**, *287*, 31794–31803. [CrossRef]

16. Jarman, K.E.; Moretti, P.A.B.; Zebol, J.R.; Pitson, S.M. Translocation of sphingosine kinase 1 to the plasma membrane is mediated by calcium- and integrin-binding protein 1. *J. Biol. Chem.* **2010**, *285*, 483–492. [CrossRef]

17. Delon, C.; Manifava, M.; Wood, E.; Thompson, D.; Krugmann, S.; Pyne, S.; Ktistakis, N.T. Sphingosine kinase 1 is an intracellular effector of phosphatidic acid. *J. Biol. Chem.* **2004**, *279*, 44763–44774. [CrossRef]

18. Pulkoski-Gross, M.J.; Jenkins, M.L.; Truman, J.-P.; Salama, M.F.; Clarke, C.J.; Burke, J.E.; Hannun, Y.A.; Obeid, L.M. An intrinsic lipid-binding interface controls sphingosine kinase 1 function. *J. Lipid Res.* **2018**, *59*, 462–474. [CrossRef]

19. Wang, J.; Knapp, S.; Pyne, N.J.; Pyne, S.; Elkins, J.M. Crystal Structure of Sphingosine Kinase 1 with PF-543. *ACS Med. Chem. Lett.* **2014**, *5*, 1329–1333. [CrossRef]

20. Wang, Z.; Min, X.; Xiao, S.-H.; Johnstone, S.; Romanow, W.; Meininger, D.; Xu, H.; Liu, J.; Dai, J.; An, S.; et al. Molecular basis of sphingosine kinase 1 substrate recognition and catalysis. *Structure* **2013**, *21*, 798–809. [CrossRef]

21. Adams, D.R.; Pyne, S.; Pyne, N.J. Sphingosine Kinases: Emerging Structure-Function Insights. *Trends Biochem. Sci.* **2016**, *41*, 395–409. [CrossRef] [PubMed]

22. Alemany, R.; van Koppen, C.J.; Danneberg, K.; ter Braak, M.; Meyer zu Heringdorf, D. Regulation and functional roles of sphingosine kinases. *Naunyn Schmiedebergs. Arch. Pharmacol.* **2007**, *374*, 413–428. [CrossRef] [PubMed]

23. ter Braak, M.; Danneberg, K.; Lichte, K.; Liphardt, K.; Ktistakis, N.T.; Pitson, S.M.; Hla, T.; Jakobs, K.H.; Meyer zu Heringdorf, D. Galpha(q)-mediated plasma membrane translocation of sphingosine kinase-1 and cross-activation of S1P receptors. *Biochim. Biophys. Acta* **2009**, *1791*, 357–370. [CrossRef] [PubMed]

24. Blankenbach, K.V.; Bruno, G.; Wondra, E.; Spohner, A.K.; Aster, N.J.; Vienken, H.; Trautmann, S.; Ferreirós, N.; Wieland, T.; Bruni, P.; et al. The WD40 repeat protein, WDR36, orchestrates sphingosine kinase-1 recruitment and phospholipase C-β activation by Gq-coupled receptors. *Biochim. Biophys. Acta Mol. Cell Biol. Lipids* **2020**, *1865*, 158704. [CrossRef] [PubMed]

25. Bruno, G.; Cencetti, F.; Bernacchioni, C.; Donati, C.; Blankenbach, K.V.; Thomas, D.; Meyer zu Heringdorf, D.; Bruni, P. Bradykinin mediates myogenic differentiation in murine myoblasts through the involvement of SK1/Spns2/S1P2 axis. *Cell. Signal.* **2018**, *45*, 110–121. [CrossRef] [PubMed]

26. Vienken, H.; Mabrouki, N.; Grabau, K.; Claas, R.F.; Rudowski, A.; Schömel, N.; Pfeilschifter, J.; Lütjohann, D.; van Echten-Deckert, G.; Meyer zu Heringdorf, D. Characterization of cholesterol homeostasis in sphingosine-1-phosphate lyase-deficient fibroblasts reveals a Niemann-Pick disease type C-like phenotype with enhanced lysosomal Ca^{2+} storage. *Sci. Rep.* **2017**, *7*. [CrossRef]

27. Kostenis, E.; Martini, L.; Ellis, J.; Waldhoer, M.; Heydorn, A.; Rosenkilde, M.M.; Norregaard, P.K.; Jorgensen, R.; Whistler, J.L.; Milligan, G. A highly conserved glycine within linker I and the extreme C terminus of G protein alpha subunits interact cooperatively in switching G protein-coupled receptor-to-effector specificity. *J. Pharmacol. Exp. Ther.* **2005**, *313*, 78–87. [CrossRef]

28. Hughes, T.E.; Zhang, H.; Logothetis, D.E.; Berlot, C.H. Visualization of a functional Galpha q-green fluorescent protein fusion in living cells. Association with the plasma membrane is disrupted by mutational activation and by elimination of palmitoylation sites, but not be activation mediated by receptors or AlF4-. *J. Biol. Chem.* **2001**, *276*, 4227–4235. [CrossRef]

29. Carman, C.V.; Parent, J.L.; Day, P.W.; Pronin, A.N.; Sternweis, P.M.; Wedegaertner, P.B.; Gilman, A.G.; Benovic, J.L.; Kozasa, T. Selective regulation of Galpha(q/11) by an RGS domain in the G protein-coupled receptor kinase, GRK2. *J. Biol. Chem.* **1999**, *274*, 34483–34492. [CrossRef]

30. Sterne-Marr, R.; Tesmer, J.J.G.; Day, P.W.; Stracquatanio, R.P.; Cilente, J.-A.E.; O'Connor, K.E.; Pronin, A.N.; Benovic, J.L.; Wedegaertner, P.B. G protein-coupled receptor Kinase 2/G alpha q/11 interaction. A novel surface on a regulator of G protein signaling homology domain for binding G alpha subunits. *J. Biol. Chem.* **2003**, *278*, 6050–6058. [CrossRef]

31. Tesmer, V.M.; Kawano, T.; Shankaranarayanan, A.; Kozasa, T.; Tesmer, J.J.G. Snapshot of activated G proteins at the membrane: The Galphaq-GRK2-Gbetagamma complex. *Science* **2005**, *310*, 1686–1690. [CrossRef] [PubMed]

32. Moepps, B.; Tulone, C.; Kern, C.; Minisini, R.; Michels, G.; Vatter, P.; Wieland, T.; Gierschik, P. Constitutive serum response factor activation by the viral chemokine receptor homologue pUS28 is differentially regulated by Galpha(q/11) and Galpha(16). *Cell. Signal.* **2008**, *20*, 1528–1537. [CrossRef] [PubMed]

33. Shankaranarayanan, A.; Boguth, C.A.; Lutz, S.; Vettel, C.; Uhlemann, F.; Aittaleb, M.; Wieland, T.; Tesmer, J.J.G. Galpha q allosterically activates and relieves autoinhibition of p63RhoGEF. *Cell. Signal.* **2010**, *22*, 1114–1123. [CrossRef] [PubMed]

34. van Unen, J.; Reinhard, N.R.; Yin, T.; Wu, Y.I.; Postma, M.; Gadella, T.W.J.; Goedhart, J. Plasma membrane restricted RhoGEF activity is sufficient for RhoA-mediated actin polymerization. *Sci. Rep.* **2015**, *5*, 14693. [CrossRef]

35. Sinnecker, D.; Schaefer, M. Real-time analysis of phospholipase C activity during different patterns of receptor-induced Ca2+ responses in HEK293 cells. *Cell Calcium* **2004**, *35*, 29–38. [CrossRef]

36. Freeman, T.C.; Black, J.L.; Bray, H.G.; Dagliyan, O.; Wu, Y.I.; Tripathy, A.; Dokholyan, N.V.; Leisner, T.M.; Parise, L.V. Identification of novel integrin binding partners for calcium and integrin binding protein 1 (CIB1): Structural and thermodynamic basis of CIB1 promiscuity. *Biochemistry* **2013**, *52*, 7082–7090. [CrossRef]

37. Black, J.L.; Harrell, J.C.; Leisner, T.M.; Fellmeth, M.J.; George, S.D.; Reinhold, D.; Baker, N.M.; Jones, C.D.; Der, C.J.; Perou, C.M.; et al. CIB1 depletion impairs cell survival and tumor growth in triple-negative breast cancer. *Breast Cancer Res. Treat.* **2015**, *152*, 337–346. [CrossRef]

38. Offermanns, S.; Zhao, L.P.; Gohla, A.; Sarosi, I.; Simon, M.I.; Wilkie, T.M. Embryonic cardiomyocyte hypoplasia and craniofacial defects in G alpha q/G alpha 11-mutant mice. *EMBO J.* **1998**, *17*, 4304–4312. [CrossRef]

39. Döll, F.; Pfeilschifter, J.; Huwiler, A. The epidermal growth factor stimulates sphingosine kinase-1 expression and activity in the human mammary carcinoma cell line MCF7. *Biochim. Biophys. Acta* **2005**, *1738*, 72–81. [CrossRef]

40. Lyon, A.M.; Taylor, V.G.; Tesmer, J.J.G. Strike a pose: Gαq complexes at the membrane. *Trends Pharmacol. Sci.* **2014**, *35*, 23–30. [CrossRef]

41. Lutz, S.; Freichel-Blomquist, A.; Yang, Y.; Rümenapp, U.; Jakobs, K.H.; Schmidt, M.; Wieland, T. The guanine nucleotide exchange factor p63RhoGEF, a specific link between Gq/11-coupled receptor signaling and RhoA. *J. Biol. Chem.* **2005**, *280*, 11134–11139. [CrossRef] [PubMed]

42. Lutz, S.; Shankaranarayanan, A.; Coco, C.; Ridilla, M.; Nance, M.R.; Vettel, C.; Baltus, D.; Evelyn, C.R.; Neubig, R.R.; Wieland, T.; et al. Structure of Galphaq-p63RhoGEF-RhoA complex reveals a pathway for the activation of RhoA by GPCRs. *Science* **2007**, *318*, 1923–1927. [CrossRef] [PubMed]

43. Watt, S.A.; Kular, G.; Fleming, I.N.; Downes, C.P.; Lucocq, J.M. Subcellular localization of phosphatidylinositol 4,5-bisphosphate using the pleckstrin homology domain of phospholipase C delta1. *Biochem. J.* **2002**, *363*, 657–666. [CrossRef] [PubMed]

44. Szymańska, E.; Sobota, A.; Czuryło, E.; Kwiatkowska, K. Expression of PI(4,5)P2-binding proteins lowers the PI(4,5)P2level and inhibits FcgammaRIIA-mediated cell spreading and phagocytosis. *Eur. J. Immunol.* **2008**, *38*, 260–272. [CrossRef] [PubMed]

45. Weernink, P.A.O.; Meletiadis, K.; Hommeltenberg, S.; Hinz, M.; Ishihara, H.; Schmidt, M.; Jakobs, K.H. Activation of type I phosphatidylinositol 4-phosphate 5-kinase isoforms by the Rho GTPases, RhoA, Rac1, and Cdc42. *J. Biol. Chem.* **2004**, *279*, 7840–7849. [CrossRef]

46. Rümenapp, U.; Schmidt, M.; Olesch, S.; Ott, S.; Eichel-Streiber, C.V.; Jakobs, K.H. Tyrosine-phosphorylation-dependent and rho-protein-mediated control of cellular phosphatidylinositol 4,5-bisphosphate levels. *Biochem. J.* **1998**, *334 Pt 3*, 625–631. [CrossRef]

47. Pitson, S.M.; Moretti, P.A.; Zebol, J.R.; Xia, P.; Gamble, J.R.; Vadas, M.A.; D'andrea, R.J.; Wattenberg, B.W. Expression of a catalytically inactive sphingosine kinase mutant blocks agonist-induced sphingosine kinase activation. A dominant-negative sphingosine kinase. *J. Biol. Chem.* **2000**, *275*, 33945–33950. [CrossRef]

48. Day, P.W.; Tesmer, J.J.G.; Sterne-Marr, R.; Freeman, L.C.; Benovic, J.L.; Wedegaertner, P.B. Characterization of the GRK2 binding site of Galphaq. *J. Biol. Chem.* **2004**, *279*, 53643–53652. [CrossRef]

49. Venkatakrishnan, G.; Exton, J.H. Identification of determinants in the alpha-subunit of Gq required for phospholipase C activation. *J. Biol. Chem.* **1996**, *271*, 5066–5072. [CrossRef]

50. Sutherland, C.M.; Moretti, P.A.B.; Hewitt, N.M.; Bagley, C.J.; Vadas, M.A.; Pitson, S.M. The calmodulin-binding site of sphingosine kinase and its role in agonist-dependent translocation of sphingosine kinase 1 to the plasma membrane. *J. Biol. Chem.* **2006**, *281*, 11693–11701. [CrossRef]

51. Inagaki, Y.; Li, P.-Y.; Wada, A.; Mitsutake, S.; Igarashi, Y. Identification of functional nuclear export sequences in human sphingosine kinase 1. *Biochem. Biophys. Res. Commun.* **2003**, *311*, 168–173. [CrossRef] [PubMed]

52. Mulders, A.C.M.; Mathy, M.-J.; Meyer zu Heringdorf, D.; ter Braak, M.; Hajji, N.; Olthof, D.C.; Michel, M.C.; Alewijnse, A.E.; Peters, S.L.M. Activation of sphingosine kinase by muscarinic receptors enhances NO-mediated and attenuates EDHF-mediated vasorelaxation. *Basic Res. Cardiol.* **2009**, *104*, 50–59. [CrossRef] [PubMed]

53. Billich, A.; Urtz, N.; Reuschel, R.; Baumruker, T. Sphingosine kinase 1 is essential for proteinase-activated receptor-1 signalling in epithelial and endothelial cells. *Int. J. Biochem. Cell Biol.* **2009**, *41*, 1547–1555. [CrossRef]

54. Böhm, A.; Flößer, A.; Ermler, S.; Fender, A.C.; Lüth, A.; Kleuser, B.; Schrör, K.; Rauch, B.H. Factor-Xa-induced mitogenesis and migration require sphingosine kinase activity and S1P formation in human vascular smooth muscle cells. *Cardiovasc. Res.* **2013**, *99*, 505–513. [CrossRef]

55. Siedlinski, M.; Nosalski, R.; Szczepaniak, P.; Ludwig-Gałęzowska, A.H.; Mikołajczyk, T.; Filip, M.; Osmenda, G.; Wilk, G.; Nowak, M.; Wołkow, P.; et al. Vascular transcriptome profiling identifies Sphingosine kinase 1 as a modulator of angiotensin II-induced vascular dysfunction. *Sci. Rep.* **2017**, *7*, 44131. [CrossRef] [PubMed]

56. Huwiler, A.; Döll, F.; Ren, S.; Klawitter, S.; Greening, A.; Römer, I.; Bubnova, S.; Reinsberg, L.; Pfeilschifter, J. Histamine increases sphingosine kinase-1 expression and activity in the human arterial endothelial cell line EA.hy 926 by a PKC-alpha-dependent mechanism. *Biochim. Biophys. Acta* **2006**, *1761*, 367–376. [CrossRef]

57. Sun, W.Y.; Abeynaike, L.D.; Escarbe, S.; Smith, C.D.; Pitson, S.M.; Hickey, M.J.; Bonder, C.S. Rapid histamine-induced neutrophil recruitment is sphingosine kinase-1 dependent. *Am. J. Pathol.* **2012**, *180*, 1740–1750. [CrossRef]

58. Pyne, N.J.; Pyne, S. Sphingosine Kinase 1: A Potential Therapeutic Target in Pulmonary Arterial Hypertension? *Trends Mol. Med.* **2017**, *23*, 786–798. [CrossRef]

59. Kostenis, E.; Pfeil, E.M.; Annala, S. Heterotrimeric Gq proteins as therapeutic targets? *J. Biol. Chem.* **2020**, *295*, 5206–5215. [CrossRef]

60. Meshcheryakova, A.; Svoboda, M.; Tahir, A.; Köfeler, H.C.; Triebl, A.; Mungenast, F.; Heinze, G.; Gerner, C.; Zimmermann, P.; Jaritz, M.; et al. Exploring the role of sphingolipid machinery during the epithelial to mesenchymal transition program using an integrative approach. *Oncotarget* **2016**, *7*, 22295–22323. [CrossRef]

61. Liu, S.-Q.; Xu, C.-Y.; Wu, W.-H.; Fu, Z.-H.; He, S.-W.; Qin, M.-B.; Huang, J.-A. Sphingosine kinase 1 promotes the metastasis of colorectal cancer by inducing the epithelial-mesenchymal transition mediated by the FAK/AKT/MMPs axis. *Int. J. Oncol.* **2019**, *54*, 41–52. [CrossRef] [PubMed]

Article

Inflammatory Conditions Disrupt Constitutive Endothelial Cell Barrier Stabilization by Alleviating Autonomous Secretion of Sphingosine 1-Phosphate

Jefri Jeya Paul [1,2,3], Cynthia Weigel [1,2,4,5], Tina Müller [1,2], Regine Heller [2,3,6], Sarah Spiegel [5] and Markus H. Gräler [1,2,3,*]

[1]   Department of Anesthesiology and Intensive Care Medicine, Jena University Hospital, 07740 Jena, Germany; jefri.jeyapaul@gmail.com (J.J.P.); Cynthia.Weigel@vcuhealth.org (C.W.); Tina.Mueller2@med.uni-jena.de (T.M.)
[2]   Center for Molecular Biomedicine, Jena University Hospital, 07745 Jena, Germany; regine.heller@med.uni-jena.de
[3]   Center for Sepsis Control and Care, Jena University Hospital, 07740 Jena, Germany
[4]   Leibniz Institute on Aging—Fritz Lipmann Institute, 07745 Jena, Germany
[5]   Department of Biochemistry and Molecular Biology, Virginia Commonwealth University School of Medicine, Richmond, VA 23298, USA; sarah.spiegel@vcuhealth.org
[6]   Institute of Molecular Cell Biology, Jena University Hospital, 07745 Jena, Germany
*   Correspondence: markus.graeler@med.uni-jena.de; Tel.: +49-3641-939-5715; Fax: +49-3641-939-5789

Received: 6 March 2020; Accepted: 7 April 2020; Published: 10 April 2020

Abstract: The breakdown of the endothelial cell (EC) barrier contributes significantly to sepsis mortality. Sphingosine 1-phosphate (S1P) is one of the most effective EC barrier-stabilizing signaling molecules. Stabilization is mainly transduced via the S1P receptor type 1 (S1PR1). Here, we demonstrate that S1P was autonomously produced by ECs. S1P secretion was significantly higher in primary human umbilical vein endothelial cells (HUVEC) compared to the endothelial cell line EA.hy926. Constitutive barrier stability of HUVEC, but not EA.hy926, was significantly compromised by the S1PR1 antagonist W146 and by the anti-S1P antibody Sphingomab. HUVEC and EA.hy926 differed in the expression of the S1P-transporter Spns2, which allowed HUVEC, but not EA.hy926, to secrete S1P into the extracellular space. Spns2 deficient mice showed increased serum albumin leakage in bronchoalveolar lavage fluid (BALF). Lung ECs isolated from Spns2 deficient mice revealed increased leakage of fluorescein isothiocyanate (FITC) labeled dextran and decreased resistance in electric cell-substrate impedance sensing (ECIS) measurements. Spns2 was down-regulated in HUVEC after stimulation with pro-inflammatory cytokines and lipopolysaccharides (LPS), which contributed to destabilization of the EC barrier. Our work suggests a new mechanism for barrier integrity maintenance. Secretion of S1P by EC via Spns2 contributed to constitutive EC barrier maintenance, which was disrupted under inflammatory conditions via the down-regulation of the S1P-transporter Spns2.

Keywords: S1P receptor; inflammation; S1P transporter; spinster homolog 2; barrier dysfunction

---

1. Introduction

Endothelial cell (EC) barriers are important intercellular structures that regulate the movement of fluids and dissolved substances into tissues. Maintaining barrier function is particularly important at sites where fluids need to be efficiently separated from tissues such as the vasculature, lymph vessels, gut, brain, and lung [1]. Several different junctional complexes are involved in barrier formation, including tight and adherens junctions, gap junctions, and desmosomes [2]. Adherens junctions are formed by cadherins and nectins and provide a mechanical linker similar to zippers, while tight junctions are formed by claudins, occludin, and junctional adhesion molecules in the transmembrane regions and perform most of the sealing functions to prevent passage of fluids and molecules [3,4]. The bioactive sphingolipid signaling molecule sphingosine 1-phosphate (S1P) and its G protein-coupled receptor S1PR1 are critical mediators of adherens junction assembly [5]. The deletion of S1PR1 in ECs or the deletion of the two known S1P-producing sphingosine kinases (SphK1 and SphK2) in hematopoietic cells and ECs of adult mice result in severe disruption of the EC barrier [6–8]. Despite this apparent phenotype, the exact mechanism of barrier maintenance by S1P is still unknown. One of the most puzzling questions is of how high S1P concentrations in circulation that are sufficient to induce activation-induced internalization and desensitization of S1PR1 are able to constitutively maintain the vascular EC barrier. Two models were proposed as potential explanations [8]: (1) the static model postulating that there is constantly sufficient S1PR1 expression on the luminal cell surface of ECs even at high S1P concentrations, due to efficient receptor recycling and (2) the dynamic model suggesting that S1PR1 is only expressed on the tissue-facing side of vascular ECs which are activated by S1P leaking through the EC barrier and subsequently induce adherens junction assembly and EC barrier stabilization. In either case, reduced S1P leakage and decreased barrier stability occur until the amount of S1P leaking through the EC barrier increases again and starts a new cycle of EC barrier formation. The validity of either of these models has not yet been verified.

The collapse of the EC barrier is a life-threatening condition and a major severity factor in sepsis [9]. The S1P concentration in circulation decreases significantly during systemic inflammation [10–13]. Whether or not this observed decrease of S1P has something to do with the vascular EC barrier collapse is not known. Previous data indicate that even low amounts of S1P in plasma are sufficient to maintain S1P and S1PR1 mediated lymphocyte circulation [14].

Here, we show that, in vitro, ECs can autonomously produce and secrete S1P, rendering their ability to maintain EC barrier formation largely independent from exogenously added S1P. The S1P transporter Spinster homolog 2 (Spns2) plays a crucial role in the proper release of S1P into the extracellular space. Our work has uncovered an important function of Spns2 in ECs to regulate barrier stability. Spns2 deficient mice demonstrated significantly reduced EC barrier formation presumably due to the lack of S1P exportation from ECs. Furthermore Spns2, but not S1PR1, was down-regulated in ECs stimulated with lipopolysaccharides (LPS) and pro-inflammatory cytokines. Inflammation-induced EC barrier breakdown due to down-regulation of Spns2 resulted in decreased S1P release. Thus, decreased exportation of S1P from ECs due to reduced expression of Spns2 may contribute to EC barrier dysfunction during inflammation. This mechanism may be particularly important in sepsis, where inflammation-induced collapse of the EC barrier significantly contributes to increased morbidity and mortality. The observed stable expression of S1PR1 and the most likely local autocrine and paracrine activity of Spns2-released S1P points to local approaches for S1P supplementation in tissues rather than systemic alteration of S1P in circulating plasma as a potential medical intervention.

2. Materials and Methods

2.1. Cell Culture

Human umbilical vein EA.hy926 cells (ATCC CRL-2922) were grown in M199 medium containing 10% fetal bovine serum (FBS; Biochrom, Berlin, Germany), 1% penicillin/streptomycin (100 U/mL, Biochrom), 0.2% glutamine (200 mM, Lonza, Basel, Switzerland), 0.2% heparin (12.5 mg/mL, Carl Roth, Karlsruhe, Germany), and 0.6% ascorbic acid (20 mM, Sigma-Aldrich, Steinheim, Germany). HUVEC were freshly isolated from human umbilical cords and grown in M199 containing 17.5% FBS, 1% penicillin/streptomycin (100 U/mL), 0.34% glutamine (200 mM), 0.2% heparin (12.5 mg/mL), 0.5% ascorbic acid (20 mM) and endothelial mitogen (5 mg/mL, Alfa Aesar, Karlsruhe, Germany). FBS was heat inactivated at 56 °C. Rat hepatoma HTC4 cells expressing human S1PR1 together with human Gαi subunit of trimeric G proteins [15] were grown in minimal essential medium (MEM) with Earle's salts (MEM Earle's medium) containing 10% FBS, 2% 100× non-essential amino acids (Biochrom),

1% 100 mM sodium pyruvate (Biochrom) and 1% penicillin/streptomycin (100 U/mL). Cells were incubated at 37 °C and 5% CO_2 in a humidified incubator (Panasonic, Hamburg, Germany).

2.2. Isolation of Primary Lung Endothelial Cells

Mice deficient for the S1P-transporter Spinster homolog 2 (Spns2) and their wild-type (wt) littermates were obtained from the NIH Knockout Mouse Project [16]. All described animal procedures were done with dead mice in accordance with the Association for Assessment and Accreditation of Laboratory Animal Care (AAALAC), Animal Welfare Assurance Number AD10000996, effective date March 26, 2017, renewed February 21, 2020 at the Virginia Commonwealth University School of Medicine, Richmond, VA, USA. Mice were killed, and after perfusion through the left ventricle with PBS supplemented with 0.1% heparin (10 U/mL), lungs were dissected from the thoracic cavity and cut into single lobes. The bronchial area was removed and the remaining tissue was digested in 15 mL conical tubes containing 5 mL of PBS with 0.5 mg/mL Liberase TL (Sigma-Aldrich) and 0.02 mg/mL DNAse 1 (Thermo Scientific, Braunschweig, Germany) for 45 min at 37 °C with shaking. The tissue was subsequently minced, further incubated for 30 min and passed through a 70 μm cell strainer (BD Biosciences, Heidelberg, Germany). Single cell suspensions were centrifuged at 300 rcf for 10 min at 4 °C. Supernatants were aspirated and cell pellets re-suspended in 90 μL of PBS containing 2 mM EDTA, 0.5% BSA, and 10 μL of CD31 MicroBeads per 107 cells (Miltenyi Biotec, Bergisch-Gladbach, Germany). The suspensions were incubated for 15 min at 4 °C and re-suspended in 500 μL PBS with 2 mM EDTA and 0.5% BSA. The labeled cells were separated with LS columns and the QuadroMACS separation system according to the manufacturer's instructions (Miltenyi Biotec). The isolated ECs were cultured in endothelial growth medium (5 × 105 cells/mL, Cell Biologics M1168) on poly-L-lysine-coated six-well plates.

2.3. cDNA Synthesis and Quantitative Polymerase Chain Reaction (qPCR)

Total RNA was extracted using the Quick-RNA Miniprep Plus kit (Zymo Research, Freiburg, Germany). cDNA was synthesized with the RevertAid first strand cDNA synthesis kit (Thermo Scientific) according to the manufacturer's instructions and diluted in nuclease-free water to a final concentration of 5 ng/μL. qPCR was performed with 8 μL cDNA, 4.4 μL of 25 mM $MgCl_2$ (VWR), 2 μL of 10× reaction buffer (VWR) with 15 mM $MgCl_2$, 0.2 μL of 10 mM dNTPs (Thermo Scientific), 0.8 μL of 5 μM 6-carboxy-X-rhodamine (ROX, Eurofins, Ebersberg, Germany), 0.048 μL of 5 U/μL of Taq-Polymerase (VWR), 3.752 μL of nuclease free water (Thermo Scientific), and 0.8 μL of TaqMan primer/probe-mix (2.5 μM each, Eurofins) for each reaction. The reaction was performed with the Mastercycler Realplex (Eppendorf, Hamburg, Germany) using the following program: initial activation at 94 °C for 10 min followed by 40 amplification cycles of denaturation at 94 °C for 10 sec and annealing and extension at 60 °C for 1 min. The internal reference dye ROX was used to normalize the fluorescent reporter signal. The expressions of genes of interest were normalized to hypoxanthine-guanine phosphoribosyltransferase (*HPRT*). Relative gene expression was calculated by the ΔCt and $2^{-\Delta\Delta Ct}$ methods [17]. Primers used for the reaction are listed in Table 1.

2.4. Agarose Gel Electrophoresis

Gel electrophoresis was performed with 1.2% agarose gels in TBE buffer according to standard protocols for 30 min at 150 V. Gels were stained with ethidium bromide and visualized with a UV trans-illuminator.

Table 1. Sequences of primers and probes used for quantitative polymerase chain reaction (qPCR) analyses.

Gene	Forward Primer	Reverse Primer	Probe (5′-56-FAM; 3′-36-TAM)
HPRT	agcctaagatgagagttc	cacagaactagaacattgata	atctggagtcctattgacatcgcc
S1PR1	agcactatatcctcttctg	tgaccaaggagtagattc	tcttcactctgcttctgctctcc
S1PR2	catcgtcatcctctgttg	agtggaacttgctgtttc	ccttctggtgctcattgcgg
S1PR3	ccaagcagaagtaaatcaag	catggagacgatcagttg	agcagcaacaatagcagccac
S1PR4	gcttctgtgtgattctgg	ccatgatcgaacttcaatg	cctctctgggcctcagtagg
S1PR5	ggaacaatgatggagatt	ggcattgtccttgataac	attccactcttacactcaattcctgag
SK1	ggcagcttccttgaacca	gcaggttcatgggtgaca	ctatgagcaggtcaccaatgaagacctcct
SK2	ccgacggcctctcagt	cctggccctgggtctta	acagtgagacctgactccttgctcctacc
SGPL1	cctagcacagaccttctgatgt	actccatgcaattagctgcca	aaggcctttgagccctactt
SPNS2	ttactggctccagcgtga	tgatcatgcccaggacag	ctgggcattgcgggtgtc

Abbreviations: FAM, 6-carboxyfluorescein; TAM, tetramethylrhodamine.

2.5. Flow Cytometry

Cells were suspended in 200 µL of 5% FBS in PBS and transferred to a 96-well round bottom plate. The plate was centrifuged at 265 rcf for 5 min. After centrifugation, 300 µL of the primary antibody specific for the human S1PR1 receptor produced in mouse (20 µg/mL, custom-made by Abmart, Berkeley Heights, NJ, USA) was added and incubated for 1 h at 4 °C. After incubation, cells were centrifuged and washed with 5% FBS in PBS, followed by 30 min incubation with 50 µL of anti-mouse Biotin-SP (1:200 dilution, Jackson Immuno Research, West Grove, PA, USA, 115–065–075). Subsequently, cells were washed and incubated with 50 µL of 40 ng/mL streptavidin-PE (Biolegend, San Diego, CA, USA, 405203) for 30 min. Cells were washed again and resuspended in 200 µL of 5% FBS and 1 µg/mL propidium iodide (BD Biosciences) in PBS. Resuspended cells were analyzed with the Accuri C6 Plus (BD Biosciences). After doublet exclusion and life-dead discrimination by propidium iodid, the mean floerescence intensity was analyzed.

2.6. Western Blot and Calcium Measurement

Cells were washed and lysed with buffer containing radioimmunoprecipitation assay buffer (RIPA, Merck, Darmstadt, Germany), 0.5 M EDTA, and protease and phosphatase inhibitor cocktail (Thermo Scientific). Samples were adjusted to 10 µg protein with the BCA protein assay kit (Thermo Scientific) and blotted to polyvinylidenfluoride (PVDF) membranes (GE Healthcare Life Sciences, Freiburg, Germany) according to standard protocols. Detection was performed with antibodies against GAPDH (1:2000 dilution, Santa Cruz Biotechnology, Dallas, TX, USA SC-166574) and VE-cadherin (1:1000 dilution, BD Biosciences 610252), anti-mouse HRP secondary antibody (1:1000 dilution, Carl Roth 47591) and SuperSignal West Pico Chemiluminescent Substrate (Thermo Scientific) with the C-Digit Blot Scanner (LI-COR Biosciences, Bad Homburg, Germany).

Calcium measurements were performed as described [18].

2.7. Electric Cell-Substrate Impedance Sensing (ECIS)

ECIS arrays (Ibidi, Gräfelfing, Germany 8W10E PET) were coated with 200 µL of 0.2% gelatin (Sigma-Aldrich) for 20 min at room temperature. Gelatin was replaced with 400 µL of medium containing 1×10^5 cells. Measurements were carried out at 6 kHz with 10 s interval. Once a monolayer was attained, 200 µL of the medium was replaced by 200 µL of starvation medium (2% FBS containing growth medium). The monolayer was stimulated 6 h later with various stimulants. Results were analyzed using ECIS software v1.2.214.0 (Ibidi) by normalizing the data points after treatment to the data point before treatment. The basal barrier function was set to 1.

2.8. Fluorescence Microscopy

Cells were fixed with 100% ice-cold methanol for 15 min at $-20\,^{\circ}$C. Coverslips were washed five times by dipping them in ice-cold HBSS containing Ca^{2+}/Mg^{2+}. Slides were blocked in 50 µL blocking buffer (5% goat serum in HBSS -including Ca^{2+}/Mg^{2+} and 0.1% saponin) for 1 h at room temperature followed by overnight incubation with 40 µL of mouse primary VE-cadherin antibody (BD Biosciences 610252) diluted 1:100 in 1% BSA in HBSS+ Ca^{2+}/Mg^{2+} +0.1% saponin at $4\,^{\circ}$C in a dark wet chamber. Slides were washed three times in washing buffer (0.1% saponin in HBSS containing Ca^{2+}/Mg^{2+}) and incubated in 40 µL of 10 µg/mL goat anti-mouse Cy3 secondary antibody (Life Technologies, Darmstadt, Germany A10521) diluted in 1% BSA in HBSS+ Ca^{2+}/Mg^{2+} +0.1% saponin at room temperature in a dark wet chamber. Slides were washed and stained with 40 µL of 300 nM DAPI (VWR) for 20 min at room temperature. Then, the slides were washed three times in ice-cold PBS and mounted using CC Mount (Sigma-Aldrich).

2.9. Measurement of S1P and Sphingosine

Lipid extraction and quantification of S1P and sphingosine by liquid chromatography coupled to triple-quadrupole mass spectrometry (LC-MS/MS) was done as described [19].

2.10. FITC-Dextran Leakage Assay

Sixty thousand primary lung ECs in 100 µL medium were seeded in the upper chamber of Transwell inserts with 0.4 µm pores (Sarstedt, Nürnbrecht, Germany) in 24-well plates and grown until confluency. The lower chamber was filled with 600 µL medium. Afterwards, the media of the upper chamber was replaced with 200 µL of media containing 2 mg/mL 70 kDa FITC-dextran (Sigma-Aldrich). The amount of FITC-dextran was measured in the medium of the lower chamber after 24 h incubation at $37\,^{\circ}$C and 5% CO_2 with the Infinite 200 plate reader (Tecan, Stadt Crailsheim, Germany) at 485 nm excitation and 530 nm emission wave lengths. The exact amount of FITC-dextran was determined using a standard curve of FITC-dextran diluted in cell culture media.

2.11. Albumin Measurement

The detection of mouse albumin in bronchoalveolar lavage fluid (BALF) was carried out with the Mouse Albumin ELISA Quantitation Set (Bethyl Laboratories, Montgomery, TX, USA) according to the manufacturer's instructions. BALF was diluted 1:20.000 in dilution buffer (50 mM Tris base, 0.14 M NaCl, 1% BSA, 0.05% Tween 20), and 100 µL were added to anti-mouse albumin antibody pre-coated 96-well plates.

2.12. Reagents

Stimuli used in this study were S1P (Sigma-Aldrich), FTY720 (Cayman Chemicals, Ann Arbor, MI, USA), FTY720-phosphate (Cayman Chemicals), W146 (Tocris, Wiesbaden-Nordenstadt, Germany), LPS (Sigma-Aldrich), TNFα (ImmunoTools, Friesoythe, Germany), and IL1β (Thermo Scientific). The anti-S1P antibody Sphingomab (LT1002) and the corresponding isotype control antibody LT1013 were kindly provided by Roger Sabbadini (LPath Inc. and San Diego State University, San Diego, CA, USA).

2.13. Statistics

Statistical analysis was performed using GraphPad Prism® Software Version 5.00 (San Diego, CA, USA). Data are presented as mean ± SEM. Unpaired two-tailed t-tests were used to compare two groups. The significance threshold was set to * $p < 0.05$, ** $p < 0.01$, and *** $p < 0.001$.

3. Results

3.1. EC Barrier Stabilizing Function of S1P and S1PR1

To investigate the role of S1P in EC barrier function, the human endothelial cell line EA.hy926 and primary HUVEC were used. EA.hy926 represents a somatic cell hybrid of HUVEC and the lung epithelial carcinoma cell line A549. Quantitative PCR demonstrated that both, HUVEC and EA.hy926 expressed mainly *S1PR1* followed by *S1PR3*, although HUVEC expressed both receptors stronger than EA.hy926 (Figure 1A). Higher expression of S1PR1 in HUVEC was confirmed by flow cytometry (FACS) using a highly specific antibody against human S1PR1 (Figure 1B). Specific staining was demonstrated by the incubation of cells with 1 µM FTY720 overnight, which led to S1PR1 internalization and consequently low cell surface staining as expected (Figure 1B). In line with these expression data, HUVEC responses were greater than those of EA.hy926 after stimulation with 100 nM S1P in intracellular calcium flux measurements (Figure 1C). However, to our surprise, EA.hy926 responses were greater in ECIS measurements after stimulation with 1 µM S1P compared to HUVEC, indicating a higher barrier stabilization in EA.hy926 (Figure 1D). Basal resistance of the EC monolayer, however, was lower in EA.hy926 compared to HUVEC (Figure 1E).

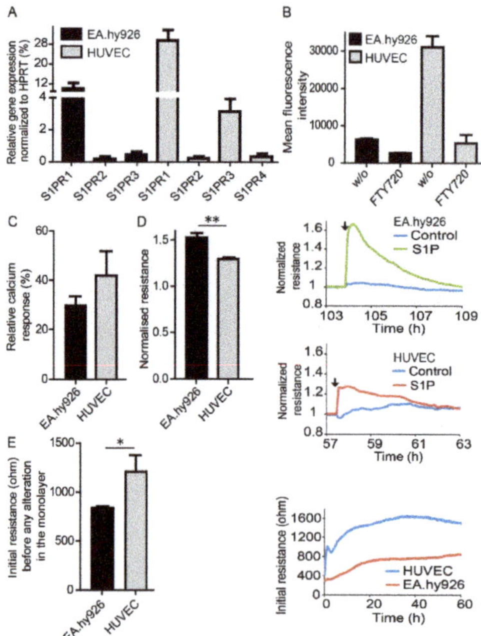

Figure 1. S1PR expression and signaling in EC. (**A**) qPCR analysis of S1PR expression in EA.hy926 and HUVEC. Dara are means ± SEM, *n* = 3. (**B**) Flow Cytometric analysis cell surface expression of S1PR1 on EC before and after treatment with 1 µM FTY720 overnight. means ± SEM, *n* = 3. (**C**) Intracellular calcium responses in EA.hy926 and HUVEC upon stimulation with 100 nM S1P. Data were normalized to the response of 10 µM ATP. Means ± SEM, *n* = 3. (**D**) Resistance following treatment with 1 µM S1P, normalized resistance values were taken at the time of the established maximum resistance after S1P treatment divided by resistance of carrier-treated control cells at the same time and are means ± SEM, *n* = 3, ** *p* <0.01, determined by two-sided Student's t-test. Line plots represent one experiment out of three with black arrows indicating the addition of S1P or vehicle at the corresponding time. (**E**) Difference in initial non-stimulated resistance of EA.hy926 and HUVEC in ECIS measurements 60 h after seeding, means ± SEM, *n* = 3, * *p* < 0.05, determined by a two-sided Student's t-test. Line plot represents one experiment out of three.

3.2. Endogenous Differences in S1P Signaling between HUVEC and EA.hy926

To explore the reason for the different behavior of HUVEC and EA.hy926 in ECIS measurements, both cells were treated with 3 µM of the S1PR1 antagonist W146. While EA.hy926 resistance was not affected by W146 treatment, HUVEC monolayers showed significantly reduced resistance by 60% in ECIS measurements, suggesting involvement of S1PR1 in constitutive basal EC barrier maintenance in HUVEC, but not in EA.hy926 (Figure 2A). A similar observation was recorded in ECIS measurements after treatment with the anti-S1P antibody Sphingomab. Sphingomab (120 µg/mL) reduced the basal resistance of the HUVEC monolayer by 30%, while EA.hy926 did not respond at all (Figure 2B). Determination of S1P in the supernatant of both cell types revealed three fold greater S1P level in HUVEC medium than EA.hy926 medium (Figure 2C). Conditioned HUVEC medium consequently provided a four-fold enhanced calcium signal in S1PR1, overexpressing rat hepatoma HTC_4 cells compared to EA.hy926 conditioned medium (Figure 2D). Conditioned medium from HUVEC induced a significant 20% increase of the measured resistance in ECIS experiments when added to EA.hy926, while conditioned medium from EA.hy926 in contrast reduced the corresponding resistance by 20% of a HUVEC monolayer (Figure 2E). HUVEC re-established their barrier integrity within hours, while the observed increased resistance in EA.hy926 after incubation with conditioned medium from HUVEC subsequently decreased further and fell below the value of HUVEC (Figure 2E).

3.3. Reversibility and VE-Cadherin Disturbance of EC Barrier Destabilization by S1PR1 Antagonism and S1P Blocking

Since HUVEC responded immediately to treatment with the S1PR1 antagonist W146 and the anti-S1P antibody Sphingomab with EC barrier disruption, we next asked if this destabilizing effect was reversible. To this end, HUVECs were applied to ECIS measurements and treated with either 3 µM W146 or 120 µg/mL Sphingomab. After decreased resistance leveled off in ECIS measurements, the medium was replaced without the addition of W146 and Sphingomab. Subsequently, the recorded resistance increased immediately and eventually reached normal levels of untreated control cells (Figure 3A). This experiment demonstrated that S1PR1 had to be constantly and persistently activated to induce the EC barrier stabilizing effect, and S1P had to be present all the time as a stimulus. Barrier stabilization was not induced by a single long-lasting activation of S1PR1, but required continuous stimulation. This was also supported by staining HUVEC monolayers for VE-cadherin expression. HUVEC showed pronounced VE-cadherin staining in the intercellular regions of monolayers, which was severely disrupted after treatment with 3 µM W146 (Figure 3B and Figure S1).

3.4. Role of Spns2 in EC Barrier Maintenance

Autonomously produced S1P by HUVEC, but not by EA.hy926 obviously contributed to constitutive basal EC barrier maintenance. qPCR data revealed that HUVEC expressed significant amounts of the S1P transporter Spns2 mRNA, while EA.hy926 were negative (Figure 4A). Monolayers of primary lung EC isolated from Spns2 deficient mice showed significantly decreased resistance values in ECIS measurements compared to those isolated from wt mice (Figure 4B). Spns2 deficient mouse lung EC also demonstrated increased leakage of fluorescein isothiocyanate (FITC) labeled dextran compared to wt mouse lung EC in Transwell cell monolayer permeability assays (Figure 4C). These results supported a significant contribution of Spns2-driven export of autonomously produced S1P for EC barrier formation ex vivo. To examine whether Spns2 also contributed to EC barrier stabilization in vivo, serum albumin leakage from circulation into the lung was measured in bronchoalveolar lavage fluid (BALF) of wt and Spns2 deficient mice. In line with ex vivo data, BALF retrieved from Spns2 deficient mice contained significantly more serum albumin compared to wt mice (Figure 4D).

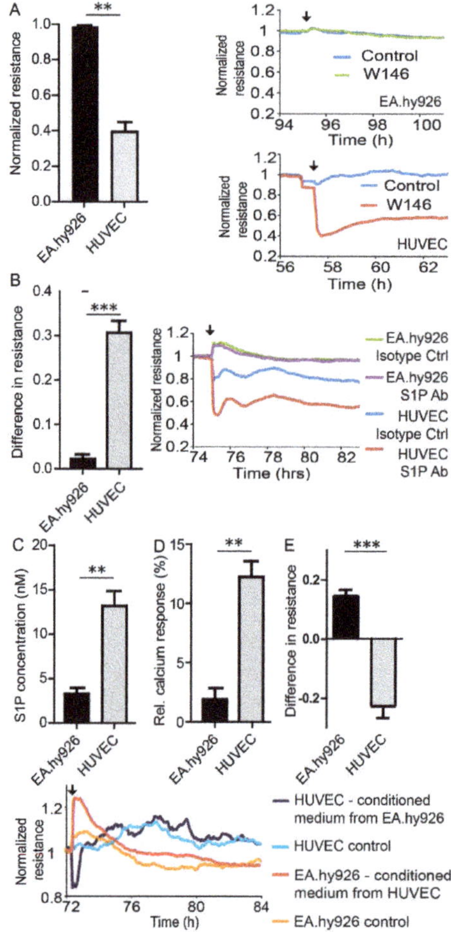

Figure 2. Comparison of S1P-signaling in HUVEC and EA.hy926. (**A**) Resistance following treatment with 3 μM of the S1PR1 antagonist W146. Normalized resistance values were taken at the time of the established maximal change of resistance after W146 treatment divided by resistance of carrier-treated control cells at the same time and are means ± SEM, $n = 3$, ** $p < 0.001$, determined by two-sided Student's t-test. Line plots represent one experiment out of three with black arrows indicating the addition of W146 or vehicle at the corresponding time. (**B**) Resistance following treatment with 120 μg/mL of the anti-S1P antibody Sphingomab. The difference in resistance is the difference between S1P-antibody treatment and isotype control antibody treatment taken at the time of maximal change of resistance after treatment. Shown are means ± SEM, $n = 3$, *** $p < 0.001$, determined by a two-sided Student's t-test. Line plot represents one experiment out of three with a black arrow indicating the addition of Sphingomab (S1P Ab) or isotype control antibody (Isotype Ctrl) at the corresponding time. (**C**) LC-MS/MS quantification of extracellular S1P production by EA.hy926 and HUVEC. (**D**) Intracellular calcium response in rat hepatoma HTC4 cells transfected with human S1PR1 and the human Gαi subunit of trimeric G proteins, stimulated with lipid extracts from EA.hy926 and HUVEC as indicated. (**E**) Resistance following medium exchange. The difference in resistance is the difference between exchange of conditioned medium and control medium at the time of maximal change of resistance after medium exchange. Line plot represents one experiment out of three with a black arrow indicating the exchange of medium at the corresponding time, controls represent unconditioned medium. (**C–E**) Data are means ± SEM, $n = 3$, ** $p < 0.01$, *** $p < 0.001$, determined by a two-sided Student's t-test.

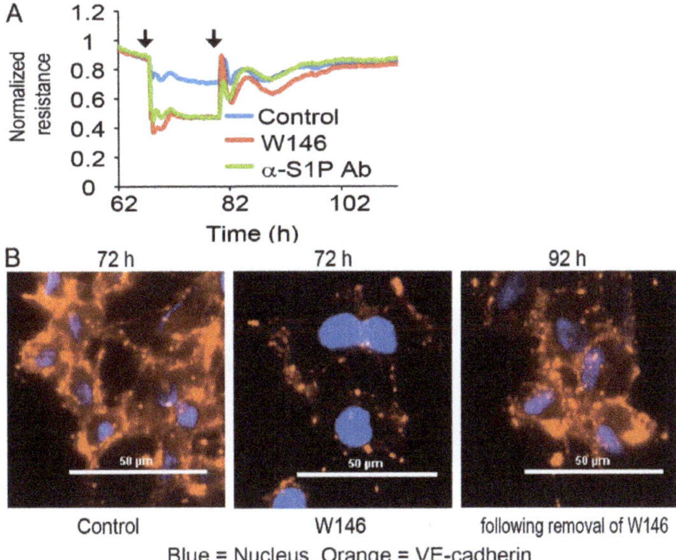

Figure 3. Dependence and reversibility of EC barrier stability in HUVEC and EA.hy926. (**A**) Resistance following treatment with 3 µM S1PR1 antagonist W146 or 120 µg/mL of anti-S1P antibody Sphingomab, followed by the removal of the added substances. Line plot represents one experiment out of three with black arrows indicating the addition and removal of W146 or Sphingomab at the corresponding time points. (**B**) Immunofluorescence staining of VE-cadherin in HUVEC after addition of 3 µM S1PR1 antagonist W146, followed by removal of the added substance. Representative images from one out of three individual experiments are shown. Pictures were taken 6 h after addition of W146 and 12 h following removal of W146.

3.5. S1P-Mediated EC Barrier Maintenance Under Inflammatory Conditions

To test the potential influence of inflammation on EC barrier formation, HUVEC and EA.hy926 monolayers were treated with a mix of the pro-inflammatory cytokines tumor necrosis factor-alpha (TNF-α) and interleukin-1beta (IL-1β) together with lipopolysaccharide (LPS). While HUVEC responded with a severe decrease of resistance in ECIS measurements upon treatment, EA.hy926 only showed a weak response (Figure 5A). S1P measurements demonstrated increased S1P levels in the supernatant of HUVEC, but not EA.hy926, which was markedly reduced after treatment with cytokines and LPS (Figure 5B). Conditioned HUVEC medium consequently reduced the calcium signal by 60% in S1PR1 overexpressing rat hepatoma HTC_4 cells after treatment with cytokines and LPS compared to non-treated HUVEC medium (Figure 5C). qPCR analyses revealed decreased expression of Spns2 in HUVEC after cytokine and LPS treatment, while expression of the S1P-degrading enzyme S1P-lyase (SGPL1), the S1P-dephosphorylating enzyme lipid phosphate phosphatase 3 (LPP3), and SphK1 was increased (Figure 5D). SphK2 and S1PR1 expression did not change (data not shown). Compared to EA.hy926, HUVEC showed pronounced VE-cadherin staining in the intercellular regions of monolayers, which was severely disrupted after treatment with cytokines and LPS (Figure 6A). In contrast to EA.hy926, VE-cadherin protein expression was reduced after treatment with cytokines and LPS (Figure 6B). Importantly, S1P was able to substantially rescue barrier disruption of HUVEC monolayer in ECIS measurements by transiently increasing the resistance (Figure 6C).

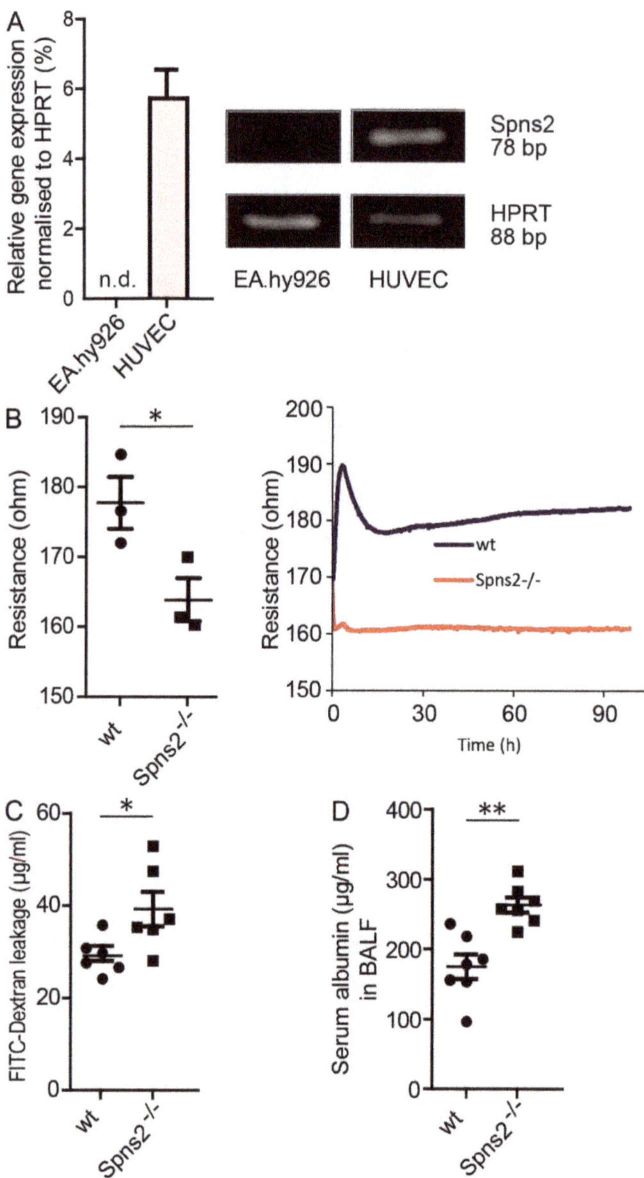

Figure 4. Role of Spns2 for EC barrier stability. (**A**) qPCR analysis of *Spns2* in EA.hy926 and HUVEC, *n* = 3, means ± SEM. Images show representative agarose gel electrophoresis signals of amplified PCR products for *HPRT* and *Spns2*. (**B**) Difference in initial non-stimulated resistance of primary lung ECs isolated from wt and Spns2 deficient mice in ECIS measurements, means ± SEM, *n* = 3, * p <0.05, determined by two-sided Student's *t*-test. Single values represent the resistance values for separate mice taken 90 h after seeding. Line plot represents one experiment out of three. (**C**) FITC-dextran leakage assay with primary lung endothelial cells isolated from wt and Spns2 deficient mice, means ± SEM, *n* = 6, * p < 0.05, determined by two-sided Student's t-test. (**D**) Serum albumin measurement in BALF isolated from wt and Spns2 deficient mice, means ± SEM, *n* = 6, ** p < 0.01, determined by a two-sided Student's *t*-test.

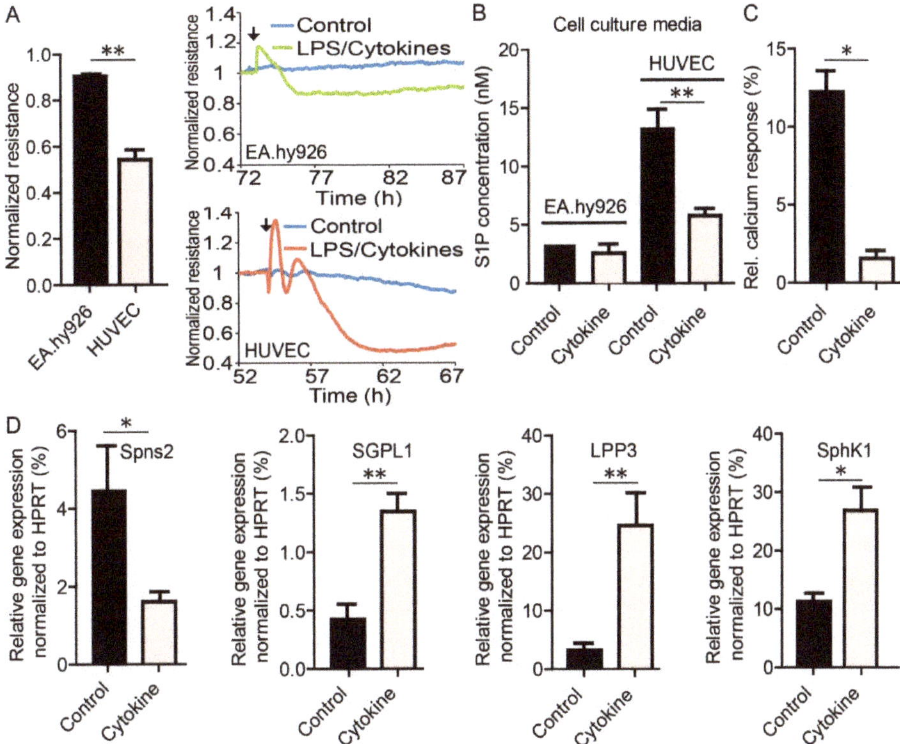

Figure 5. Influence of LPS and cytokines on EC barrier stabilization. (**A**) Resistance following treatment with a mix of LPS and cytokines (50 ng/mL IL1β, 50 ng/mL TNFα, 1 μg/mL LPS). Normalized resistance values are resistance of cells 12 h after LPS/cytokine treatment divided by resistance of carrier-treated control cells at the same time and are means ± SEM, $n = 3$, ** $p < 0.01$, determined by a two-sided Student's *t*-test. Line plots represent one experiment out of three with black arrows indicating the addition of LPS and cytokines or vehicle at the corresponding time. (**B**) LC-MS/MS quantification of extracellular S1P production by EA.hy926 and HUVEC after stimulation with LPS and cytokines or vehicle control. (**C**) Intracellular calcium response in rat hepatoma HTC4 cells transfected with human S1PR1 and the human Gαi subunit of trimeric G proteins, stimulated with lipid extracts from HUVEC without or with LPS and cytokines as indicated. (**D**) qPCR analysis of *Spns2*, *SGPL1*, *LPP3*, and *SphK1* expression in cytokine mix treated HUVEC. (**B–D**) Data are means ± SEM, $n = 3$, * $p < 0.05$, ** $p < 0.01$, determined by two-sided Student's *t*-test.

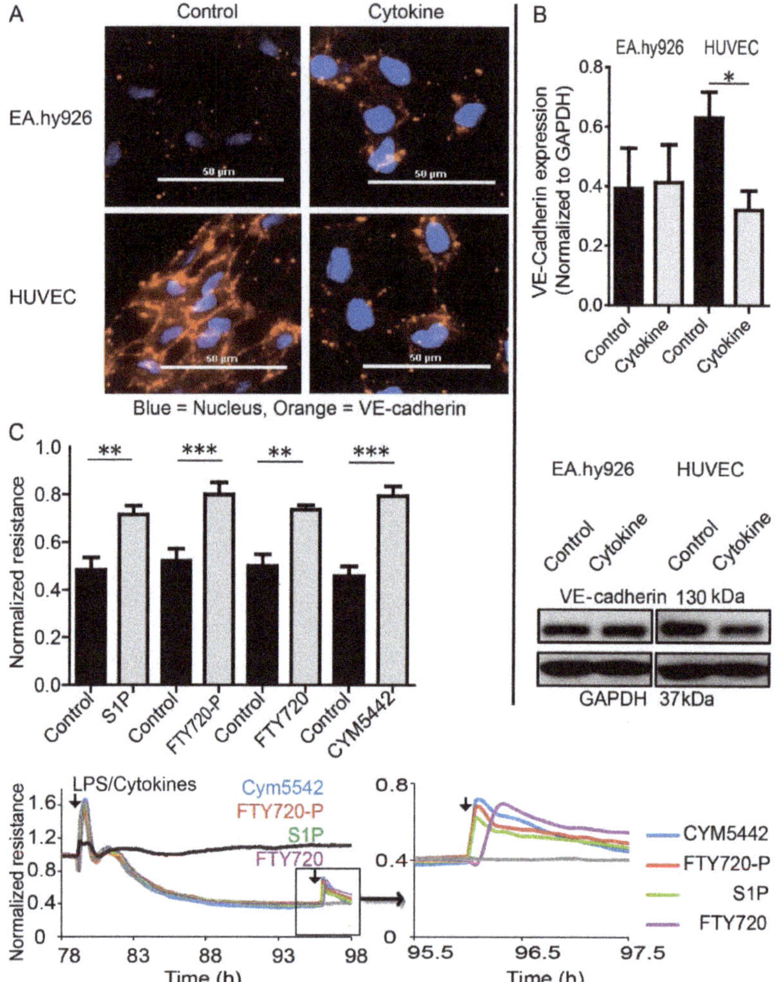

Figure 6. Role of S1P and S1PRs in EC barrier destabilization induced by LPS and cytokines. (**A**) Immunofluorescence staining of VE-cadherin and Western blot analysis of VE-cadherin and GAPDH in EA.hy926 and HUVEC after stimulation with LPS and cytokines or vehicle. Representative images from one out of three individual experiments are shown, size bar = 50 µm. (**B**) Western blot quantification (top) and representative Western blot (bottom) of VE-cadherin expression in EA.hy926 and HUVEC after stimulation with LPS and cytokines or vehicle, means ± SEM, $n = 3$, * $p < 0.05$, determined by two-sided Student's t-test. (**C**) Resistance following treatment with a mix of LPS and cytokines (50 ng/mL IL1β, 50 ng/mL TNFα, 1 µg/mL LPS), and re-stimulated with 1 µM of the S1PR1 agonist CYM5442, 1 µM of the S1PR1,3,4,5 agonist FTY720-phosphate and its non-phosphorylated precursor FTY720, and 1 µM S1P. Line plots represent one experiment out of three with black arrows indicating the addition of stimuli at the corresponding time points. The dark grey line represents an unstimulated control, the light grey line represents a control stimulated with LPS and cytokines without second stimulation. Bar graph represents means ± SEM, $n = 3$, ** $p < 0.01$, *** $p < 0.001$, determined by two-sided Student's *t*-test. Normalized resistance values were taken before (controls) and after treatment with S1P, FTY720-P, FTY720, and CYM5442 at the time of the established maximal change of resistance of cells divided by resistance of cells before LPS/cytokine treatment.

4. Discussion

S1P is present in the circulation at high nM up to μM concentrations. Recent studies suggested that S1P in blood is particularly important for EC barrier maintenance [8]. Data presented in this study, however, demonstrate that autonomously produced S1P by EC is important for basal constitutive maintenance of barrier function. This is also consistent with previous studies that focused on the role of S1P in the circulation, but used inducible knockout strategies that would also delete the only S1P-producing enzymes SphK1 and SphK2 in EC [20]. Deletion of both SphK1 and SphK2 in EC prevented the autonomous production and release of S1P in EC. Furthermore, the important role of autonomously produced S1P in EC was previously demonstrated by the treatment of HUVEC monolayer with the S1PR1 antagonist W146 prior to addition of platelets in transendothelial electrical resistance (TEER) measurements, which also induced a significant reduction of resistance [20]. While the role of S1PR1 in EC barrier formation and maintenance was investigated many times, the source of S1P required for S1PR1 mediated EC barrier stabilization is still under debate. Our data support the autonomous contribution of EC to produce their own barrier-stabilizing S1P.

S1P is produced intracellularly and needs to be transported out of the cell to act as an extracellular ligand for S1PRs. The main S1P transporter expressed in EC is Spns2 [21,22]. Our data indicate that Spns2-deficient EC suffer from a compromised barrier function due to the defective export of S1P and low extracellular S1P concentrations, which resembled the decreased circulating levels of S1P in mice lacking Spns2. Isolated primary lung EC established very quickly a stable resistance base line as a measure for barrier integrity, lacking the common slow increase in resistance in the first couple of hours of this assay that we observed with EA.hy926 and HUVEC. One reason for this unusual behavior could be the lack of cell growth and division, and seeding of these cells in quantities sufficient to rapidly establish a confluent cell layer to compensate for their observed growth arrest. EC barrier destabilization in Spns2 deficient mice was not observed in a previous study using Evans Blue dye to investigate EC barrier leakage [16]. A possible reason could be different experimental approaches. In this study, we found that endogenous serum albumin was significantly increased in BALF of Spns2-deleted mice compared to wt mice, which reflects the steady-state of barrier stability formed over a long period of time. In contrast, Evans Blue leakage is a single event monitored over a very brief period of 90 min, which may not be sufficient to detect differences in basal EC barrier disturbances. Our ex vivo data was consistent with our in vivo data obtained in three different experimental setups with primary lung EC that all confirmed a disturbed EC barrier function after depletion of Spns2.

Deficiency of S1P in circulation contributes to detrimental effects during inflammation [8,20]. Although the EC barrier stabilizing function of S1P is well accepted, a contributing role of S1P production, transportation, and signaling in inflammation-induced EC barrier breakdown is a novel observation of this study. Particularly, the demonstration that cytokine and LPS-induced down-regulation of Spns2 in EC contributed significantly to lower basal extracellular S1P levels and consequent EC barrier disruption, has not been reported before. Thus far, many studies, including ours, observed an up-regulation of SphK1 during infection, and it therefore could be considered to be an inflammatory kinase [23,24]. The down-regulation of Spns2 in EC might be an effective response to support the infiltration of leukocytes at sites of local infection. In the event of a systemic infection, however, reduced expression of Spns2 in endothelial cells likely contributes to a global collapse of the vascular EC barrier leading to septic shock. Treatment of ECs with inflammatory stimuli did not compromise the expression of S1PR1 on EC, which opens the possibility of using S1PR1 agonists for EC barrier stabilization. Previous studies investigated the influence of S1PR1 agonists on EC barrier protection in various different disease models with mixed results. While some studies showed beneficial effects [7,25], others reported they were ineffective [26]. Thus far, even successful approaches with S1PR1 agonists were not very effective. These results are consistent with our data, which showed incomplete rescue of the EC barrier under inflammatory conditions, probably due to concomitant cytokine and LPS-induced down-regulation of VE-cadherin, which is an essential player in EC barrier formation as well [3].

The contribution of autonomously produced and secreted S1P by EC also implicates S1PR1 and EC barrier function. Since levels of S1P released by EC with 4–15 nM are much lower than the typical concentrations measured in plasma with 200–1000 nM, it is unlikely that activation occurs at the plasma-facing side of EC. EC barrier stabilizing stimulation supposedly occurs at the tissue-facing side of EC, and strategies to increase S1P in tissues such as inhibition of the S1P-metabolizing enzyme SGPL1 may be more promising for future medical interventions than gross application of S1PR1 agonists [27]. SGPL1 inhibitors have already been tested in rheumatoid arthritis [28]. Additional studies are required to evaluate the full potential of targeting S1P signaling and metabolism for EC barrier stabilization.

5. Conclusions

We confirmed that S1P is a major EC barrier-stabilizing factor, predominantly via S1PR1 stimulation. The constitutive production of S1P by EC and its release in the local environment by Spns2 rather than systemic S1P-levels in plasma were important for basal EC barrier stabilization. Inflammatory stimuli resulted in EC barrier disruption due to the down-regulation of the S1P-transporter Spns2, while the expression of S1PR1 was not altered. Based on our data, inflammation-induced EC barrier disruption may be prevented by local application of S1P or S1PR1 agonists, e.g., via permanent inhalation in the lung or by inhibition of cellular S1P-degradation in tissues with S1P-lyase inhibitors to compensate for the reduced release of S1P from EC.

Supplementary Materials: The following are available online at http://www.mdpi.com/2073-4409/9/4/928/s1, Figure S1: Dependence and reversibility of VE-cadherin assembly in HUVEC. Immunofluorescence staining of VE-cadherin in HUVEC after addition of 3 μM S1PR1 antagonist W146, followed by removal of the added substance. Representative images from three individual experiments are shown. Pictures were taken 6 h after addition of W146 and 12 h following removal of W146.

Author Contributions: Conceptualization, M.H.G.; Methodology, J.J.P., C.W., T.M., and R.H.; Formal analysis, J.J.P.; Investigation, J.J.P., C.W., and T.M.; Resources, R.H., S.S., and M.H.G.; Writing—Original Draft, M.H.G.; Writing—Review and Editing, J.J.P., C.W., T.M., S.S., R.H., and M.H.G.; Visualization, J.J.P. and M.H.G.; Supervision, M.H.G.; Project Administration, M.H.G.; Funding Acquisition, S.S., R.H., and M.H.G. All authors have read and agreed to the published version of the manuscript.

Funding: This research was funded by the Center for Sepsis Control and Care (CSCC), the Jena School for Microbial Communication (JSMC), and the National Institutes of Health, Grant R01GM043880 (to S.S.). We also acknowledge support by the German Research Foundation and the Open Access Publication Fund of the Thueringer Universitaets- und Landesbibliothek Jena Projekt-Nr. 433052568.

Acknowledgments: We thank Elke Teuscher for the preparation of HUVEC, Roger Sabbadini (LPath Inc. and San Diego State University, San Diego, CA, USA) for kindly providing the anti-S1P antibody Sphingomab and the corresponding isotype control antibody LT1013, and Mareike Lipinski for technical assistance.

Conflicts of Interest: The authors declare no conflict of interest.

References

1. Stevens, T.; Garcia, J.G.; Shasby, D.M.; Bhattacharya, J.; Malik, A.B. Mechanisms regulating endothelial cell barrier function. *Am. J. Physiol. Lung Cell. Mol. Physiol.* **2000**, *279*, L419–L422. [CrossRef] [PubMed]

2. Vasileva, E.; Citi, S. The role of microtubules in the regulation of epithelial junctions. *Tissue Barriers* **2018**, *6*, 1539596. [CrossRef] [PubMed]

3. Campbell, H.K.; Maiers, J.L.; DeMali, K.A. Interplay between tight junctions & adherens junctions. *Exp. Cell Res.* **2017**, *358*, 39–44. [PubMed]

4. Hartsock, A.; Nelson, W.J. Adherens and tight junctions: Structure, function and connections to the actin cytoskeleton. *Biochim. Biophys. Acta* **2008**, *1778*, 660–669. [CrossRef]

5. McVerry, B.J.; Garcia, J.G. Endothelial cell barrier regulation by sphingosine 1-phosphate. *J. Cell. Biochem.* **2004**, *92*, 1075–1085. [CrossRef]

6. Allende, M.L.; Yamashita, T.; Proia, R.L. G-protein-coupled receptor s1p1 acts within endothelial cells to regulate vascular maturation. *Blood* **2003**, *102*, 3665–3667. [CrossRef]

7. Burg, N.; Swendeman, S.; Worgall, S.; Hla, T.; Salmon, J.E. Sphingosine 1-phosphate receptor 1 signaling maintains endothelial cell barrier function and protects against immune complex-induced vascular injury. *Arthritis Rheumatol.* **2018**, *70*, 1879–1889. [CrossRef]

8. Camerer, E.; Regard, J.B.; Cornelissen, I.; Srinivasan, Y.; Duong, D.N.; Palmer, D.; Pham, T.H.; Wong, J.S.; Pappu, R.; Coughlin, S.R. Sphingosine-1-phosphate in the plasma compartment regulates basal and inflammation-induced vascular leak in mice. *J. Clin. Invest.* **2009**, *119*, 1871–1879. [CrossRef]

9. Schnoor, M.; Garcia Ponce, A.; Vadillo, E.; Pelayo, R.; Rossaint, J.; Zarbock, A. Actin dynamics in the regulation of endothelial barrier functions and neutrophil recruitment during endotoxemia and sepsis. *Cell. Mol. Life Sci.* **2017**, *74*, 1985–1997. [CrossRef]

10. Coldewey, S.M.; Benetti, E.; Collino, M.; Pfeilschifter, J.; Sponholz, C.; Bauer, M.; Huwiler, A.; Thiemermann, C. Elevation of serum sphingosine-1-phosphate attenuates impaired cardiac function in experimental sepsis. *Sci. Rep.* **2016**, *6*, 27594. [CrossRef]

11. Frej, C.; Linder, A.; Happonen, K.E.; Taylor, F.B.; Lupu, F.; Dahlback, B. Sphingosine 1-phosphate and its carrier apolipoprotein m in human sepsis and in Escherichia coli sepsis in baboons. *J. Cell. Mol. Med.* **2016**, *20*, 1170–1181. [CrossRef]

12. Gomes, L.; Fernando, S.; Fernando, R.H.; Wickramasinghe, N.; Shyamali, N.L.; Ogg, G.S.; Malavige, G.N. Sphingosine 1-phosphate in acute dengue infection. *PLoS ONE* **2014**, *9*, e113394. [CrossRef] [PubMed]

13. Winkler, M.S.; Nierhaus, A.; Holzmann, M.; Mudersbach, E.; Bauer, A.; Robbe, L.; Zahrte, C.; Geffken, M.; Peine, S.; Schwedhelm, E.; et al. Decreased serum concentrations of sphingosine-1-phosphate in sepsis. *Crit. Care* **2015**, *19*, 372. [CrossRef] [PubMed]

14. Pappu, R.; Schwab, S.R.; Cornelissen, I.; Pereira, J.P.; Regard, J.B.; Xu, Y.; Camerer, E.; Zheng, Y.W.; Huang, Y.; Cyster, J.G.; et al. Promotion of lymphocyte egress into blood and lymph by distinct sources of sphingosine-1-phosphate. *Science* **2007**, *316*, 295–298. [CrossRef] [PubMed]

15. Graler, M.H.; Goetzl, E.J. The immunosuppressant fty720 down-regulates sphingosine 1-phosphate g-protein-coupled receptors. *FASEB J.* **2004**, *18*, 551–553. [CrossRef] [PubMed]

16. Mendoza, A.; Breart, B.; Ramos-Perez, W.D.; Pitt, L.A.; Gobert, M.; Sunkara, M.; Lafaille, J.J.; Morris, A.J.; Schwab, S.R. The transporter Spns2 is required for secretion of lymph but not plasma sphingosine-1-phosphate. *Cell Rep.* **2012**, *2*, 1104–1110. [CrossRef]

17. Livak, K.J.; Schmittgen, T.D. Analysis of relative gene expression data using real-time quantitative PCR and the 2(-delta delta c(t)) method. *Methods* **2001**, *25*, 402–408. [CrossRef]

18. Sensken, S.C.; Staubert, C.; Keul, P.; Levkau, B.; Schoneberg, T.; Graler, M.H. Selective activation of g alpha i mediated signalling of s1p3 by fty720-phosphate. *Cell. Signal.* **2008**, *20*, 1125–1133. [CrossRef]

19. Bode, C.; Graler, M.H. Quantification of sphingosine-1-phosphate and related sphingolipids by liquid chromatography coupled to tandem mass spectrometry. *Methods Mol. Biol.* **2012**, *874*, 33–44.

20. Gazit, S.L.; Mariko, B.; Therond, P.; Decouture, B.; Xiong, Y.; Couty, L.; Bonnin, P.; Baudrie, V.; Le Gall, S.M.; Dizier, B.; et al. Platelet and erythrocyte sources of s1p are redundant for vascular development and homeostasis, but both rendered essential after plasma s1p depletion in anaphylactic shock. *Circ. Res.* **2016**, *119*, e110–e126. [CrossRef]

21. Fukuhara, S.; Simmons, S.; Kawamura, S.; Inoue, A.; Orba, Y.; Tokudome, T.; Sunden, Y.; Arai, Y.; Moriwaki, K.; Ishida, J.; et al. The sphingosine-1-phosphate transporter spns2 expressed on endothelial cells regulates lymphocyte trafficking in mice. *J. Clin. Invest.* **2012**, *122*, 1416–1426. [CrossRef] [PubMed]

22. Kawahara, A.; Nishi, T.; Hisano, Y.; Fukui, H.; Yamaguchi, A.; Mochizuki, N. The sphingolipid transporter Spns2 functions in migration of zebrafish myocardial precursors. *Science* **2009**, *323*, 524–527. [CrossRef] [PubMed]

23. Nagahashi, M.; Yamada, A.; Katsuta, E.; Aoyagi, T.; Huang, W.C.; Terracina, K.P.; Hait, N.C.; Allegood, J.C.; Tsuchida, J.; Yuza, K.; et al. Targeting the sphk1/s1p/s1pr1 axis that links obesity, chronic inflammation, and breast cancer metastasis. *Cancer Res.* **2018**, *78*, 1713–1725. [CrossRef] [PubMed]

24. Vettorazzi, S.; Bode, C.; Dejager, L.; Frappart, L.; Shelest, E.; Klassen, C.; Tasdogan, A.; Reichardt, H.M.; Libert, C.; Schneider, M.; et al. Glucocorticoids limit acute lung inflammation in concert with inflammatory stimuli by induction of sphk1. *Nat. Commun.* **2015**, *6*, 7796. [CrossRef]

25. Dong, J.; Wang, H.; Zhao, J.; Sun, J.; Zhang, T.; Zuo, L.; Zhu, W.; Gong, J.; Li, Y.; Gu, L.; et al. Sew2871 protects from experimental colitis through reduced epithelial cell apoptosis and improved barrier function in interleukin-10 gene-deficient mice. *Immunol. Res.* **2015**, *61*, 303–311. [CrossRef]

26. Flemming, S.; Burkard, N.; Meir, M.; Schick, M.A.; Germer, C.T.; Schlegel, N. Sphingosine-1-phosphate receptor-1 agonist sew2871 causes severe cardiac side effects and does not improve microvascular barrier breakdown in sepsis. *Shock* **2018**, *49*, 71–81. [CrossRef]

27. Hemdan, N.Y.; Weigel, C.; Reimann, C.M.; Graler, M.H. Modulating sphingosine 1-phosphate signaling with dop or fty720 alleviates vascular and immune defects in mouse sepsis. *Eur. J. Immunol.* **2016**, *46*, 2767–2777. [CrossRef]

28. Fleischmann, R. Novel small-molecular therapeutics for rheumatoid arthritis. *Curr. Opin. Rheumatol.* **2012**, *24*, 335–341. [CrossRef]

Article

Opposing Roles of S1P$_3$ Receptors in Myocardial Function

Dina Wafa [1,*], Nóra Koch [1], Janka Kovács [1], Margit Kerék [1], Richard L. Proia [2], Gábor J. Tigyi [1,3], Zoltán Benyó [1] and Zsuzsanna Miklós [1,*]

[1]   Institute of Translational Medicine, Semmelweis University, 1094 Budapest, Hungary; kochnori@gmail.com (N.K.); kovacsjankee@gmail.com (J.K.); margit.nagy9@gmail.com (M.K.); gtigyi@uthsc.edu (G.J.T.); benyo.zoltan@med.semmelweis-univ.hu (Z.B.)

[2]   National Institute of Diabetes and Digestive and Kidney Diseases (NIDDK), National Institues of Health, Bethesda, MD 20892, USA; richard.proia@nih.gov

[3]   Department of Physiology, University of Tennessee Health Science Center, Memphis, TN 38163, USA

*   Correspondence: dina.wafa.93@gmail.com (D.W.); miklos.zsuzsanna@med.semmelweis-univ.hu (Z.M.)

Received: 15 May 2020; Accepted: 22 July 2020; Published: 24 July 2020

Abstract: Sphingosine-1-phosphate (S1P) is a lysophospholipid mediator with diverse biological function mediated by S1P$_{1-5}$ receptors. Whereas S1P was shown to protect the heart against ischemia/reperfusion (I/R) injury, other studies highlighted its vasoconstrictor effects. We aimed to separate the beneficial and potentially deleterious cardiac effects of S1P during I/R and identify the signaling pathways involved. Wild type (WT), S1P$_2$-KO and S1P$_3$-KO Langendorff-perfused murine hearts were exposed to intravascular S1P, I/R, or both. S1P induced a 45% decrease of coronary flow (CF) in WT-hearts. The presence of S1P-chaperon albumin did not modify this effect. CF reduction diminished in S1P$_3$-KO but not in S1P$_2$-KO hearts, indicating that in our model S1P$_3$ mediates coronary vasoconstriction. In I/R experiments, S1P$_3$ deficiency had no influence on postischemic CF but diminished functional recovery and increased infarct size, indicating a cardioprotective effect of S1P$_3$. Preischemic S1P exposure resulted in a substantial reduction of postischemic CF and cardiac performance and increased the infarcted area. Although S1P$_3$ deficiency increased postischemic CF, this failed to improve cardiac performance. These results indicate a dual role of S1P$_3$ involving a direct protective action on the myocardium and a cardiosuppressive effect due to coronary vasoconstriction. In acute coronary syndrome when S1P may be released abundantly, intravascular and myocardial S1P production might have competing influences on myocardial function via activation of S1P$_3$ receptors.

Keywords: sphingosine-1-phosphate; ischemia/reperfusion; cardioprotection; vasoconstriction; coronary flow; myocardial function; myocardial infarct; albumin

1. Introduction

Ischemic heart disease, including acute coronary syndrome (ACS), is a major cause of death worldwide [1]. ACS is the sudden loss of adequate blood perfusion to the heart, most commonly initiated by the rupture of an atherosclerotic plaque and consequent activation of blood coagulation. This process results in thrombotic occlusion of the coronary artery causing cardiac tissue damage [2]. Urgent reestablishment of blood perfusion to the affected area is crucial to minimizing ischemic tissue injury. Besides the therapeutic time window, the success of reperfusion depends on several other factors such as vascular response to pathophysiological events happening prior to and during thrombus formation. Platelet activation might be relevant in this context as it releases numerous vasoactive mediators which might have an impact on the dynamics and severity of ischemic injury. Sphingosine-1-phosphate (S1P) is one of these many mediators [3–8].

S1P is a sphingolipid mediator which is produced by a wide variety of cells [9]. In vivo, albumin and APO-M in HDL are the most recognized carriers of S1P in blood plasma, which have also been reported to modulate the actions of S1P [10,11]. S1P actions include regulation of diverse physiological and pathophysiological processes such as inflammation, autoimmunity, and neurodegeneration [12,13]. In the cardiovascular (CV) system, activated platelets synthesize and release S1P in large amounts [3–8], and S1P has been reported to play a role in regulating vascular tone [14,15], atherogenesis, cardiac remodeling, and cardioprotection [16–18]. To date, five different G protein-coupled receptors belonging to the endothelial differentiation gene (EDG) family have been identified as specific S1P receptors ($S1P_{1-5}$) [19,20]. From these, $S1P_1$, $S1P_2$ and $S1P_3$ receptors are expressed abundantly in the CV system and have been reported to mediate CV actions of S1P [16].

S1P has been attributed with cardioprotective effects against ischemia-reperfusion (I/R) injury by several research groups [18,21–27]. The key enzymes in S1P synthesis, sphingosine-kinase 1 and 2 (SphK1 and 2), have been implicated in the ischemia-induced increased release of S1P from cardiomyocytes as well as in mediating the beneficial effects of ischemic pre- and post-conditioning [18,23,24]. Combined deletion of $S1P_2$ and $S1P_3$ receptors increased the infarcted area and enhanced apoptotic cell death after I/R, suggesting that activation of these receptors is cardioprotective [25].

Preischemic S1P treatment has also been reported to decrease infarct size in ex vivo experimental models [24,26]. It has already been shown by Theilmeier and colleagues that HDL and S1P directly protect the heart against I/R injury via the $S1P_3$ receptor in vivo [27]. However, in ACS when S1P is released in substantial amounts from platelets and endothelial cells in blood plasma, it might bind to carriers other than HDL. Several studies have highlighted the vasoconstrictive effects of S1P in various vessel beds from multiple species including the coronaries: S1P had a constrictive effect on isolated porcine pulmonary artery rings [28], in canine, rat, murine, and leporine basilar and middle cerebral arteries [29,30], in rat portal veins [31] and in canine coronaries [32]. S1P administration to the coronary perfusate has been shown to diminish coronary flow (CF) in Langendorff-perfused rat hearts [33]. This effect was attenuated by pharmacological inhibition of $S1P_3$ receptors in the same experiment [33]. Another study conducted on coronary smooth muscle cells raised the potential involvement of $S1P_2$ receptors in the vasoconstrictor response elicited by S1P [34]. Beside its actions on CF, S1P exerts other short-term cardiac effects including generation of arrhythmias and negative inotropy [35–38].

The cardiac effects of S1P reported in I/R injury are controversial. Activation of S1P receptors seems to be cardioprotective, whereas the acute effects of S1P to reduce CF and cardiac contractility are expected to interfere with successful post-ischemic recovery. Moreover, $S1P_2$ and $S1P_3$ receptors have been shown to be involved in both mechanisms. In ACS, when S1P is released in large amounts from activated platelets, its favorable and potentially deleterious effects might clash with one another. In the present study, we aim to delineate how these opposing S1P actions actually affect postischemic cardiac injury after a non-fatal ischemic insult.

For this purpose, we conducted ex vivo experiments in isolated murine hearts mounted on the Langendorff-system. First, we mimicked ACS-related massive S1P release into the blood in order to characterize its coronary effects and consequences on heart function. In these experiments we used albumin as an S1P chaperone and also in subsequent experiments Krebs buffer without added chaperone protein as the vehicle. Second, using S1P receptor gene knockout (KO) mouse models, we aimed to identify receptors involved in these cardiac effects. Third, to understand the role of $S1P_3$ receptor in cardioprotection, we applied a non-fatal I/R protocol in $S1P_3$ deficient hearts. Finally, the complete sequence of ACS was modeled with an initial exposure of the coronaries to S1P as it occurs during plaque rupture and platelet activation, followed by 20 min of complete ischemia, during which the myocardial S1P-producing machinery can be activated, and concluding with 120 min reperfusion representing successful reopening of the coronary artery in a clinical setting. With this approach, we were able to separate the consequences of intravascular and myocardial S1P-related effects during ACS and also to evaluate their combined effects.

2. Materials and Methods

2.1. Animals

All experiments reported here were performed in hearts of 130–150-day-old male mice. Animals were bred and housed in the animal facility at Semmelweis University, kept in a 12/12-h light/dark cycle and with free access to water and food. C57BL/6 (WT) mice were bred from breeding pairs obtained from Charles River Laboratories (Isaszeg, Hungary). To answer our specific questions, $S1P_2$-KO and $S1P_3$-KO animals along with wild-type littermates on C57BL/6 genetic background were tested [39]. All procedures were carried out according to the guidelines of the Hungarian Law of Animal Protection (28/1998) and were approved by the Government Office of Pest County (Permission number: PEI/001/820-2/2015).

2.2. Isolated Perfused Heart Experiments

General anesthesia was induced by intraperitoneal injection of 40 mg/kg pentobarbital, followed by thoracotomy and isolation of the heart. The isolated heart was mounted in a Langendorff apparatus (Experimetria Ltd., Budapest) and perfused at constant 80 mmHg pressure with modified Krebs-Henseleit buffer (118 mM NaCl, 4.3 mM KCl, 25 mM $NaHCO_3$, 1.2 mM $MgSO_4$, 1.2 mM KH_2PO_4, 0.5 mM NaEDTA, 2.0 mM $CaCl_2$, 11 mM glucose, 5 mM pyruvate (pH 7.4) - all purchased from Sigma-Aldrich, Budapest, Hungary) [40–43]. The solution was continuously gassed with 95% O_2 and 5% CO_2 at 37 °C. During the experiment, the heart was surrounded by a thermally-regulated chamber filled with Krebs-Henseleit buffer.

CF was continuously monitored with a transit-time flow meter placed into the inflow line (Transonic 2PXN flow probe, Transonic Systems Inc., Ithaca, NY, USA). In order to measure left ventricular pressure (LVP), a fluid-filled balloon catheter connected to a pressure transducer was inserted into the ventricle to maintain diastolic pressure at 8 mmHg.

Devices were connected to a computer and data were recorded and analyzed by the Haemosys software (Experimetria Ltd., Budapest, Hungary). Left ventricular developed pressure (LVDevP) was calculated as the difference between peak systolic and minimum diastolic (LVDiastP) pressures. The positive and negative maximum values of the first derivative of the LVP (+dLVP/dt$_{max}$, −dLVP/dt$_{max}$) were determined as indices of left ventricular contractile and lusitropic performance, respectively.

2.3. Experimental Protocol

After cannulation of the isolated heart, a 30-min stabilization period was allowed. Subsequently, baseline data were recorded and S1P (D-erythro-sphingosine-1-phosphate, Avanti Inc., 10^{-6} M) or vehicle was infused to the perfusion line for 5 min. One mg S1P was dissolved in 263 μL 0.3 N NaOH. This solution (10^{-2} M) was further diluted with Krebs solution to give the required concentration in the perfusate.

First, to characterize S1P effects on CF we performed dose-response experiments with S1P in the concentration range of 10^{-9}–10^{-5} M. S1P was administered in the presence of S1P chaperon human serum albumin (HSA) (Sigma Aldrich, Budapest, Hungary, Cat.No: A3782, lyophilized powder, fatty acid free, globulin free, >99%) or diluted directly in chaperone-free Krebs buffer. The molar S1P to HSA ratio was 1:2 at every concentration [44]. In these experiments we applied S1P in cumulative doses, the next S1P dose was added to the previous dose after the response had reached its maximum. The biological effects of the applied vehicles were tested in separate experiments, and they proved to be without an effect.

In order to answer our main questions, two different experimental protocols were tested (Supplementary Materials Figure S1). To understand the effects of S1P under stable baseline conditions, a 5-min S1P (or vehicle) infusion was applied that was followed by a 20-min washout period. Whereas in the I/R-injury protocol, the 5-min S1P (or vehicle) infusion was followed by a 20-min global ischemia that was brought about by complete cessation of perfusion. At the end of the ischemic period, perfusion was restarted and reperfusion was maintained for 2 h. In these protocols, S1P was delivered in albumin free Krebs solution at a concentration of 10^{-6} M. The S1P concentration applied was defined as a dose approximating its ED_{50} value ($1.17 * 10^{-6}$ M in Krebs solution).

2.4. Measurement of Infarct Size

In the I/R experiments, hearts were removed from the apparatus after the 2-h reperfusion period and placed into a $-20\,^\circ$ C freezer for at least 15 min. The left ventricle of the frozen heart was cut into ~1 mm thick slices (4 to 6 slices per heart). To visualize the infarcted area of the heart, the slices were then incubated in a phosphate buffer containing 1% triphenyltetrazolium (TTC) (Sigma-Aldrich) for 20 min at a temperature of 37 °C. The TTC powder was diluted in a two-part phosphate buffer system at a pH of 7.4 and the slices were fixed in 10% formalin for 15 min [45]. Photos of the TTC-stained slices were captured using a stereomicroscope equipped with a high-resolution digital camera (Rasband, W.S.) and analyzed using Image-J software (National Institutes of Health, Bethesda, MD, USA). The area at risk, defined as the total area consisting of the pale plus red parts and the infarcted pale area, was measured and relative infarct size was calculated as a percentage of the area at risk.

2.5. Statistical Analysis

Results are presented as mean ± standard error of the mean (SEM). In order to compare time series data between 2 experimental groups, we used two-way repeated measurement ANOVA and Dunnett's *post hoc* test for multiple comparisons. Comparison of data acquired from the 4 experimental groups (WT/S1P$_3$-KO vs. vehicle/S1P infusion) was performed by two-way ANOVA and Sidak's *post hoc* test. To compare the maximal effects of S1P infusion, unpaired *t*-test was applied. To determine the total perfusion loss during S1P infusion, area over the curve (AOC) was calculated. In dose-response experiments non-linear regression analysis was used to find the best fit and ED_{50} values. To compare dose-responses between 2 experimental groups, comparison of Fits analysis of the statistical software was applied. All statistical analyses were performed using GraphPad Prism 7.0 (San Diego, CA, USA) and $p < 0.05$ was considered as statistically significant.

3. Results

3.1. Dose-Dependent Effects of Intravascular S1P on CF Administered with or without S1P-Chaperon Albumin

To characterize the effect of intravascular S1P on CF, we carried out dose-response experiments with and without albumin as a chaperone. When administered without a carrier, S1P elicited a dose-dependent CF reduction in isolated hearts with an ED_{50} value of 1.17×10^{-6} M. Therefore, the S1P in further experiments was applied at 1 microM—a dose close to its ED_{50} value. The coronary effect of S1P was similar in the presence of albumin, however the ED_{50} value slightly shifted to a smaller concentration range (1.85×10^{-7} M) though it was not statistically significant ($p = 0.12$, $F = 2.46$) (Figure 1). The maximal reduction in CF was also indistinguishable between groups regardless whether S1P was applied carrier-free or with albumin as vehicle.

3.2. Effects of Intravascular S1P Exposition on CF and Heart Function

To investigate the effects of a robust S1P release on CF and cardiac function, 10^{-6} M S1P or its vehicle was administered to the perfusate of isolated WT murine hearts for 5 min. Administration of S1P reduced CF by 44 ± 3% (Figure 2A). This remarkable decrease started at the beginning of the S1P infusion and continued progressively during the 5 min. During the 20-min wash-out period, CF did not return to the baseline level and remained at a significantly lower value ($p < 0.0001$).

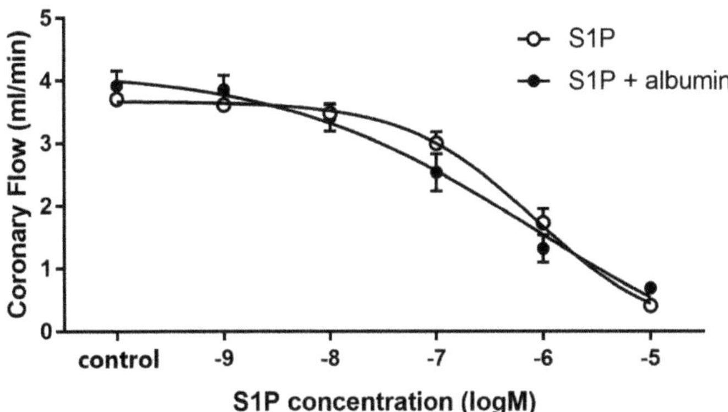

Figure 1. Dose-dependent effects of S1P on coronary flow of isolated murine hearts infused alone or in the presence of S1P-carrier albumin. In these experiments S1P was applied in a range of 10^{-9} to 10^{-5} M in cumulative doses without (S1P) or in the presence of its carrier, human serum albumin (S1P + albumin), and its effects on coronary flow were investigated. Albumin was present in a concentration twice that of S1P. ED_{50} values were 1.17×10^{-6} M (S1P) and 1.85×10^{-7} M (S1P + albumin). Mean ± SEM; $n = 9$; 8. Non-linear regression analysis and comparison of Fits using GraphPad Prism 7.0.

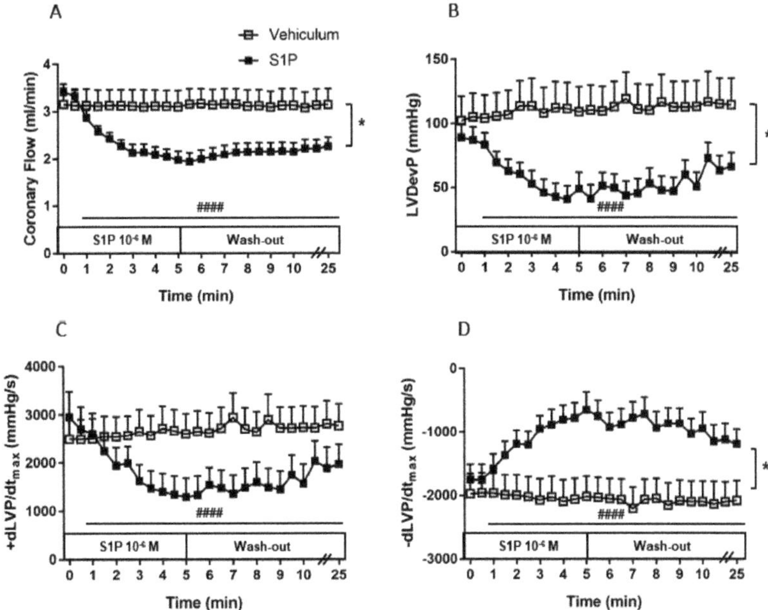

Figure 2. Effects of S1P on coronary flow (CF) (**A**), left ventricular developed pressure (LVDevP) (**B**), $+dLVP/dt_{max}$ (**C**) and $-dLVP/dt_{max}$ (**D**) of isolated mouse hearts. S1P (10^{-6} M) or its vehicle was administered to the perfusate of isolated wild-type (WT) murine hearts for 5 min. The infusion was followed by a 20-min wash-out period. Administration of S1P resulted in a remarkable decrease in CF, which prevailed throughout the infusion and the wash-out period ($p < 0.0001$). CF reduction compromised left ventricular contractile performance as evidenced by a concomitant decrease in LVDevP, $+dLVP/dt_{max}$ and $-dLVP/dt_{max}$ ($p < 0.0001$). Mean ± SEM; $n = 6$, 9; #### $p < 0.0001$ vs. baseline (pre-infusion value), * $p < 0.05$ vs. vehicle; two-way repeated measurement ANOVA and Dunnett's *post hoc* test.

CF reduction induced by S1P coexisted with compromised left ventricular contractile performance, which is indicated by a 54 ± 9% drop in LVDevP (Figure 2B) and by the markedly decreased +dLVP/dt$_{max}$, and −dLVP/dt$_{max}$ values ($p < 0.0001$) (Figure 2C–D). The vehicle did not affect either CF or other measured heart function parameters (Figure 2A–D).

Earlier studies suggested that S1P might affect coronaries via S1P$_2$ and S1P$_3$ receptors [33,34]. Therefore, we aimed to identify which of these receptors mediate(s) the effect of S1P on the CF. For this purpose, we perfused S1P into the isolated hearts of S1P$_2$ (Figure 3) and S1P$_3$ (Figure 4) KO mice following the experimental protocol described above.

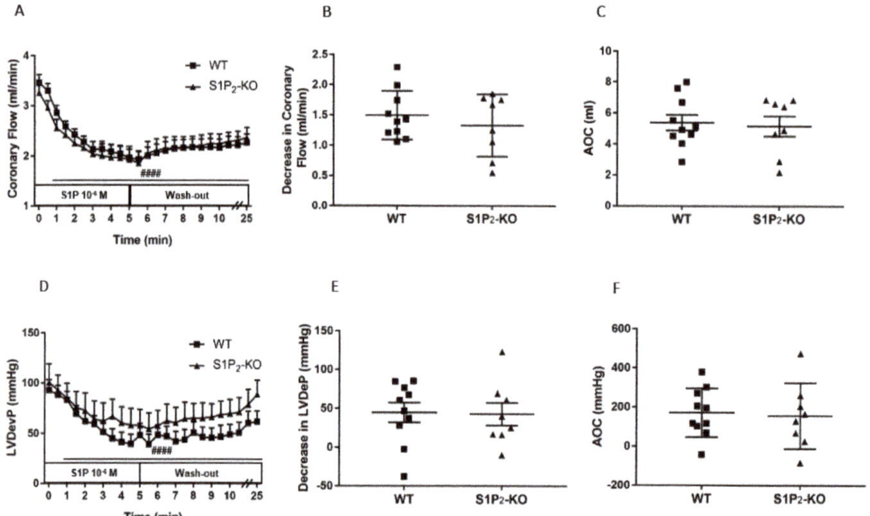

Figure 3. Effects of S1P on coronary flow (CF) (**A–C**) and left ventricular developed pressure (LVDevP) (**D–F**) of hearts isolated from wild-type (WT) and S1P$_2$ knock-out (KO) mice. S1P (10^{-6} M) was administered to the perfusate of isolated WT and S1P$_2$-KO murine hearts for 5 min. The infusion was followed by a 20-min wash-out period. CF and LVDevP were monitored during the entire experiment (panels A and D). Maximal decrease in CF and LVDevP compared to preinfusion baseline are shown in panels B and E. Values of area over the curve (AOC) during S1P infusion are shown in panels C and F. In S1P$_2$-KO hearts S1P-induced CF and LVDevP reduction was similar to that observed in WT hearts. Mean ± SEM; $n = 10, 8$; #### $p < 0.0001$ vs. baseline (preinfusion value) in both groups, two-way repeated measurement ANOVA followed by Dunnett's *post hoc* test.

The CF-reducing effect of S1P developing in S1P$_2$-deficient mice was similar to that of WT littermates (Figure 3A–C). The drop of the LVDevP was also similar in the two groups (Figure 3D–F), with no statistically significant difference.

In S1P$_3$-KO hearts, the CF-reducing effect of S1P was markedly diminished compared to WT mice (Figure 4A). There was a significant difference in the maximal effects: CF was dropped by 1.95 ± 0.33 mL/min in WT and only by 0.93 ± 0.10 mL/min in S1P$_3$-KO mice (Figure 4B). The AOC used as an index for total perfusion loss during the infusion period showed similar decrease. During the 5-min S1P infusion, the total perfusion loss was 8.56 ± 1.60 mL in WT vs. 3.70 ± 0.57 mL in S1P$_3$-KO mice (Figure 4C).

The decrease in left ventricular contractile performance upon S1P infusion was also attenuated in S1P$_3$-KO mice (Figure 4D): both the maximal drop in LVDevP (Figure 4E) and the area over the LVDevP curve used as a measure of loss of contractile activity (Figure 4F) were significantly reduced compared to WT controls.

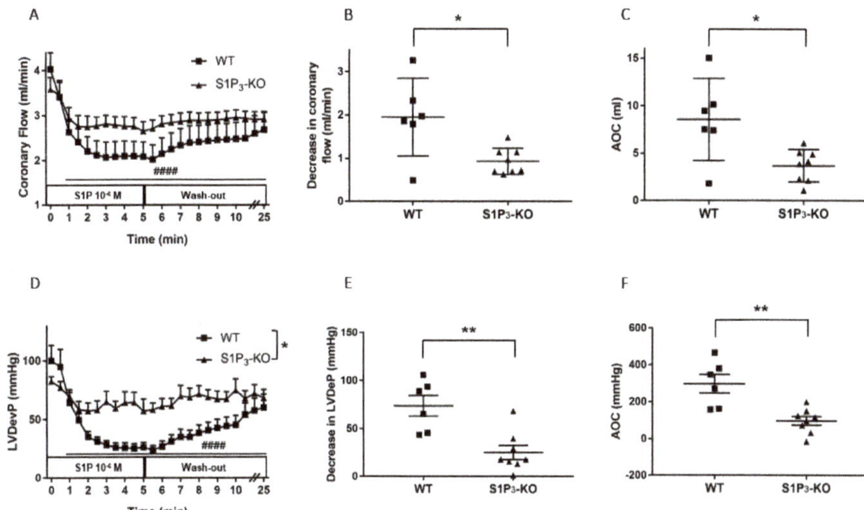

Figure 4. Effects of S1P on coronary flow (CF) (**A–C**) and left ventricular developed pressure (LVDevP) (**D–F**) of hearts isolated from wild-type (WT) and S1P$_3$ knock-out (KO) mice. S1P (10^{-6} M) was administered to the perfusate of isolated WT and S1P$_3$-KO murine hearts for 5 min. The infusion was followed by a 20-min wash-out period. CF and LVDevP are shown in panels A and D. Maximal decrease in CF and LVDevP compared to preinfusion baseline are shown in panels B and E. Values of area over the curve (AOC) during S1P infusion are shown in panels C and F. In S1P$_3$-KO hearts, the S1P-induced CF and LVDevP reduction was significantly reduced. Mean ± SEM; n = 6, 8; #### $p < 0.0001$ vs. baseline (preinfusion value); * $p < 0.05$, ** $p < 0.01$ vs. WT; two-way repeated measurement ANOVA and Dunnett's *post hoc* test (**A,D**) and unpaired *t*-test (**B–F**).

3.3. Role of Myocardial S1P$_3$ Receptor Activation in I/R Injury

To better understand the apparent contradiction between the widely reported cardioprotective and observed cardiosuppressive effects of S1P, we aimed to separate the myocardial and coronary actions of S1P in a model of I/R injury.

First, we investigated the effects of potential S1P$_3$ receptor activation during I/R in the absence of intravascularly administered S1P. WT and S1P$_3$-KO hearts were exposed to an I/R protocol, CF and myocardial function were monitored during reperfusion. CF did not differ significantly between the WT and S1P$_3$-KO mice (Figure 5A1). In contrast, parameters describing myocardial performance showed marked differences. The lack of S1P$_3$ resulted in a far worse postischemic functional recovery as evidenced by the drop of the LVDevP (Figure 5B1), decreased +dLVP/dt$_{max}$, and −dLVP/dt$_{max}$ (Figure 5C1,D1), and elevated LVDiastP (Figure 5E1).

These results indicate that S1P$_3$ receptors play a beneficial role in preventing ischemia-induced myocardial dysfunction, most probably by activation from S1P generated locally by the tissues of the ischemic heart. However, this myocardial S1P release did not induce S1P$_3$-mediated coronary vasoconstriction as observed in the previous experiments with intravascular S1P administration.

Interestingly, lack of S1P$_2$ receptors did not influence any of these functional parameters nor the infarct size in this I/R model (Figure S2.)

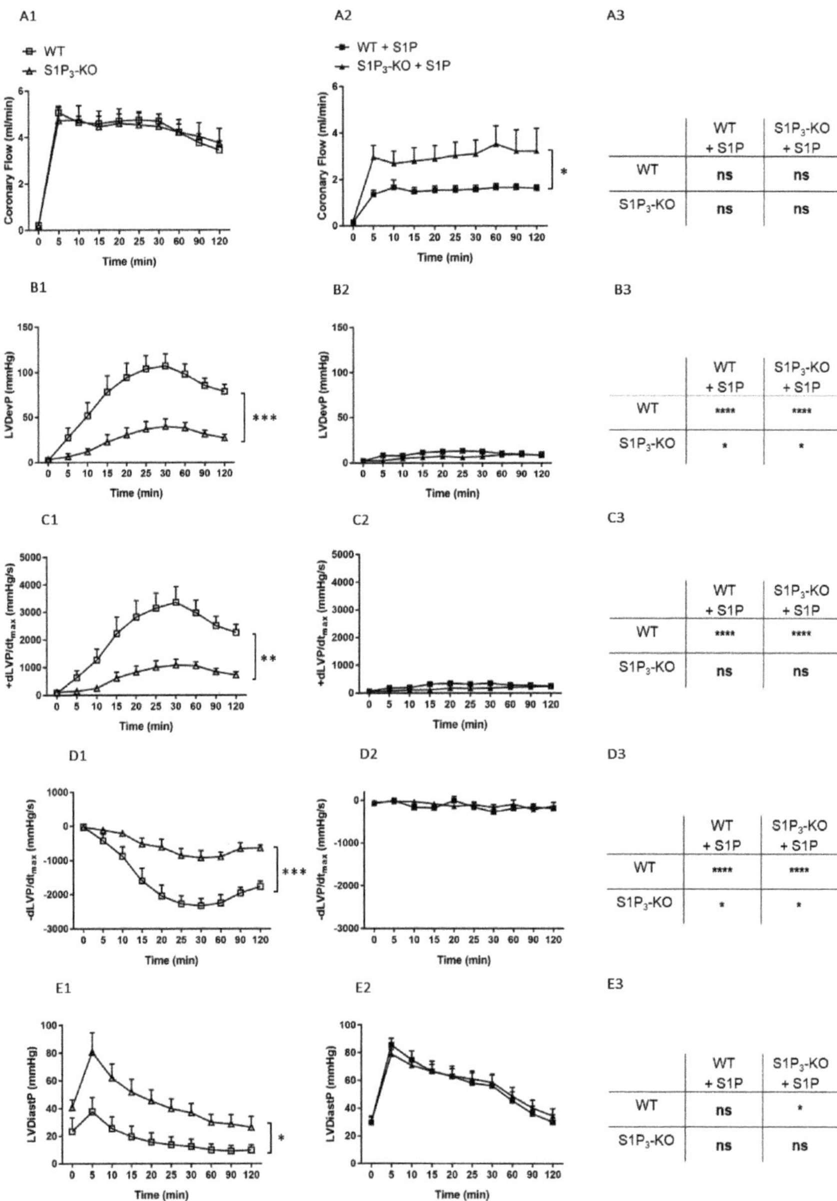

Figure 5. Postischemic coronary flow (CF) (**A**), left ventricular developed pressure (LVDevP) (**B**), +dLVP/dt$_{max}$ (**C**), −dLVP/dt$_{max}$ (**D**) and left ventricular diastolic pressure (LVDiastP) (**E**) in isolated WT and S1P$_3$ knock-out (KO) mouse hearts without (left panels: **A1–E1**) or with S1P administration (middle panels: **A2–E2**) for 5 min to the perfusate at 10^{-6} M before the induction of a 20-min ischemia followed by a 120-min reperfusion period. The right panels (**A3–E3**) demonstrate statistical comparison of the parameters captured at the end of the reperfusion period. Mean ± SEM; $n = 6, 8, 7, 7$; * $p < 0.05$, ** $p < 0.01$, *** $p < 0.001$, **** $p < 0.0001$, with two-way repeated measurement ANOVA and Dunnett's *post hoc* test in the graphs and two-way ANOVA followed by Sidak's *post hoc* test in the table insets.

3.4. Effects of Preischemic Intravascular S1P Exposure on I/R Injury

Next, we investigated the role of intravascular S1P by administering S1P to the perfusion solution before I/R at a concentration of 10^{-6} M for 5 min. Under these conditions, CF returned to a significantly higher value during reperfusion in the S1P$_3$-KO hearts (Figure 5A2) indicating S1P$_3$-mediated coronary vasoconstriction. Postischemic myocardial function failed to return during the reperfusion without any difference between the two groups (Figure 5B2–E2).

In order to determine the effects of S1P infusion on postischemic CF and cardiac performance, results obtained by the two experimental protocols were compared (see the table panels in Figure 4). S1P-exposed WT hearts showed a marked reduction in post-ischemic functional myocardial recovery as compared to WT or S1P$_3$-KO hearts without S1P administration (Figure 5B3–E3). Furthermore, the difference in the functional parameters between WT and S1P$_3$-KO hearts was not statistically significant (B2–E2 table insets in Figure 5.).

Finally, we determined whether alterations in myocardial function were reflected in the irreversible ischemic damage of cardiomyocytes. TTC staining revealed that without S1P administration, the relative infarct size was larger in S1P$_3$-KO (10.72 ± 2.93%) than in WT (1.12 ± 0.37%) hearts (Figure 6A,C). In the S1P-exposed groups, the infarcts were substantially larger, but they did not differ between S1P$_3$-KO and WT hearts (Figure 6B,D).

Comparing the size of the infarcted area in S1P-pretreated (Figure 6B,D) with the untreated groups (Figure 5A,C), we detected a marked increase in the size of the infarcted myocardium as a result of S1P administration (Figure 6E).

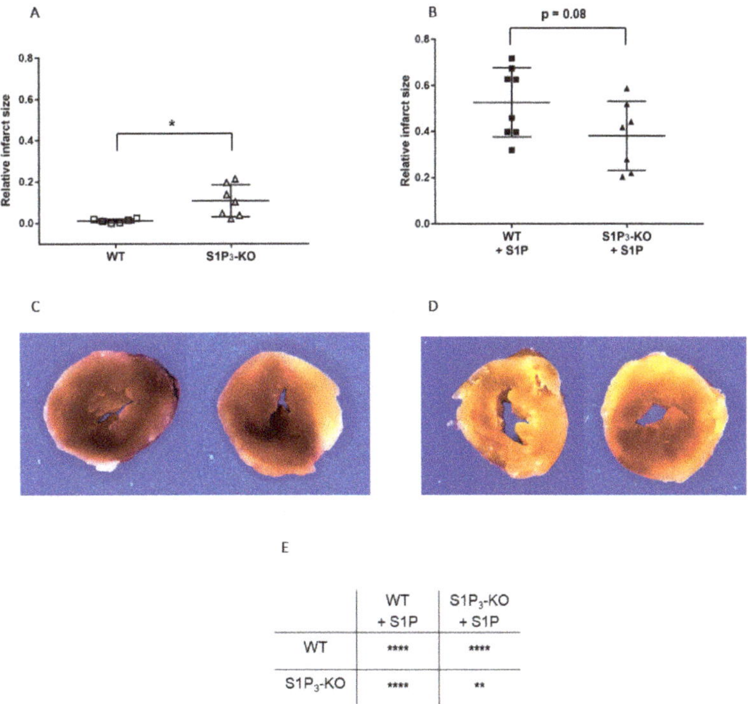

	WT + S1P	S1P$_3$-KO + S1P
WT	****	****
S1P$_3$-KO	****	**

Figure 6. Relative infarct size (**A,B**) and representative sections (**C,D**) from hearts subjected to ischemia/reperfusion without (A & C) or with (B & D) 10^{-6} M S1P infusion. Mean ± SEM; $n = 6, 8, 7, 7$; * $p < 0.05$, ** $p < 0.01$, **** $p < 0.0001$, with unpaired t-test (**A,B**) or two-way ANOVA and Sidak's multiple comparison test (**E**).

4. Discussion

In the present ex vivo study, three different experimental protocols were tested in order to understand the complexity of S1P-induced alterations of cardiac function in ACS and also to determine the involvement of S1P receptor subtypes in mediating these effects. First, we focused on intravascular S1P release, which occurs at the onset of ACS when plaque rupture initiates platelet activation. We found that S1P caused a 44 ± 3% reduction of heart perfusion and simultaneous suppression of myocardial contractility (54 ± 9% decrease in LVDevP). Both effects were attenuated in hearts of S1P$_3$-KO mice, indicating a major role of S1P$_3$ in signaling. In the second part of our study, we focused on the effects of S1P receptor activation within the heart in response to ischemia. Under these conditions, hearts of S1P$_3$-KO mice exhibited worse postischemic contractile recovery (82 ± 3% vs. 52 ± 3% decrease in LVDevP compared to preischemic baseline value in S1P$_3$-KO and WT mice, respectively) and larger infarct size (11 ± 3% vs. 1 ± 3% in S1P$_3$-KO and WT mice, respectively) than WT hearts, indicating that ischemia-induced myocardium-related S1P actions are cardioprotective via activation of S1P$_3$. Finally, we proposed to model the complex scenario of ACS, when intravascular and myocardial S1P release may occur simultaneously and influence cardiac function. Under these conditions, WT hearts showed limited coronary perfusion without any sign of postischemic functional recovery. In S1P$_3$-KO hearts, coronary reflow was better (Figure 5A2), but this failed to improve cardiac function (Figure 5B2–E2) or to reduce infarct size (Figure 6B) compared to WT. These observations indicate that although S1P$_3$-mediated vasoconstriction contributes to the deleterious no-reflow phenomenon, elimination of this effect in S1P$_3$-KO hearts does not moderate I/R injury because it also abolishes the benefits of S1P$_3$-mediated cardioprotection.

The major sources of S1P in blood plasma are red blood cells, platelets, and endothelial cells [5,9]. Sphingosine kinase is highly active in platelets and synthesizes S1P from sphingosine taken up from plasma and produces it in the outer leaflet of the platelet plasma membrane [4]. Platelets store S1P abundantly and release it upon activation [3,7,8]. In ACS, when blood clotting is activated by the rupture of an atherosclerotic plaque, substantial amount of S1P might be released to the circulation [6]. S1P has been reported to have vasoconstrictor and endothelium-dependent vasodilator actions in different vascular beds [14,46,47]. For instance, S1P was shown to have a constrictor effect in isolated porcine pulmonary artery rings [28], in canine, rat, murine and leporine basilar and middle cerebral arteries [29,30], in rat portal veins [31], and in canine coronaries [32]. Whereas S1P increased NO production in cultured HUVEC [48] and in bovine lung microvascular endothelial cells [49], and HDL, a carrier of S1P was shown to cause endothelium-dependent vasodilation in aortic rings of rats and mice mediated via S1P$_3$ activation [50]. However, despite its potential pathophysiological relevance, only a few of these studies have investigated the effects of S1P on the coronaries, and none of them have attempted to relate it to myocardial function. In our study, we found that S1P causes dose-dependent reduction in CF of Langendorff-perfused murine hearts. This observation is in agreement with earlier reports that also ascribed vasoconstrictor effects to S1P in the coronaries and other vascular beds [33,34,51]. Murakami et al. reported dose-dependent S1P-induced CF reduction in rat hearts in a similar experimental model [33]. When we delivered S1P in the presence of albumin, the coronary effect of S1P remained unchanged (Figure 1.). Although, a slight, nonsignificant shift in the ED$_{50}$ to lower concentration was observed potentially indicating that albumin may enhance S1P coronary effects by protecting it from degradation by phosphatases in the vessels of isolated hearts.

One of the main aims of our study was to (1) mimic the effect of robust S1P exposure of the coronary arteries that might occur in ACS upon thrombotic platelet activation, and (2) explore its effects on coronary perfusion, and (3) on heart function. For this purpose, we administered S1P to the coronary perfusate of isolated murine hearts at 1 microM, a concentration that might easily occur in a thrombotic coronary artery [5,6,52,53], and was close to the ED$_{50}$ value, we defined (Figure 1). This produced a remarkable decrease in CF (Figure 2A). This observation is in agreement with earlier studies, which also ascribed vasoconstrictor effects to S1P in coronaries and other vascular beds [33,34,51]. Murakami et al. reported dose-dependent S1P-induced CF reduction in rat hearts in a similar experimental setting [33]. The S1P-induced flow deprivation in our study was associated with a significant decline in cardiac performance, which was evidenced by decreased LVDevP, +dLVP/dt$_{max}$ and −dLVP/dt$_{max}$ (Figure 2B–D). This might be primarily attributed to CF reduction. However, direct negative inotropic effect of S1P on cardiomyocytes reported by earlier studies might also play a role [25]).

The cellular actions of S1P are attributed to the presence of five specific G protein-coupled S1P receptors [19,20]. Among these, S1P$_1$, S1P$_2$, and S1P$_3$ receptors are expressed abundantly in the CV system [16]. Detailed description of S1P signaling in coronaries is not available in the literature. However, a few studies provide evidence that S1P$_2$ or S1P$_3$ might play a role in the regulation of heart function. In a recent study, we reported a dominant role of S1P$_2$ in S1P-induced enhancement of vasoconstrictor stimuli in the circulation [54]. Therefore, in the present study we aimed to characterize the role of these two receptors in mediating CF reduction by S1P. Using an S1P$_3$-KO mouse model, we showed that the S1P$_3$ receptor plays a relevant role in mediating S1P-induced CF reduction, because the absence of this receptor diminishes significantly the CF reducing effect of S1P (Figure 3C). This observation confirms the findings of Murakami *et al.*, who proposed the role of S1P$_3$ in coronary constriction using the S1P$_3$ receptor antagonist TY-52156 in a similar experimental setting [33]. Levkau et al. found that S1P decreases myocardial perfusion in vivo and this effect was absent in S1P$_3$-KO mice [51]. Other investigators proposed the role of S1P$_2$ receptors because S1P induced constriction in human coronary smooth muscle cells that was attenuated by the S1P$_2$ antagonist, JTE-013 [34]. However, in our experiments S1P$_2$-KO mice did not reproduce these pharmacological observations. We acknowledge that this does not necessarily mean that the S1P$_2$ has no role in regulating coronary vessel tone, because it might be that S1P$_2$ also activates pathways in the heart which cause coronary dilation, and these and the direct vasoconstrictor effects in smooth muscle cells canceled out each other in our experiments. However, this putative mechanism requires further investigation. Moreover, S1P$_2$ activation might also sensitize the smooth muscle to other vasoconstrictor stimuli, as has been shown in the systemic circulation [54].

S1P is frequently implicated in cardioprotection [21,22,55,56]. Indeed, numerous studies have shown that it decreases the infarcted area and apoptotic cell death after I/R injury, and that it plays a role in the mechanism of ischemic pre- and post-conditioning [18,23–27]. However, myocardial function has not yet been evaluated in detail in these previous studies, although the involvement of S1P$_2$ and S1P$_3$ receptors has already been suggested [27,33,34]. This protective effect has been inferred from experiments, through the use of fundamentally different methodological approaches. In most of these studies, inhibition of S1P signaling in ischemia was achieved by using S1P receptor gene-deficient models or the pharmacological inhibition of SphK1 and SphK2 enzymes, which made I/R injury more severe and/or reduced the benefits of ischemic pre- and post-conditioning. These observations suggest that S1P signaling is stimulated in ischemia most likely by locally generated S1P released from the heart tissue.

The other approach introduced intravascular S1P administration into the coronary blood flow before an ischemic insult. Although this experimental setting can be considered as a relevant model for studying S1P effects in ACS, inasmuch as S1P infusion mimics S1P release during thrombus formation, whereas the flow cessation models thrombotic occlusion, only a few investigators have explored S1P effects this way, and they only assessed tissue damage without monitoring postischemic heart function. Nevertheless, these studies consistently reported a decrease in the infarcted area [24,27]. This is surprising considering that S1P has several short-term effects in the heart by reducing CF and causing negative inotropy that might be detrimental to postischemic contractile recovery [35–38].

Our current study was designed to combine these approaches in the context of S1P receptor signaling. Our choice of focus on S1P$_3$ signaling was motivated by the results of our experiments shown in Figures 3 and 4 which highlight that the short-term cardiac effects of S1P are mediated in large part by S1P$_3$. However, S1P$_2$, which is the other receptor proposed to participate in cardioprotection [25], did not have a major role. Moreover, the exposure of S1P$_2$-KO hearts to our I/R protocol produced similar functional and tissue injury to that observed in control hearts (Figure S2), showing that this receptor has no detectable role in cardioprotection in our experimental setting.

First, we aimed to clarify whether intrinsic activation of $S1P_3$ signaling during ischemia was protective in our experimental setting. Our results showed that, in the absence of $S1P_3$, murine hearts were more susceptible to a 20-min global ischemia. This was indicated by weaker contractile recovery during the 2-h reperfusion period (Figure 5B1–D1), higher postischemic end-diastolic pressure (Figure 5E1) an indicator of more severe myocardial ischemic contracture, and increased infarct size (Figure 6A,C). These observations are in agreement with other studies which also implicated the participation of $S1P_3$ signaling in cardioprotection against I/R injury [25–27]. Notably, the severe functional and morphological injury in $S1P_3$-KO hearts developed despite a relatively maintained CF, which approached the preischemic value and was not worse than that of WT hearts during the reperfusion period (Figure 5A1). The observation that CF during reperfusion was similar in WT and $S1P_3$-KO hearts indicates that vascular $S1P_3$ was not exposed to S1P levels sufficient to induce $S1P_3$-mediated vasoconstriction. This is not surprising if we consider that the perfusion fluid was free of exogenously added S1P.

In our study we also investigated the effects of S1P on I/R injury by the other approach described in the literature, where S1P was administered to the coronary circulation before ischemia. After S1P pretreatment, we applied a non-fatal ischemia protocol and followed up the recovery of cardiac function upon reperfusion. This model allowed for the exploration of S1P actions which may take place in a complex ACS scenario in which simultaneous intravascular and myocardial S1P release may occur. Preischemic S1P infusion can be considered as simulation of platelet-derived S1P release in ACS [3–8], whereas the ischemia protocol as S1P release from the myocardium [23,24]. Furthermore, our experiments investigated S1P effects more broadly than previous studies. In addition to determining infarct size, we also assessed postischemic cardiac function. We found that preischemic S1P exposure exacerbated ischemic injury. After an ischemic insult which is supposed to be non-fatal, the infarcted tissue extended to a large part of the myocardium and restitution of contractile activity was hardly observed in WT hearts. The latter was indicated by extremely low LVDevP, $+dLVP/dt_{max}$, and $-dLVP/dt_{max}$ values (Figure 5) that failed to approach preischemic levels during reperfusion, although CF partly recovered. Comparing ischemic injury of S1P pretreated and non-treated WT hearts, infarct size was significantly larger (Figure 6E), whereas LVDevP, $+dLVP/dt_{max}$ and -$dLVP/dt_{max}$ (Figure 5B3–D3) values were significantly lower at the end of the 2-h reperfusion period. Interestingly, some researchers observed a decrease in infarct size after preischemic S1P treatment [24,26,27]. This difference might be explained by differences in methodology because the infusion time and concentration of S1P preinfusion were slightly different [21,24,26]. A possible explanation can be that our infusion protocol, where we applied S1P at 10^{-6} M, might have caused sustained desensitization of cardiomyocytes to S1P. Several studies have shown in different cell types that S1P causes rapid desensitization of S1P receptors which persists for hours [57,58]. It might well be that in our experimental setting the combination of sustained vasoconstriction, which is potentially detrimental to postischemic recovery, and the loss of $S1P_3$ mediated activation of prosurvival and antiapoptotic protective pathways due to desensitization, results in enhanced I/R injury. Whereas, in experimental settings, where lower doses are used (0.1 micromolar), $S1P_3$ mediated protection is more active and dominates over effects of more moderate perfusion loss.

Because we observed that the coronary effects of S1P are in part mediated by $S1P_3$ (Figure 4), we also explored the effects of preischemic S1P exposure on ischemic damage in $S1P_3$-KO hearts. Although preischemic CF and the function of $S1P_3$-KO hearts were better (data not shown, also cf. Figure 4), their functional recovery was as weak, and infarct size as large, as those of WT hearts. Interestingly, although CF in $S1P_3$-KO hearts returned close to the preischemic value, this relatively better perfusion did not provide any benefit for cardiac performance. All these results indicate, that although the absence of $S1P_3$ receptors might mitigate the detrimental effects of preischemic intravascular S1P exposure by decreasing the CF-reducing effect of S1P and allow for better reflow during reperfusion, the concomitant loss of $S1P_3$-mediated cardioprotection obliterates this potential benefit. Therefore, the $S1P_3$ receptor seems to mediate two opposing S1P actions in the heart, as schematically shown in Figure 7.

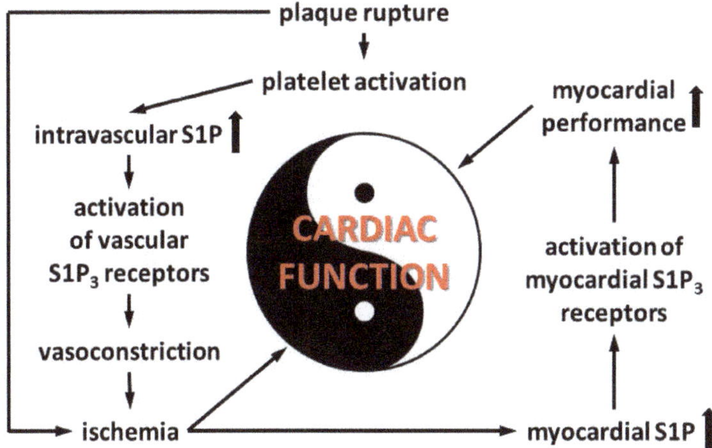

Figure 7. Events in acute coronary syndrome related to $S1P_3$-mediated alterations of cardiac function.

5. Conclusions

In this study, using isolated perfused murine hearts, we designed experimental models to simulate and explore the actions of S1P release in ACS. First, we described the effects of intravascular S1P exposure which occur during thrombus formation; and second, the effects of S1P within the heart in response to myocardial ischemia in a separate experimental paradigm. Finally, we investigated the combined effects of simultaneous intravascular and myocardial S1P effects, which can be a real-life scenario in ACS. Intracoronary administration of S1P caused a substantial decrease in CF and heart function. Using S1P receptor KO mouse models, we established that $S1P_2$ has only a minor role, whereas $S1P_3$ is a key determinant in this effect. In I/R experiments, the postischemic functional recovery was weakened and the ratio of the infarcted area was increased in $S1P_3$-KO hearts, confirming the cardioprotective role of this receptor subtype. Preischemic intravascular S1P administration worsened the recovery of cardiac function and increased infarct size both in WT and $S1P_3$-KO hearts, although coronary hypoperfusion was attenuated by $S1P_3$ deficiency. These findings highlight that S1P has opposing effects in the myocardium: S1P released from the ischemic myocardium seems to be cardioprotective, whereas S1P acting via the coronary circulation is deleterious to the heart. Moreover, both of these effects are in a large part mediated by $S1P_3$ receptor. These results taken together suggest that in clinical situations, when thrombotic coronary occlusion causes cardiac ischemia, the released S1P might compromise postischemic recovery due to its unfavorable coronary effects, which might outweigh the presumed cardioprotective effects of S1P produced by the ischemic myocardium. Clearly, further studies are warranted using in vivo and ex vivo models to obtain a better understanding of the (patho)physiological actions of S1P in ACS.

Supplementary Materials: The following are available online at http://www.mdpi.com/2073-4409/9/8/1770/s1, Figure S1: Experimental protocols of isolated heart experiments, Figure S2: Postischemic coronary flow (CF) (A), left ventricular developed pressure (LVDevP)(B), +dLVP/dtmax (C), −dLVP/dtmax (D) and left ventricular diastolic pressure (LVDiastP) (E) in isolated WT and $S1P_2$ knock-out (KO) mouse hearts.

Author Contributions: Conceptualization, D.W., G.J.T., Z.B., Z.M.; Validation, Z.M.; Formal analysis, D.W., N.K., J.K., Z.M.; Investigation, D.W., N.K., J.K., M.K., Z.M.; Resources, R.L.P.; Writing—original draft preparation, D.W., Z.B., Z.M.; Writing—review and editing, D.W., R.L.P., G.J.T., Z.B., Z.M.; Visualization, D.W.; Supervision, G.J.T., Z.B., Z.M.; Project administration, Z.M.; Funding acquisition, Z.M., Z.B. All authors have read and agreed to the published version of the manuscript.

Funding: The research was funded by the Hungarian National Research, Development and Innovation Office (K-112964, K-125174 and NVKP_16-1-2016-0042) as well as by the Higher Education Institutional Excellence Program of the Ministry of Human Capacities in Hungary, within the framework of the Molecular Biology thematic program of the Semmelweis University and supported by the EFOP-3.6.3-VEKOP-16-2017-00009 grant.

Acknowledgments: The authors are grateful to Ildikó Murányi for expert technical assistance as well as to Nathan Tipton and Erzsébet Fejes for critically reading the manuscript.

Conflicts of Interest: The authors declare no conflict of interest.

References

1. Mahmood, S.S.; Levy, D.; Vasan, R.S.; Wang, T.J. The Framingham Heart Study and the epidemiology of cardiovascular disease: A historical perspective. *Lancet* **2013**, *383*, 999–1008. [CrossRef]
2. Makki, N.; Brennan, T.M.; Girotra, S. Acute coronary syndrome. *J. Intensive Care Med.* **2015**, *30*, 186–200. [CrossRef] [PubMed]
3. Ulrych, T.; Böhm, A.; Polzin, A.; Daum, G.; Nüsing, R.M.; Geisslinger, G.; Hohlfeld, T.; Schror, K.; Rauch, B.H. Release of sphingosine-1-phosphate from human platelets is dependent on thromboxane formation. *J. Thromb. Haemost.* **2011**, *9*, 790–798. [CrossRef] [PubMed]
4. Tani, M.; Sano, T.; Ito, M.; Igarashi, Y.; Moschetta, A.; Xu, F.; Hagey, L.R.; Van Berge-Henegouwen, G.P.; Van Erpecum, K.J.; Brouwers, J.F.; et al. Mechanisms of sphingosine and sphingosine 1-phosphate generation in human platelets. *J. Lipid Res.* **2005**, *46*, 2458–2467. [CrossRef] [PubMed]
5. Yatomi, Y.; Igarashi, Y.; Yang, L.; Hisano, N.; Qi, R.; Asazuma, N.; Satoh, K.; Ozaki, Y.; Kume, S. Sphingosine 1-phosphate, a bioactive sphingolipid abundantly stored in platelets, is a normal constituent of human plasma and serum. *J. Biochem.* **1997**, *121*, 969–973. [CrossRef]
6. Ono, Y.; Kurano, M.; Ohkawa, R.; Yokota, H.; Igarashi, K.; Aoki, J.; Tozuka, M.; Yatomi, Y. Sphingosine 1-phosphate release from platelets during clot formation: Close correlation between platelet count and serum sphingosine 1-phosphate concentration. *Lipids Heal. Dis.* **2013**, *12*, 20. [CrossRef]
7. Yatomi, Y.; Ohmori, T.; Rile, G.; Kazama, F.; Okamoto, H.; Sano, T.; Satoh, K.; Kume, S.; Tigyi, G.; Igarashi, Y.; et al. Sphingosine 1-phosphate as a major bioactive lysophospholipid that is released from platelets and interacts with endothelial cells. *Blood* **2000**, *96*, 3431–3438. [CrossRef]
8. Yatomi, Y.; Ruan, F.; Hakomori, S.; Igarashi, Y. Sphingosine-1-phosphate: A platelet-activating sphingolipid released from agonist-stimulated human platelets. *Blood* **1995**, *86*, 193–202. [CrossRef] [PubMed]
9. Venkataraman, K.; Lee, Y.-M.; Michaud, J.; Thangada, S.; Ai, Y.; Bonkovsky, H.L.; Parikh, N.S.; Habrukowich, C.; Hla, T. Vascular Endothelium As a Contributor of Plasma Sphingosine 1-Phosphate. *Circ. Res.* **2008**, *102*, 669–676. [CrossRef]
10. Blaho, V.A.; Galvani, S.; Engelbrecht, E.; Liu, C.; Swendeman, S.L.; Kono, M.; Proia, R.L.; Steinman, L.; Han, M.H.; Hla, T. HDL-bound sphingosine-1-phosphate restrains lymphopoiesis and neuroinflammation. *Nature* **2015**, *523*, 342–346. [CrossRef]
11. Ruiz, M.; Okada, H.; Dahlbäck, B. HDL-associated ApoM is anti-apoptotic by delivering sphingosine 1-phosphate to S1P1 & S1P3 receptors on vascular endothelium. *Lipids Heal. Dis.* **2017**, *16*, 36. [CrossRef]
12. Tsai, H.-C.; Han, M.H. Sphingosine-1-Phosphate (S1P) and S1P Signaling Pathway: Therapeutic Targets in Autoimmunity and Inflammation. *Drugs* **2016**, *76*, 1067–1079. [CrossRef] [PubMed]
13. Wang, G.; Bieberich, E. Sphingolipids in neurodegeneration (with focus on ceramide and S1P). *Adv. Boil. Regul.* **2018**, *70*, 51–64. [CrossRef] [PubMed]
14. Igarashi, J.; Michel, T. Sphingosine-1-phosphate and modulation of vascular tone. *Cardiovasc. Res.* **2009**, *82*, 212–220. [CrossRef] [PubMed]
15. Kerage, D.; Brindley, D.; Hemmings, D.G. Review: Novel insights into the regulation of vascular tone by sphingosine 1-phosphate. *Placenta* **2014**, *35*, S86–S92. [CrossRef]
16. Peters, S.L.; Alewijnse, A. Sphingosine-1-phosphate signaling in the cardiovascular system. *Curr. Opin. Pharmacol.* **2007**, *7*, 186–192. [CrossRef]
17. Siess, W. Athero- and thrombogenic actions of lysophosphatidic acid and sphingosine-1-phosphate. *Biochim. Et Biophys. Acta (Bba) Bioenerg.* **2002**, *1582*, 204–215. [CrossRef]
18. Karliner, J.S. Sphingosine kinase and sphingosine 1-phosphate in the heart: A decade of progress. *Biochim. Et Biophys. Acta (Bba) Mol. Cell Boil. Lipids* **2013**, *1831*, 203–212. [CrossRef]

19. Mutoh, T.; Rivera, R.; Chun, J. Insights into the pharmacological relevance of lysophospholipid receptors. *Br. J. Pharmacol.* **2012**, *165*, 829–844. [CrossRef]

20. Tigyi, G.J. Lysophospholipid Receptors. In *Encyclopedia of Biological Chemistry*, 2nd ed.; John Wiley & Sons, Inc.: Hoboken, NJ, USA, 2013; pp. 1–3.

21. Xiang, S.Y.; Ouyang, K.; Yung, B.S.; Miyamoto, S.; Smrcka, A.V.; Chen, J.; Brown, J.H. PLC, PKD1, and SSH1L Transduce RhoA Signaling to Protect Mitochondria from Oxidative Stress in the Heart. *Sci. Signal.* **2013**, *6*, ra108. [CrossRef]

22. Karliner, J.S.; Honbo, N.; Summers, K.; Gray, M.O.; Goetzl, E.J. The Lysophospholipids Sphingosine-1-Phosphate and Lysophosphatidic Acid Enhance Survival during Hypoxia in Neonatal Rat Cardiac Myocytes. *J. Mol. Cell. Cardiol.* **2001**, *33*, 1713–1717. [CrossRef]

23. Jin, Z.-Q.; Goetzl, E.J.; Karliner, J.S. Sphingosine Kinase Activation Mediates Ischemic Preconditioning in Murine Heart. *Circulation* **2004**, *110*, 1980–1989. [CrossRef] [PubMed]

24. Vessey, D.A.; Kelley, M.; Li, L.; Huang, Y.; Zhou, H.-Z.; Zhu, B.Q.; Karliner, J.S. Role of sphingosine kinase activity in protection of heart against ischemia reperfusion injury. *Med. Sci. Monit.* **2006**, *12*, 318–324.

25. Means, C.K.; Xiao, C.-Y.; Li, Z.; Zhang, T.; Omens, J.H.; Ishii, I.; Chun, J.; Brown, J.H. Sphingosine 1-phosphate S1P2 and S1P3 receptor-mediated Akt activation protects against in vivo myocardial ischemia-reperfusion injury. *Am. J. Physiol. Circ. Physiol.* **2007**, *292*, H2944–H2951. [CrossRef] [PubMed]

26. Yung, B.S.; Brand, C.S.; Xiang, S.Y.; Gray, C.B.B.; Means, C.K.; Rosen, H.; Chun, J.; Purcell, N.H.; Brown, J.H.; Miyamoto, S. Selective coupling of the S1P3 receptor subtype to S1P-mediated RhoA activation and cardioprotection. *J. Mol. Cell. Cardiol.* **2016**, *103*, 1–10. [CrossRef] [PubMed]

27. Theilmeier, G.; Schmidt, C.; Herrmann, J.; Keul, P.; Schäfers, M.; Herrgott, I.; Mersmann, J.; Larmann, J.; Hermann, S.; Stypmann, J.; et al. High-Density Lipoproteins and Their Constituent, Sphingosine-1-Phosphate, Directly Protect the Heart Against Ischemia/Reperfusion Injury In Vivo via the S1P 3 Lysophospholipid Receptor. *Circulation* **2006**, *114*, 1403–1409. [CrossRef]

28. Hsiao, S.-H.; Constable, P.D.; Smith, G.W.; Haschek, W.M. Effects of Exogenous Sphinganine, Sphingosine, and Sphingosine-1-Phosphate on Relaxation and Contraction of Porcine Thoracic Aortic and Pulmonary Arterial Rings. *Toxicol. Sci.* **2005**, *86*, 194–199. [CrossRef]

29. Tosaka, M.; Okajima, F.; Hashiba, Y.; Saito, N.; Nagano, T.; Watanabe, T.; Kimura, T.; Sasaki, T. Sphingosine 1-phosphate contracts canine basilar arteries in vitro and in vivo: Possible role in pathogenesis of cerebral vasospasm. *Stroke* **2001**, *32*, 2913–2919. [CrossRef]

30. Salomone, S.; Yoshimura, S.-I.; Reuter, U.; Foley, M.; Thomas, S.S.; Moskowitz, M.A.; Waeber, C. S1P3 receptors mediate the potent constriction of cerebral arteries by sphingosine-1-phosphate. *Eur. J. Pharmacol.* **2003**, *469*, 125–134. [CrossRef]

31. Ikeda, H.; Nagashima, K.; Yanase, M.; Tomiya, T.; Arai, M.; Inoue, Y.; Tejima, K.; Nishikawa, T.; Watanabe, N.; Omata, M.; et al. Sphingosine 1-phosphate enhances portal pressure in isolated perfused liver via S1P2 with Rho activation. *Biochem. Biophys. Res. Commun.* **2004**, *320*, 754–759. [CrossRef]

32. Sugiyama, A.; Yatomi, Y.; Ozaki, Y.; Hashimoto, K. Sphingosine 1-phosphate induces sinus tachycardia and coronary vasoconstriction in the canine heart. *Cardiovasc. Res.* **2000**, *46*, 119–125. [CrossRef]

33. Murakami, A.; Takasugi, H.; Ohnuma, S.; Koide, Y.; Sakurai, A.; Takeda, S.; Hasegawa, T.; Sasamori, J.; Konno, T.; Hayashi, K.; et al. Sphingosine 1-Phosphate (S1P) Regulates Vascular Contraction via S1P3 Receptor: Investigation Based on a New S1P3 Receptor Antagonist. *Mol. Pharmacol.* **2010**, *77*, 704–713. [CrossRef] [PubMed]

34. Ohmori, T.; Yatomi, Y.; Osada, M.; Kazama, F.; Takafuta, T.; Ikeda, H.; Ozaki, Y. Sphingosine 1-phosphate induces contraction of coronary artery smooth muscle cells via S1P. *Cardiovasc. Res.* **2003**, *58*, 170–177. [CrossRef]

35. Means, C.K.; Miyamoto, S.; Chun, J.; Brown, J.H. S1P1 receptor localization confers selectivity for G i-mediated cAMP and contractile responses. *J. Biol. Chem.* **2008**, *283*, 11954–11963. [CrossRef]

36. Landeen, L.K.; Dederko, D.A.; Kondo, C.S.; Hu, B.S.; Aroonsakool, N.; Haga, J.H.; Giles, W.R. Mechanisms of the negative inotropic effects of sphingosine-1-phosphate on adult mouse ventricular myocytes. *Am. J. Physiol. Circ. Physiol.* **2008**, *294*, H736–H749. [CrossRef] [PubMed]

37. Sanna, M.G.; Liao, J.; Jo, E.; Alfonso, C.; Ahn, M.-Y.; Peterson, M.S.; Webb, B.; Lefebvre, S.; Chun, J.; Gray, N.; et al. Sphingosine 1-Phosphate (S1P) Receptor Subtypes S1P1 and S1P3, Respectively, Regulate Lymphocyte Recirculation and Heart Rate. *J. Boil. Chem.* **2004**, *279*, 13839–13848. [CrossRef] [PubMed]

38. Forrest, M.; Sun, S.-Y.; Hajdu, R.; Bergstrom, J.; Card, D.; Doherty, G.; Hale, J.; Keohane, C.; Meyers, C.; Milligan, J.; et al. Immune Cell Regulation and Cardiovascular Effects of Sphingosine 1-Phosphate Receptor Agonists in Rodents Are Mediated via Distinct Receptor Subtypes. *J. Pharmacol. Exp. Ther.* **2004**, *309*, 758–768. [CrossRef] [PubMed]

39. Kono, M.; Mi, Y.; Liu, Y.; Sasaki, T.; Allende, M.L.; Wu, Y.-P.; Yamashita, T.; Proia, R.L. The Sphingosine-1-phosphate Receptors S1P1, S1P2, and S1P3 Function Coordinately during Embryonic Angiogenesis. *J. Boil. Chem.* **2004**, *279*, 29367–29373. [CrossRef]

40. Kemecsei, P.; Miklós, Z.; Bíró, T.; Marincsák, R.; Tóth, B.I.; Komlodi-Pasztor, E.; Barnucz, E.; Mirk, E.; Van Der Vusse, G.J.; Ligeti, L.; et al. Hearts of surviving MLP-KO mice show transient changes of intracellular calcium handling. *Mol. Cell. Biochem.* **2010**, *342*, 251–260. [CrossRef]

41. Miklós, Z.; Kemecsei, P.; Biro, T.; Marincsak, R.; Tóth, B.I.; Buijs, J.; Benis, E.; Drozgyik, A.; Ivanics, T. Early cardiac dysfunction is rescued by upregulation of SERCA2a pump activity in a rat model of metabolic syndrome. *Acta Physiol.* **2012**, *205*, 381–393. [CrossRef]

42. Skrzypiec-Spring, M.; Grotthus, B.; Szelag, A.; Schulz, R. Isolated heart perfusion according to Langendorff—Still viable in the new millennium. *J. Pharmacol. Toxicol. Methods* **2007**, *55*, 113–126. [CrossRef]

43. Lacza, Z.; Dézsi, L.; Káldi, K.; Horvath, E.M.; Sándor, P.; Benyó, Z. Prostacyclin-mediated compensatory mechanism in the coronary circulation during acute NO synthase blockade. *Life Sci.* **2003**, *73*, 1141–1149. [CrossRef]

44. Liliom, K.; Sun, G.; Bünemann, M.; Virág, T.; Nusser, N.; Baker, D.L.; Wang, D.A.; Fabian, M.J.; Brandts, B.; Bender, K.; et al. Sphingosylphosphocholine is a naturally occurring lipid mediator in blood plasma: A possible role in regulating cardiac function via sphingolipid receptors. *Biochem. J.* **2001**, *355*, 189–197. [CrossRef] [PubMed]

45. Nachlas, M.M.; Shnitka, T.K. Macroscopic Identification of Early Myocardial Infarcts by Alterations in Dehydrogenase Activity. *Am. J. Pathol.* **1963**, *42*, 379–405.

46. Hemmings, D.G.; Xu, Y.; Davidge, S.T. Sphingosine 1-phosphate-induced vasoconstriction is elevated in mesenteric resistance arteries from aged female rats. *Br. J. Pharmacol.* **2004**, *143*, 276–284. [CrossRef] [PubMed]

47. Hemmings, D.G.; Hudson, N.K.; Halliday, D.; O'Hara, M.; Baker, P.N.; Davidge, S.T.; Taggart, M. Sphingosine-1-Phosphate Acts via Rho-Associated Kinase and Nitric Oxide to Regulate Human Placental Vascular Tone1. *Boil. Reprod.* **2006**, *74*, 88–94. [CrossRef]

48. Wilkerson, B.A.; Argraves, K.M. The role of sphingosine-1-phosphate in endothelial barrier function. *Biochim. Et Biophys. Acta (Bba) Mol. Cell Boil. Lipids* **2014**, *1841*, 1403–1412. [CrossRef] [PubMed]

49. Morales-Ruiz, M.; Lee, M.-J.; Zöllner, S.; Gratton, J.-P.; Scotland, R.; Shiojima, I.; Walsh, K.; Hla, T.; Sessa, W.C. Sphingosine 1-Phosphate Activates Akt, Nitric Oxide Production, and Chemotaxis through a GiProtein/Phosphoinositide 3-Kinase Pathway in Endothelial Cells. *J. Boil. Chem.* **2001**, *276*, 19672–19677. [CrossRef]

50. Nofer, J.-R.; Van Der Giet, M.; Tolle, M.; Wolinska, I.; Lipinski, K.V.W.; Baba, H.A.; Tietge, U.J.; Gödecke, A.; Ishii, I.; Kleuser, B.; et al. HDL induces NO-dependent vasorelaxation via the lysophospholipid receptor S1P3. *J. Clin. Investig.* **2004**, *113*, 569–581. [CrossRef]

51. Levkau, B.; Hermann, S.; Theilmeier, G.; Van Der Giet, M.; Chun, J.; Schober, O.; Schäfers, M. High-Density Lipoprotein Stimulates Myocardial Perfusion In Vivo. *Circulation* **2004**, *110*, 3355–3359. [CrossRef]

52. Deutschman, D.H.; Carstens, J.S.; Klepper, R.L.; Smith, W.S.; Page, M.; Young, T.R.; Gleason, L.A.; Nakajima, N.; Sabbadini, R.A. Predicting obstructive coronary artery disease with serum sphingosine-1-phosphate. *Am. Hear. J.* **2003**, *146*, 62–68. [CrossRef]

53. Okajima, F. Plasma lipoproteins behave as carriers of extracellular sphingosine 1-phosphate: Is this an atherogenic mediator or an anti-atherogenic mediator? *Biochim. Et Biophys. Acta (Bba) Bioenerg.* **2002**, *1582*, 132–137. [CrossRef]

54. Panta, C.R.; Ruisanchez, É.; Móré, D.; Dancs, P.T.; Balogh, A.; Fülöp, Á.; Kerék, M.; Proia, R.L. Sphingosine-1-phosphate enhances α1-adrenergic vasoconstriction via S1P2–G12/13–ROCK mediated signaling. *Int. J. Mol. Sci.* **2019**, *20*, 6361. [CrossRef] [PubMed]

55. Karliner, J.S. Sphingosine Kinase and Sphingosine 1-Phosphate in Cardioprotection. *J. Cardiovasc. Pharmacol.* **2009**, *53*, 189–197. [CrossRef] [PubMed]

56. Knapp, M. Cardioprotective role of sphingosine-1-phosphate. *J. Physiol. Pharmacol. Off. J. Pol. Physiol. Soc.* **2011**, *62*, 601–607.

57. Xin, C.; Ren, S.; Pfeilschifter, J.; Huwiler, A. Heterologous desensitization of the sphingosine-1-phosphate receptors by purinoceptor activation in renal mesangial cells. *Br. J. Pharmacol.* **2004**, *143*, 581–589. [CrossRef]

58. Gräler, M.H.; Goetzl, E.J. The immunosuppressant FTY720 down-regulates sphingosine 1-phosphate G protein-coupled receptors. *Faseb J.* **2004**, *18*, 551–553. [CrossRef]

Article

Cystic Fibrosis Defective Response to Infection Involves Autophagy and Lipid Metabolism

Alessandra Mingione [1,†], Emerenziana Ottaviano [2,†], Matteo Barcella [2], Ivan Merelli [3], Lorenzo Rosso [4], Tatiana Armeni [5], Natalia Cirilli [6], Riccardo Ghidoni [1,7], Elisa Borghi [2] and Paola Signorelli [1,*]

[1]   Biochemistry and Molecular Biology Laboratory, Health Science Department, University of Milan, San Paolo Hospital, 20142 Milan, Italy; alessandra.mingione@unimi.it (A.M.); riccardo.ghidoni@unimi.it (R.G.)
[2]   Laboratory of Clinical Microbiology, Health Science Department, University of Milan, San Paolo Hospital, 20142 Milan, Italy; emerenziana.ottaviano@unimi.it (E.O.); matteobarcella84@gmail.com (M.B.); elisa.borghi@unimi.it (E.B.)
[3]   National Research Council of Italy, Institute for Biomedical Technologies, Via Fratelli Cervi 93, 20090 Milan, Italy; ivan.merelli@itb.cnr.it
[4]   Health Sciences Department, University of Milan, Thoracic surgery and transplantation Unit, Fondazione IRCCS Ca Granda Ospedale Maggiore Policlinico, 20122 Milan, Italy; lorenzo.rosso@unimi.it
[5]   Department of Clinical Sciences, Section of Biochemistry, Biology and Physics, Polytechnic University of Marche, 60131 Ancona, Italy; t.armeni@staff.univpm.it
[6]   Cystic Fibrosis Referral Care Center, Mother-Child Department, United Hospitals Le Torrette, 60126 Ancona, Italy; natalia.cirilli@ospedaliriuniti.marche.it
[7]   "Aldo Ravelli" Center for Neurotechnology and Experimental Brain Therapeutics, via Antonio di Rudinì 8, 20142 Milan, Italy
*   Correspondence: paola.signorelli@unimi.it
†   The authors contributed equally to this work.

Received: 26 June 2020; Accepted: 4 August 2020; Published: 6 August 2020

Abstract: Cystic fibrosis (CF) is a hereditary disease, with 70% of patients developing a proteinopathy related to the deletion of phenylalanine 508. CF is associated with multiple organ dysfunction, chronic inflammation, and recurrent lung infections. CF is characterized by defective autophagy, lipid metabolism, and immune response. Intracellular lipid accumulation favors microbial infection, and autophagy deficiency impairs internalized pathogen clearance. Myriocin, an inhibitor of sphingolipid synthesis, significantly reduces inflammation, promotes microbial clearance in the lungs, and induces autophagy and lipid oxidation. RNA-seq was performed in *Aspergillusfumigatus*-infected and myriocin-treated CF patients' derived monocytes and in a CF bronchial epithelial cell line. Fungal clearance was also evaluated in CF monocytes. Myriocin enhanced CF patients' monocytes killing of *A. fumigatus*. CF patients' monocytes and cell line responded to infection with a profound transcriptional change; myriocin regulates genes that are involved in inflammation, autophagy, lipid storage, and metabolism, including histones and heat shock proteins whose activity is related to the response to infection. We conclude that the regulation of sphingolipid synthesis induces a metabolism drift by promoting autophagy and lipid consumption. This process is driven by a transcriptional program that corrects part of the differences between CF and control samples, therefore ameliorating the infection response and pathogen clearance in the CF cell line and in CF peripheral blood monocytes.

Keywords: cystic fibrosis; sphingolipids; autophagy; myriocin; *Aspergillus fumigatus*

1. Introduction

Cystic fibrosis (CF) is a hereditary disease associated with different classes of mutations in the Cystic Fibrosis Transmembrane conductance Regulator (CFTR), a chloride/carbonate channel. A deletion of phenylalanine 508 (F508) affects more than 70% of patients and results in unfoldedproteins accumulation, causing a proteinopathy responsible for inflammation, impaired autophagy, and altered lipid metabolism [1–5]. CF is a multiple-organ disease characterized by life-threatening chronic inflammation and recurrent infection of the lungs, where defective immune response and altered mucus viscosity and acidity pave the way to persistent microbial colonization [6]. Indeed, CF is characterized by ineffective clearance of pathogens and reduced killing of internalized microbes, both in pulmonary airways epithelia and in macrophages [7–9]. Moreover, recurrent antibacterial therapies favor the insurgence of drug resistance and, by altering the microbiological milieu, promote colonization by opportunistic pathogens. *Aspergillus fumigatus* is the most prevalent filamentous fungus in the respiratory tract of CF patients, contributing to lung deterioration. Approximately 35% of CF patients are infected by *A. fumigatus*, and researchers are still investigating ways to prevent its colonization [10], since CFTR dysfunction itself has been directly associated with a reduced clearance ability of *A. fumigatus* conidia by CF airway epithelia [8,11].

CF cells suffer from defective autophagy, which is related to altered proteostasis and chronic inflammation [12–14]. Autophagy is a conserved cell-autonomous stress response, dedicated to the breakdown of cellular material and to cell content recycling. Apart from its homeostatic role and its crucial activity in stress conditions, autophagy is actively involved in pathogen clearance, offering the cell an extremely efficient defensive response. A specialized form of autophagy called xenophagy involves the recognition and clearance of foreign particles and pathogens. Pattern recognition receptors (PRRs), upon antigen engagement, recruit microtubule-associated protein 1B light chain-3 (LC3) and successively other components of the canonical autophagy pathway, therefore enabling the binding of foreign particles that are already contained within a single membraned phagosome or endosome and their addressing to lysosomes [15].

The ability to detoxify and/or assimilate host lipids is a crucial aspect of the infection process [16,17], particularly in fungal infection [18]. In addition, the mobilization of internal lipid stores to release growth substrates, as well as lipid-mediated cellular signaling, contribute to the pathogen invasion outcome [18]. The activation of lipid catabolism via transcriptional activity of transcription factor EB/peroxisome proliferator-activated receptor α (TFEB/PPARα) potentiates macrophage response to infection [19].

Several studies reported an altered lipid homeostasis in blood and peripheral tissues in CF patients [1,2,20–24], and demonstrated the pathological role of the inflammatory ceramide accumulation in CF airways [4,25]. In a CF mouse model and in bronchial epithelial cells, our group previously demonstrated that hampering ceramide accumulation by the sphingolipid synthesis inhibitor myriocin (Myr) reduces inflammation and ameliorates the response against microbial infection in vivo and in vitro by promoting cell-killing ability [8,25,26]. Moreover, the inhibition of ceramide synthesis in CF cells activates TFEB, a master regulator of the stress response. Indeed, TFEB is responsible for enhancing autophagy and lipid oxidation via the activation of lipid-metabolism-related transcription factors such as PPARs and FOXO, thus reducing the overall lipid content [5].

Here we studied the myriocin modulation of gene expression profile and its effect in promoting conidial killing in CF patients' monocytes infected with *A. fumigatus*. Next, we demonstrated that *A. fumigatus* infection triggers a different expression profile in CF bronchial epithelial cells compared to healthy control cells. We further investigated the effect of Myr treatment on the expression of genes involved in inflammation and in the autophagic response to infection. Our data suggest that regulating the altered lipid metabolism could represent a possible therapeutic strategy in ameliorating CF disease.

2. Materials and Methods

2.1. Cells and Treatments

IB3-1 cells (named CF cells), an adeno-associated virus-transformed human bronchial epithelial cell line derived from a CF patient (ΔF508/W1282X) and provided by LGC Promochem (US), were grown in LHC-8 medium supplemented with 10% fetal bovine serum (FBS) and 1% penicillin/streptomycin at 37 °C and 5% CO_2. Human lung bronchial epithelial cells 16HBE14o- (named CTRL cells), originally developed by Dieter C. Gruenert, were provided by Luis J. Galietta, (Telethon Institute of Genetics and Medicine—TIGEM, Napoli) and cultured, as recommended, in Minimum Essential Medium (MEM) Earle's salt, supplemented with 10% FBS and 1% penicillin/streptomycin at 37 °C and 5% CO_2. Myriocin (Myr) treatments were performed at a concentration of 50 μM for the indicated time lengths in 100 mm dishes plated at 1×10^5 cells/each. At least triplicate samples for each experiment were performed.

2.2. Aspergillus Fumigatus Culture

The reference strain *A. fumigatus* Af293 (ATCC MYA-4609, CBS 101355) was used in the study. Frozen conidia were streaked out from glycerol stocks stored at −80 °C on fresh Potato Dextrose Agar (PDA) slant and incubated at 30 °C for 72–96 h, until sporulation. Conidia were then harvested, suspended in 0.01% Tween-20, washed in phosphate-buffered saline (PBS), and counted by hemocytometer.

2.3. Cells Infection

We plated 1×10^6 cells, either CF or CTRL, in 100 mm Petri dishes and grew them overnight in 5 mL of appropriate medium. The next day, cells were pre-treated with 50 μM Myr for 1 h and then infected with *A. fumigatus* conidia with a multiplicity of infection (MOI) of 1:100 for 1 h. Cells were washed three times in PBS to eliminate non-engulfed conidia and incubated for 3 h with 5 mL medium containing or not 50 μM Myr. Cells were collected at this time, or after 12 h, for subsequent analysis.

For the killing assay, patients' monocytes were pretreated with Myr (50 μM) or medium alone for 1 h at 37 °C and then infected with *A. fumigatus* conidia (MOI 1:1). One hour after infection, cells were washed twice with PBS to remove non-internalized conidia, and incubated for a further 4 h to allow conidia killing. Cells were then harvested, counted, and lysed by osmotic shock to recover live internalized conidia. Cellular lysis was confirmed by microscopy. The supernatant was mixed vigorously, serially diluted in PBS, and plated onto Sabouraud Dextrose Agar (SAB). Plates were incubated at 37 °C, and colony forming units (CFU) were counted after 48 and 72 h of growth.

2.4. PBMC Isolation and Infection

We collected 43 blood samples from CF patients referred to the Ancona Cystic Fibrosis Centre, 22 homozygotes for the F508del mutation of the *CFTR* gene and 21 compound heterozygotes with one F508del allele (as shown in Supplementary Table S1) (United Hospital Le Torrette, Ancona, Italy, CE Regionale Marche -CERM-, protocol number 2016 0606OR).

Peripheral blood mononuclear cells (PBMCs) were freshly isolated using Leucosep protocol and stored at −80 °C until use. To avoid the influence of anti-CD14-coated microbeads selection on cytokine production, we isolated monocytes by plastic adherence [27]. Briefly, 1×10^6 PBMCs per well were seeded into 6-well plates (Corning Inc. Costar, New York, NY, USA) and allowed to adhere in a 5% CO_2 incubator at 37 °C for 2 h in 2 mL Roswell Park Memorial Institute (RPMI) 1640 medium supplemented with 10% FBS and 1% penicillin/streptomycin. Free-floating cells were then removed by washing and the adhering monocytes were used for *A. fumigatus* infection as described above for cell lines.

2.5. qRT-PCR

One microgram of purified RNA was reverse transcribed to cDNA. The amplification was performed for the following target genes: *IL-1B, CXCL8, IL-10, NOD2, TLR2, TLR7, TBK1, OPTN, TFEB, LAMP2a, TP53INP1, TMEM59, HIGD1A*. Relative mRNA expression of target genes was normalized to the endogenous *GAPDH* control gene and represented as fold change versus control, calculated by the comparative CT method ($\Delta\Delta$CT Method). The analysis was performed by referring to control values that did not significantly differentiate (triplicate samples, and their standard deviation value divided by their mean value was <1). Supplementary Figure S1 shows the $\Delta\Delta$CT analysis of the target genes in basal condition, with CF cells versus CTRL cells. Real-time PCR was performed by SYBR Premix Ex Taq™ II (Takara); primer sequences are available on request.

2.6. Statistical Analysis

Data are expressed as mean ± SE, calculated from experimental replicates. Data significance was evaluated by two-way ANOVA followed by Bonferroni correction ($p < 0.05$) or Wilcoxon *t*-test (for paired samples), as specified in the figure legends. Statistical analysis was performed by GraphPad InStat software (La Jolla, CA, USA) and graph illustrations were generated by GraphPad Prism software (La Jolla, CA, USA).

2.7. RNA Extraction and Sequencing

Total RNA was isolated from harvested cells and patients' samples with the ReliaPrep™ Miniprep RNA extraction system (Promega), according to the manufacturer instructions.

Sequencing was performed on Illumina NextSeq using the SMART-Seq protocol for the preparation of the libraries, obtaining an average of 15 million single-end reads per sample. All sequences were of fixed length 75 bp, and of a high quality for a single base.

Twenty four samples were analyzed: 12 derived from a cell line of CF pulmonary epithelium, and 12 from a cell line of healthy pulmonary epithelium (CTRL). For each group, the following conditions were analyzed: (1) basal condition (CTRL or CF); (2) infection with *A. fumigatus* (Asp); (3) myriocin treatment (Myr); (4) infection and treatment (Asp-Myr). Three biological replicates were used for each condition.

RNA was also extracted from a selection of *A. fumigatus*-infected monocytes from 11 homozygous and 9 heterozygous patients. For each subject, we analyzed 2 conditions: (1) samples infected with *A. fumigatus* (untreated); (2) samples infected with *A. fumigatus* and treated with Myr (treated).

2.8. Exploratory Data Analysis

In order to evaluate samples variability, we applied principal component analysis (PCA) as a tool to make assumptions. In particular, we performed PCA starting from regularized log-transformed counts matrices created by the rlog function implemented in the DESeq2 package [28].

2.9. RNA-Seq Data Analysis

Raw single-end reads were aligned to the human reference genome (GRCh38) using STAR [29], and only uniquely mapping reads were considered for downstream analyses. Reads were assigned to genes with featureCounts [30], using the Gencode primary assembly v.31 gene transfer file (GTF) as reference annotation for the genomic features. Raw counts matrices were then processed with the R/Bioconductor differential gene expression analysis packages Deseq2 and EdgeR [31] following standard workflows. In particular, for the monocyte dataset we set up a paired analysis, modeling gene counts using the following design formula: ~patient + condition. For cell lines, due to the high variability between and within cell lines, we split the dataset into specific subsets. In particular, in both CTRL and CF cells, we evaluated the effect of *A. fumigatus* (Asp) infection and of Myr with and without infection. Genes with adjusted *p*-values less than 0.01 (cell lines) and 0.05 (monocytes dataset) were considered differentially expressed (DEGs). Downstream analyses, including gene set enrichment analysis (GSEA) and over-representation analysis (ORA), were performed with the ClusterProfiler R/Bioconductor package [32] using a list of databases including Gene Ontology (GO), KEGG, Reactome pathway, and the Molecular Signatures Database (MsigDB). Enriched terms with an adjusted *p*-value < 0.05 were considered statistically significant. Charts and images were produced using the ggplot2 R package.

We performed protein–protein interaction (PPI) networks functional enrichment analysis using string-db [33] web site. In particular, starting from the default setting we applied the following changes: removal of text-mining flag from active interaction sources, selection of highest confidence interaction scores, and hiding of disconnected nodes from graphical output.

3. Results

3.1. Myriocin Transcriptionally Regulates and Ameliorates the Response to A. fumigatus Infection in CF-Patients-Derived Monocytes

In order to study the effects of Myr on the immune response against fungal infection in CF, monocytes derived from the PBMCs of CF patients bearing homozygous or heterozygous ΔF508 CFTR mutation were infected in vitro with *A. fumigatus* and either treated or not treated with Myr. Myr treatment significantly increased *A. fumigatus* killing compared with untreated infected CF monocytes (Figure 1).

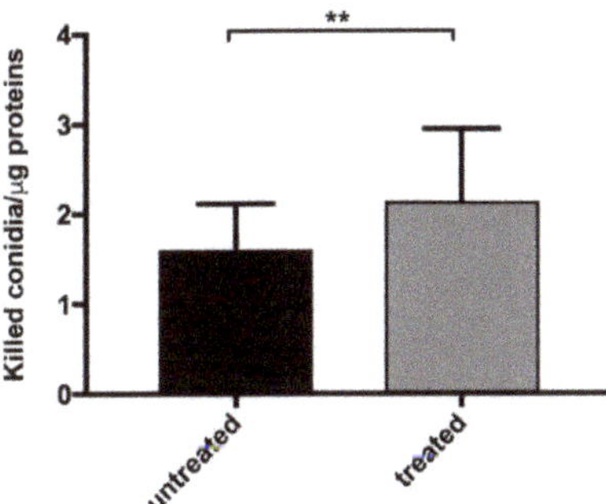

Figure 1. Myriocin (Myr) treatment improved monocyte killing ability. *Aspergillus fumigatus* conidia were added to patients' monocytes, untreated or treated with Myr, and cells incubated for 1 h to allow conidia internalization. Non-internalized conidia were then washed, and monocytes were incubated for a further 4 h before undergoing lysis. Live conidia were counted by plating cell lysate on solid medium. To normalize the results, data are expressed as killed conidia/µg protein. Myr partially rescued cystic fibrosis (CF) monocytes' killing ability (** *p* < 0.01 Wilcoxon *t*-test).

Exploratory data analysis showed a patient-specific bias that we took into account by using a paired differential expression model, as described in the Methods section. After correcting for batch effect (patient), we clearly observed a separation on the second principal component (Figure 2A). Differential gene expression analysis identified 1460 DEGs between Myr-treated and untreated cells, in particular 563 upregulated and 896 downregulated genes in Myr-treated cells (Figure 2B).

A

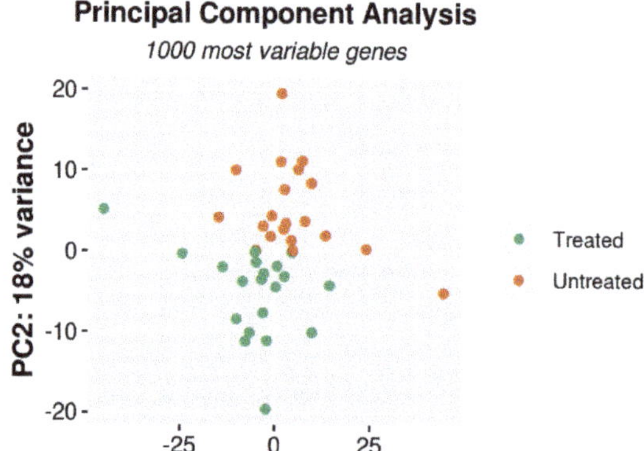

B

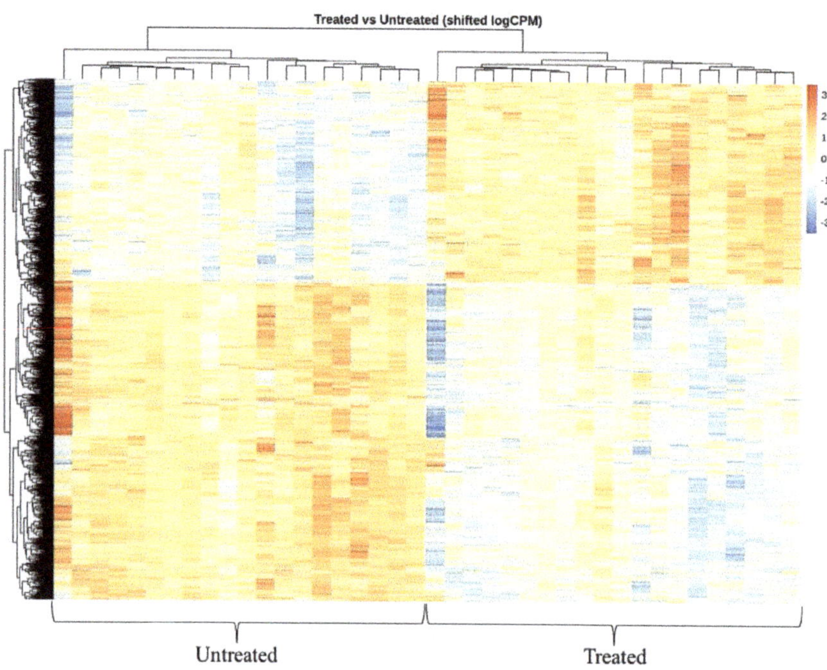

Figure 2. PCA and differential gene expression (DGE) induced by myriocin treatment in CF-patients-derived monocytes. (**A**) PCA was calculated considering the 1000 most variable genes across the dataset. Log-normalized counts were "shifted" for patient in order to take into account the experimental design adopted for DGE analysis. (**B**) Heatmap of DEGs with adjusted p-value < 0.05 resulted from the comparison between treated and untreated samples. The data used for creating this image were logCPM-corrected for patient batch and row-scaled (heatmap package).

Starting from the DEG list, we performed over-representation analysis using different gene sets, including Gene Ontology (GO), Kyoto Encyclopedia of Genes and Genomes (KEGG), and Molecular Signatures Database (MSigDB) terms, in order to identify the presence of enriched functional categories.

In particular, we observed categories related to CF disease as inflammation, autophagy, lipid metabolism, and infection (Figure 3A). We then analyzed the proportion of differentially upregulated and downregulated genes and highlighted the most significantly modulated ones. We noticed a significantly higher number of upregulated genes under Myr treatment (Figure 3B).

A

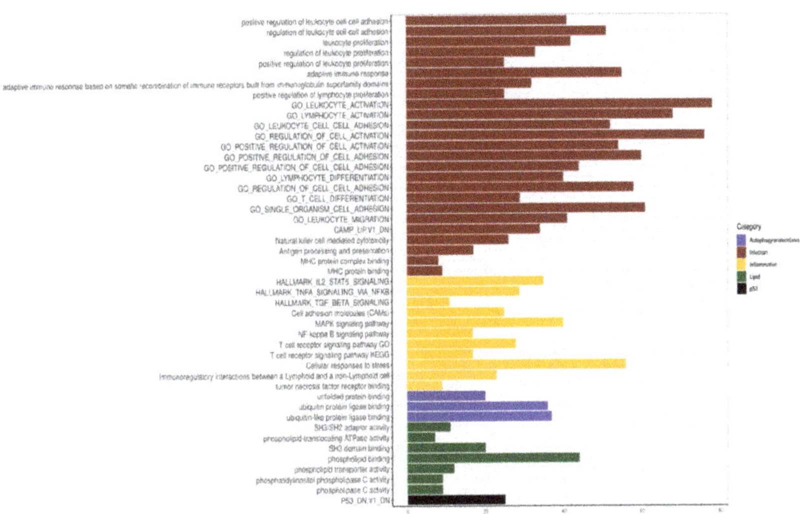

B

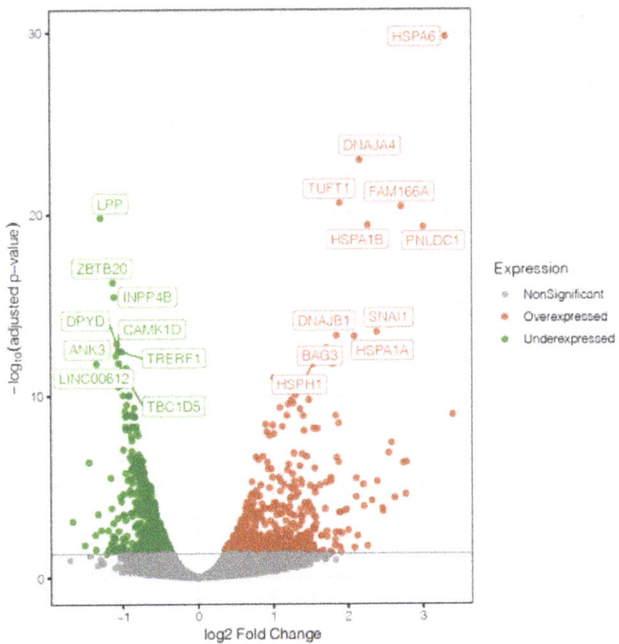

Figure 3. *Cont.*

C

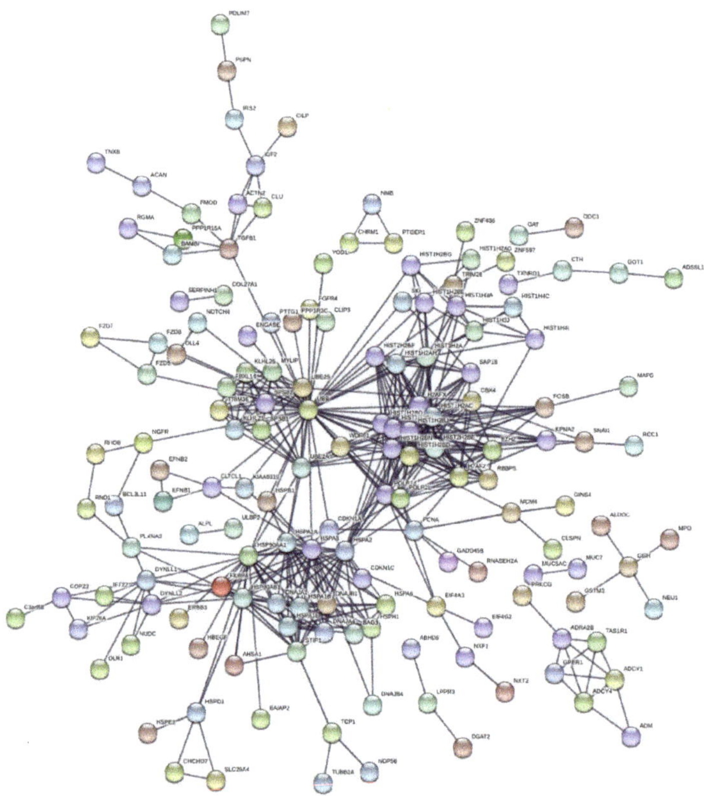

Figure 3. (**A**) CF-related over-represented terms. DEGs obtained from Myr-treated and untreated *A. fumigatus*-infected CF monocytes; the comparisons were used for performing over-representation analysis (ORA) using GO, KEGG, and MSigDB terms. Here are shown the enriched terms (adjusted *p*-value < 0.05) matching the top functional categories related to CF including autophagy, infection, inflammation, and lipid metabolism. On the *x*-axis, the number of DEGs belonging to the enriched term. (**B**) Volcano plot showing DEGs distribution between up- and downregulated genes with labels for the top 20 significant DEGs. On the *x*-axis, the log2Fold change, and on the *y*-axis the adjusted *p*-value (−log10). (**C**) STRING analysis of up regulated DEGs in treated versus untreated comparison.

Moreover, we performed protein–protein interaction (PPI) networks and functional enrichment analysis (STRING) of both up- and downregulated DEGs. Interestingly, we noticed that PPI network analysis of upregulated genes highlighted the presence of two main clusters of interacting proteins enriched in histone molecules and heat-shock proteins, respectively (Figure 3C).

By studying the most significant DEGs, we observed that genes belonging to the above-mentioned clusters are related to the ontologies indicated in Figure 3A and to the two main clusters of interacting proteins (Figure 3C). Moreover, their activity is relevant and novel in CF infection. A number of genes found to be upregulated belong to the histone cluster family and their corresponding proteins, which can locate within the cytosol and are recognized for exerting antimicrobial activity [34–40] (Table 1).

Table 1. Cluster of histone protein DEGs between Myr-treated and untreated CF-patients-derived monocytes.

Symbol	Log2FoldChange	Padj	Description	Ensembl
HIST1H2BH	1.86	8.07×10^{-5}	histone cluster 1 H2B family member h	ENSG00000275713
HIST1H2BG	1.56	0.001394033	histone cluster 1 H2B family member g	ENSG00000273802
HIST1H2BO	1.54	0.021985759	histone cluster 1 H2B family member o	ENSG00000274641
HIST1H2BE	1.53	0.007600047	histone cluster 1 H2B family member e	ENSG00000274290
HIST1H2BJ	1.29	0.002758228	histone cluster 1 H2B family member j	ENSG00000124635
HIST1H2BD	1.21	0.001620018	histone cluster 1 H2B family member d	ENSG00000158373
HIST1H4C	1.14	0.004057245	histone cluster 1 H4 family member c	ENSG00000197061
H2AFX	1.04	0.009827979	H2A histone family member X	ENSG00000188486
HIST1H3A	1.02	0.05944192	histone cluster 1 H3 family member a	ENSG00000275714
HIST2H2BE	0.88	0.028428384	histone cluster 2 H2B family member e	ENSG00000184678
H2AFZ	0.63	0.044623207	H2A histone family member Z	ENSG00000164032
HIST1H3J	0.93	0.204175587	histone cluster 1 H3 family member j	ENSG00000197153
HIST1H4I	0.72	0.065534653	histone cluster 1 H4 family member i	ENSG00000276180
HIST1H2AC	0.47	0.318228769	histone cluster 1 H2A family member c	ENSG00000180573

To note that Myr also induced the upregulation of several genes encoding for heat-shock proteins: Hsp70, Hsp90, Hsp40 (DnaJ), and Hsp-interacting proteins (Table 2), as already highlighted by PPI network analysis (Figure 3C).

Table 2. Cluster of heat-shock protein DEGs between Myr-treated and untreated CF-patients-derived monocytes.

Symbol	Log2FoldChange	Padj	Description	Ensembl
HSPA6	3.41	2.65×10^{-9}	heat shock protein family A (Hsp70) member 6	ENSG00000173110
DNAJA4	2.71	2.25×10^{-16}	DnaJ heat shock protein family (Hsp40) member A4	ENSG00000140403
HSPA1B	2.71	1.19×10^{-10}	heat shock protein family A (Hsp70) member 1B	ENSG00000204388
HSPA1A	2.64	3.02×10^{-13}	heat shock protein family A (Hsp70) member 1A	ENSG00000204389
DNAJB1	2.47	8.60×10^{-13}	DnaJ heat shock protein family (Hsp40) member B1	ENSG00000132002
BAG3	2.14	1.79×10^{-11}	BCL2 associated athanogene 3	ENSG00000151929
HSPA2	2.00	1.54×10^{-5}	heat shock protein family A (Hsp70) member 2	ENSG00000126803
SERPINH1	1.94	3.78×10^{-9}	serpin family H member 1	ENSG00000149257
HSPD1	1.68	3.78×10^{-9}	heat shock protein family D (Hsp60) member 1	ENSG00000144381
ZFAND2A	1.60	1.29×10^{-9}	zinc finger AN1-type containing 2A	ENSG00000178381
HSPE1	1.40	7.08×10^{-5}	heat shock protein family E (Hsp10) member 1	ENSG00000115541
HSP90AA1	1.35	1.13×10^{-6}	heat shock protein 90 alpha family class A member 1	ENSG00000080824
DNAJB4	1.13	5.80×10^{-7}	DnaJ heat shock protein family (Hsp40) member B4	ENSG00000162616
HSPA1L	1.13	8.07×10^{-5}	heat shock protein family A (Hsp70) member 1 like	ENSG00000204390
DNAJA1	0.90	6.78×10^{-5}	DnaJ heat shock protein family (Hsp40) member A1	ENSG00000086061
HSP90AB1	0.82	0.002280106	heat shock protein 90 alpha family class B member 1	ENSG00000096384
HSPA8	0.53	0.058998158	heat shock protein family A (Hsp70) member 8	ENSG00000109971

Such proteins are directly involved in chaperone-mediated autophagy, which sustains the autophagic activity. In addition, we identified an increased transcription of the *PNPLA3* gene that codes for a lipase supporting lipid storage degradation in autophagolysosomal compartment [41].

Moreover, Myr inducedthe upregulation of LC3IIB (*MAP1LC3B2*), which belongs to the LC3B family, required for cargo and autophagy protein recruitment and for autophagosome nucleation [42].

3.2. CF and Control Airways Epithelial Cell Lines Exhibit a Different Transcriptional Response to Infection

Chronic inflammation contributes to defective pathogen clearance in CF airways, and epithelial cells offer the first barrier to microbial infection by exerting pathogens' uptake and killing. In order to better understand the therapeutic effect of sphingolipid synthesis inhibition in CF, we studied the transcriptomic profile of CF bronchial epithelial cells in the early phase of infection (4 h). Hence, we performed an RNA-sequencing analysis of CF and CTRL bronchial epithelial cells after infection with *A. fumigatus*.

As shown in Figure 4A, infected CF cells exhibited a stronger transcriptional response to fungal infection compared to CTRL cells, considering both fold changes and differentially expressed genes (DEGs). In response to infection, we identified 3954 DEGs in CTRL cells and 5109 DEGs in CF cells (Figure 4B).

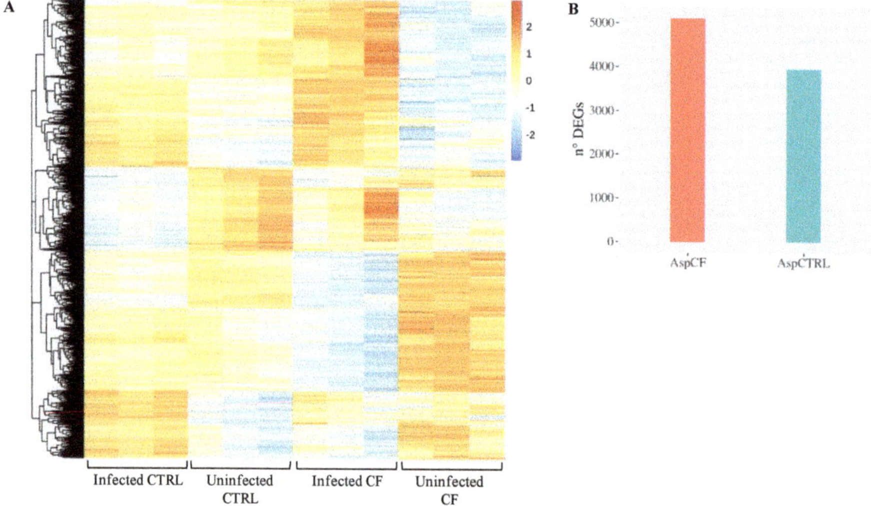

Figure 4. Transcriptional response to infection. (**A**) The heatmap represents the expression (rlog row-scaled) of DEGs obtained in the AspCTRL and AspCF comparisons. (**B**) DEGs numbers. AspCTRL: DEGs between infected and uninfected CTRL cells; AspCF: DEGs between infected and uninfected CF cells.

3.3. Myriocin Treatment Modulates CF Expression Profile under Infection

Next, we evaluated the effect of myriocin on gene expression in CF and CTRL cells, both in basal and in *A. fumigatus* infection conditions. We observed a significantly higher number of DEGs in Myr-treated CF (MyrCF, 1629 DEGs) than in CTRL cells (MyrCTRL 55 DEGs) (Figure 5A). Myr treatment in *A. fumigatus*-infected CF cells (MyrAspCF/Asp) significantly modulated only 62 DEGs versus 240 DEGs in infected CTRL cells (MyrAspCTRL/Asp) (Figure 5B). This could be related to an attenuated transcriptional effect of the compound due to the massive transcriptional change induced by *A. fumigatus* infection in CF cells (AspCF), which was significantly stronger than in CTRL cells (AspCTRL) (see above, Figure 4).

To get insight from the obtained data, we created Venn diagrams showing the common DEGs, deriving from the three different comparisons: (i) infected CTRL cells versus uninfected (AspCTRL); (ii) infected CF cells versus uninfected (AspCF); (iii) infected and Myr-treated CF cells versus uninfected and untreated (MyrAspCF).

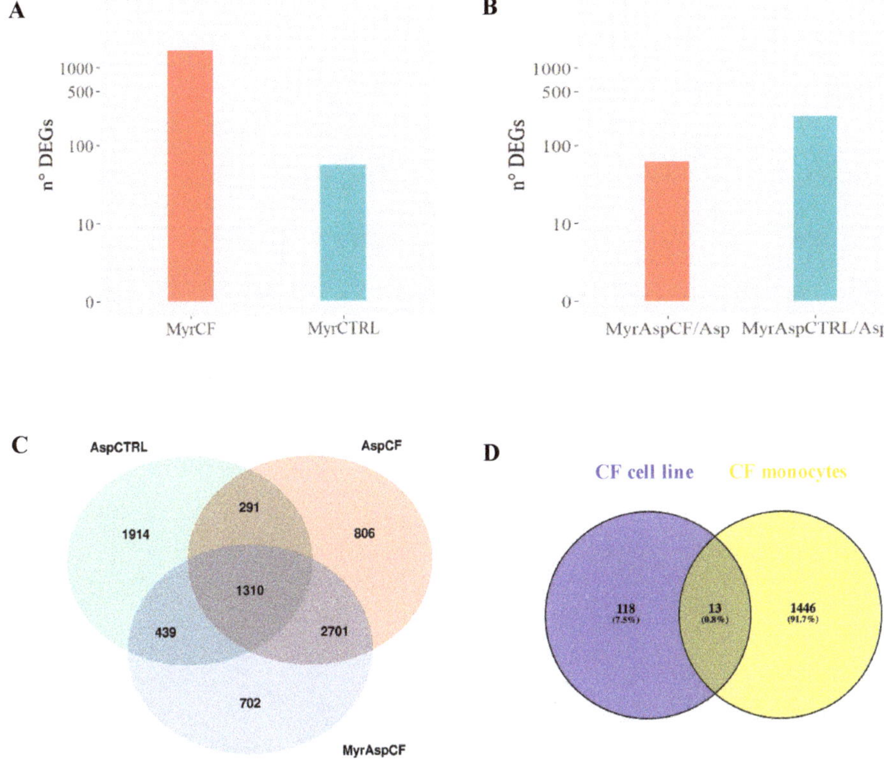

Figure 5. Effect of myriocin on CF and CTRL cells in infected and infection-free environments. (**A**) MyrCF indicates DEGs between Myr-treated CF cells and untreated CF cells; MyrCTRL indicates DEGs between treated CTRL cells and untreated CTRL cells; (**B**) MyrAspCF/Asp indicates DEGs between Myr-treated infected CF cells and untreated infected CF cells; MyrAspCTRL/Asp indicates DEGs between Myr-treated infected CTRL cells and untreated infected CTRL cells. (**C**) Venn diagram of CRTL and CF cells in response to infection (AspCTRL and AspCF) and infected CF cells in response to treatment (MyrAspCF). MyrAspCF: DEGs between Myr-treated infected cells and uninfected untreated CF cells. (**D**) Venn diagram of CF cells and CF monocytes: DEGs between Myr-treated infected CF cells and Myr-treated infected monocytes from CF patients.

We found 292 genes that were differentially expressed in both CTRL and CF cells in response to fungal infection (AspCTRL and AspCF). When CF cells were treated with Myr (MyrAspCF), the number of common regulated genes increased up to 439, indicating that Myr drives CF response to infection and partially restores the CTRL expression profile (Figure 5C). Finally, we compared the genes that found to be regulated by Myr in infected airways epithelial CF cells (MyrAspCF) and in infected CF patients' derived monocytes (treated-infected versus untreated-infected) (Figure 5D). Regardless of the different origins of the cells, we found a common upregulation of HSP90AA1 and ZFAS1, two genes belonging to heat shock and zinc fingers proteins families, involved in chaperone-mediated autophagy (previously described, Table 2, Section 3.1, [40,43,44]). In addition, the expression of another 11 genes was commonly modulated, although at minor extent in respect to the above discussed.

3.4. Myriocin Activates Gene Sets Involved in Inflammation, Infection, Autophagy/Proteostasis, and Lipid Metabolism in CF Bronchial Epithelial Infected Cells

In order to evaluate the effect of Myr on infected CF cells, we performed GSEA using logFC pre-ranked list of MyrAspCF/Asp DEGs, thus evaluating only Myr-related transcriptional activities and excluding infection-induced transcriptional modification. Results indicate that the Myr treatment modulates the expression of genes involved in autophagy/proteostasis, lipid metabolism, and in response to infection and inflammation. In particular, as highlighted in Figure 6, Myr treatment upregulated the genes involved in autophagy/proteostasis and lipid metabolism, whereas it downregulated genes related to inflammation and infection processes.

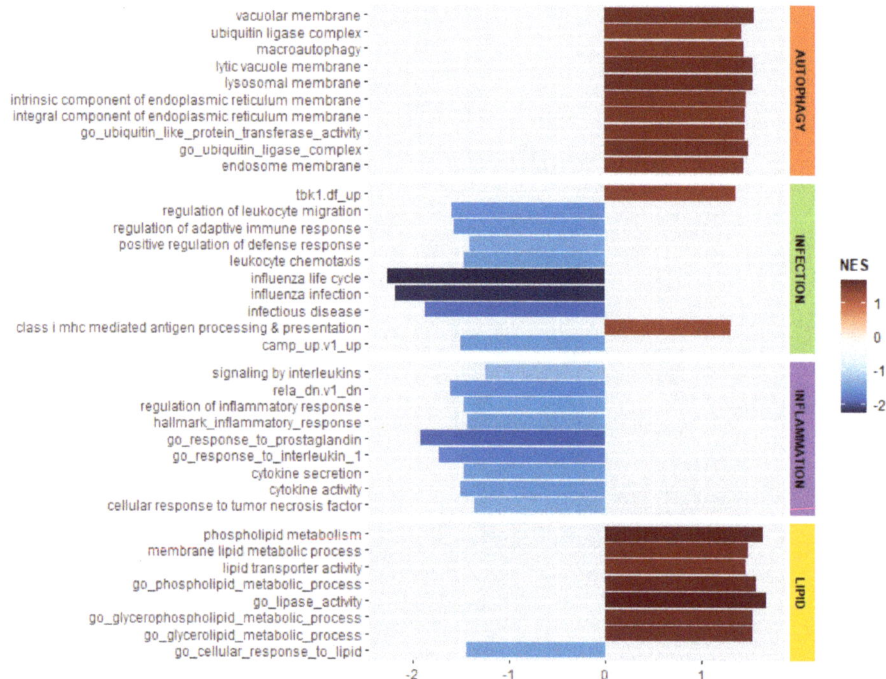

Figure 6. Enriched gene set enrichment analysis (GSEA) terms—functional categories. This chart shows the results of GSEA analysis on functional categories related to CF pathology using pre-ranked logFC list genes from MyrAsp/Asp CF analysis. In red, the terms with positive normalized enriched score (NES), upregulated by myriocin, whereas in blue the terms with negative NES, downregulated by myriocin. GSEA terms include GO, MutSigDB, and KEGG pathway terms.

Considering a significant adjusted *p*-value threshold of 0.05, we observed 139 DEGs in infected CF cells in response to treatment. Among such DEGs we identified specific genes related to the molecular processes that are regulated by Myr, as evidenced by GSEA. Other than increasing IL1β, a primary cytokine in CF disease and its defective response to infection [45,46], Myr upregulated the expression of the transmembrane protein TMEM59, which induces the LC3 labelling of endosomal vesicles, stimulating their fusion with lysosomes, in response to autophagy and xenophagy [47]. Moreover, Myr upregulated the p53-inducible protein 1 (*TP53INP1*) gene, which also promotes autophagy by interacting with autophagy-related protein family (ATG) [48]. The expression of SNX14, involved in the regulation of autophagy and lipid metabolism, was significantly increased by Myr [49,50], as well as that of hypoxia inducible gene 1 (*HIGD1A*), involved in oxidative stress and lipotoxicity protection [51]. At the same time, Myr downregulated ACACA, the rate-limiting enzyme regulating de novo fatty acid synthesis, whose increased activity has been associated with inflammation and CFTR deficiency [52,53]. We next validated the upregulation of three of the above reported genes by RT-PCR and demonstrated the increase of their expression in infected and Myr-treated CF cells (Figure 7).

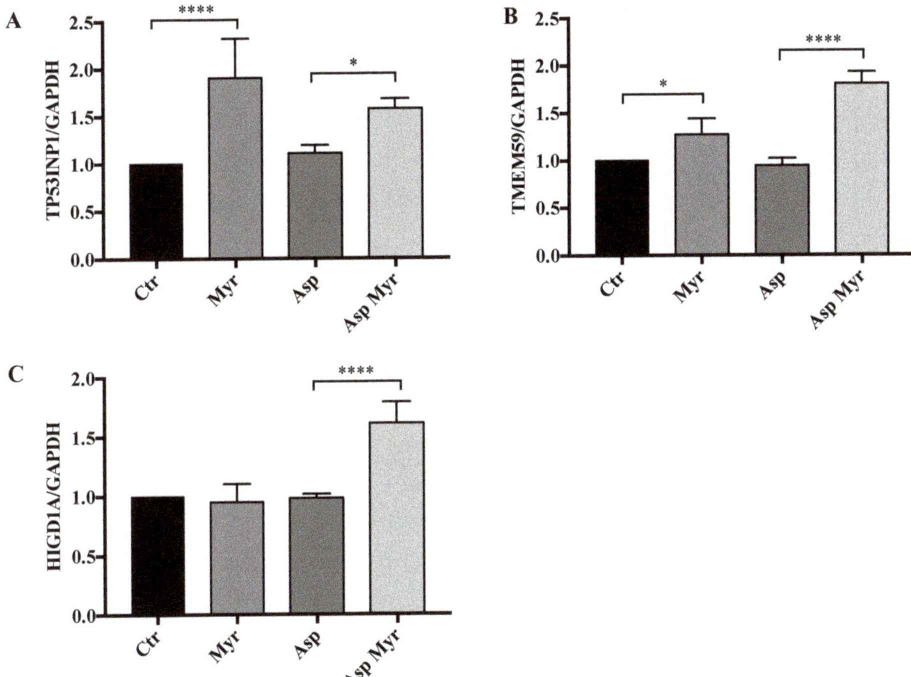

Figure 7. RNA-seq validation by RT-PCR quantification of the expression of genes involved in response to infection in CF cells: (**A**) transmembrane protein TMEM59; (**B**) p53-inducible protein 1, *TP53INP1*; (**C**) hypoxia inducible gene 1, *HIGD1A*. *GAPDH* was used as a housekeeping gene. Data, derived for triplicate samples, are expressed as mean ± SE (* $p < 0.05$; **** $p < 0.0001$); two-way ANOVA followed by Bonferroni correction was used for all data.

In view of our previous results on Myr transcriptional effects in CF bronchial epithelial cells [5,8,25], we evaluated a delayed expression of specific marker genes involved in the inflammatory process and in the response against microbial infection in CF cells infected with *A. fumigatus*, treated or untreated with Myr. By real-time PCR (RT-PCR), we proved that the expression of pro-inflammatory interleukin-1β (*IL-1β*) and *IL-8* chemokines, known to be upregulated in CF [25], were significantly reduced by Myr, in both basal conditions and in *A. fumigatus*-infected CF cells, whereas the expression of anti-inflammatory *IL-10* was upregulated by the compound (Figure 8A–C).

Pathogen recognition receptors (PRRs) are responsible for the identification of antigens, and can mediate their lysosomal clearance. We observed that Myr treatment enhanced the expression of the *NOD2*, *TLR2*, and *TLR7* in infected CF cells (Figure 8D–F). Myr reduction of sphingolipid synthesis regulates lipid metabolism by enhancing fatty acids oxidation and reducing the overall cell amount of glycerolipids and cholesterols [5]. Myr's action on lipid-energy homeostasis is sensed as a stress that drives TFEB activation and transcriptional activities that sustain lipid consumption and autophagy induction [5]. We observed an increased expression of *TFEB* in infected CF bronchial epithelial cells treated with Myr compared to untreated infected cells (Figure 8G). TBK1 phosphorylation and interaction with optineurin (OPTN) promote autophagy-mediated pathogen clearance, namely xenophagy [54–56]. We observed that *A. fumigatus* infection reduced the expression of *TBK1* while

Myr treatment significantly rescued it (Figure 8H). Myr was also able to increase *OPTN* expression in infected CF cells (Figure 8I). Finally, Myr induced a significant increase of *Lamp2a* expression, which was reduced in *A. fumigatus*-infected CF cells, confirming the TFEB and the autophagy-related increase in lysosome formation (Figure 8L).

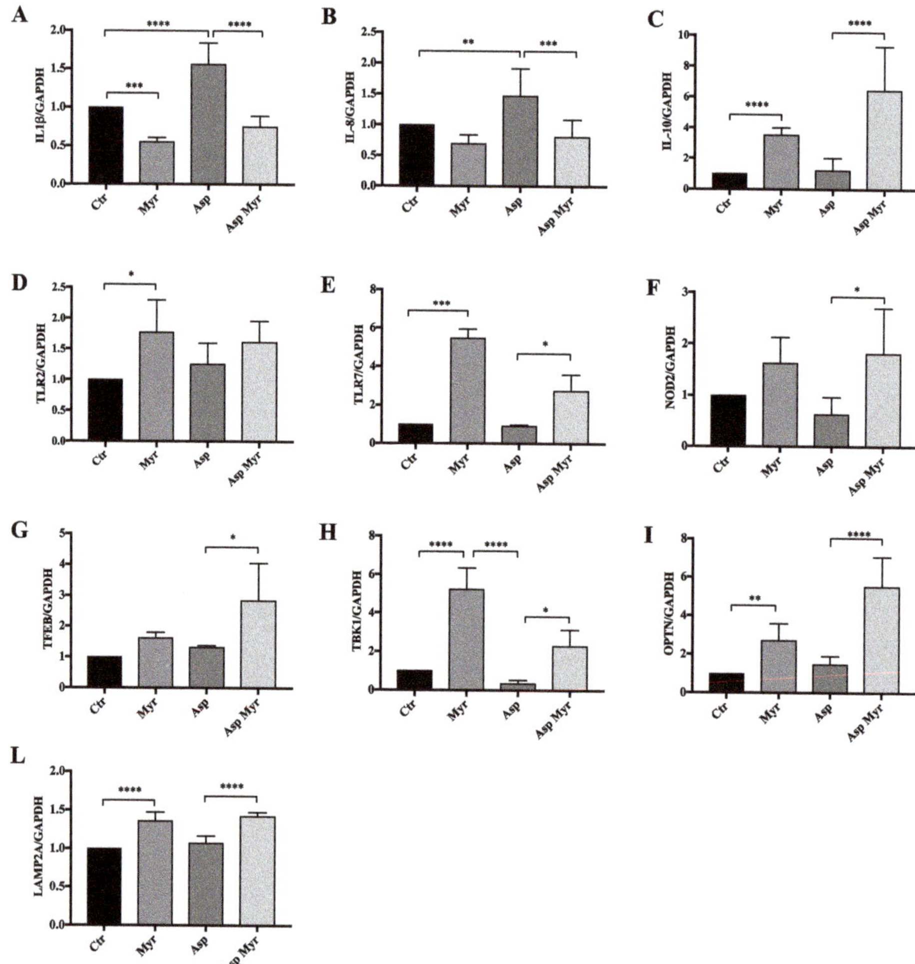

Figure 8. Quantification of the expression of genes involved in inflammation, response to infection, and autophagy: (**A**) pro-inflammatory *IL1β* interleukin; (**B**) pro-inflammatory *IL-8* chemokine; (**C**) anti-inflammatory *IL10* interleukin; (**D–F**) pathogen recognition receptors (PRRs): *NOD2*, *TLR2*, and *TLR7*; (**G**) *TFEB*; (**H**) *TBK1*; (**I**) *OPTN*; (**L**) *LAMP2a* by qRT-PCR in *A. fumigatus*-infected and uninfected CF cells, treated and untreated with Myr (12 h after infection). *GAPDH* was used as a housekeeping gene. Data, derived for triplicate samples, are expressed as mean ± SE (* $p < 0.05$; ** $p < 0.01$; *** $p < 0.001$; **** $p < 0.0001$); two-way ANOVA followed by Bonferroni correction was used for all data.

4. Discussion

Our study demonstrates that the defective response to infection in CF is related to a dysfunction in autophagy, inflammation resolution, and lipid metabolism, which are known to be caused by mutated CFTR [1–5]. Dyslipidemia has been associated with CF disease, and it is characterized by a reduced absorption and increased synthesis of lipids [1,2,53,57–62]. Moreover, cholesterol and the inflammatory lipid ceramide have been shown to accumulate in CF peripheral organs, in particular at the airways level [4,23,26,63–66]. Although altered lipid metabolism is a common feature of chronic inflammatory diseases, the contribution of lipids in CF pathophysiology is still to be fully elucidated. We previously demonstrated that Myr—a specific inhibitor of sphingolipid de novo synthesis—reduces the accumulation of not only ceramide but also most lipid species in CF cells [5]. By impairing sphingolipid synthesis, Myr induces the activation of a stress response that initiates with the TFEB-induced transcriptional program, sustained by PPARs and FOXOs transcription factors, aimed at increasing lipid oxidation and promoting autophagy [5]. Its modulatory action results in an overall decline of CF hyperinflammation, as a consequence of a reduced expression of pro-inflammatory cytokines, and in an ameliorated defensive response to infection, driven by the increase in xenophagy-activating PRR expression [8,25], known to be downregulated in CF [67].

Monocytes play a crucial role in pathogen eradication and CF prominent susceptibility to recurrent infections, largely relying on altered monocytes response. Therefore, we studied peripheral-blood-derived monocytes from CF patients, bearing either homozygous or compound heterozygous ΔF508 mutation of CFTR, by in vitro infection with *A. fumigatus*. Myr treatment shaped a significantly different transcriptional process in infected monocytes, modulating the expression of genes involved in inflammation, response to infection, lipid metabolism, and autophagy. Among differentially expressed genes, we identified two functional clusters in terms of significant response to Myr that have not previously been associated to CF (Tables 1 and 2). Myr upregulated the expression of several proteins belonging to the histone family. Other than fundamental components of eukaryotic chromatin, histones and histone fragments display antimicrobial activities [34], either by being secreted and reacting against extracellular pathogens or by accumulating in the cytosol and binding intracellular infectious agents [38]. Cytosolic histones elude proteolysis thanks to the binding to lipid cytosolic storage [35], are released upon interaction with pathogens, and are processed to act as antimicrobial peptides against bacteria and fungi [36,37,39]. The observation that infection increases this cytosolic histone-related fraction [35] supports our hypothesis that, possibly by modulating lipid metabolism, Myr promotes histones transcription in infected cells in order to boost their cytosolic pool, which is endowed with antimicrobial activity. Moreover, Myr upregulated the expression of a number of heat-shock proteins (HSP) belonging to the 70 family (Hsp70), 90 family (Hsp90), and DnaJ family (Hsp40). HSPs take part in the response to infection by receptor-mediated activation of the innate immune response and by participation in the antigen presentation for the adaptive immune response [68,69], which is also defective in CF [6]. Hsp70 and Hsp90 play an important role in chaperone-mediated autophagy (CMA) and pathogen recognition [44,70,71]. Proteins degraded by CMA are identified in the cytosol by a chaperone complex which includes Hsp70 [72]. Upon binding to the target, Hsp70 [73] and Hsp90 [44] interact with the lysosome-associated membrane protein type 2A (LAMP-2A), which we observed to be upregulated by Myr. DNAJ proteins (Hsp40) regulate Hsp70 chaperones by stimulating ATP hydrolysis [74]. Thus, Myr's action on HSPs is aimed at sustaining innate and adaptive immunity as well as autophagy-related unfolded proteins and pathogens clearance in infected CF monocytes. Moreover, Myr increased the expression of the *PNPLA3* gene, which encodes for a lipase responsible for the mobilization of intracellular fat storage [75] and degradation in the autophagolysosomes. This process, named lipophagy, is directly promoted by TFEB activation [41]. Hence, Myr might sustain both classic autophagy, as we previously demonstrated [5], and chaperone-mediated autophagy, driving not only the autophagic pathway, but also TFEB-induced lysosome biogenesis, as revealed by increased *Lamp2a* expression. In agreement with previous data suggesting that Myr enhances killing ability in an *A. fumigatus*-infected CF bronchial epithelial cell line [8], we observed a significant increase of conidia killing in *A. fumigatus*-infected CF monocytes. This latter observation indicates that reducing inflammatory lipid accumulation and promoting autophagy is an effective therapeutic approach to correcting proteinopathy stress, mainly related to ΔF508 mutation of CFTR, and to restoring an effective response against infection.

Next, we extended our previously published data by investigating the effect of sphingolipid synthesis inhibition in the first phase of microbial infection in the airways. In order to better understand whether Myr could modulate CF response to infection, we analyzed the whole-transcriptome modification induced by *A. fumigatus* in a CF bronchial epithelial cell line and a healthy control counterpart, either treated or untreated with Myr. Our data provide clear evidence that a profound modification in gene expression, triggered by infection, differentiates CF from healthy cells, suggesting that the disease itself causes higher sensitivity to infection. Indeed, healthy cells responded to the same stimulus by engaging a smaller number of genes. This evidence is in line with CF patients' defective ability to elicit a coordinated defensive response against infection. Indeed, we observed a stronger modulation of CF cells' transcriptional activity by Myr treatment, whereas the impact on control cells was milder. These data are in agreement with our previous observation on the effects of Myr treatment on inflamed cells or organs with respect to their healthy counterparts [5,25,76]. We suggest that sphingolipid de novo synthesis is enhanced under stress conditions such as proteinopathy and inflammation. In a homeostatic state, Myr inhibition of the sphingolipid synthesis rate-limiting enzyme serine palmitoyl transferase (SPT) may be buffered by a consequent spontaneous modulation of Nogo and ORMs, two enzymes that normally control SPT activity [77,78]. When stress or inflammation upregulate the sphingolipid synthesis pathway, the effect of Myr is more pronounced and is perceived as an alarm signal that drives defense response by moving the metabolic resources via lipid mobilization and autophagy. During an infectious process, CF cells' transcriptional activities are perturbed to a higher extent than control cells. Nonetheless, Myr treatment increased the number of genes that were commonly regulated by infection in CF and healthy cells, thus driving the CF phenotype closer to that of control cells. This confirms the hypothesis that inhibition of sphingolipid synthesis activates a profound transcriptional modification aimed at inducing stress tolerance and resilience as well as pathogen resistance in CF. This concept is crucial in any chronic inflammatory disease, in which the possibility of counteracting the cellular stress is the required therapeutic effect. Our data indicate that the Myr-induced effects on infected CF cells involve genes mediating inflammation, immune response, autophagy, and lipid metabolism. Among genes whose expression was significantly affected by Myr in CF cells during infection, we identified a few that are related to the significantly regulated pathways and that may indicate novel targets for CF therapy. A specific form of LC3-associated phagocytosis can be activated by TLR signaling during the phagocytosis of fungal and bacterial pathogens: LC3 binds to the cytosolic side of the TLR-induced endosome, which evolves into autophagosomes [68]. Mediating this mechanism, TMEM59 and TMEM166 interact with ATG16L1, become incorporated into phagosomes' membranes, driving the activation and lipidation of LC3, and thus the maturation into autophagosomes [69]. Similarly, SNX14 proteins generally localize to endosomes' membranes via the PX domain, which binds to phosphoinositides (PtdIns), and this is reported to promote autophagy [49,50]. The expression of both SNX14 and TMEM59 was enhanced by Myr upon infection, suggesting that the compound drives a pathogen-clearance-related autophagy. Moreover, altered SNX14 function has been associated with reduced cholesterol esters, suggesting an alteration in neutral lipid metabolism and possibly reduced lipid delivery to droplets storage [50]. Accordingly, Myr reduced the expression of the gene encoding for acetyl-CoA carboxylase (ACACA), the rate-limiting enzyme involved in fatty acids synthesis and regulated in opposition to their oxidation. ACACA has been associated with inflammation [52] and most notably, its activity is enhanced in CFT-deficient cells [53], in line with the increased sphingolipid synthesis in CF chronic inflammation [5,25]. Myr treatment also increased the expression of hypoxia-inducible gene domain family member 1A (*HIGD1a*). Its gene product was recently reported to protect cells from hypoxia and from lipotoxicity related to fat oxidation impairment. Indeed, HIGD1a decreases oxygen radical production, helping to maintain a normal mitochondrial function [51]. A reduced autophagy paired with lipid accumulation because of decreased oxidation rate was already observed in CF cells and rescued by Myr [5]. Thus, Myr could possibly enhance

SNX14 to induce autophagy and lipid consumption, and it may upregulate *HIGD1a* to reduce the oxidative stress caused by proteinopathy and lipid accrual in infected CF cells. Therefore, Myr is able to directly target specific genes whose functions have not been previously associated to CF, but that sustain its pathophysiology.

In addition, lipid synthesis inhibition is a strategy to fight microbial infections that are known to take advantage of cell lipid storage, and ACACA inhibitors have been designed to develop antimicrobial strategies in infectious diseases [79].

Finally, we investigated the possibility of Myr-induced transcriptional activation of TFEB and of downstream target genes in *A. fumigatus*-infected CF bronchial epithelial cells. As expected, TFEB transcription was increased by Myr, suggesting that lipid catabolism and autophagy are enhanced even under infection conditions. A growing body of evidence highlighted that TBK1-OPTN signaling is pivotal for the initiation and resolution of the innate immune responses, and in eliciting autophagy [55,56,80]. Myr significantly increased the expression of these two proteins, both in uninfected and infected cells, thus promoting internalized pathogen clearance. To note, Myr upregulated the expression of *LAMP2A* (lysosomal-associated membrane protein 2A), encoding for a membrane glycoprotein involved in autophagy, which is directly promoted by TFEB induction of lysosomes biogenesis.

5. Strength and Limitations of the Study

The strength of the present study is that it presents data deriving from a significant number of CF patients, variable for sex and age, affected by ΔF508 mutation of the *CFTR* gene (homozygous and compound heterozygous). The limitations include the fact that part of the study was conducted on epithelial cell lines, and further studies on primary epithelial cells are needed to corroborate the obtained data. Moreover, we cannot exclude that patient PBMC storage and in vitro culturing could impact the cell transcriptional response. Next, although the used *Aspergillus fumigatus* strain (Af293) is a clinical strain derived from invasive pulmonary aspergillosis, a genome-sequenced biofilm producer, the in vitro infection procedure may not completely resemble in vivo infection, and the host response may differ in terms of intensity and timing of gene expression modulation. Similarly, we cannot predict whether Myr administration would be efficient in modulating chronic infections in patients. From a technical point of view, gene expression was carried out by normalizing data with a unique housekeeping gene (*GAPDH*), whereas the use of three different housekeeping genes would ensure the highest reliability of transcriptional evaluation.

6. Conclusions

CF bronchial epithelial cells display a profound difference in the transcriptional profile compared with their normal counterpart, and infection sharpens this diversity. We demonstrated that by modulating the biosynthesis of sphingolipids by Myr, a metabolism drift occurs which is aimed at fueling energy to act against stress, and includes autophagy and lipid consumption. This is driven by a transcriptional program that significantly modifies cellular phenotype and amends part of the differences between CF and control, therefore ameliorating the response against infection and pathogens clearance in both a CF cell line [8] and monocytes. We speculate that lipid metabolism is deeply altered in CF patients, possibly due to the chronic and systemic inflammation associated with the disease. Lipids accumulation, or simply deregulation in their storage/consumption, may contribute to the defective immunity of CF patients. Literature evidence reports dyslipidemia in CF patients, measured both in blood and peripheral organs [2,81,82]. We conclude that more attention should be devoted to the alteration of lipid metabolism in CF.

Supplementary Materials: The following are available online at http://www.mdpi.com/2073-4409/9/8/1845/s1. Figure S1: Quantification of the expression of genes involved in inflammation, response to infection, and autophagy in CTRL and CF brochial epithelial cells; Table S1: CF patients cohort.

Author Contributions: Conceptualization, A.M., P.S. Investigation, A.M., E.O., E.B., P.S. Formal analysis, M.B., I.M. Drafting of the manuscript, A.M., E.O. Supervision, R.G., E.B., P.S. Writing—review and editing, A.M., E.O., E.B., P.S. Resources, L.R., T.A., N.C. Funding acquisition, P.S. All authors have read and agreed to the published version of the manuscript.

Funding: This research was funded by the Italian Cystic Fibrosis Foundation, Grant FFC#11-2016.

Acknowledgments: We thank the Italian Cystic Fibrosis Foundation for their financial support of this work. E.O. was supported by the PhD program in Molecular and Translational Medicine of the Università degli Studi di Milano, Milan.

Conflicts of Interest: There are no conflict of interest.

References

1. Gelzo, M.; Sica, C.; Elce, A.; Dello Russo, A.; Iacotucci, P.; Carnovale, V.; Raia, V.; Salvatore, D.; Corso, G.; Castaldo, G. Reduced absorption and enhanced synthesis of cholesterol in patients with cystic fibrosis: A preliminary study of plasma sterols. *Clin. Chem. Lab. Med.* **2016**, *54*, 1461–1466. [CrossRef] [PubMed]

2. Figueroa, V.; Milla, C.; Parks, E.J.; Schwarzenberg, S.J.; Moran, A. Abnormal lipid concentrations in cystic fibrosis. *Am. J. Clin. Nutr.* **2002**, *75*, 1005–1011. [CrossRef] [PubMed]

3. Luciani, A.; Villella, V.R.; Esposito, S.; Gavina, M.; Russo, I.; Silano, M.; Guido, S.; Pettoello-Mantovani, M.; Carnuccio, R.; Scholte, B.; et al. Targeting autophagy as a novel strategy for facilitating the therapeutic action of potentiators on DeltaF508 cystic fibrosis transmembrane conductance regulator. *Autophagy* **2012**, *8*, 1657–1672. [CrossRef] [PubMed]

4. Teichgraber, V.; Ulrich, M.; Endlich, N.; Riethmuller, J.; Wilker, B.; De Oliveira-Munding, C.C.; van Heeckeren, A.M.; Barr, M.L.; von Kurthy, G.; Schmid, K.W.; et al. Ceramide accumulation mediates inflammation, cell death and infection susceptibility in cystic fibrosis. *Nat. Med.* **2008**, *14*, 382–391. [CrossRef] [PubMed]

5. Mingione, A.; Dei Cas, M.; Bonezzi, F.; Caretti, A.; Piccoli, M.; Anastasia, L.; Ghidoni, R.; Paroni, R.; Signorelli, P. Inhibition of Sphingolipid Synthesis as a Phenotype-Modifying Therapy in Cystic Fibrosis. *Cell. Physiol. Biochem. Int. J. Exp. Cell. Physiol. Biochem. Pharmacol.* **2020**, *54*, 110–125.

6. Bruscia, E.M.; Bonfield, T.L. Innate and Adaptive Immunity in Cystic Fibrosis. *Clin. Chest Med.* **2016**, *37*, 17–29. [CrossRef]

7. Zhang, S.; Shrestha, C.L.; Kopp, B.T. Cystic fibrosis transmembrane conductance regulator (CFTR) modulators have differential effects on cystic fibrosis macrophage function. *Sci. Rep.* **2018**, *8*, 17066. [CrossRef]

8. Caretti, A.; Torelli, R.; Perdoni, F.; Falleni, M.; Tosi, D.; Zulueta, A.; Casas, J.; Sanguinetti, M.; Ghidoni, R.; Borghi, E.; et al. Inhibition of ceramide de novo synthesis by myriocin produces the double effect of reducing pathological inflammation and exerting antifungal activity against *A. fumigatus* airways infection. *Biochim. Biophys. Acta* **2016**, *1860*, 1089–1097. [CrossRef]

9. Leveque, M.; Le Trionnaire, S.; Del Porto, P.; Martin-Chouly, C. The impact of impaired macrophage functions in cystic fibrosis disease progression. *J. Cyst. Fibros.* **2017**, *16*, 443–453. [CrossRef]

10. Tracy, M.C.; Moss, R.B. The myriad challenges of respiratory fungal infection in cystic fibrosis. *Pediatr. Pulmonol.* **2018**, *53*, S75–S85. [CrossRef]

11. Chaudhary, N.; Datta, K.; Askin, F.B.; Staab, J.F.; Marr, K.A. Cystic fibrosis transmembrane conductance regulator regulates epithelial cell response to Aspergillus and resultant pulmonary inflammation. *Am. J. Respir. Crit. Care Med.* **2012**, *185*, 301–310. [CrossRef] [PubMed]

12. Bodas, M.; Vij, N. Adapting Proteostasis and Autophagy for Controlling the Pathogenesis of Cystic Fibrosis Lung Disease. *Front. Pharmacol.* **2019**, *10*, 20. [CrossRef] [PubMed]

13. Luciani, A.; Villella, V.R.; Esposito, S.; Brunetti-Pierri, N.; Medina, D.L.; Settembre, C.; Gavina, M.; Raia, V.; Ballabio, A.; Maiuri, L. Cystic fibrosis: A disorder with defective autophagy. *Autophagy* **2011**, *7*, 104–106. [CrossRef] [PubMed]

14. Esposito, S.; Tosco, A.; Villella, V.R.; Raia, V.; Kroemer, G.; Maiuri, L. Manipulating proteostasis to repair the F508del-CFTR defect in cystic fibrosis. *Mol. Cell. Pediatr.* **2016**, *3*, 13. [CrossRef]

15. Evans, R.J.; Sundaramurthy, V.; Frickel, E.M. The Interplay of Host Autophagy and Eukaryotic Pathogens. *Front. Cell Dev. Biol.* **2018**, *6*, 118. [CrossRef]

16. Rameshwaram, N.R.; Singh, P.; Ghosh, S.; Mukhopadhyay, S. Lipid metabolism and intracellular bacterial virulence: Key to next-generation therapeutics. *Future Microbiol.* **2018**, *13*, 1301–1328. [CrossRef]

17. Walpole, G.F.W.; Grinstein, S.; Westman, J. The role of lipids in host-pathogen interactions. *IUBMB Life* **2018**, *70*, 384–392. [CrossRef]

18. Keyhani, N.O. Lipid biology in fungal stress and virulence: Entomopathogenic fungi. *Fungal Biol.* **2018**, *122*, 420–429. [CrossRef]

19. Kim, Y.S.; Lee, H.M.; Kim, J.K.; Yang, C.S.; Kim, T.S.; Jung, M.; Jin, H.S.; Kim, S.; Jang, J.; Oh, G.T.; et al. PPAR-alpha Activation Mediates Innate Host Defense through Induction of TFEB and Lipid Catabolism. *J. Immunol.* **2017**, *198*, 3283–3295. [CrossRef]

20. Ollero, M.; Astarita, G.; Guerrera, I.C.; Sermet-Gaudelus, I.; Trudel, S.; Piomelli, D.; Edelman, A. Plasma lipidomics reveals potential prognostic signatures within a cohort of cystic fibrosis patients. *J. Lipid Res.* **2011**, *52*, 1011–1022. [CrossRef]

21. Ollero, M. Methods for the study of lipid metabolites in cystic fibrosis. *J. Cyst. Fibros. Soc.* **2004**, *3* (Suppl. 2), 97–98. [CrossRef] [PubMed]

22. Ziobro, R.; Henry, B.; Edwards, M.J.; Lentsch, A.B.; Gulbins, E. Ceramide mediates lung fibrosis in cystic fibrosis. *Biochem. Biophys. Res. Commun.* **2013**, *434*, 705–709. [CrossRef] [PubMed]

23. Fang, D.; West, R.H.; Manson, M.E.; Ruddy, J.; Jiang, D.; Previs, S.F.; Sonawane, N.D.; Burgess, J.D.; Kelley, T.J. Increased plasma membrane cholesterol in cystic fibrosis cells correlates with CFTR genotype and depends on de novo cholesterol synthesis. *Respir. Res.* **2010**, *11*, 61. [CrossRef] [PubMed]

24. Del Ciampo, I.R.; Sawamura, R.; Fernandes, M.I. Cystic fibrosis: From protein-energy malnutrition to obesity with dyslipidemia. *Iran. J. Pediatr.* **2013**, *23*, 605–606. [PubMed]

25. Caretti, A.; Bragonzi, A.; Facchini, M.; De Fino, I.; Riva, C.; Gasco, P.; Musicanti, C.; Casas, J.; Fabrias, G.; Ghidoni, R.; et al. Anti-inflammatory action of lipid nanocarrier-delivered myriocin: Therapeutic potential in cystic fibrosis. *Biochim. Biophys. Acta* **2014**, *1840*, 586–594. [CrossRef]

26. Caretti, A.; Vasso, M.; Bonezzi, F.T.; Gallina, A.; Trinchera, M.; Rossi, A.; Adami, R.; Casas, J.; Falleni, M.; Tosi, D.; et al. Myriocin treatment of CF lung infection and inflammation: Complex analyses for enigmatic lipids. *Naunyn-Schmiedeberg Arch. Pharmacol.* **2017**, *390*, 775–790. [CrossRef]

27. Elkord, E.; Williams, P.E.; Kynaston, H.; Rowbottom, A.W. Human monocyte isolation methods influence cytokine production from in vitro generated dendritic cells. *Immunology* **2005**, *114*, 204–212. [CrossRef]

28. Love, M.I.; Huber, W.; Anders, S. Moderated estimation of fold change and dispersion for RNA-seq data with DESeq2. *Genome Biol.* **2014**, *15*, 550. [CrossRef]

29. Dobin, A.; Davis, C.A.; Schlesinger, F.; Drenkow, J.; Zaleski, C.; Jha, S.; Batut, P.; Chaisson, M.; Gingeras, T.R. STAR: Ultrafast universal RNA-seq aligner. *Bioinformatics* **2013**, *29*, 15–21. [CrossRef]

30. Liao, Y.; Smyth, G.K.; Shi, W. featureCounts: An efficient general purpose program for assigning sequence reads to genomic features. *Bioinformatics* **2014**, *30*, 923–930. [CrossRef]

31. Robinson, M.D.; McCarthy, D.J.; Smyth, G.K. edgeR: A Bioconductor package for differential expression analysis of digital gene expression data. *Bioinformatics* **2010**, *26*, 139–140. [CrossRef] [PubMed]

32. Yu, G.; Wang, L.G.; Han, Y.; He, Q.Y. clusterProfiler: An R package for comparing biological themes among gene clusters. *Omics J. Integr. Biol.* **2012**, *16*, 284–287. [CrossRef] [PubMed]

33. Szklarczyk, D.; Gable, A.L.; Lyon, D.; Junge, A.; Wyder, S.; Huerta-Cepas, J.; Simonovic, M.; Doncheva, N.T.; Morris, J.H.; Bork, P.; et al. STRING v11: Protein-protein association networks with increased coverage, supporting functional discovery in genome-wide experimental datasets. *Nucleic Acids Res.* **2019**, *47*, D607–D613. [CrossRef] [PubMed]

34. Hirsch, J.G. Bactericidal action of histone. *J. Exp. Med.* **1958**, *108*, 925–944. [CrossRef]

35. Anand, P.; Cermelli, S.; Li, Z.; Kassan, A.; Bosch, M.; Sigua, R.; Huang, L.; Ouellette, A.J.; Pol, A.; Welte, M.A.; et al. A novel role for lipid droplets in the organismal antibacterial response. *Elife* **2012**, *1*, e00003. [CrossRef]

36. Kim, H.S.; Yoon, H.; Minn, I.; Park, C.B.; Lee, W.T.; Zasloff, M.; Kim, S.C. Pepsin-mediated processing of the cytoplasmic histone H2A to strong antimicrobial peptide buforin I. *J. Immunol.* **2000**, *165*, 3268–3274. [CrossRef]

37. Watson, K.; Edwards, R.J.; Shaunak, S.; Parmelee, D.C.; Sarraf, C.; Gooderham, N.J.; Davies, D.S. Extra-nuclear location of histones in activated human peripheral blood lymphocytes and cultured T-cells. *Biochem. Pharmacol.* **1995**, *50*, 299–309. [CrossRef]

38. Zlatanova, J.S.; Srebreva, L.N.; Banchev, T.B.; Tasheva, B.T.; Tsanev, R.G. Cytoplasmic pool of histone H1 in mammalian cells. *J. Cell Sci.* **1990**, *96 Pt 3*, 461–468.

39. De Lucca, A.J.; Heden, L.O.; Ingber, B.; Bhatnagar, D. Antifungal properties of wheat histones (H1-H4) and purified wheat histone H1. *J. Agric. Food Chem.* **2011**, *59*, 6933–6939. [CrossRef]

40. Hu, B.; Zhang, Y.; Jia, L.; Wu, H.; Fan, C.; Sun, Y.; Ye, C.; Liao, M.; Zhou, J. Binding of the pathogen receptor HSP90AA1 to avibirnavirus VP2 induces autophagy by inactivating the AKT-MTOR pathway. *Autophagy* **2015**, *11*, 503–515. [CrossRef]

41. Negoita, F.; Blomdahl, J.; Wasserstrom, S.; Winberg, M.E.; Osmark, P.; Larsson, S.; Stenkula, K.G.; Ekstedt, M.; Kechagias, S.; Holm, C.; et al. PNPLA3 variant M148 causes resistance to starvation-mediated lipid droplet autophagy in human hepatocytes. *J. Cell. Biochem.* **2019**, *120*, 343–356. [CrossRef] [PubMed]

42. Tanida, I.; Ueno, T.; Kominami, E. LC3 and Autophagy. *Methods Mol. Biol.* **2008**, *445*, 77–88. [PubMed]

43. Hu, F.; Shao, L.; Zhang, J.; Zhang, H.; Wen, A.; Zhang, P. Knockdown of ZFAS1 Inhibits Hippocampal Neurons Apoptosis and Autophagy by Activating the PI3K/AKT Pathway via Up-regulating miR-421 in Epilepsy. *Neurochem. Res.* **2020**. [CrossRef] [PubMed]

44. Bandyopadhyay, U.; Kaushik, S.; Varticovski, L.; Cuervo, A.M. The chaperone-mediated autophagy receptor organizes in dynamic protein complexes at the lysosomal membrane. *Mol. Cell. Biol.* **2008**, *28*, 5747–5763. [CrossRef] [PubMed]

45. Levy, H.; Murphy, A.; Zou, F.; Gerard, C.; Klanderman, B.; Schuemann, B.; Lazarus, R.; Garcia, K.C.; Celedon, J.C.; Drumm, M.; et al. IL1B polymorphisms modulate cystic fibrosis lung disease. *Pediatr. Pulmonol.* **2009**, *44*, 580–593. [CrossRef]

46. Gillette, D.D.; Shah, P.A.; Cremer, T.; Gavrilin, M.A.; Besecker, B.Y.; Sarkar, A.; Knoell, D.L.; Cormet-Boyaka, E.; Wewers, M.D.; Butchar, J.P.; et al. Analysis of human bronchial epithelial cell proinflammatory response to Burkholderia cenocepacia infection: Inability to secrete il-1beta. *J. Biol. Chem.* **2013**, *288*, 3691–3695. [CrossRef] [PubMed]

47. Boada-Romero, E.; Letek, M.; Fleischer, A.; Pallauf, K.; Ramon-Barros, C.; Pimentel-Muinos, F.X. TMEM59 defines a novel ATG16L1-binding motif that promotes local activation of LC3. *EMBO J.* **2013**, *32*, 566–582. [CrossRef]

48. Nowak, J.; Archange, C.; Tardivel-Lacombe, J.; Pontarotti, P.; Pebusque, M.J.; Vaccaro, M.I.; Velasco, G.; Dagorn, J.C.; Iovanna, J.L. The TP53INP2 protein is required for autophagy in mammalian cells. *Mol. Biol. Cell* **2009**, *20*, 870–881. [CrossRef]

49. Akizu, N.; Cantagrel, V.; Zaki, M.S.; Al-Gazali, L.; Wang, X.; Rosti, R.O.; Dikoglu, E.; Gelot, A.B.; Rosti, B.; Vaux, K.K.; et al. Biallelic mutations in SNX14 cause a syndromic form of cerebellar atrophy and lysosome-autophagosome dysfunction. *Nat. Genet.* **2015**, *47*, 528–534. [CrossRef] [PubMed]

50. Bryant, D.; Liu, Y.; Datta, S.; Hariri, H.; Seda, M.; Anderson, G.; Peskett, E.; Demetriou, C.; Sousa, S.; Jenkins, D.; et al. SNX14 mutations affect endoplasmic reticulum-associated neutral lipid metabolism in autosomal recessive spinocerebellar ataxia 20. *Hum. Mol. Genet.* **2018**, *27*, 1927–1940. [CrossRef] [PubMed]

51. Li, T.; Xian, W.J.; Gao, Y.; Jiang, S.; Yu, Q.H.; Zheng, Q.C.; Zhang, Y. Higd1a Protects Cells from Lipotoxicity under High-Fat Exposure. *Oxid. Med. Cell. Longev.* **2019**, *2019*, 6051262. [CrossRef] [PubMed]

52. Idrovo, J.P.; Yang, W.L.; Jacob, A.; Corbo, L.; Nicastro, J.; Coppa, G.F.; Wang, P. Inhibition of lipogenesis reduces inflammation and organ injury in sepsis. *J. Surg. Res.* **2016**, *200*, 242–249. [CrossRef] [PubMed]

53. Mailhot, G.; Ravid, Z.; Barchi, S.; Moreau, A.; Rabasa-Lhoret, R.; Levy, E. CFTR knockdown stimulates lipid synthesis and transport in intestinal Caco-2/15 cells. *Am. J. Physiol. Gastrointest. Liver Physiol.* **2009**, *297*, G1239–G1249. [CrossRef]

54. Weidberg, H.; Elazar, Z. TBK1 mediates crosstalk between the innate immune response and autophagy. *Sci. Signal.* **2011**, *4*, pe39. [CrossRef] [PubMed]

55. Kuo, C.J.; Hansen, M.; Troemel, E. Autophagy and innate immunity: Insights from invertebrate model organisms. *Autophagy* **2018**, *14*, 233–242. [CrossRef]

56. Nozawa, T.; Sano, S.; Minowa-Nozawa, A.; Toh, H.; Nakajima, S.; Murase, K.; Aikawa, C.; Nakagawa, I. TBC1D9 regulates TBK1 activation through Ca^{2+} signaling in selective autophagy. *Nat. Commun.* **2020**, *11*, 770. [CrossRef]

57. Rhodes, B.; Nash, E.F.; Tullis, E.; Pencharz, P.B.; Brotherwood, M.; Dupuis, A.; Stephenson, A. Prevalence of dyslipidemia in adults with cystic fibrosis. *J. Cyst. Fibros.* **2010**, *9*, 24–28. [CrossRef]

58. Ishimo, M.C.; Belson, L.; Ziai, S.; Levy, E.; Berthiaume, Y.; Coderre, L.; Rabasa-Lhoret, R. Hypertriglyceridemia is associated with insulin levels in adult cystic fibrosis patients. *J. Cyst. Fibros.* **2013**, *12*, 271–276. [CrossRef]

59. Freedman, S.D.; Blanco, P.G.; Zaman, M.M.; Shea, J.C.; Ollero, M.; Hopper, I.K.; Weed, D.A.; Gelrud, A.; Regan, M.M.; Laposata, M.; et al. Association of cystic fibrosis with abnormalities in fatty acid metabolism. *N. Engl. J. Med.* **2004**, *350*, 560–569. [CrossRef]

60. Olveira, G.; Dorado, A.; Olveira, C.; Padilla, A.; Rojo-Martinez, G.; Garcia-Escobar, E.; Gaspar, I.; Gonzalo, M.; Soriguer, F. Serum phospholipid fatty acid profile and dietary intake in an adult Mediterranean population with cystic fibrosis. *Br. J. Nutr.* **2006**, *96*, 343–349. [CrossRef]

61. Peretti, N.; Roy, C.C.; Drouin, E.; Seidman, E.; Brochu, P.; Casimir, G.; Levy, E. Abnormal intracellular lipid processing contributes to fat malabsorption in cystic fibrosis patients. *Am. J. Physiol. Gastrointest. Liver Physiol.* **2006**, *290*, G609–G615. [CrossRef] [PubMed]

62. Dogliotti, E.; Vezzoli, G.; Nouvenne, A.; Meschi, T.; Terranegra, A.; Mingione, A.; Brasacchio, C.; Raspini, B.; Cusi, D.; Soldati, L. Nutrition in calcium nephrolithiasis. *J. Transl. Med.* **2013**, *11*, 109. [CrossRef] [PubMed]

63. White, N.M.; Jiang, D.; Burgess, J.D.; Bederman, I.R.; Previs, S.F.; Kelley, T.J. Altered cholesterol homeostasis in cultured and in vivo models of cystic fibrosis. *Am. J. Physiol. Lung Cell. Mol. Physiol.* **2007**, *292*, L476–L486. [CrossRef] [PubMed]

64. Desbenoit, N.; Saussereau, E.; Bich, C.; Bourderioux, M.; Fritsch, J.; Edelman, A.; Brunelle, A.; Ollero, M. Localized lipidomics in cystic fibrosis: TOF-SIMS imaging of lungs from Pseudomonas aeruginosa-infected mice. *Int. J. Biochem. Cell Biol.* **2014**, *52*, 77–82. [CrossRef]

65. Gentzsch, M.; Choudhury, A.; Chang, X.B.; Pagano, R.E.; Riordan, J.R. Misassembled mutant DeltaF508 CFTR in the distal secretory pathway alters cellular lipid trafficking. *J. Cell Sci.* **2007**, *120*, 447–455. [CrossRef]

66. Gentzsch, M.; Chang, X.B.; Cui, L.; Wu, Y.; Ozols, V.V.; Choudhury, A.; Pagano, R.E.; Riordan, J.R. Endocytic trafficking routes of wild type and DeltaF508 cystic fibrosis transmembrane conductance regulator. *Mol. Biol. Cell* **2004**, *15*, 2684–2696. [CrossRef] [PubMed]

67. Chillappagari, S.; Venkatesan, S.; Garapati, V.; Mahavadi, P.; Munder, A.; Seubert, A.; Sarode, G.; Guenther, A.; Schmeck, B.T.; Tummler, B.; et al. Impaired TLR4 and HIF expression in cystic fibrosis bronchial epithelial cells downregulates hemeoxygenase-1 and alters iron homeostasis in vitro. *Am. J. Physiol. Lung Cell. Mol. Physiol.* **2014**, *307*, L791–L799. [CrossRef] [PubMed]

68. Bolhassani, A.; Agi, E. Heat shock proteins in infection. *Clin. Chim. Acta Int. J. Clin. Chem.* **2019**, *498*, 90–100. [CrossRef]

69. Stewart, G.R.; Young, D.B. Heat-shock proteins and the host-pathogen interaction during bacterial infection. *Curr. Opin. Immunol.* **2004**, *16*, 506–510. [CrossRef] [PubMed]

70. Osterloh, A.; Breloer, M. Heat shock proteins: Linking danger and pathogen recognition. *Med. Microbiol. Immunol.* **2008**, *197*, 1–8. [CrossRef]

71. Dokladny, K.; Myers, O.B.; Moseley, P.L. Heat shock response and autophagy–cooperation and control. *Autophagy* **2015**, *11*, 200–213. [CrossRef] [PubMed]

72. Chiang, H.L.; Terlecky, S.R.; Plant, C.P.; Dice, J.F. A role for a 70-kilodalton heat shock protein in lysosomal degradation of intracellular proteins. *Science* **1989**, *246*, 382–385. [CrossRef] [PubMed]

73. Cuervo, A.M.; Dice, J.F. A receptor for the selective uptake and degradation of proteins by lysosomes. *Science* **1996**, *273*, 501–503. [CrossRef] [PubMed]

74. Hartl, F.U.; Bracher, A.; Hayer-Hartl, M. Molecular chaperones in protein folding and proteostasis. *Nature* **2011**, *475*, 324–332. [CrossRef] [PubMed]

75. Tardelli, M.; Bruschi, F.V.; Trauner, M. The role of metabolic lipases in the pathogenesis and management of liver disease. *Hepatology* **2020**. [CrossRef] [PubMed]

76. Bonezzi, F.; Piccoli, M.; Dei Cas, M.; Paroni, R.; Mingione, A.; Monasky, M.M.; Caretti, A.; Riganti, C.; Ghidoni, R.; Pappone, C.; et al. Sphingolipid Synthesis Inhibition by Myriocin Administration Enhances Lipid Consumption and Ameliorates Lipid Response to Myocardial Ischemia Reperfusion Injury. *Front. Physiol.* **2019**, *10*, 986. [CrossRef]

77. Sasset, L.; Zhang, Y.; Dunn, T.M.; Di Lorenzo, A. Sphingolipid De Novo Biosynthesis: A Rheostat of Cardiovascular Homeostasis. *Trends Endocrinol. Metab. TEM* **2016**, *27*, 807–819. [CrossRef]

78. Cai, L.; Oyeniran, C.; Biswas, D.D.; Allegood, J.; Milstien, S.; Kordula, T.; Maceyka, M.; Spiegel, S. ORMDL proteins regulate ceramide levels during sterile inflammation. *J. Lipid Res.* **2016**, *57*, 1412–1422. [CrossRef]

79. Chen, L.; Duan, Y.; Wei, H.; Ning, H.; Bi, C.; Zhao, Y.; Qin, Y.; Li, Y. Acetyl-CoA carboxylase (ACC) as a therapeutic target for metabolic syndrome and recent developments in ACC1/2 inhibitors. *Expert Opin. Investig. Drugs* **2019**, *28*, 917–930. [CrossRef]

80. Ammanathan, V.; Mishra, P.; Chavalmane, A.K.; Muthusamy, S.; Jadhav, V.; Siddamadappa, C.; Manjithaya, R. Restriction of intracellular Salmonella replication by restoring TFEB-mediated xenophagy. *Autophagy* **2019**, 1–14. [CrossRef]

81. Khoury, T.; Asombang, A.W.; Berzin, T.M.; Cohen, J.; Pleskow, D.K.; Mizrahi, M. The Clinical Implications of Fatty Pancreas: A Concise Review. *Dig. Dis. Sci.* **2017**, *62*, 2658–2667. [CrossRef] [PubMed]

82. Tham, R.T.; Heyerman, H.G.; Falke, T.H.; Zwinderman, A.H.; Bloem, J.L.; Bakker, W.; Lamers, C.B. Cystic fibrosis: MR imaging of the pancreas. *Radiology* **1991**, *179*, 183–186. [CrossRef] [PubMed]

MDPI
St. Alban-Anlage 66
4052 Basel
Switzerland
Tel. +41 61 683 77 34
Fax +41 61 302 89 18
www.mdpi.com

Cells Editorial Office
E-mail: cells@mdpi.com
www.mdpi.com/journal/cells